ORIGIN OF MATTER AND EVOLUTION OF GALAXIES

To learn more about the AIP Conference Proceedings, including the Conference Proceedings Series, please visit the webpage
http://proceedings.aip.org/proceedings

ORIGIN OF MATTER AND EVOLUTION OF GALAXIES

International Symposium on Origin of Matter and Evolution of Galaxies 2005: New Horizon of Nuclear Astrophysics and Cosmology

Tokyo, Japan   8 – 11 November 2005

EDITORS

S. Kubono
Center for Nuclear Study (CNS), University of Tokyo

W. Aoki
T. Kajino
National Astronomical Observatory (NAO), Japan

T. Motobayashi
Nishina Center for Accelerator-Based Science, RIKEN, Japan

K. Nomoto
Department of Astronomy, University of Tokyo

SPONSORING ORGANIZATIONS
The Graduate University for Advanced Studies - (GUAS)
National Astronomical Observatory - (NAO)
Center for Nuclear Study, University of Tokyo - (CNS, Tokyo)
Nishina Center for Accelerator-Based Science, RIKEN, Japan
Department of Astronomy, Graduate School of Science, University of Tokyo
Japan Society for the Promotion of Science

Melville, New York, 2006
AIP CONFERENCE PROCEEDINGS ■ VOLUME 847

Editors:

S. Kubono
Center for Nuclear Study
University of Tokyo
Wako Branch at RIKEN
Hirosawa 2-1, Wako,
Saitama 351-0198
Japan

E-mail: kubono@cns.s.u-tokyo.ac.jp

W. Aoki
National Astronomical Observatory
2-21-1 Osawa, Mitaka
Tokyo 181-8588
Japan

E-mail: aoki.wako@nao.ac.jp

T. Kajino
National Astronomical Observatory
2-21-1 Osawa, Mitaka
Tokyo 181-8588
Japan

E-mail: kajino@nao.ac.jp

T. Motobayashi
Nishina Center for Accelerator-Based Science,
RIKEN
Hirosawa 2-1, Wako,
Saitama 351-0198
Japan

E-mail: motobaya@riken.jp

K. Nomoto
Department of Astronomy
University of Tokyo
7-3-1 Hongo, Bunkyo-ku
Tokyo 113-0033
Japan

E-mail: nomoto@astron.s.u-tokyo.ac.jp

Authorization to photocopy items for internal or personal use, beyond the free copying permitted under the 1978 U.S. Copyright Law (see statement below), is granted by the American Institute of Physics for users registered with the Copyright Clearance Center (CCC) Transactional Reporting Service, provided that the base fee of $23.00 per copy is paid directly to CCC, 222 Rosewood Drive, Danvers, MA 01923, USA. For those organizations that have been granted a photocopy license by CCC, a separate system of payment has been arranged. The fee code for users of the Transactional Reporting Services is: 0-7354-0342-2/06/$23.00

© 2006 American Institute of Physics

Permission is granted to quote from the AIP Conference Proceedings with the customary acknowledgment of the source. Republication of an article or portions thereof (e.g., extensive excerpts, figures, tables, etc.) in original form or in translation, as well as other types of reuse (e.g., in course packs) require formal permission from AIP and may be subject to fees. As a courtesy, the author of the original proceedings article should be informed of any request for republication/reuse. Permission may be obtained online using Rightslink. Locate the article online at http://proceedings.aip.org, then simply click on the Rightslink icon/"Permission for Reuse" link found in the article abstract. You may also address requests to: AIP Office of Rights and Permissions, Suite 1NO1, 2 Huntington Quadrangle, Melville, NY 11747-4502, USA; Fax: 516-576-2450; Tel.: 516-576-2268; E-mail: rights@aip.org.

L.C. Catalog Card No. 2006928553
ISBN 0-7354-0342-2
ISSN 0094-243X

Printed in the United States of America

CONTENTS

Preface... xiii
List of Committees... xv
Photograph of the Participants................................... xvii

1. INVITED AND ORAL TALKS

1.1 OPENING SESSION

Opening Address.. 3
    K. Kodaira
Opening Address.. 5
    T. Kajino

1.2 BIG-BANG COSMOLOGY AND PARTICLE ASTROPHYSICS

Supernovae and Dark Energy...................................... 9
    P. Nugent
On the Origin of Dark Energy in Brane World Cosmology... 15
    K. Ichiki, K. Umezu, T. Kajino, G. J. Mathews, R. Nakamura, and
    M. Yahiro
Lithium Abundances in Halo Subgiants........................ 21
    D. Yong, B. W. Carney, W. Aoki, A. McWilliam, and W. J. Schuster
Big-Bang Nucleosynthesis: Lithium Problems and Scalar-Tensor Theories of Gravity.. 25
    A. Coc
Experimental Study of Photonuclear Reactions Relevant to Nuclear Astrophysiscs.. 31
    T. Shima
Determination of the Astrophysical $^8Li(n, \gamma)^9Li$ Reaction Rate from the Measurement of $^2H(^8Li, ^9Li)^1H$ Reaction... 37
    Z. Li, W. Liu, X. Bai, B. Guo, G. Lian, S. Yan, B. Wang, S. Zeng, Y. Lu,
    J. Su, Y. Chen, K. Wu, N. Shu, and T. Kajino

1.3 MOST METAL-DEFICIENT STARS: OBSERVATION AND THEORY

Stellar Abundances—First Generation to Solar............... 45
    J. E. Norris
An Abundance Study of the Most Iron-Poor Star HE1327-2326 with Subaru/HDS... 53
    W. Aoki, A. Frebel, N. Christlieb, J. E. Norris, T. C. Beers, T. Minezaki,
    P. S. Barklem, S. Honda, M. Takada-Hidai, M. Asplund, S. G. Ryan,
    S. Tsangarides, K. Eriksson, A. Steinhauer, C. P. Deliyannis, K. Nomoto,
    M. Y. Fujimoto, H. Ando, Y. Yoshii, and T. Kajino

Nuclesosynthetic Signatures of Pop.III Survivors and the Origin of HE0107-5240 and HE1327-2326 59
    T. Suda, T. Nishimura, N. Iwamoto, M. Aikawa, M. Y. Fujimoto, and I. Iben, Jr.

The Abundance Pattern and Formation of Extremely Metal-Poor Stars 65
    H. Umeda, N. Tominaga, N. Iwamoto, K. Nomoto, and K. Maeda

Rotating Massive Stars @ Very Low Z: High C & N Production 71
    R. Hirschi

1.4 COSMIC AND GALACTIC CHEMICAL EVOLUTION AND STRUCTURE FORMATION

The Chemical Evolution of the Milky Way: From Light to Heavy Elements 79
    F. Matteucci

Chemical Evolution in the Milky Way Disk 87
    B. Nordström

Galactic Chemical Evolution with Heavy Metals Produced by the First Generation Stars 92
    Y. Ishimaru, S. Wanajo, W. Aoki, S. G. Ryan, and N. Prantzos

1.5 EXPLOSIVE NUCLEOSYNTHESIS IN SUPERNOVAE

Nucleosynthesis in Core Collapse Supernovae 99
    M. Limongi and A. Chieffi

Light Element Production in Type Ic Supernovae 105
    T. Shigeyama, K. Nakamura, S. Wanajo, and S. Inoue

Nucleosythesis and Emission Processes in Aspherical Supernovae 111
    K. Maeda

1.6 WEAK INTERACTION AND NEUTRINO PHYSICS

Results from KamLAND 119
    K. Inoue

Neutrino Magnetic Moment 128
    A. B. Balantekin

The Effect of Neutrino Oscillations on Supernova Light Element Synthesis 134
    T. Yoshida, T. Kajino, H. Yokomakura, K. Kimura, A. Takamura, and D. H. Hartmann

Supernova Detection via a Network of Neutral Current Spherical TPC's 140
    J. D. Vergados and Y. Giomataris

**DM Search by Studying X-Rays Following WIMP
Nuclear Interactions** 147
    H. Ejiri, C. C. Moustakidis, and J. D. Vergados

1.7 SUPERNOVAE, NEUTRON STARS AND HIGH DENSITY MATTER

**Constraints on the Dense Matter Equation of State
from Observations** 155
    J. M. Lattimer
**Recent Developments in Neutron Star Thermal Evolution Theories
and Observation** 163
    S. Tsuruta
Microscopic Origin of the Magnetic Field in Compact Stars 171
    T. Tatsumi

1.8 SUPERNOVA EXPLOSION MECHANISM

Core Collapse Supernovae: Modeling Requirements and Surprises 179
    A. Mezzacappa, J. M. Blondin, O. E. B. Messer, and S. W. Bruenn
Three-Dimensional Modeling of Type Ia Supernova Explosions 190
    F. K. Röpke and W. Hillebrandt
Gravitational Collapse of Massive Stars 196
    S. Yamada

1.9 NEUTRON-CAPTURE AND r-PROCESS NUCLEOSYNTHESIS

Abundance of Heavy Elements in Extremely Metal-Poor Stars 205
    P. François, E. Depagne, V. Hill, M. Spite, F. Spite, B. Plez, T. C. Beers,
    B. Barbuy, R. Cayrel, J. Anderson, P. Bonifacio, P. Molaro, B. Nordström,
    and F. Primas
Radioactive Beams and Exploding Stars at ORNL 213
    M. S. Smith
**Subaru/HDS Studies of r-Process Elements in Metal-Poor Stars from
near UV-Spectra** 221
    S. Honda, W. Aoki, Y. Ishimaru, S. Wanajo, and S. G. Ryan
Origin of the Main r-Process Elements 227
    K. Otsuki, J. Truran, M. Wiescher, J. Gorres, G. Mathews, D. Frekers,
    A. Mengoni, A. Bartlett, and J. Tostevin

1.10 NEUTRON-CAPTURE AND s-PROCESS NUCLEOSYNTHIESIS

**Neutron Capture in Massive Stars—The Challenge of the
Weak s Process** 235
    M. Heil and F. Käppeler

1.11 NUCLEAR ASTROPHYSICS AND NUCLEOSYNTHESIS (I)

Electron Screening in Metallic Environments: A Plasma of the Poor Man 245
    C. Rolfs

Study of Astrophysical (α, n) Reactions Using Light-Neutron Rich Radioactive Nuclear Beams 249
    H. Ishiyama, Y. Watanabe, N. Imai, Y. Hirayama, H. Miyatake, M. Tanaka, N. Yoshikawa, S. Jeong, Y. Fuchi, I. Katayama, T. Nomura, T. Ishikawa, S. K. Das, Y. Mizoi, T. Fukuda, T. Hashimoto, K. Nishio, S. Mitsuoka, H. Ikezoe, M. Matsuda, S. Ichikawa, and T. Shimoda

Indirect Techniques in Nuclear Astrophysics 255
    A. M. Mukhamedzhanov, G. Rogachev, L. D. Blokhintsev, S. Brown, V. Burjan, S. Cherubini, V. Z. Gol'dberg, B. F. Irgaziev, E. Johnson, K. Kemper, V. Kroha, A. Momotyuk, R. G. Pizzone, B. Roeder, S. Romano, C. Spitaleri, R. E. Tribble, and A. Tumino

Trojan Horse Method: Recent Experiments 263
    S. Cherubini, C. Spitaleri, V. Crucilla, M. Gulino, M. La Cognata, L. Lamia, R. G. Pizzone, S. Romano, S. Tudisco, A. Tumino, A. Mukhamedzhanov, L. Trache, R. Tribble, C. Rolfs, and S. Typel

Determination of S_{17} from ^{8}B Breakup by Means of the Method of Continuum-Discretized Coupled-Channels 269
    K. Ogata, S. Hashimoto, Y. Iseri, M. Kamimura, and M. Yahiro

Proton Resonance Scattering of ^{7}Be 275
    H. Yamaguchi, A. Saito, J. J. He, Y. Wakabayashi, G. Amadio, H. Fujiakwa, S. Kubono, L. H. Khiem, Y. K. Kwon, M. Niikura, T. Teranishi, S. Nishimura, Y. Togano, N. Iwasa, and K. Inafuku

Coulomb Dissociation of ^{27}P for study of ^{26}Si(p,γ)^{27}P Reaction 281
    Y. Togano, T. Gomi, T. Motobayashi, Y. Ando, N. Aoi, H. Baba, K. Demichi, Z. Elekes, N. Fukuda, Z. Fülöp, U. Futakami, H. Hasegawa, Y. Higurashi, K. Ieki, N. Imai, M. Ishihara, K. Ishikawa, N. Iwasa, H. Iwasaki, S. Kanno, Y. Kondo, T. Kubo, S. Kubono, M. Kunibu, K. Kurita, Y. U. Matsuyama, S. Michimasa, T. Minemura, M. Miura, H. Murakami, T. Nakmura, M. Notani, S. Ota, A. Saito, M. Serata, S. Shimoura, T. Sugimoto, E. Takeshita, S. Takeuchi, K. Ue, K. Yamada, Y. Yanagisawa, K. Yoneda, and A. Yoshida

1.12 X-RAYS, GAMMA RAYS, AND COSMIC RAYS

Studies of Isotopic Abundances through Gamma-Ray Lines 289
    R. Diehl

Probing Galactic ^{26}Al with Exotic Ion Beams 298
    A. A. Chen (*On behalf of the DRAGON and CRIB Collaborations*)

The Abundance of Live ^{60}Fe in the Early Solar System 304
    S. Tachibana, G. R. Huss, N. T. Kita, G. Shimoda, and Y. Morishita

1.13 METEORITIC ABUNDANCES

Presolar Graphite from the Murchison Meteorite: Imprint of Nucleosynthesis and Grain Formation 311
    S. Amari, R. Gallino, M. Limongi, and A. Chieffi
Stardusts in Meteorites—Precursors of Planets 319
    H. Yurimoto
Eu Isotopic Analyses of SiC Grains from the Murchison Meteorite 324
    K. Terada, T. Yoshida, N. Iwamoto, W. Aoki, and I. S. Williams

1.14 NUCLEAR ASTROPHYSICS AND NUCLEOSYNTHESIS (II)

Composition of the Innermost Core Collapse Supernova Ejecta and the νp-Process 333
    C. Fröhlich, M. Liebendörfer, G. Martínez-Pinedo, F.-K. Thielemann,
    E. Bravo, N. T. Zinner, W. R. Hix, K. Langanke, A. Mezzacappa, and
    K. Nomoto
Universality of the p-Process Nucleosynthesis in Supernova Explosions and Scaling Laws for p- and s-Process Nuclei in the Solar System Abundances 339
    T. Hayakawa, N. Iwamoto, T. Kajino, T. Shizuma, H. Umeda, and
    K. Nomoto
The rp-Process in Core-Collapse Supernovae 345
    S. Wanajo
Elastic α-Scattering on Proton Rich Nuclei at Astrophysically Relevant Energies 351
    Z. Fülöp, D. Galaviz, G. Gyürky, G. G. Kiss, Z. Máté, P. Mohr,
    T. Rauscher, E. Somorjai, and A. Zilges

2. POSTERS

R-Matrix and Potential Model Extrapolations for the NACRE Update and Extension Project 359
    M. Aikawa, K. Arai, M. Katsuma, K, Takahashi, M. Arnould, and
    H. Utsunomoiya
Study of Inelastic Contribution in the ^{7}Be+p Scattering Experiment at CRIB 362
    G. Amadio, H. Yamaguchi, J. J. He, A. Saito, Y. Wakabayashi,
    H. Fujikawa, S. Kubono, L. H. Khiem, Y. K. Kwon, T. Teranishi,
    S. Nishimura, M. Niikura, Y. Togano, N. Iwasa, and S. Inafuku
Neutrino Flavor Changing Neutral Currents and Stellar Collapse 365
    P. S. Amanik and G. M. Fuller
Low-Metallicity Lead Stars: Comparison between Theory and Observations 368
    S. Bisterzo, R. Gallino, O. Straniero, W. Aoki, S. Ryan, and T. C. Beers

Cosmic History of Star Formation and Metal Production 371
    F. Calura and F. Matteucci

A New Measurement of the ^{8}Li$(\alpha,n)^{11}$B Reaction for Astrophysical Interest. 374
    S. K. Das, T. Fukuda, Y. Mizoi, H. Ishiyama, H. Miyatake, Y. X. Watanabe,
    Y. Hirayama, M. H. Tanaka, N. Yoshikawa, S. C. Jeong, Y. Fuchi,
    I. Katayama, T. Nomura, T. Ishikawa, K. Nakai, T. Hashimoto,
    S. Mitsuoka, K. Nishio, P. K. Saha, M. Matsuda, S. Ichikawa, H. Ikezoe,
    T. Furukawa, H. Izumi, T. Shimoda, and T. Sasaqui

Oxygen in Very Metal-Poor Stars. 377
    É. Depagne and F. Primas

Finite-Size Effects on the Hadron-Quark Mixed Phase 380
    T. Endo, T. Maruyama, S. Chiba and T. Tatsumi

Search for α-Enhanced Stars in the Spectroscopic SDSS Stellar Database 383
    M. Franchini, C. Morossi, P. D. Marcantonio, M. L. Malagnini, and
    M. Chavez

Nucleosynthesis inside Magnetically-Driven Jets in a Gamma-Ray Burst 386
    S. Fujimoto, M. Hashimoto, K. Kotake, and S. Yamada

Radioactive Elements in Stellar Atmospheres. 389
    V. Gopka, A. Yushchenko, S. Goriely, A. Shavrina, and Y. W. Kang

Missing Mass in Galaxies in Dynamic Universe Model of Cosmology (Part 3) 392
    S. N. P. Gupta

Weak-Coupling Structure of Proton Resonant States in ^{23}Al Studied with RI Beam at CNS 395
    J. J. He, S. Kubono, T. Teranishi, M. Notani, S. Michimasa, H. Baba,
    S. Nishimura, M. Nishimura, Y. Yanagisawa, N. Hokoiwa, M. Kibe,
    Y. Gono, J. Y. Moon, J. H. Lee, C. S. Lee, H. Iwasaki, and S. Kato

Cosmological Solutions to the Discrepancy among the Light Elements Abundances and WMAP 400
    K. Ichikawa

Origin of Cosmological Magnetic Fields 403
    K. Ichiki, K. Takahashi, H. Ohno, H. Hanayama, and N. Sugiyama

Neutrino Emission from Type Ia Supernovae. 406
    K. Iwamoto and T. Kunugise

Explosive Nucleosynthesis in Different Y_e Conditions 409
    N. Iwamoto, H. Umeda, K. Nomoto, N. Tominaga, F.-K. Thielemann, and
    W. R. Hix

Solar Neutrino Fluxes Using the Exponential S-Factor. 412
    H. A. Kassim, I. A. Jalil, and N. Yusof

MSW Effect in Supernova-Shock Propagation. 415
    S. Kawagoe, T. Kajino, K. Yoshihara, H. Suzuki, K. Sumiyoshi, and
    S. Yamada

Neutrino-Induced Hydrogen Burning 418
    C. T. Kishimoto and G. M. Fuller

Multigroup Flux-Limited Diffusion Neutrino Transport Simulations
for Magnetized and Rotating Core-Collapse Supernovae.................... 421
 K. Kotake, N. Ohnishi, S. Yamada, and K. Sato

The p-Process in the Carbon Deflagration Model for Type Ia
Supernovae and Chronology of the Solar System Formation 424
 M. Kusakabe, N. Iwamoto, and K. Nomoto

Current Progress of Nuclear Astrophysics Experiments at CIAE............. 427
 W. Liu, Z. Li, J. Su, X. Bai, Y. Wang, G. Lian, B. Guo, S. Zeng, S. Yan,
 B. Wang, N. Shu, and Y. Chen

Two-Dimensional Simulation of Core-Collapse Supernovae: Role of
Anisotropic Neutrino Radiation on Explosion Dynamics 430
 H. Madokoro, T. Shimizu, and Y. Motizuki

Time-Variable Complex Metal Absorption Lines in the Quasar
HS1603+3820 433
 T. Misawa, M. Eracleous, J. C. Charlton, and A. Tajitsu

Calibration of Lick Indices from SDSS Spectra........................... 436
 C. Morrosi, M. Franchini, P. Di Marcantonio, M. L. Malagnini, and
 M. Chavez

Self-Bound Object with Kaon Condensates as a Baryonic
Dark Matter 439
 T. Muto

The Metal Enrichment of Galaxies and Galaxy Clusters in the Cold
Dark Matter Universe 442
 M. Nagashima, C. G. Lacey, T. Okamoto, C. M. Baugh, C. S. Frenk, and
 S. Cole

Explosive Nucleosynthesis Inside/Outside of the Jet Launched by a
Collapsar *(Abstract)* 445
 S. Nagataki, A. Mizuta, and K. Sato

Light Elements Produced by Nitrogen-Rich Type Ic Supernovae 446
 K. Nakamura, S. Inoue, S. Wanajo, T. Suzuki, and T. Shigeyama

Constraints on Brans-Dicke Cosmology with a Varying Λ Term Due
to the Big-Bang Nucleosynthesis and WMAP................................ 449
 R. Nakamura, M. Hashimoto, and K. Arai

Heavy Element Nucleosynthesis in the MHD Jet Explosions of
Core-Collapse Supernovae.. 452
 N. Nishimura, M. Hashimoto, S. Fujimoto, K. Kotake, and S. Yamada

Neutron-Capture Nucleosynthesis in the He-Flash Convective Zone in
Extremely Metal-Poor Stars 455
 T. Nishimura, N. Iwamoto, T. Suda, M. Aikawa, M. Y. Fujimoto,
 and I. Iben, Jr.

Nucleosynthesis by Type Ia Supernova for Different Metallicity.......... 458
 T. Ohkubo, H. Umeda, K. Nomoto, and T. Yoshida

SNe Feedback and the Formation of Elliptical Galaxies................... 461
 A. Pipino and F. Matteucci

Zinc Abundances in Metal-Poor Stars..................................... 464
 Y. Saito, M. Takada-Hidai, Y. Takeda, S. Honda, and M. Katsumata

Astrophysics at RIA (ARIA) Working Group 467
 M. S. Smith, H. Schatz, F. X. Timmes, M. Wiescher, and U. Greife

Computational Infrastructure for Nuclear Astrophysics.................470
 M. S. Smith, E. J. Lingerfelt, J. P. Scott, C. D. Nesaraja, W. R. Hix,
 K. Chae, H. Koura, R. A. Meyer, D. W. Bardayan, J. C. Blackmon, and
 M. W. Guidry

Fate of Core-Collapse Supernovae: Formation of Neutron Star and
Black Hole..473
 K. Sumiyoshi, H. Suzuki, and S. Yamada

Magnetorotational Collapse of Very Massive Stars: Formation of Jets
and Black Holes..476
 Y. Suwa, T. Takiwaki, K. Kotake, and K. Sato

Neutrino-Nucleus Reactions Induced by Supernova Neutrinos..............479
 T. Suzuki, S. Chiba, O. Iwamoto, and T. Kajino

Chemical Evolution of Sulfur in the Metallicity Range of
$-4<[Fe/H]<+0.5$...482
 M. Takada-Hidai, M. Katsumata, and Y. Saito

Aspherical Ejecta of Type Ia Supernovae Inferred from High
Velocity Features..485
 M. Tanaka, P. A. Mazzali, K. Maeda, and K. Nomoto

Population III Core-Collapse Supernova Yields and Extremely
Metal-Poor Star Abundance Pattern...............................488
 N. Tominaga, H. Umeda, and K. Nomoto

Electric Field Strength of Coherent Radio Emission in Rock Salt
Concerning Ultra High-Energy Neutrino Detection....................491
 Y. Watanabe, M. Chiba, O. Yasuda, Y. Shibasaki, T. Kamijo, Y. Chikashige,
 T. Kon, Y. Shimizu, A. Amano, Y. Takeoka, S. Ninomiya, and S. Mori

The Extraction of Fractions of the Resonant Component from
Analyzing Powers in $^6Li(d, \alpha)^4He$ and $^6Li(d,p_0)^7Li$ Reactions at Very
Low Incident Energies...494
 M. Yamaguchi, Y. Tagishi, Y. Aoki, T. Iizuka, T. Nagatomo, T. Shinba,
 N. Yoshimaru, Y. Yamato, T. Katabuchi, and M. Tanifuji

Study on the Dominant Reaction Path in Nucleosynthesis during
Stellar Evolution by Means of the Monte Carlo Method.................497
 K. Yamamoto, K. Hashizume, T. Wada, M. Ohta, T. Nishimura,
 M. Y. Fujimoto, K. Kato, T. Suda, and M. Aikawa

Seven-Layer Supernova Mixtures Reproducing Isotopic Ratios of
Presolar Grains..500
 T. Yoshida

Accretion in Sirius Binary System..............................503
 A. Yushchenko and V. Gopka

The Exponential S-Factor in the PP Chain........................506
 N. Yusof, I. A. Jalil, and H. A. Kassim

List of Participants..509
Scientific Program..521
Author Index...525

PREFACE

The International Symposium on Origin of Matter and Evolution of Galaxies OMEG05 was held on November 8 – 11 at the Koshiba Hall of the University of Tokyo. This is a series of nuclear astrophysics international symposia being held roughly every other year since 1988. The first meeting was very much promoted by the strong interest of radioactive nuclear beams in nuclear physics under the title of Heavy Ion Physics and Nuclear Astrophysical Problems.

This symposium was hosted by five institutions: National Astronomical Observatory (NAO), The Graduate University for Advanced Studies (GUAS), Department of Astronomy, Graduate School of Science, University of Tokyo (Tokyo), Center for Nuclear Study, University of Tokyo (CNS), and RIKEN Accelerator Research Facility (RIKEN).

The symposium was attended by about 150 people, including about 50 people from outside Japan. Because of the special fund provided from GAUS under the conference name of "SUBARU Astronomy for the Establishment of Cosmo-Nuclear Astrophysics", we placed an emphasis this time on observational works and cosmology. As for nuclear physics, it also was a good occasion. The third-generation RI Beam facility RIBF is just about to deliver the first beam, and low energy RI beams have become available at the CNS RI Beam in-flight separator CRIB at the University of Tokyo and also at Tokai RI ACcelerator facility TRIAC by the KEK-JAEA collaboration.

We are very grateful to the administration officers of the host institutes, students and young researchers for their services provided to this Symposium, and especially to Ms. Miwa Yamaguchi, Sayuri Nagano and Shiho Okada for the secretarial assistance.

We apologize for the slow publication of this book. We instituted a peer review process for all manuscripts, and thus it took a little longer to finalize everything, but we hope the delay will be forgiven as this process assures the quality of our accomplishments.

April, 2006

S. Kubono (Editor in Chief)
W. Aoki, T. Kajino, T. Motobayashi, K. Nomoto,

List of Committees

Local organizing committee

W. Aoki* (NAO/GUAS: Co-chair)
S. Honda (NAO)
S. Kawanomoto (NAO)
T. Motobayashi (RIKEN/Tokyo: Co-chair)
K. Nomoto (Tokyo: Co-chair)
Y. Takeda (NAO/GUAS)
* Scientific Secretary

M.Y. Fujimoto (Hokkaido)
T. Kajino (NAO/GUAS/Tokyo: Chair)
S. Kubono (CNS, Tokyo: Co-chair)
K. Noguchi (NAO/GUAS)
T. Shigeyama* (Big Bang Center, Tokyo)
K. Yoneda* (RIKEN)

International advisory committee

H. Ando (NAO/Tokyo)
R. Boyd (Ohio State/NSF)
R. Gallino (Torino)
K. Koyama (Kyoto)
C.S. Lee (Korea)
G. Mathews (Notre Dame)
Y. Nagai (Osaka)
N. Prantzos (Paris)
K. Sato (Tokyo)
J. Silk (Cambridge)
Y. Suzuki (Tokyo)
C. Spitaleri (Catania)
J. Truran (Chicago)
Y. Yoshii (Tokyo)

A. Balantekin (Wisconsin)
L. Buchmann (TRIUMF)
W. Haxton (Washington, Seattle)
K. Langanke (GSI)
W.P. Liu (CIAE, Beijing)
A. Mengoni (IAEA, Vienna/CERN)
P. Nissen (Aarhus)
C. Rolfs (Bochum)
H. Schatz (Michigan State)
C. Sneden (Texas)
M. Smith (Oak Ridge)
F.-K. Thielemann (Basel)
M. Wiescher (Notre Dame)

Host institutes

The Graduate University for Advanced Studies (GUAS)
National Astronomical Observatory (NAO), Japan
Center for Nuclear Study, University of Tokyo (CNS, Tokyo)
The Accelerator Research Facility, the Institute of Physical and Chemical Research (RIKEN)
Department of Astronomy, Graduate School of Science, University of Tokyo (Tokyo)

Supported by

The Graduate University for Advanced Studies (GUAS)
National Astronomical Observatory (NAO)
Center for Nuclear Study, University of Tokyo (CNS, Tokyo)
The Accelerator Research Facility, the Institute of Physical and Chemical Research (RIKEN)
Department of Astronomy, Graduate School of Science, University of Tokyo (Tokyo)
Japan Society for the Promotion of Science

1. INVITED and ORAL TALKS

1.1 OPENING SESSION

Opening Address

Keiichi KODAIRA
President of the Graduate University for Advanced Studies (SOKENDAI)

Dear Symposium Participants, Dear Colleagues, Ladies and Gentlemen,

It is a great pleasure for me to welcome you and to open this symposium on the Origin of Matter and Evolution of Galaxies, which shall illuminate the new horizon of Nuclear Astrophysics and Cosmology.
I feel highly honored as the President of the Graduate University for Advanced Studies, SOKENDAI, to host this symposium.
In recent years, many of theoretical scenarios in the field of Nuclear Astrophysics and Cosmology became the subjects of solid observational and/or experimental examination. The large telescopes on the ground, such as Subaru, VLT, Gemini, Keck, revealed the details of physical phenomena in the remote universe, like supernova explosions. At the same time they enabled us to take closer looks into nearby faint objects, such as to study the chemical composition of most metal-poor stars.
The telescopes in the space orbits, observing cosmic gamma-ray or X-ray, provided us with dynamic views of the high-energy cosmic events, which turned out to be rich in information to improve our knowledge about the Matter and Universe.
Not necessary to mention, the experimental efforts to detect the solar and cosmic neutrinos were highly rewarded, to develop a new field of astrophysics, neutrino astronomy. Today's venue of this symposium, the KOSHIBA Hall, carries the name of Dr. Masatoshi KOSHIBA, one of the Nobel Prize Laureates in physics 2004, whose group first clearly confirmed the detection of the neutrinos of the cosmic origin, emitted by the SN1987A in the Large Magellanic Cloud. I remember his talk at the Astronomy Seminar just a few months before his retirement from the University of Tokyo, spring 1987. He told us the performance of the KAMIOKANDE, the predecessor of the SUPER-KAMIOKANDE, which had been just completed under his initiative. He showed us a map of Cherenkov photon distribution,
with dots densely covering the diagram. He indicated that it would be highly challenging to pick up a few of real solar or cosmic events from thousands of the background shots.
You know, however, already, lucky enough for him, just a month before his retirement,
the sharp, spiky signal of the supernovae neutrino-burst arrived to the earth after traveling over 150 thousands years from the Large Magellanic Cloud.
Many experimental challenges are further going on. Ground-based Gamma-ray Telescopes and Gravitational-Wave Detector Arrays are certainly new assets to widen the horizon of the nuclear astrophysics. High-power laser technology and current nuclear-fusion

plasma experiments are opening the access to the Experimental Astrophysics, which is closely related to the topics of this symposium. The quick development of computer science also contributed to the high-grade refinement of the theoretical prediction and interpretation in nuclear astrophysics.

One says that the science in the 21st century may be focused onto the Universe and the Life. Both subjects involve the question of its origin and evolution.

In the life science, we observe the various structure scales, starting from molecules like DNA, RNA, protein, through organelles and cells, tissues and organs, and individual beings and species. The origin and the evolution of the living world is still full of mysteries, but DNA is regarded now as one of the important indicators of evolution, by adopting the concept of the molecular clock. You cannot, however, look back into the past, except for studying the fossils. As for the universe and the cosmology, there are different scales of matter, starting from elementary particles, atoms and molecules, dust and grains, stars and galaxies, clusters of galaxies, and large cosmic structures. The chemical composition in this case serves something like DNA as a kind of important indicator of evolution, by adopting the concept of the chemical evolution basing upon the nuclear astrophysics. In the cosmology, you regard the non-convective envelopes of low-mass stars as cosmic fossils, in addition you can look back into the past by observing remote galaxies and intergalactic matter there.

The evolution of the universe, at the first glance, seems to be more straightforward to comprehend than the evolution of the life, for the basic processes are controlled by "Physics"; that is, each step of the evolution seems to be more deterministic, and to be commanded by smaller amount of ordered information.

Probably, today, we are standing pretty near to the position to comprehend the whole story of the evolution of the universe, but the recent introduction of the "Dark Matter" and the "Dark Energy" in particular, may cast a doubt on such optimism.

I wish that the discussions in this symposium may lighten up the horizon of the nuclear astrophysics, and bring us a step closer to the full understanding of the origin of matter and the evolution of the galaxy world.

Now, let me allow to say some words about my university, SOKENDAI. SOKENDAI is a unique system of research schools composed of 18 academic research institutes; among them NAOJ, KEK, NIFS, ISAS, NIPR, etc. SOKENDAI has only doctor courses, and accepts students from any country in the world. I would be most happy, if you, the participants to this symposium take a look at the SOKENDAI pamphlet available at the information desk, and encourage your young successors to come to SOKENDAI.

Finally I wish that this symposium will be stimulating for you, and that you enjoy this 4-day meeting in Tokyo,

Thank you for your attention.

2005.11.8

Opening Address

Toshitaka Kajino

*National Astronomical Observatory and The Graduate University for Advanced Studies
Department of Astronomy, Graduate School of Science, University of Tokyo*

Dear Distinguished Guests, Participants,
Dear Colleagues, Ladies and Gentlemen,

On behalf of the Organizing Committee, I heartily welcome you all especially those who came from foreign countries. It is my honor to address an opening address as the chairman of this conference on the Origin of Matter and Evolution of Galaxies -- New Frontier of Nuclear Astrophysics and Cosmology.

This is a series of international conferences started in 1988 which was one year after the SN1987 had appeared in Large Magellanic Cloud. The SN1987 was the celestial gift as an event of the century. Its brightness literally shed lights on all scientists. Many astronomers and astrophysicists have enjoyed a success in understanding and modeling the explosion mechanism and explosive nucleosynthesis in massive stars. Mysteries also still remain: An example is a question why X-ray pulsar has not yet been found for more than 18 years although the proto-neutron star once formed and 99% of its gravitational energy was released as intensive flux of energetic neutrinos from the proto-neutron star. This neutrino burst was detected in KAMIOKANDE and several other observatories in 1987. Neutrino is a tiny particle but provides very profound information of rich and deep structure of the stars, the galaxies and the Universe. Standing at the dawn of new science with the SN1987, we all shared the same feeling that the next decades should be very exciting to find more surprises in astronomy, cosmology, astrophysics and fundamental physics. We then decided to organize an international conference on nuclear astrophysics and cosmology, which was the 1st conference held in 1988. Professor Koshiba was awarded Nobel Prize in Physics in 2002 for his pioneering work of opening the new field of neutrino astronomy. We are happy today to bring ourselves together here in the Koshiba Memorial Hall which was built for celebrating his great scientific achievement.

Our interdisciplinary field has further advanced and developed since then in both depth and width. In the recent ten years nuclear astrophysics had a renaissance by producing and controlling new radioactive nuclear beams with very short lives in order to simulate explosive nuclear burning processes in the laboratory system, which were thought to realize only in celestial events like supernovae or in the Big-Bang. We are also able to look back the entire history of the evolving Universe observationally by using gigantic telescopes like SUBARU, VLT, KECK and the Hubble Space Telescope. One of the big surprises found in recent years is that we know little about the invisible part of the Universe. We need dark matter and dark energy to make an accelerating universal expansion, which is suggested by several cosmological observations, but we do not know what they are. These dark components in the Universe manifest themselves only through the interactions with ordinary luminous matter, baryons. Only when we will definitely detect them in direct manner and find their true nature, we can conclude that these dark components are the real entity of our Universe. Towards this goal, therefore, we should get to the bottom of the origin of ordinary matter, i.e. the standard elementary particles and atomic nuclides, as precisely as possible because the known ordinary matter forms beautiful hierarchy in the Universe, taking the keys to approach invisible mass and energy. For our common goal and purpose to reach the truth in nature, though we are working in different neighboring fields, we should promote tight collaboration and interactions. It is our spirit of this series of conferences to bring together all scientists and exchange expertise among them who come from interdisciplinary field of science which includes astronomy and astrophysics, cosmology, cosmic-ray physics, particle physics, nuclear physics, meteoritic science, cosmo-chemistry, and many others.

Finally, let me express how important role the young generation would play in such interdisciplinary scientific field in the future or even today. We are very happy to find many graduate students and junior scientists as well as established senior scientists in this conference. I wish all of you stimulate and inspire one another with exchanging lots of new ideas through active and constructive discussions. I hope you enjoy this conference and your stay in Tokyo for a week!

Thank you for your attention.

1.2 BIG-BANG COSMOLOGY and PARTICLE ASTROPHYSICS

Supernovae and Dark Energy

Peter Nugent

Lawrence Berkeley National Laboratory, 1 Cyclotron Rd, MS 50F-1650, Berkeley, CA, 94720, USA

Abstract.
Astronomers have begun to measure the fundamental parameters of cosmology through the observation of very distant Type Ia supernovae. Over the past decade more than 300 spectroscopically confirmed high-redshift supernovae have been discovered. These supernovae are used as standardized candles to measure the history of the expansion of the universe. Under the current standard model for cosmology these measurements indicate the presence of a heretofore unknown dark energy causing a recent acceleration in the expansion of the universe.

At this time supernova measurements of the cosmological parameters are no longer limited by statistical uncertainties, rather systematic uncertainties are the dominant source of error. These include the effects of evolution (further back in time do the supernovae behave the same way?), the effect of intergalactic dust on the brightness of the supernovae and the relationship between supernovae and their environments. Here I present exciting new developments in the field of cosmology using Type II-P supernovae as standardized candles and the prospect of using them to independently measure the cosmological parameters.

Keywords: Supernovae, Cosmology
PACS: 97.60.Bw, 98.80.-k

INTRODUCTION

The recent discovery of a cosmic acceleration based on the analysis of the Hubble diagram of Type Ia supernovae [SNe Ia; 1, 2] has far-reaching implications for our understanding of the Universe. While indirect evidence for the acceleration can be deduced from a combination of studies of the cosmic microwave background and large scale structure [3, 4, 5] distance measurements to supernovae provide a valuable direct and model independent tracer of the evolution of the expansion scale factor necessary to constrain the nature of the proposed dark energy. The mystery of dark energy lies at the crossroads of astronomy and fundamental physics: the former is tasked with measuring its properties and the latter with explaining its origin.

Systematic uncertainties (rather than statistical errors) may soon limit SN Ia measurements of the expansion rate at $z \sim 0.5$ [see 6, 7, for recent analyses]. A largely unexplored source of potential bias is evolution in the progenitor properties and/or the SN explosion. While several programs are underway to measure, test, and constrain SN Ia systematics [8, 9], it is highly desirable to consider independent tests of the cosmology where both the underlying physics and susceptibility to bias and evolution are different.

As cosmological probes, SNe II have lagged behind their brighter and better calibrated cousins, SNe Ia, but their potential has improved significantly as a result of several recent studies. Baron et al. [10], Baron et al. [11], Mitchell et al. [12], and two doctoral theses [13, 14], have utilized new samples of SNe II and demonstrated that a subset, the plateau SNe II-P, are particularly promising as distance indicators. From an astrophysical standpoint, SNe II-P hold three advantages over SNe Ia as cosmological probes: (1) Their progenitor stars are well understood, [15, 16, 17], (2) the physics of their atmospheres, dominated by hydrogen, is much simpler to understand and model [10], and (3) while fainter, they are more abundant per unit volume [18, 19]. The two main disadvantages are that they are on average 1.2 magnitudes fainter in the optical than SNe Ia and that all distance measurements currently based on SNe II-P require a reasonable-quality spectrum of the event.

SNE II-P

Unlike the other members of the core-collapse supernova family, SNe II-P maintain a massive hydrogen envelope prior to explosion. From analyses of their optical light curves and spectra [e.g., 20], they evidently suffer little subsequent interaction with the surrounding medium afterwards as well – they are the result of the putative red supergiant exploding into a near-vacuum. Recent results from spectropolarimetric studies also suggest that, at least during the plateau epoch, the ejecta and electron-scattering photosphere are quite spherical (see Leonard and Filippenko [21] and references therein).

There are now two different approaches used to measure the distance to SNe II: the spectral expanding atmosphere method [SEAM; see 10], the descendent of the traditional expanding photosphere method [EPM; see 22, 23], and the standardized candle method implemented by Hamuy and Pinto [24], Hamuy [25], hereafter HP02 and H03 respectively. Each provides an independent way to achieve distances to SNe IIP, the former based on theoretical modeling while the latter is a completely empirical approach.

The approach advocated by HP02 is a particularly significant development since it is motivated by sound physical principles. In more luminous supernovae, the hydrogen recombination front is maintained at higher velocities, pushing the photosphere farther out in radius. When a SN II-P is on the plateau phase, a period which lasts for around 100 days, a strong correlation is expected and observed between the velocity of the weak Fe II lines near 5000 Å (which nicely track the electron-scattering photosphere) and the luminosity. Extinction corrections are based on a variety of different methods. Application of this empirical correlation to 24 SNe IIP in the Hubble flow, as seen in Fig. 1 (H03), yields a Hubble diagram in I-band with a scatter of 0.29 magnitudes (15% in distance).

IMPROVEMENTS

Recently we have created a modified version of this technique where we have simultaneously combined both the extinction correction, using the restframe $V - I$ color on the plateau, and the Fe II velocity-luminosity correction to arrive at a simple correlation between these parameters and the absolute brightness.

$$M_I = -\alpha \log_{10}(V_{FeII}/5000) - 1.36[(V-I) - (V-I)_0] + M_{I_0} \qquad (1)$$

Here, as in H03, we have adopted the relative SBF distance scale [26]. In this fit we have employed the standard relationship between the $V-I$ colors for a dust law with $R_V = 3.1$ for a SN II-P at day 50 ($A_I = 1.36*E_{V-I}$). As is done in the SN Ia studies [27] for the color stretch-relationship, we have adopted a ridge-line, unextinguished $(V-I)_0$ color for SN II-P of 0.53 magnitudes. The exact choice is irrelevant since this term and the M_{I_0} term are degenerate. Using this technique we can produce a Hubble diagram in restframe I-band with a scatter of only 0.28 magnitudes for those SNe II-P in the Hubble flow ($cz > 3000$ km/s), similar to that found in H03. The scatter for all SN II-P is reduced from 1.11 to 0.52 magnitudes. Crucially, we find there is no advantage in using reddening estimators based on late-time color measures or on detailed modeling of the spectroscopic data. In addition, based upon both modeling and data, we have found that at sufficiently late epochs (> 40 days) the velocity of H-β correlates nicely with the velocity of the Fe II features. This feature is slightly stronger than the Fe II features and is more readily identifiable at higher redshift. This work has provided the necessary breakthrough to make SNe II-P valuable cosmological probes at high-redshift.

A COSMOLOGICAL HUBBLE DIAGRAM BASED ON A SAMPLE OF HIGH REDSHIFT SNE II-P

In the previous section, we have shown that extinction corrections using $V-I$ colors during the plateau phase at day 50 and concurrent expansion velocities determined using a variety of prominent absorption lines can be used to generalize and extend the important distance determination method initially proposed by HP02. As SNe II-P are now being found to redshifts $z \simeq 0.4$ in rolling searches such as the SNLS [8], we can now apply our method to these distant supernovae to independently verify the cosmic acceleration.

Starting in 2003, we began using the Low Resolution Imaging Spectrometer [LRIS; Oke et al. [28]], a double-arm spectrograph mounted on the 10-m Keck I telescope, to observe SNLS-discovered supernovae. Although our primary program initially targeted SNe Ia [9], during the first two years we also successfully studied five moderate-redshift ($0.1 < z < 0.3$) SNe II-P, identified from the SNLS rolling search via their light curves. These events were typically observed five times per lunation in $g'r'i'z'$ [7] with coverage starting before explosion and extending throughout the plateau phase.

The SNLS represents an impressive 5-year commitment to deliver various wide-field products using the MegaPrime $1\times1°$ camera. The "Deep Synoptic Survey" involves repeat imaging of four square-degree fields in $griz$ on every third dark night per lunation for 6 lunations per year. Supernova candidates are always available in at least two separate fields. The huge advantage of the SNLS is its continuous monitoring of the same deep fields. Previous time-specific campaigns could often be ruined by poor weather on only a single night during search or follow-up.

To calculate the restframe $V-I$ colors during the plateau phase, we first made an estimate of the explosion date of the supernova conservatively based on the midpoint between the last non-detection and the first detection of the supernova in any filter during the rolling search. We then interpolated the $g'r'i'z'$ colors at day $50(1+z)$. These colors were used to "warp" a day 50 template SN II-P spectrum, redshifted appropriately,

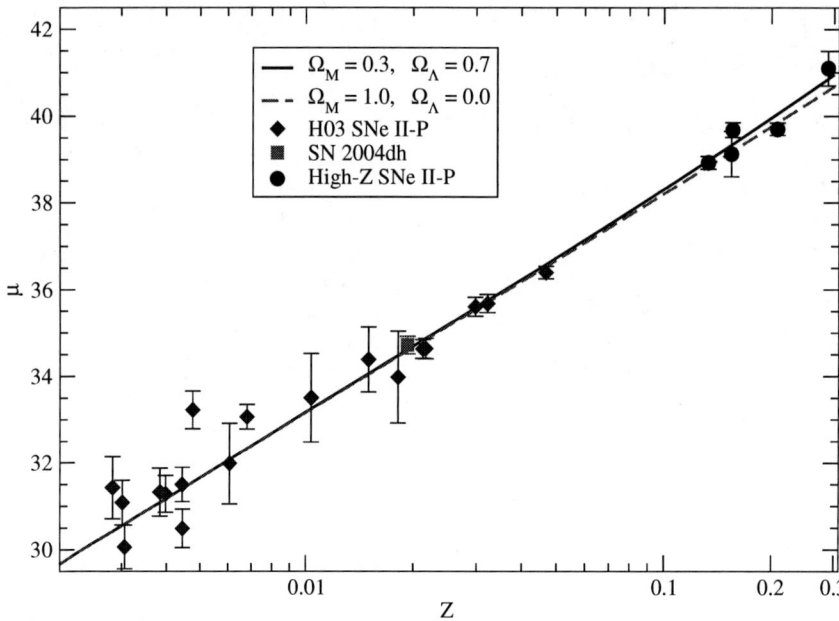

FIGURE 1. The Hubble diagram for both the local SNe II-P (diamonds and square) and the high-redshift SNLS SNe II-P (circles) observed spectroscopically with Keck+LRIS. The observed scatter for the supernova in the Hubble flow is 0.26 magnitudes with a reduced χ^2 of 1.4, which is indicates a small amount of intrinsic uncertainty. To understand the current power of this technique we have over-plotted two differing Hubble lines for a flat cosmology with $\Omega_M = 1$ and 0.3.

following the protocol described in Nugent et al. [29] with the exception that here we use spline fits to the underlying colors to adjust the template rather than a reddening law. This spectrum was then de-redshifted and the restframe $V - I$ color of the supernova was calculated. The Fe II velocities were measured via a cross-correlation technique, extrapolated to restframe day 50 using a power-law fit to the velocity evolution $V(50) = V(t) * (t/50)^{0.464 \pm 0.017}$. The full analysis can be seen in Nugent et al. [30]. The Hubble diagram is presented in Figure 1. The observed scatter was 0.26 magnitudes for all supernovae in the Hubble flow and the reduced χ^2 is 1.4 suggesting a small amount of intrinsic dispersion.

CONCLUSIONS

Given multi-color lightcurves ($g'r'i'z'$) provided by SNLS, we conclude that this method *as-is* can be practically applied, using existing instrumentation, to $z = 0.3$. Over this redshift range we could detect the cosmic acceleration at >95% level, independent of any other constraints on the cosmological parameters, with $\simeq 12$ additional high-redshift SNe II-P coupled with the five low-redshift CCCP SNe II-P observed this past year. Extending these measurements to $z = 0.5$ for the SNLS supernovae would require *J*-band imaging with *HST/NICMOS* for average luminosity SNe II-P. However, due to the wide dispersion in their apparent luminosity, a large fraction of this distribution could be observed in the infrared with 8-10 meter ground-based telescopes.

Exploring the utility of measuring distances to SNe II-P has potential benefits well beyond simply verifying, independently, the acceleration seen at redshifts $z < 1$. Several plausible models for the time evolution of the dark energy require distance measures to $z \simeq 2$ and beyond [31]. At such high redshifts, other cosmological probes may become less effective than at $z \leq 1$. Weak lensing, for example, will suffer from the loss of suitable lenses, and, while evidence at $z < 1$ suggests some fraction of SNe Ia explode with very short delay-times and hence will be abundant at high-redshift [18, 32], the efficiency and metallicity dependence of the SN Ia progenitor sysstem is still in doubt and may curtail their production [e.g. 33]. However, current models for the cosmic star-formation history predict an abundant source of SNe II at these epochs and future facilities, such as the proposed *JDEM* telescope, in concert with *JWST* and/or future 30-m telescopes such as TMT, could potentially use SNe II-P to determine distances at these very high redshifts.

ACKNOWLEDGMENTS

I wish to thank the organizers of OMEGA05 for a very enjoyable conference. This research used resources of the National Energy Research Scientific Computing Center, which is supported by the Office of Science of the U.S. Department of Energy under Contract No. DE-AC03-76SF00098. We thank them for a generous allocation of computing time.

REFERENCES

1. A. Riess, et al., *AJ* **116**, 1009 (1998).
2. S. Perlmutter, et al., *ApJ* **517**, 565 (1999).
3. G. Efstathiou, S. Moody, J. A. Peacock, W. J. Percival, C. Baugh, J. Bland-Hawthorn, T. Bridges, R. Cannon, S. Cole, M. Colless, C. Collins, W. Couch, G. Dalton, R. de Propris, S. P. Driver, R. S. Ellis, C. S. Frenk, K. Glazebrook, C. Jackson, O. Lahav, I. Lewis, S. Lumsden, S. Maddox, P. Norberg, B. A. Peterson, W. Sutherland, and K. Taylor, *MNRAS* **330**, L29–L35 (2002).
4. C. L. Bennett, M. Halpern, G. Hinshaw, N. Jarosik, A. Kogut, M. Limon, S. S. Meyer, L. Page, D. N. Spergel, G. S. Tucker, E. Wollack, E. L. Wright, C. Barnes, M. R. Greason, R. S. Hill, E. Komatsu, M. R. Nolta, N. Odegard, H. V. Peiris, L. Verde, and J. L. Weiland, *ApJS* **148**, 1–27 (2003).
5. D. J. Eisenstein, M. Blanton, I. Zehavi, N. Bahcall, J. Brinkmann, J. Loveday, A. Meiksin, and D. Schneider, *ApJ* **619**, 178–192 (2005).

6. R. A. Knop, et al., *ApJ* **598**, 102 (2003).
7. P. Astier, et al., *A&A* (2005), accepted.
8. M. Sullivan, et al., in *1604-2004: Supernovae as Cosmological Lighthouses*, edited by S. B. M. Turatto, W. R. J. Shea, and L. Zampieri, ASP, Padua, Italy, 2005, p. 1.
9. R. S. Ellis, et al. (2006), in preparation.
10. E. Baron, P. E. Nugent, D. Branch, and P. H. Hauschildt, *ApJ* **616**, 91 (2004).
11. E. Baron, P. Nugent, D. Branch, P. H. Hauschildt, D. Leonard, A. V. Filippenko, M. Turatto, and E. Cappellaro, *ApJ* **586**, 1199 (2003).
12. R. Mitchell, et al., *ApJ* **574** (2002).
13. M. Hamuy, Ph.D. thesis, University of Arizona (2002).
14. D. C. Leonard, Ph.D. thesis, University of California, Berkeley (2000).
15. A. Heger, C. L. Fryer, S. E. Woosley, N. Langer, and D. H. Hartmann, *ApJ* **591**, 288–300 (2003).
16. W. Li, S. D. Van Dyk, A. V. Filippenko, and J.-C. Cuillandre, *PASP* **117**, 121–131 (2005).
17. S. J. Smartt, J. R. Maund, G. F. Gilmore, C. A. Tout, D. Kilkenny, and S. Benetti, *MNRAS* **343**, 735–749 (2003).
18. F. Mannucci, M. della Valle, N. Panagia, E. Cappellaro, G. Cresci, R. Maiolino, A. Petrosian, and M. Turatto, *A&A* **433**, 807–814 (2005).
19. E. Cappellaro, M. Riello, G. Altavilla, M. T. Botticella, S. Benetti, A. Clocchiatti, J. I. Danziger, P. Mazzali, A. Pastorello, F. Patat, M. Salvo, M. Turatto, and S. Valenti, *A&A* **430**, 83–93 (2005).
20. N. N. Chugai, "Supernovae with dense circumstellar winds," in *Circumstellar Media in Late Stages of Stellar Evolution*, 1994, pp. 148–+.
21. D. C. Leonard, and A. V. Filippenko, in *1604-2004: Supernovae as Cosmological Lighthouses*, edited by S. B. M. Turatto, W. R. J. Shea, and L. Zampieri, ASP, Padua, Italy, 2005, p. 100.
22. R. P. Kirshner, and J. Kwan, *ApJ* **193**, 27 (1974).
23. B. P. Schmidt, Ph.D. thesis, Harvard University (1993).
24. M. Hamuy, and P. A. Pinto, *ApJ* **566**, 63 (2002).
25. M. Hamuy, in *I.A.U. Colloquium 192: Supernovae (10 years of SN1993J)*, edited by J. Marcaide, and K. Weiler, Springer Verlag, Valencia, Spain, 2003.
26. J. L. Tonry, J. P. Blakeslee, E. A. Ajhar, and A. Dressler, *ApJ* **530**, 625–651 (2000).
27. J. Guy, P. Astier, S. Nobili, N. Regnault, and R. Pain, *A&A* **443**, 781 (2005).
28. J. B. Oke, J. G. Cohen, M. Carr, J. Cromer, A. Dingizian, F. H. Harris, S. Labrecque, R. Lucinio, W. Schaal, H. Epps, and J. Miller, *PASP* **107**, 375–+ (1995).
29. P. Nugent, A. Kim, and S. Perlmutter, *PASP* **114**, 803–819 (2002).
30. P. Nugent, et al., *ApJ* (2006), in press.
31. E. V. Linder, and D. Huterer, *PhRvD* **67**, 081303–+ (2003).
32. M. Sullivan, et al. (2006), in preparation.
33. C. Kobayashi, T. Tsujimoto, K. Nomoto, I. Hachisu, and M. Kato, *ApJ* **503**, L155 (1998).

On the Origin of Dark Energy in Brane World Cosmology

Kiyotomo Ichiki[*,†], Ken-ichi Umezu[*,**], Toshitaka Kajino[*], Grant J. Mathews[‡], Riou Nakamura[§] and Masanobu Yahiro[§]

[*]*National Astronomical Observatory, Mitaka, Tokyo 181-8588, JAPAN*
[†]*Research Fellows of Japan Society for the Promotion of Science*
[**]*Graduate University for Advanced Studies, 2-21-1 Osawa, Mitaka, Tokyo 181-8588, Japan*
[‡]*Center for Astrophysics, Department of Physics, University of Notre Dame, Notre Dame, IN 46556, U.S.A.*
[§]*Department of Physics, Graduate School of Science, Kyushu University, 6-10-1 Hakozaki, Higashi-ku, Fukuoka 812-8581, Japan*

Abstract. We investigate the cosmological evolution in a brane-world scenario in which the bulk is not empty. Rather, exchange of mass-energy between the bulk and the bane is allowed. The expansion history of the universe and the evolution of matter fields on the brane are then modified due to this exchange. We show that the flow of matter realizes an accelerating expansion of the brane, which mimics the dark energy. We investigate the constraints from various cosmological observations on the flow of matter from the bulk into the brane. Interestingly, it is possible to have a $\Lambda = 0$ cosmology to an observer in the brane which satisfies standard cosmological constraints including the CMB temperature fluctuations, Type Ia supernovae at high redshift, and the matter power spectrum.

Keywords: dark energy, brane world cosmology
PACS: 98.80.-k,95.36.+x,98.80.Cq

INTRODUCTION

The origin of the dark energy responsible for the accelerating expansion of the universe is one of the biggest mysteries in modern cosmology [1]. The simplest explanation is that of a cosmological constant, or a vacuum energy in the form of a "quintessence" scalar field slowly evolving along an effective potential. In this contribution, we consider an alternative mechanism by which the observed cosmic acceleration could be produced even without the need to invoke dark energy and its associated complexities. Specifically, we explore models in which the cosmic acceleration is driven [2, 3, 4, 5, 6, 7] by the flow of dark matter from a higher dimension (the bulk) into our three-space (the brane).

This study is also in part motivated by the currently popular view that our universe could be a sub-manifold embedded in a higher-dimensional space-time. As a practical phenomenological model, a thin three-brane (Randall-Sundrum brane) embedded in a five-dimensional anti-de Sitter space AdS_5 has been proposed [8] in which to explore such higher dimensional physics. In such a picture, the extra (bulk) dimension can be infinite and the observed 3-dimensional fields are represented by zero modes of bulk fields in the domain-wall background. These modes are localized and thus behave like 3-dimensional mass-less fields. In this domain-wall scenario, physical matter fields are dynamically confined to this sub-manifold, while gravity can reside in the extra bulk dimension.

The stability of massive matter fields has been analyzed [9] in this scenario where it was shown that such massive particles are metastable on the brane. From the viewpoint of an observer in the 4-dimensional space time, these massive particles appear to propagate for some time in three spatial dimensions and then disappear into the fifth dimension. This disappearance of massive particles from the 3-brane constitutes an energy flow from the brane to the bulk. The cosmological constraints on such disappearing matter have been studied in Ref. [10].

If the massive particles can also exist in the bulk, it becomes possible to consider the inverse flow from the bulk into our three-brane. In this contribution, we build a model with such mass-energy exchange, in which the flow from the bulk to the brane provides the present observed acceleration [2, 3, 5, 4, 6, 7] of the universe. As noted in Ref. [10] a heavy ($\sim$TeV) dark-matter candidate such as the lightest supersymmetric partner is likely to have the largest tunneling rate between the brane and bulk. Hence, in the present work we mainly consider the exchange between the bulk and brane involving a growth of the cold dark-matter component on the brane. This model [which we refer to as growing cold dark matter (GCDM)] is, thus, an alternative to the standard Λ plus cold dark matter (SΛCDM) cosmology to an observer on the 3-brane.

We test this model by analyzing the observations of Type Ia supernovae at high redshift, the temperature fluctuation spectrum of the cosmic microwave background (CMB), and the matter power spectrum.

MODEL WITH ENERGY EXCHANGE BETWEEN THE BULK AND BRANE: GCDM MODEL

We begin with the five-dimensional Einstein equations $G_{AB} + \Lambda_5 g_{AB} = \kappa^2 T_{AB}$, where $\kappa^2 = M^{-3}$ is the 5-dimensional gravitation constant with M the 5-dimensional Planck mass, and Λ_5 is taken to be negative responsible for AdS$_5$ bulk space. In a frame for which the three-brane is at rest we can decompose the energy-momentum tensor into two parts.

$$T_{AB} = \delta(y) T_{AB}^{BRANE} + T_{AB}^{DM} , \qquad (1)$$

where the δ-function identifies the location of the brane at $y = 0$ to which the standard-model particles are assumed to be confined. $^{(BRANE)}T_B^A$ corresponds to the usual three-density and pressure, ρ and p, of ordinary relativistic and non-relativistic particles plus the tension τ on the brane. $^{(BRANE)}T_B^A = \text{diag}(-\tau - \rho, -\tau + p, -\tau + p, -\tau + p, 0)$. The energy momentum tensor T_{AB}^{DM} is further decomposed into usual dark matter on the brane, plus the bulk components $^{(DM-BULK)}T_B^A$ as

$$^{(DM)}T_B^A = \delta(y)\text{diag}(-\bar{\rho}, \bar{p}, \bar{p}, \bar{p}, 0) + {}^{(DM-BULK)}T_B^A. \qquad (2)$$

where, $^{(DM-BULK)}T_5^0 \sim (\hat{\rho} + \hat{p})U_5$, represents the matter-energy flow from the bulk to the brane, while $^{(DM-BULK)}T_5^5 = (\hat{\rho} + \hat{p})U^5 U_5 + \hat{p}$, represents a bulk pressure in the limit of vanishing U_5. When a fluid is static, T_5^5 represents a pressure, it is then natural to set the pressure to be zero. Hereafter, we shall neglect T_5^5 in our analysis. But T_5^5 is in general non-vanishing for a moving flow. We leave the case with non-vanishing T_5^5 in the future investigation.

The cosmological equations of motion with brane-bulk energy exchange have previously been formulated in Refs. [2, 3, 5, 4, 6, 7]. The covariant derivative of the energy-momentum tensor T_{AB}^{BRANE} leads to the usual energy conservation condition for various components i of normal matter on the brane;

$$\dot{\rho}_i + 3(1+w_i)H\rho_i = 0 , \qquad (3)$$

plus a new condition for the dark matter which takes into account the flow of matter from the brane world,

$$\dot{\bar{\rho}} + 3(1+\bar{w})H\bar{\rho} = -T, \qquad (4)$$

where $T \equiv 2T_5^0$ is the discontinuity of the (0,5) component of $^{(DM-BULK)}T_B^A$ at $y=0$. The (0,0) component of the Einstein equation produces a modified Friedmann cosmology to an observer on the brane:

$$H^2 = \frac{\dot{a}^2}{a^2} = \frac{8}{3}\pi G_N(\rho+\bar{\rho}) + \Lambda_4 + \frac{\kappa^4}{36}(\rho+\bar{\rho})^2 + \chi , \qquad (5)$$

while the (5,5) component leads to an equation for the dark radiation term χ,

$$\dot{\chi} + 4\frac{\dot{a}}{a}\chi = \frac{\kappa^4}{18}(\rho+\bar{\rho}+\tau)T . \qquad (6)$$

In the limit of an empty bulk and no exchange between the bulk and the brane, $T_5^0 = 0$, the quantity χ varies with scale factor on the brane as C/a^4. Thus, χ reduces to the standard dark-radiation term when all matter fields are confined to the brane.

For the purposes of scaling, we adopt [2, 3] the usual parameterization of the EOS for matter in the bulk and parameterize the 0-5 component of the bulk dark-matter energy-momentum tensor as

$$(DM-BULK)T_5^0 = -\frac{\alpha}{2}\left(\frac{\rho_{cr}}{a^q}\right) \times H_0 , \qquad (7)$$

where we have assumed a constant transition rate of matter between the bulk and the brane absorbed into the dimensionless parameter α which can be either positive or negative. q can be written $q = 3(1+\hat{w})$, where $\hat{w} = \hat{p}/\hat{\rho}$ is the usual equation of state parameter with, $\hat{w} = 0$ for normal cold dark matter, while $\hat{w} = 1/3$ for relativistic matter. A value of $\hat{w} = -1$ corresponds to a vacuum energy, while $\hat{w} = -2/3$ for a string-like topological defect.

Figure 1 illustrates the evolution of various components on the brane in a simple $\Lambda = k = 0$ cosmology. This cosmology separates into three characteristic epochs on this figure. First, the usual early radiation dominated epoch ($a < 10^{-4}$). Second, a dark-matter dominated epoch ($10^{-4} < a < 10^{-1}$). Third, the dark matter and dark radiation dominated accelerating epoch ($a > 0.5$). During the first and second epochs, the dark radiation component evolves as $\rho_{DR} \propto a^{2-q}$ and $\rho_{DR} \propto a^{3/2-q}$, respectively.

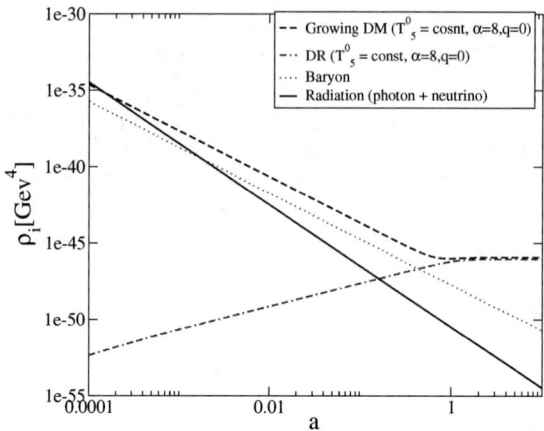

FIGURE 1. Evolution of various energy densities as a function of scale factor in a $\Lambda = k = 0$ growing cold-dark-matter model. The curves labeled as "Growing DM" shows the dark radiation component. Note that the dark radiation content is plotted as its absolute value. This quantity is actually negative in these models.

Eventually, in the third region the cosmology of interest to this paper emerges. The dark radiation component dominates. In the case of $q \approx 0$ it becomes a constant dark-energy density. Hence, the dark radiation associated with in-flowing cold dark matter leads the cosmic acceleration without the need for a cosmological constant on the right hand side of Eq. (5) as long as matter in the bulk has the right equation of state. In what follows we show that the fits to observations based upon these two models are indistinguishable from each other and from the best standard $S\Lambda CDM$ model.

OBSERVATIONAL CONSTRAINTS

Having defined the cosmology of interest we now analyze the various cosmological constraints as a test of this hypothesis by solving equations numerically. In particular, we examine the magnitude-redshift relation for type Ia supernovae (SNIa), the cosmic microwave background (CMB) and the matter power spectrum $P(k)$. Parameters summarizing the best fits to these constraints are given in Table 1 for models with and without growing cold dark matter. The $S\Lambda CDM$ fits are consistent with the usually inferred cosmological parameters (e.g. [11]). An important constraint on the GCDM model is that the sum of $\Omega_{DM} + \Omega_{DR}$ be approximately equivalent to the sum of $\Omega_{DM} + \Omega_{\Lambda}$ in the standard cosmology. This is evidenced by the fourth column of Table 1. Note, however, that the magnitude of Ω_{DM} and Ω_{DR} individually can be quite large in the GCDM model, as long as $\Omega_{tot} \approx 1$. The last column indicates that GCDM model can give a equivalent fit with the standard $S\Lambda CDM$ cosmology. Details can be found in Ref. [12].

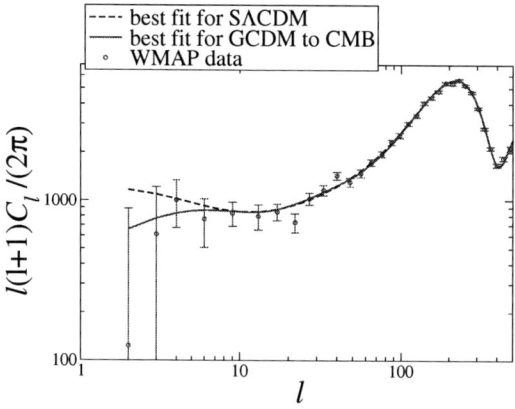

FIGURE 2. CMB angular power spectrum with and without GCDM compared with observational data from WMAP. The dashed line corresponds to the best fit SΛCDM model. The solid line shows GCDM best-fit models with the cosmological constant equal to zero, and $q = 2.92$, $\alpha = 2.14$. This figure demonstrates how GCDM modifies the CMB power spectrum.

TABLE 1. Parameter sets for various fits. α and q are parameters for inflowing matter defined in eq.(7), Ω_i denotes energy density of i component in critical density units, h is Hubble parameter, z_{re} is reionization redshift, n_s is spectral index of scalar perturbations, τ is optical depth of Compton scattering, b is bias parameter to fit the matter power spectra, and χ_r^2 represents reduced chi-squared.

GCDM / SΛCDM	α	q	Ω_{m+dr} / $\Omega_{m+\Lambda}$	Ω_m	Ω_{dr} / Ω_Λ	h	z_{re}	n_s	τ	b	χ_r^2
SNIa Only											
GCDM	11.0	0.006	0.93	3.31	-2.38	0.58	-	-	-	-	1.23
SΛCDM	-	-	0.97	0.26	0.71	0.71	-	-	-	-	1.24
CMB Only											
GCDM	2.14	2.92	0.93	1.91	-0.98	0.64	29.1	1.18	0.53	-	1.02
SΛCDM	-	-	0.94	0.23	0.71	0.71	14.9	0.97	0.13	-	1.01
SN + CMB											
GCDM	8.45	0.023	0.95	3.14	-2.19	0.71	15.0	0.97	0.13	-	1.04
SΛCDM	-	-	0.96	0.25	0.71	0.70	13.3	0.96	0.11	-	1.04
All											
GCDM	8.33	0.037	0.95	3.05	-2.44	0.71	15.3	0.98	0.14	2.1	1.03
SΛCDM	-	-	0.95	0.24	0.71	0.70	13.7	0.97	0.12	1.05	1.03

SUMMARY AND DISCUSSION

We have considered models in which the apparent cosmological constant derives from the energy exchange of cold dark matter from the bulk dimension to the brane. The energy momentum tensor in this extra dimensional brane cosmology leads naturally to terms resembling a cosmological constant at the present time. If such energy exchange occurs our universe accelerates without the need to invoke a cosmological constant on the 3-brane.

We find that such GCDM exchanges are consistent with observations including the supernova magnitude-redshift relation, temperature fluctuations in the CMB, and the matter power-spectrum data. This cosmology is even slightly preferred as it fits better the suppression of the CMB power spectrum at low multipoles as shown in figure (2). We have thus demonstrated that this cosmology represents an alternative model to the SΛCDM cosmology for an observer on the 3-brane. A consistent fit to the observational constraints, however, requires that the EOS parameter for matter in the bulk be small $q \approx 0.0$ as shown in the third column of Table 1. This EOS is consistent with a need for an AdS_5 geometry in the bulk.

A peculiar feature of the present best fit models, is the fact that the true value of Ω_{DM} is much larger than in the standard cosmology, though its gravitational effect is canceled by the dark-radiation contribution. Indeed, the key constraint is that $\Omega_{DM} + \Omega_{DR} \approx \Omega_{DM} + \Omega_{\Lambda}$ as evidenced in the fourth column of Table 1.

One consequence of such a large and growing dark-matter contribution is the need for a somewhat large bias parameter. On the other hand, such a large dark matter content suggests new observational tests of this cosmology. For example, if the dark-matter content at the present time is much larger than in a SΛCDM model, then direct terrestrial measurements of the total density of cold dark-matter particles could indicate a much higher density than expected based upon their mass and gravitation effect. Another test of this cosmology is that there should be a suppression of the matter power spectrum on the scale of the horizon compared to a SΛCDM cosmology [12].

REFERENCES

1. P. M. Garnavich, et al., Astrophys. J., **509**, 74 (1998); S. Perlmutter, et al. Nature, **391**, 51 (1998).
2. E. Kiritsis, G. Kofinas, N. Tetradis, T. N. Tomaras and V. Zarikas, JHEP, **02**, 035 (2003).
3. N. Tetradis, Phys. Lett., **B569**, 1 (2003).
4. P. S. Apostolopoulos and N. Tetradis, Class. Quant. Grav., **21**, 4781 (2004).
5. Y.S. Myung, J.Y. Kim, , Class. Quant. Grav., **20**, L169 (2003).
6. N. Tetradis, Class. Quant. Grav., **21**, 5221 (2004).
7. P. S. Apostolopoulos and N. Tetradis, Phys. Rev., **D71**, 043506 (2005).
8. L. Randall and R. Sundrum, Phys. Rev. Lett. **83**, 3370 (1999); **83**, 4690 (1999).
9. S. L. Dubovsky, V. A. Rubakov and P. G. Tinyakov, Phys. Rev. D 62, 105011 (2000).
10. K. Ichiki, P. M. Garnavich, T. Kajino, G. J. Mathews, and M. Yahiro, Phys. Rev. D 68, 083518 (2003).
11. D. Spergel, et al. (*WMAP Collaboration*, Astrophys. J. Suppl., **148**, 175 (2003).
12. K. i. Umezu, K. Ichiki, T. Kajino, G. J. Mathews, R. Nakamura and M. Yahiro, arXiv:astro-ph/0507227.

Lithium Abundances in Halo Subgiants

David Yong*, Bruce W. Carney*, Wako Aoki†, Andrew McWilliam** and William J. Schuster‡

*Department of Physics & Astronomy, University of North Carolina, Chapel Hill, NC 27599, USA
†National Astronomical Observatory, Mitaka, 181-8588 Tokyo, Japan
**Observatories of the Carnegie Institution of Washington, Pasadena, CA 91101, USA
‡Observatorio Astronómico Nacional, UNAM, Mexico

Abstract. We present lithium abundances for 28 halo subgiants based on high resolution, high signal-to-noise ratio spectra. Excluding the known lithium-rich subgiant BD +23 3912, the maximum abundances are log ε(Li) = 2.35. While subgiants evolve from stars hotter than the main sequence turn-off with shallower convection zones that may have depleted lithium to a lesser degree, lithium abundances in halo subgiants are not in agreement with the primordial value as predicted from standard big bang nucleosynthesis combined with recent results from WMAP.

Keywords: Stellar structure, interiors, evolution, nucleosynthesis, ages; Abundances, chemical composition; Giant and subgiant stars
PACS: 97.10.Cv; 97.10.Tk; 97.20.Li

INTRODUCTION

The stellar abundances of lithium continue to have important consequences in various areas of modern astronomy. Lithium is a fragile element that is destroyed at several million degrees such that the measured photospheric compositions provide important tests of stellar evolution theory, specifically for non-standard mixing processes [1, 2]. Since lithium may be synthesized in different stellar sites (novae, asymptotic giant branch stars, Type II supernovae) as well as in the interstellar medium via cosmic ray spallation, the measured abundances can be used to study Galactic chemical evolution [3]. Most importantly, lithium is one of the few isotopes produced in significant quantities during big bang nucleosynthesis. Therefore, stellar lithium abundances play an important role for cosmology. Indeed, when the first stellar abundances of lithium were measured in halo dwarfs, the observed constant abundance, log ε(Li) = 2.05, was assumed to be the primordial value produced in the big bang [4].

In standard big bang nucleosynthesis, the only free parameter is the baryon to photon ratio $\eta = n_b/n_\gamma$ which can also be expressed as the fraction of the critical density Ω_b provided by baryons ($\Omega_b = 3.652 \times 10^7 \eta/h^2$) where h is the Hubble constant in units of 100 km s^{-1} Mpc^{-1}. The WMAP satellite has observed the cosmic microwave background from which the cosmological parameters have been estimated to an unprecedented accuracy. Spergel et al. [5] found a value $\Omega_b h^2$ = 0.0224 ± 0.0009 from which standard big bang nucleosynthesis predicts a primordial lithium abundance log ε(Li) = 2.65 [6]. This represents a 0.5 dex difference (corresponding to a factor of 3) between the predicted lithium abundance and the measured abundance on the Spite plateau. Recent measurements of lithium isotope ratios suggest that there may be a plateau for ^{6}Li at a level

1000 times greater than predicted from standard BBN [7]. Such a discovery heightens the discrepancy between the observed and predicted primordial abundances for lithium.

The debate continues over whether or not the observed lithium abundances are primordial or if a combination of factors have conspired to uniformly reduce the primordial abundances in metal-poor stars with a range in mass, effective temperature, and convection zone depth. It is worth re-iterating that if metal-poor stars have depleted their primordial lithium abundances, then 75% of the original lithium must have been destroyed in almost all metal-poor main sequence stars. Trends with effective temperature and/or metallicity have been identified with surface lithium abundances decreasing with decreasing temperature [8]. This suggests that the thinner convection zones of hotter stars may better preserve the original lithium abundances. Therefore, subgiants may be the best place to search for the highest, presumably primordial, lithium abundances. Subgiants have evolved from stars hotter than the main sequence turn-off, with consequently shallower convection zones, and so are more likely to have retained their lithium. King et al. [9] identified a subgiant, BD +23 3912, whose Li abundance log ε(Li) = 2.60 is twice as high as the Spite plateau. Furthermore, Charbonnel and Primas [10] have also found statistically signficant, but small, lithium abundance differences between main sequence stars and subgiants.

OBSERVATIONS AND ANALYSIS

A sample of candidate subgiants were identified in the following three ways. A. If trigonometric parallaxes were available, then metal-poor subgiants are readily identifiable from a color-magnitude diagram. B. If spectroscopic estimates for effective temperature and surface gravity were available, metal-poor subgiants occupy a clear regime in parameter space. C. If Strömgren photometry was available, metal-poor subgiants can easily be selected in a plot of $(b-y)_0$ versus c_0.

High resolution spectroscopic observations were obtained using the echelle spectrograph on the 4-meter telescope at the Kitt Peak National Observatory, the Magellan Inamori Kyocera Echelle spectrograph (MIKE) on the Magellan telescopes, and the High Dispersion Spectrograph (HDS) on the the Subaru telescope. One dimensional wavelength calibrated normalized spectra were produced in the standard way using using the IRAF[1] package of programs. The resolving powers ranged from $R \equiv \lambda/\Delta\lambda = 30,000$ to 55,000. The signal-to-noise ratios (S/N) ranged from 100 to 400 per resolution element. Such a combination of high resolving power and S/N ensured that uncertainties in the measured equivalent widths of the lithium line are insignificant contributors to the overall error budget.

The stellar parameters were determined in the following way. Following Charbonnel & Primas (2005), the effective temperature was set using the color-temperature relations of Alonso et al. [11, 12]. The mean difference between the "dwarf" − "giant" calibration

---

[1] IRAF is distributed by the National Optical Astronomy Observatories, which are operated by the Association of Universities for Research in Astronomy, Inc., under cooperative agreement with the National Science Foundation.

was 1 ± 4 K ($\sigma = 23$ K). We adopted the "dwarf" calibration temperatures. Surface gravities were taken from the Yonsei-Yale isochrones assuming an age of 10 Gyr and [α/Fe] = +0.3 [13]. [Fe/H] and reddening estimates were determined using the formulae given by Schuster and Nissen [14]. The model atmospheres were calculated using ATLAS9 [15] with convective overshoot switched off. The lithium abundances were derived using the stellar line analysis program MOOG [16] in combination with the measured equivalent width for the 6707Å lithium line. Non-local thermodynamic equilibrium (NLTE) corrections to the lithium abundance were calculated using the interpolation code made available by Carlsson et al. [17].

RESULTS AND CONCLUSIONS

The lithium abundances are presented in Figure 1. There are 10 stars in common with Charbonnel and Primas [10]. The mean difference THIS STUDY − CHARBONNEL & PRIMAS is 0.01 ± 0.03 ($\sigma = 0.11$). Such a similarity is expected given that the stellar parameters are derived using essentially identical techniques.

In the lower panel of Figure 1, there are 6 stars with unusually low lithium abundances compared to other stars at the same metallicity. However, consideration of the upper panel in Figure 1 shows that these stars are cool and the lower lithium abundances can be understood as a natural consequence of the larger convective envelopes in such cool stars.

At a given effective temperature, there appears to be a scatter in lithium abundances that exceeds the measurement uncertainty. One possibility is that for a fixed effective temperature, the depth of the convective envelope is a function of metallicity. Another possibility is that the stars may have slightly different rotational velocities with more rapidly rotating stars having depleted additional lithium.

The most lithium-rich star in the sample is BD +23 3912. Note that Charbonnel and Primas [10] find another subgiant, HD 160617 to be unusually lithium-rich log ε(Li) = 2.52. Apart from these two stars, the remainder of the sample show lithium abundances below log ε(Li) = 2.35. That is, the measured lithium abundances in subgiants do not approach the primordial values expected from standard big bang nucleosynthesis.

ACKNOWLEDGMENTS

We thank the National Science Foundation for support through grants grants AST 96-19381, AST 99-88156, and AST 03-05431 to the University of North Carolina.

REFERENCES

1. M. H. Pinsonneault, C. P. Deliyannis, and P. Demarque, *Astrophysical Journal Supplements* **78**, 179–203 (1992).
2. S. Vauclair, and C. Charbonnel, *Astronomy & Astrophysics* **295**, 715 (1995).
3. S. G. Ryan, T. Kajino, T. C. Beers, T. K. Suzuki, D. Romano, F. Matteucci, and K. Rosolankova, *Astrophysical Journal* **549**, 55–71 (2001).
4. F. Spite, and M. Spite, *Astronomy & Astrophysics* **115** (1982).

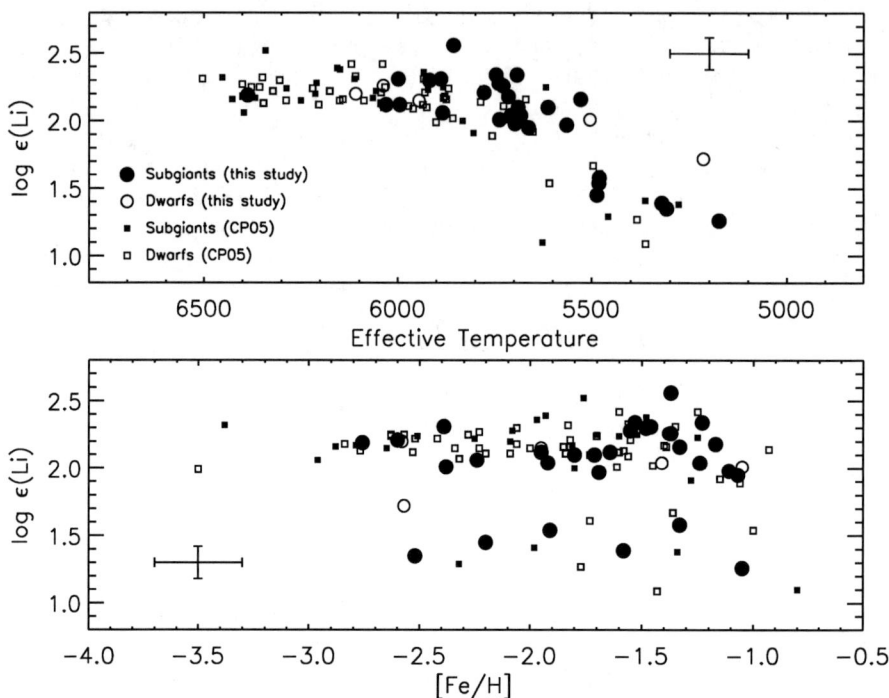

FIGURE 1. Lithium abundances for program stars in this study and in the Charbonnel & Primas (2005; CP05) study.

5. D. N. Spergel, L. Verde, H. V. Peiris, E. Komatsu, M. R. Nolta, C. L. Bennett, M. Halpern, G. Hinshaw, N. Jarosik, A. Kogut, M. Limon, S. S. Meyer, L. Page, G. S. Tucker, J. L. Weiland, E. Wollack, and E. L. Wright, *Astrophysical Journal Supplements* **148**, 175–194 (2003).
6. A. Coc, E. Vangioni-Flam, P. Descouvemont, A. Adahchour, and C. Angulo, *Astrophysical Journal* **600**, 544–552 (2004).
7. M. Asplund, D. L. Lambert, P. E. Nissen, F. Primas, and V. V. Smith, *Astrophysical Journal submitted (astro-ph/0510636)* (2005).
8. S. G. Ryan, J. E. Norris, and T. C. Beers, *Astrophysical Journal* **523**, 654–677 (1999).
9. J. R. King, C. P. Deliyannis, and A. M. Boesgaard, *Astronomical Journal* **112**, 2839 (1996).
10. C. Charbonnel, and F. Primas, *Astronomy & Astrophysics* **442**, 961–992 (2005).
11. A. Alonso, S. Arribas, and C. Martinez-Roger, *Astronomy & Astrophysics* **313**, 873–890 (1996).
12. A. Alonso, S. Arribas, and C. Martínez-Roger, *Astronomy & Astrophysics Supplements* **140**, 261–277 (1999).
13. P. Demarque, J.-H. Woo, Y.-C. Kim, and S. K. Yi, *Astrophysical Journal Supplements* **155**, 667–674 (2004).
14. W. J. Schuster, and P. E. Nissen, *Astronomy & Astrophysics* **221**, 65–77 (1989).
15. R. Kurucz, *ATLAS9 Stellar Atmosphere Programs and 2 km/s grid. Kurucz CD-ROM No. 13.* Cambridge, Mass.: Smithsonian Astrophysical Observatory, 1993. **13** (1993).
16. C. Sneden, *Astrophysical Journal* **184**, 839 (1973).
17. M. Carlsson, R. J. Rutten, J. H. M. J. Bruls, and N. G. Shchukina, *Astronomy & Astrophysics* **288**, 860–882 (1994).

Big–Bang Nucleosynthesis: lithium problems and scalar–tensor theories of gravity

Alain Coc

*Centre de Spectrométrie Nucléaire et de Spectrométrie de Masse,
Bâtiment 104, F–91405 Orsay Campus, France*

Abstract. The observations of the anisotropies of the Cosmic Microwave Background (CMB) radiation, by the WMAP satellite, has provided a determination of the baryonic density of the Universe ($\Omega_b h^2$) with an unprecedented precision. Using this value, the primordial abundances of the light elements can be calculated in the framework of the Standard Big–Bang Nucleosynthesis model (SBBN). While the agreement is excellent for D and good for ^4He, there is a difference of a factor of ≈ 3 for ^7Li. In addition, in a few halo stars, ^6Li has also been observed at a level well above SBBN predictions. To enable a more reliable calculation of these ^7Li and ^6Li yields, two nuclear reactions important for the nucleosynthesis of ^7Li and ^6Li have been studied experimentally: $D(\alpha,\gamma)^6Li$ and $^7Be(d,p)2\alpha$. Even though, the lithium primordial production is not well understood, BBN can be used to constrain theories beyond the standard model, for instance, scalar-tensor theories of gravity.

Keywords: Cosmology, Big-Bang Nucleosynthesis, Nuclear Astrophysics, Abundances, Tensor–Scalar gravity
PACS: 26.35.+c;25.45.-z;04.50.+h

INTRODUCTION

Standard Big–Bang Nucleosynthesis can be considered as a parameter free model now that the nuclear reaction rates, the number of neutrino families and the baryonic density of the Universe have been independently determined. In particular, the value $\Omega_b h^2 = 0.0224 \pm 0.0009$ has been extracted from the observations of the anisotropies of the Cosmic Microwave Background (CMB) radiation, by the WMAP satellite[14]. With this very precise value of the baryonic density and the main nuclear reaction rates obtained from an R–matrix analysis of experimental data[7], it should be possible to calculate precisely the abundance of the light isotopes. When compared to primordial abundances deduced from observations, the agreement is excellent for D, good for ^4He but there is a discrepancy of a factor of ≈ 3 for lithium and of orders of magnitude for ^6Li. To exclude any nuclear solutions to this lithium problems, two reactions, whose rates were uncertain at BBN energies, have been the subject of recent experiments: $^7Be(d,p)2\alpha$ (^7Li nucleosynthesis) and $D(\alpha,\gamma)^6Li$ (^6Li production).

Considering only ^4He and D, BBN can be used to constrain non-standard models. For instance, it can put constraints on scalar–tensor theories of gravity, natural extensions of General Relativity, at redshifts of $z \sim 10^8$, complementing those obtained from CMB anisotropies ($z \approx 1000$) and the Solar system ($z = 0$).

THE LITHIUM PROBLEMS

Since the discovery of the "Spite plateau", lithium observations in halo stars have been used to deduce the primordial ^7Li abundance. Observations by Ryan et al.[16] have led to the (95% c.l.) value of Li/H = $(1.23^{+0.68}_{-0.32}) \times 10^{-10}$, taking into account various sources of systematic uncertainties including stellar depletion and models of stellar atmospheres. Even though ^7Li is a fragile isotope, easily destroyed, it seems difficult to imagine that it has been significantly depleted in these halo stars because i) of the small scatter of the observed abundances and ii) the observation of the even more fragile ^6Li in some of these stars. This is comforted by the very recent observations (2–σ) by Asplund et al.[1] of ^6Li in nine halo stars: the total lithium display a very thin plateau while the $^6Li/^7Li$ ratio is $\sim$0.05. Figure 1 displays the ^7Li primordial abundance deduced by Ryan et al.[16], a conservative range of ^6Li abundances based on Asplund et al.[1] observations, the results of SBBN calculations as a function of $\Omega_b h^2$ and the WMAP result[14].

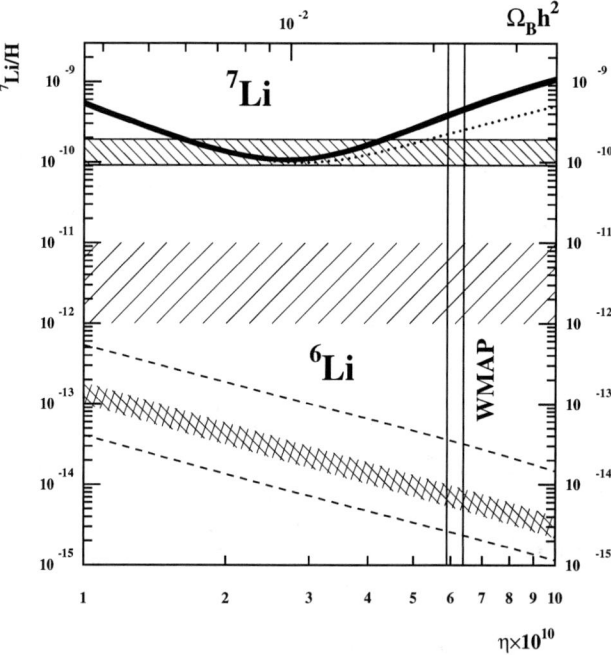

FIGURE 1. Abundances of the lithium isotopes: comparison between observations (hatched horizontal areas) and SBBN calculations with different reaction rates (provisional, see text). The vertical stripe shows the baryonic density as determined by WMAP.

In this figure, the width of the ^7Li curve is the result of a Monte–Carlo calculation[4] representing the effect of nuclear uncertainties on the rates as provided by Descouvemeont et al analysis[7]. The dashed lines represent the uncertainties on ^6Li production when using the upper and lower limit for the D$(\alpha, \gamma)^6Li$ rate provided by the NACRE compilation[2]. (This reaction is considered the main source of uncertainty in ^6Li production.) The figure shows that at the WMAP baryonic density, the ^7Li SBBN yield is higher by a factor of $\approx$3 from the Ryan et al. observations[16] and that despite large uncertainties both in SBBN yields and primordial abundance, there is an ever larger discrepancy for ^6Li. However, before proceeding any further, it is important to clarify some nuclear physics aspects.

^7Li Big–Bang Nucleosynthesis

It is well known that the valley shaped curve representing Li/H as a function of $\Omega_b h^2$ is due to two modes of ^7Li production. One, at low $\Omega_b h^2$ is produced ^7Li directly via $^3H(\alpha,\gamma)^7Li$ while ^7Li destruction comes from $^7Li(p,\alpha)^4He$. The other one, at high $\Omega_b h^2$, leads to the formation of ^7Be through $^3He(\alpha,\gamma)^7Be$ while ^7Be destruction by $^7Be(n,p)^7Li$ is inefficient because of the lower neutron abundance at high density. (^7Be later decays to ^7Li.) Since the WMAP results point toward the high $\Omega_b h^2$ region, a peculiar attention should be paid to ^7Be synthesis. In particular, the ^7Be+d reactions could be an alternative to $^7Be(n,p)^7Li$ for the destruction of ^7Be, by compensating the scarcity of neutrons at high $\Omega_b h^2$. An increase of the $^7Be(d,p)2^4He$ reaction rate by factors of 100 would alleviate the discrepancy (dotted line in Figure 1). The rate for this reaction can be traced to an estimate by Parker[13] who assumed for the astrophysical S–factor a constant value of 10^5 keV-barn based on the single experimental data available[10]. However, *no experimental data for this reaction was available at energies relevant to ^7Be Big Bang nucleosynthesis*, taking place when the temperature has dropped below 10^9 K. An experiment[3] was performed using a $^7Be^{1+}$ radioactive beam at the lowest energy of 5.545 MeV provided by the CYCLONE110 cyclotron at the CYCLONE RIB facility at Louvain-la-Neuve (Belgium). This energy was degraded down to 1.710 MeV using a 6 μm Mylar foil situated at about 50 cm from the target. The target consisted of a 200 μg/cm^2 $(CD_2)_n$ self-supporting foil. The reaction products were detected using a stack of two silicon strip detectors covering an angular range of $\theta_{lab} = 7° - 17°$. With such a set-up, we were able to investigate the center-of-mass energy ranges between 1.2 and 0.96 MeV (for a beam energy of 5.545 MeV) and between 0.38 and 0.15 MeV (for 1.710 MeV). High energy protons corresponding to the ground state and the first excited state in ^8Be were not completely stopped in the $\Delta E_1 - \Delta E_2$ telescope, while protons corresponding to other higher excited states in ^8Be were stopped. α particles, recoil and scattered particles were completely stopped in ΔE_1. This experiment showed that the cross–section is not higher than extrapolated at BBN energies but is in fact is smaller than estimated. by Parker. Hence, ^7Be destruction by $^7Be(d,p)2\alpha$ remains negligible and the ^7Li discrepancy cannot be resolved in that way. The main contribution to the ^7Li yield uncertainty at WMAP baryonic density comes from the $^3He(\alpha,\gamma)^7Be$ reaction rate where systematic uncertainties are important. Indeed, the cross section could even be slightly higher by 20–30%[15] incrasing the discrepancy.

^6Li Big–Bang Nucleosynthesis

The D$(\alpha,\gamma)^6$Li reaction is the main path for ^6Li SBBN production while destruction proceeds from the ^{6}Li$(p,\alpha)^3$He. Both rates are available in the NACRE[2] compilation. While the latter reaction rate is reasonably known at BBN energies, the former suffers from an uncertainty of more than one order of magnitude. This is reflected in the large uncertainty on SBBN ^6Li yield depicted on Figure 1 (dashed lines). The only D$(\alpha,\gamma)^6$Li data available at BBN energies have been measured indirectly via the Coulomb dissociation technique[11]. However, there is a discrepancy between these data at the lower (BBN) energies on the one hand and theoretical extrapolations from higher energies were direct measurements have been performed on the other hand. The upper and lower rates fond in NACRE originate from this difference between theory and experiment. A new Coulomb dissociation experiment was performed recently at GSI[9] that provided data over a wide energy range from the high energy region where direct measurements are available down to the BBN region. These results are now in good agreement both with direct measurement and low energy theoretical extrapolations. The corresponding reaction rate will be obtained by an R–matrix calculation. At present, using a provisional reaction rate together with estimated uncertainties, we obtained the ^6Li yield depicted in Figure 1 (crossed area) and, as a preliminary value, ^{6}Li/H$\approx 7\times 10^{-15}$ ($\pm \approx 40\%$) i.e. more than two orders of magnitudes below observed values. This estimated uncertainty also include those on the ^{6}Li$(p,\alpha)^3$He reaction rate and on $\Omega_b h^2$. Other potentially ^6Li producing reactions have a negligible contribution because they have negative Q-value (^{7}Li$(p,d)^6$Li and ^{4}He$(t,n)^6$Li) or a too low cross section (^{3}He$(t,\gamma)^6$Li). For instance, multiplying the ^{3}He$(t,\gamma)^6$Li reaction rate[8] by a factor of 1000 would only increase the ^6Li yield by a factor of ≈ 7 at WMAP baryonic density. Using a more realistic factor would not affect significantly ^6Li production as shown by Fukugita and Kajino[8].

SCALAR–TENSOR THEORIES OF GRAVITY

Even though, the lithium primordial production is not well understood, BBN together with ^4He and D observations can be used to constrain theories beyond the standard model. As an example, we can mention scalar-tensor theories of gravity. These theories involve two free functions describing the coupling of the scalar field to matter and its self-interaction potential. They are motivated by high-energy theories trying to unifying gravity with other interactions which generically involve a scalar field in the gravitational sector, in particular, in superstring theories. The action in the Jordan frame (where length, time, cross–sections, ... have their usual meaning) is written as (field + matter) :

$$S = \int \frac{d^4x}{16\pi G_*} \sqrt{-g_*} \left[R_* - 2g_*^{\mu\nu} \partial_\mu \varphi_* \partial_\nu \varphi_* - 4V(\varphi_*) \right] + S_m[A^2(\varphi_*)g^*_{\mu\nu}; \psi] \quad (1)$$

The scalar field φ_* is introduced together with the two functions $A(\varphi_*)$ (coupling) and $V(\varphi_*)$ (potential). If the coupling has a minimum, the solutions are attracted toward General Relativity (GR) verified to a high level of accuracy at present time. Recent work[12] in the framework of scalar–tensor theories, investigate the possibility that with well chosen potential, coupling and initial conditions, the ^4He, D and ^7Li BBN abundances could *all* be re-conciliated with observations. Given the new ^6Li problem we considered that lithium isotopes were not sufficiently understood to constrain new physics. This is why we preferred to consider, following Damour and Pichon[6], a more general model and only ^4He and D for constraining the theories.

In the vicinity of the minimum of the coupling, one can use the parametrization: $a(\varphi_*) \equiv \ln(A(\varphi_*)) = \frac{1}{2}\beta\varphi_*^2$. BBN ($z \approx 10^8 - 10^9$) can supplement the constraints given by the solar system ($z=0$), the CMB ($z \approx 1000$), on these parameters: β and $\alpha_0 \equiv \partial \ln A / \partial \varphi_*|_{z=0}$.

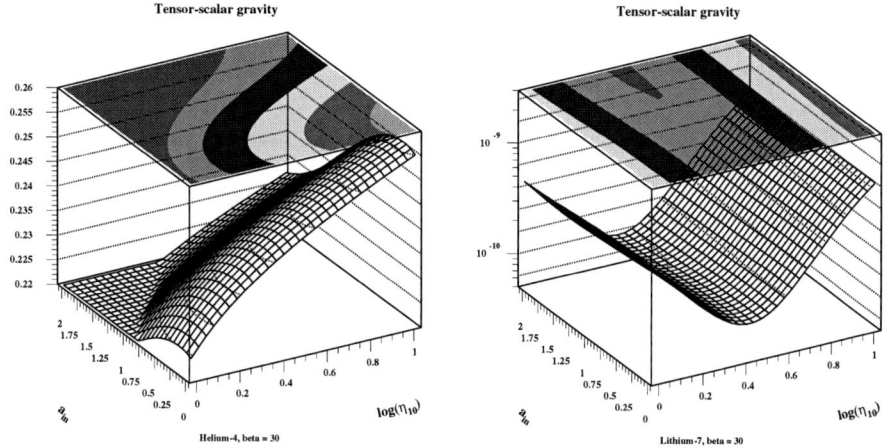

FIGURE 2. ^4He and ^7Li abundances as functions of $\ln(\eta_{10})$ [$\Omega_b h^2 = 0.00365 \times \eta_{10}$] and initial parameter (a_{in} see text) for the scalar field and for $\beta=30$.

Varying the action (1) give the modified Friedmann equation together with the scalar field evolution equation that we numerically integrated from deep in the radiation era $z \sim 10^{12}$ until present. The rate of expansion of the universe is modified during BBN resulting in different light isotope yields relative to SBBN. Figure 2 displays, for $\beta = 30$, the abundances of ^4He and ^7Li as a function of the baryon to photon numbers ratio and the initial scalar field parameter $a_{in} \equiv a[\varphi_*(z \sim 10^{12})]$. It shows that, (within the usual scale of SBBN yields) ^4He is very sensitive to a_{in} while ^7Li remains almost unchanged. This can be easily explained as ^4He abundance is mostly defined by the n/p ratio at freeze–out ($T \approx 10^{10}$ K) while nucleosynthesis of the other light elements occurs later ($T \approx 10^9 - 10^8$ K). Between these two phases, the quadratic coupling that induces a strong drive toward GR, have been effective and the results are close to SBBN. To provide limits on the parameters β and α_0 from BBN and the Solar–System, thousands of models have been calculated for the justified range of parameters $0 < a_{in} < 3$ and $0.1 < \beta < 100$[5].

ACKNOWLEDGMENTS

This contribution summarizes works performed within several collaborations. My warmest thanks go to Elisabeth Vangioni for a continuous collaboration on BBN, to Carmen Angulo and Faïirouz Hammache who led the two nuclear physics experiments and to Jean–Philippe Uzan and Keith Olive for the theoretical aspects of scalar–tensor theories. I am also indebted to the participants to the ^{7}Be(d,p)2α experiment, E. Casarejos, M. Couder, P. Demaret, P. Leleux, F. Vanderbist, J. Kiener, V. Tatischeff, T. Davinson, A.S. Murphy, N.L. Achouri, N.A. Orr, D. Cortina-Gil, P. Figuera, B.R. Fulton and I. Mukha and to the D$(\alpha,\gamma)^6$Li experiment, D. Galaviz, K. Sümmerer, S. Typel, F. Attallah, M. Caamano, D. Cortina, H. Geissel, M. Hellestrőm, N. Iwasa, J. Kiener, P. Koczon, B. Kohlmeyer, E. Schwab, K. Schwarz, F. Shümann, P. Senger, O. Sorlin, V. Tatischeff, J.P. Thibaud, F. Uhlig, A. Wagner and A. Walus.

REFERENCES

1. M. Asplund, D. Lambert, P.E. Nissen, F. Primas, and V. Smith, submitted to Astrophys. J., astro-ph/0510636.
2. C. Angulo, M. Arnould, M. Rayet, P. Descouvemont, D. Baye, et al. (NACRE), Nucl. Phys., **A656** 3–183 (1999).
3. C. Angulo, E. Casarejos, M. Couder, P. Demaret, P. Leleux, et al., Astrophys. J. Lett., **630**, L105–L108 (2005) astro-ph/0508454.
4. A. Coc, E. Vangioni-Flam, P. Descouvemont, A. Adahchour, and C. Angulo, Astrophys. J., **600** 544–552 (2004).
5. A. Coc, K. Olive, J.-P. Uzan and E. Vangioni, submitted to *Phys. Rev. D*
6. T. Damour and B. Pichon, Phys. Rev.,**D59**, 123502 (1999).
7. P. Descouvemont, A. Adahchour, C. Angulo, A. Coc, and E. Vangioni–Flam, At. Data Nucl. Data Tables, **88**, 203–236 (2004).
8. M. Fukugita and T. Kajino, Phys. Rev., **D42**, 4251–4253 (1990).
9. F. Hammache, D. Galaviz, K. Sümmerer, S. Typel, F. Attallah, et al., in "Cross section measurements of the Big–Bang nucleosynthesis reaction D$(\alpha,\gamma)^6$Li by coulomb dissociation of ^6Li," in *International Conference on Frontiers in Nuclear Structure*, edited by R. Julin and S. Harissopulos, AIP Conference Proceedings, American Institute of Physics, New York, *in press*
10. R.W. Kavanagh, Nucl. Phys., **18**, 492 (1960)
11. J. Kiener, H.J. Gils, H. Rebel, S. Zagromski, G. Gsottschneider, N. Heide, H. Jelitto, J. Wentz and G. Baur, Phys. Rev., **C44**, 2195–2208 (1991). .
12. J. Larena, J.-M. Alimi, and A. Serna, astro-ph/0511693.
13. P.D. Parker, Astrophys. J., **175**, 261 (1972).
14. D.N. Spergel, L. Verde, H.V. Peiris, E. Komatsu, M.R. Nolta et al., Astrophys. J. S.,**148**, 175–194 (2003) astro-ph/0302209.
15. C. Rolfs, these proceedings.
16. S.G. Ryan,T.C. Beers, K.A. Olive, B.D. Fields, and J.E. Norris, Astrophys. J., **530**, L57–L60 (2000).

Experimental Study Of Photonuclear Reactions Relevant To Nuclear Astrophysics

Tatsushi Shima

Research Center for Nuclear Physics, Osaka University
10-1, Mihogaoka, Ibaraki, Osaka 567-0047, Japan

Abstract. Photonuclear reactions play crucial roles in various processes of nucleosynthesis occurring in the universe. A laser-Compton backscattered γ-ray beam and an active target technique will be a promising tool for precise measurements of low-energy photon-induced nuclear reactions relevant to nucleosynthesis.

Keywords: nucleosynthesis, photonuclear reaction, laser-Compton backscattering

PACS: 21.45.+v, 23.20.-g, 24.30.Cz, 25.20.-x, 26.30.+k, 26.35.+c

INTRODUCTION

The nuclear radiative capture reactions at low energies play essential roles both in the primordial nucleosynthesis and in stellar nuclear burning processes, and accurate data of those reactions are indispensable for quantitative analysis of those processes of nucleosynthesis. So far those nuclear data have been extensively accumulated mainly by measuring radiative capture reactions. The photonuclear reactions, the inverse processes of the capture reactions, will be useful to check and/or improve the accuracies of the data from capture measurements. Also they can provide unique information about the reaction rates of many-body fusion processes like $\alpha(\alpha n,\gamma)^9 Be$, $^4He(2\alpha,\gamma)^{12}C$, etc.

Precise data of photonuclear reactions are also demanded for investigation of explosive nucleosynthesis such as p-process and γ-process, because those processes are considered to be induced by (γ,p) and/or (γ,n) reactions due to high-temperature radiative environments [1,2].

Photonuclear reactions give us important information on neutrino-induced nuclear reactions with weak neutral current, which are considered to be influential both for dynamics of supernova explosions and for the associating neutrino-process nucleosynthesis, since the operators of the electromagnetic interaction have analogous structure to those of weak interaction with neutral current [3].

Owing to recent developments of monochromatic γ-ray sources and low-energy particle detectors, high-precision experiments on nuclear reactions induced by low-energy γ-rays have become really feasible. In this paper recent experimental work on photodisintegration of ^{4}He below 30 MeV will be presented, and future prospects of nuclear astrophysics studies with real photon beams will be shown.

PHOTODISINTEGRATION OF ^{4}HE

Neutrino-nucleus interactions have been considered to play important roles in Type-II supernova explosions (SNes) [4]. Theoretical studies of the Type-II SN dynamics have suggested that the heating of the outgoing shockwave by delayed neutrinos is important to satisfy the kinetic energy needed for successful explosions [5,6]. A recent model calculation of the supernova explosion suggests that the region where the shockwave may stall is dominated by ^{4}He [7], and therefore the neutrino-inelastic scattering of ^{4}He is expected to make an important contribution to the dynamics of Type-II SNe. The neutrino-inelastic scattering reaction of ^{4}He is also considered to play important roles in production of light element as well as heavy one via the rapid process of nucleosynthesis in the neutrino-driven wind of Type-II SNe [6,8,9,10]. To study the above effects of neutrino-nucleus interactions, the responses of nuclei to weak interaction are to be known. Nuclear responses to weak charged currents have been studied using experimental information on analogous processes of nuclear Gamow-Teller interactions [11]. On the other hand, nuclear responses to weak neutral current have been less understood mainly due to poor accuracies of existing experimental data for analogous photodisintegration processes. The neutrino-induced spallation reaction of ^{4}He by neutral current has been estimated in the energy region of astrophysical importance by means of a microscopic theory based on the Lorentz integral transform method [12], suggesting 5~15% enhancement of the reaction rate compared to an old estimation using a general property of the giant dipole resonance (GDR) of light nuclei [8] as shown in Fig. 1.

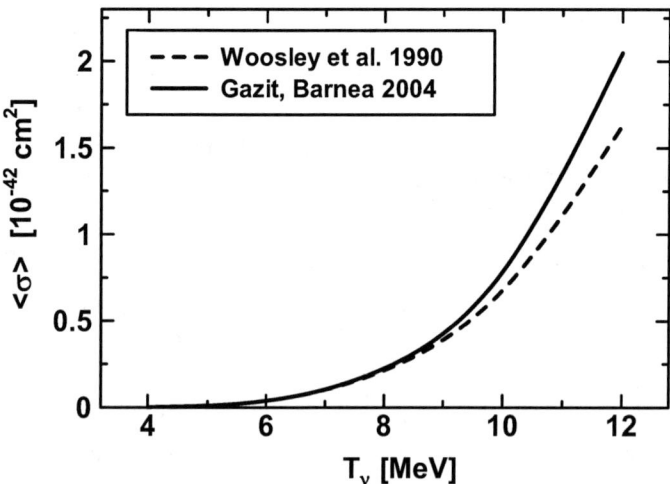

FIGURE 1. Theoretical reaction rate of the neutrino-inelastic reaction of ^{4}He as a function of the neutrino temperature. The solid line and the dashed line denote the reaction rates calculated with the Lorentz integral transform method [12] and the property of GDR of light nuclei [8], respectively.

To test the theoretical calculation it is useful to compare experimental data of the photodisintegration cross sections with the values estimated with the same theoretical method. Therefore it is important to measure the photodisintegration cross sections of ^{4}He accurately in the energy range of between a few and a few tenth of mega electron volts. Experimentally the photodisintegration cross sections of ^{4}He have been measured in the energy range from 20 to 215 MeV using quasi-monoenergetic photon beams and/or bremsstrahlung photon beams (see the references in [13]). The ^{3}H(p,γ)^{4}He and ^{3}He(n,γ)^{4}He reactions have also been studied in order to determine their inverse reactions of ^{4}He(γ,p)^{3}H and ^{4}He(γ,n)^{3}He reactions, respectively. Above 35~40 MeV most of the existing data agree with each other. However, there appear to be discrepancies as large as 50~100% in the peak region of 25~26 MeV, where the GDR peak has been supposed to exist. It has been discussed that the discrepancies are due to experimental reasons such as the backgrounds caused by low-energy g-rays contained in the bremsstrahlung photon beams, difficulties in detection of low-energy fragments from the photodisintegrations, inaccuracies in determination of the effective thickness of the target, and so on. Hence it is highly required to accurately measure these cross sections with use of a monoenergetic photon beam and a detector for fragments with an efficiency of ~100% and a solid angle of ~4π. To realize such a measurement, we employed a laser-Compton backscattered (LCS) γ-ray beam and a time projection chamber (TPC) containing helium gas as an active target. The experimental set up is schematically drawn in Fig. 2.

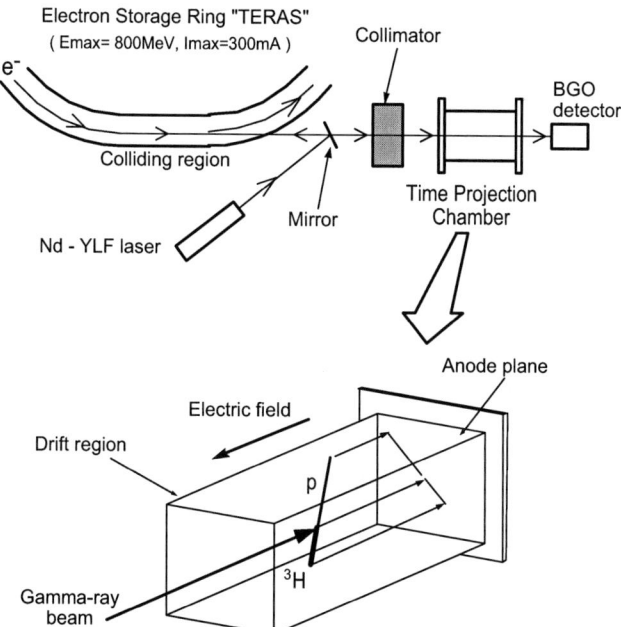

FIGURE 2. Experimental setup for the measurement of the photodisintegration of ^{4}He at AIST.

The incident γ-ray was generated by head-on collision of a laser light with a high-energy electron beam circulating in the storage ring TERAS at AIST [14], and was introduced into the TPC. The TPC was operated with the counter gas made of the mixture of helium gas and methane gas, and therefore served as an active target, which enabled us to detect the charged fragments from the photodisintegrations of ^{4}He and ^{12}C in the counter gas with the detection efficiency of ~100% and the solid angle of 4π. Fig. 3 shows an example of the track of the charged fragments from the ^{4}He(γ,p)^{3}H reaction observed with the TPC. As shown in Fig. 3, the reaction events can be unambiguously identified. The intensity and the energy distribution of the incident γ-ray were accurately measured with a BGO detector and a NaI(Tl) detector, respectively. The detail of the experimental method is described elsewhere [13,15].

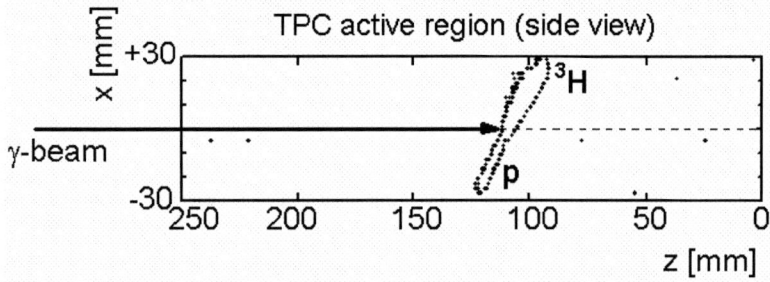

FIGURE 3. Example of the tracks of the charged fragments from the ^{4}He(γ,p)^{3}H reaction.

Fig. 4 shows the measured cross section of the ^{4}He(γ,p)^{3}H and ^{4}He(γ,n)^{3}He reactions in comparison with the previous data and the recent theoretical calculations. The present result is consistent with the previous data obtained with use of monochromatic γ-ray beams. On the other hand, it is significantly smaller than the data measured by means of bremsstrahlung photons, and that may suggest some systematic effect depending on the type of the photon beam. In the comparison with the theories, the present cross section of the ^{4}He(γ,n)^{3}He reaction is in rather good agreement with the one obtained with the Faddeev-type calculation [16], while the present ^{4}He(γ,p)^{3}H and ^{4}He(γ,n)^{3}He cross sections are smaller than the ones calculated with the Lorentz integral transform method [17]. Note that those calculations took into account only the contribution of the two-body central force in nucleon-nucleon potentials. More recently, it has been shown that the tensor force and the three-nucleon forces can significantly hinder the transition amplitude below about 30 MeV [18]. Therefore a new calculation for four-nucleon systems with more realistic nuclear potentials will be interesting.

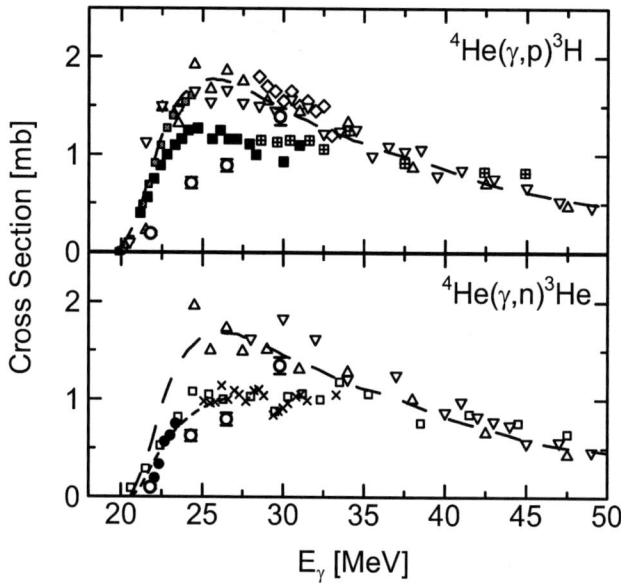

FIGURE 4. Two-body photodisintegration cross sections of ^{4}He. The open circles denote the present data, while the other symbols indicate the previous data (see the references in [13]). Upper panel; ^{4}He(γ,p)^{3}H, open upward triangles; Gorbunov, open downward triangles; Arkatov *et al.*, crossed squares; Bernabei *et al.*, filled squares; Feldman *et al.*, open diamonds; Hoorebeke *et al.*, gray squares; Hahn *et al.*. Lower panel; ^{4}He(γ,n)^{3}He, open upward triangles; Gorbunov, open downward triangles; Arkatov *et al.*, open squares; Berman *et al.*, diagonal crosses; Ward *et al.*, filled circles; Komar *et al.*. The error bars of the previous data are not shown for clarity. The long-dashed curves are the cross sections calculated using the LIT method with the MTI-III potential by Quaglioni *et al.*[17]. The short-dashed curve represents the calculated (γ,n) cross section based on the AGS formalism by Ellerkmann *et al.*[16].

NEW EXPERIMENTAL PLAN

One of the main advantages of the present method is high accuracy in determination of absolute cross sections of the photodisintegration reactions. Since the statistical and systematic errors in the present measurement are evaluated to be 6~10% and ~3%, respectively, the accuracy of the present method is governed by the observed reaction yields. Therefore we are going to perform a new experiment using new LCS γ-ray beams with an intensity of ~10^6 photons/s, which is about two orders of magnitude higher than that of AIST. Such intense LCS γ-ray beams are now available at several facilities in the world, e.g. the NewSUBARU facility of the University of Hyogo in Japan [19]. High-intensity γ-ray beams are expected to enable us to perform high-precision measurements of various photonuclear reactions at low energies. For example, the cross section of the D(p,γ)^{3}H reaction below a few hundred keV is one of the important nuclear inputs for the big-bang nucleosynthesis (BBN) calculations [20-22], and many experimental efforts have been devoted to determine it accurately [23]. However, a discrepancy of about 20% still exists between the previous experimental data, leading to one of the major error sources in the BBN calculations. The measurement of the inverse ^{3}He(γ,p)d reaction will provide experimental information independent of the possible sources of the systematic errors in the radiative capture measurements.

CONCLUSION

A quasi-monochromatic LCS γ-ray beam in combination with a time projection chamber with an active target will be a promising tool for experimental study of the nuclear reactions relevant to nuclear astrophysics. It will provide reliable data for low-energy photonuclear reactions going not only to two-body final states but also to many-body ones.

ACKNOWLEDGMENTS

The author would like to thank Profs. Y. Nagai, T. Kajino, H. Utsunomiya and S. Miyamoto for fruitful discussions. This work was supported in part by Grant-in-Aid for Specially Promoted Research of the Japan Ministry of Education, Science, Sports and Culture and in part by Grant-in-Aid for Scientific Research of the Japan.

REFERENCES

1. M. Arnould and S. Goriely, *Phys. Rep.* **384**, 1-84 (2003).
2. T. Rauscher, *Phys. Rev.* **C73**, 015804 (2006).
3. H. Ejiri, *Phys. Rep.* **338**, 265-351 (2000).
4. K. Langanke, *Rev. Mod. Phys.* **75**, 819-862 (2003).
5. W. C. Haxton, *Phys. Rev. Lett.* **60**, 1999-2002 (1988).
6. S. E. Woosley et al., *Astrophys. J.* **433**, 229-246 (1994).
7. K. Sumiyoshi et al., *Astrophys. J.* **629**, 922-932 (2005).
8. S. E. Woosley, D. H. Hartmann, R. D. Hoffman, and W. C. Haxton, *Astrophys. J.* **356**, 272-301 (1990).
9. T. Yoshida, M. Terasawa, T. Kajino, and K. Sumiyoshi, *Astrophys. J.* **600**, 204-213 (2004).
10. A. Heger et al., *Phys. Lett.* **B606**, 258-264 (2005).
11. Y. -Z. Qian, W. C. Haxton, K. Langanke, and P. Vogel, Phys. Rev. C55, 1532-1544 (1997).
12. D. Gazit and N. Barnea, *Phys. Rev.* **C70**, 048801 (2004).
13. T. Shima et al., *Phys. Rev.* **C72**, 044004 (2005).
14. H. Ohgaki et al., *IEEE Trans. Nucl. Sci.* **38**, 386 (1991).
15. T. Kii, T. Shima, T. Baba, and Y. Nagai, *Nucl. Instr. Meth. in Phys. Res.* **A552**, 329-343 (2005).
16. W. Sandhas et al., *Nucl. Phys.* **A631**, 210c-229c (1998).
17. S. Quaglioni, W. Leidemann, G. Orlandini, N. Barnea, and V. D. Efros, *Phys. Rev.* **C69**, 044002 (2004).
18. T. Myo (private communication).
19. K. Aoki et al., *Nucl. Instr. Meth. in Phys. Res.* **A516**, 228-236 (2004).
20. M. S. Smith, L. H. Kawano, and R. A. Malaney, *Astrophys. J. Suppl.* **85**, 219-247 (1993).
21. A. Coc, E. V-Flam, P. Descouvemont, A. Adahchour, and C. Angulo, *Astrophys. J.* **600**, 544-552 (2004).
12. R. H. Cyburt, *Phys. Rev.* **D70**, 023505 (2004).
23. C. Casella et al. (LUNA collaboration), *Nucl. Phys.* **A706**, 203-216 (2002).

Determination of the astrophysical $^8Li(n,\gamma)^9Li$ reaction rate from the measurement of $^2H(^8Li,^9Li)^1H$ reaction

Zhihong Li*, Weiping Liu*, Xixiang Bai*, Bing Guo*, Gang Lian*,
Shengquan Yan*, Baoxiang Wang*, Sheng Zeng*, Yun Lu*, Jun Su*,
Yongshou Chen*, Kaisu Wu*, Nengchuan Shu* and Toshitaka Kajino[†,**]

China Institute of Atomic Energy, Beijing 102413, P. R. China
[†]*National Astronomical Observatory of Japan, Mitaka, Tokyo 181-8588, Japan*
[**]*Department of Astronomy, University of Tokyo, Bunkyo-ku, Tokyo 113-0033 Japan*

Abstract. The cross section of the $^8Li(n,\gamma)^9Li$ reaction has attracted much attention for many years because of its importance in nuclear astrophysics. The single particle spectroscopic factor, $S_{1,3/2}$ for the ground state of $^9Li = {}^8Li \otimes n$ derived from the transfer reaction of $^8Li(d,p)^9Li$ was used to calculate the direct radiative capture reaction rates of $^8Li(n,\gamma)^9Li$ at energies of astrophysical interest. The present result shows that the $^8Li(n,\gamma)^9Li$ direct capture reaction may play an important role in the astrophysical environments of inhomogeneous big bang and type II supernovae.

Keywords: Transfer reaction, Spectroscopic factor, Radiative capture, Nucleosynthesis
PACS: 25.60.Je, 21.10.Jx, 25.40.Lw, 26.50.+x

INTRODUCTION

In the astrophysical environments of inhomogeneous big bang and type II supernovae, the main pathways towards heavier nuclides are the reaction chains $^1H(n,\gamma)^2H(n,\gamma)^3H(d,n)^4He(t,\gamma)^7Li(n,\gamma)^8Li(\alpha,n)^{11}B$ and $^7Li(n,\gamma)^8Li(n,\gamma)^9Li(\beta^-)^9Be$, in which $^8Li(\alpha,n)^{11}B$ and $^8Li(n,\gamma)^9Li$ are the key reactions to bridge the stability gap at mass number A = 8 and may affect the abundances of Li, Be, B and C [1, 2].

Up to now, considerable effort has been devoted to experimentally determining the $^8Li(\alpha,n)^{11}B$ cross section as described in Ref. [3] and references therein. However, a large uncertainty still remains for the $^8Li(n,\gamma)^9Li$ cross section. There were some microscopic and systematic calculations of this reaction that deviated by an order of magnitude [4, 5, 6, 7, 8, 9]. Direct measurement of the $^8Li(n,\gamma)^9Li$ reaction is impossible because no neutron target exists and the half life of ^8Li is too short (838 ms) as a target. The only experimental information was obtained from two Coulomb dissociation measurements [10, 11] that presented the different upper limits. It is therefore highly needed to measure the $^8Li(n,\gamma)^9Li$ cross section through an independent approach. A practicable method is to extract the direct capture cross section for the $^8Li(n,\gamma)^9Li$ reaction using the direct capture model [12, 13] and spectroscopic factor deduced from the angular distribution of the transfer reaction $^8Li(d,p)^9Li$.

EXPERIMENTAL RESULTS

The measurement of the ^{8}Li$(d,p)^9$Li angular distribution was performed using the secondary beam facility GIRAFFE [14, 15] built at the HI-13 tandem accelerator of China Institute of Atomic Energy. A 44 MeV ^{7}Li primary beam from the tandem accelerator impinged on a deuterium gas cell at 1.6 atm pressure to produce the ^{8}Li ions through the ^{2}H(^{7}Li,^{8}Li)^{1}H reaction. The front and rear windows of the gas cell were 1.9 mg/cm^2 thick Havar foils. Following the magnetic separation and focus with a dipole and a quadrupole doublet, the 39 MeV secondary ^{8}Li beam was delivered. After the collimation using two 3 mm apertures, the ^{8}Li beam was directed onto a deuterated polyethylene $(CD_2)_n$ foil in thickness of 1.5 mg/cm^2 to study the ^{2}H(^{8}Li,^{9}Li)^{1}H reaction. The experimental details were described in Ref. [16, 17]. Here, we only gave the angular distribution in center of mass frame for the ^{8}Li$(d,p)^9$Li (ground state) reaction, as shown in Fig. 1.

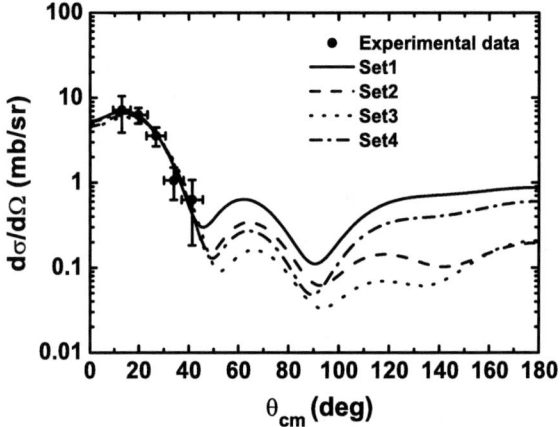

FIGURE 1. Measured angular distribution of ^{8}Li$(d,p)^9$Li$_{g.s.}$ at E_{cm} = 7.8 MeV, together with DWBA calculations using different optical potential parameters.

The spins and parities of ^8Li and ^9Li (ground state) are 2^+ and $3/2^-$, respectively. The $^8Li(d,p)^9Li_{g.s.}$ cross section involves two components corresponding to $(l=1, j=3/2)$ and $(l=1, j=1/2)$ transfers. We neglected the j = 1/2 component in the calculations because its contribution was predicted to be less than 5% [13, 18]. The differential cross section can then be simplified as

$$(\frac{d\sigma}{d\Omega})_{exp} = S_d S_{1,3/2} \sigma_{1,3/2}(\theta), \qquad (1)$$

where $(\frac{d\sigma}{d\Omega})_{exp}$ and $\sigma_{l,j}(\theta)$ denote the measured and DWBA differential cross sections, respectively. $S_{1,3/2}$ and S_d are spectroscopic factors for the ground state of $^9Li = {}^8Li \otimes n$ and $d = p \otimes n$, respectively. By knowing the value of S_d, the $S_{1,3/2}$ can be extracted by normalizing DWBA differential cross sections to the experimental data.

The finite-range Distorted Wave Born Approximation (DWBA) code PTOLEMY [19] was used to compute the angular distribution. The four sets of optical potential parameters used in the calculations were listed in Table 1, all the entrance channel potentials were taken from Ref. [20], the exit channel ones from Refs. [20] and [21], respectively. Fig. 1 presents the normalized angular distributions for four sets of optical potential parameters. For each set, five spectroscopic factors with different statistical errors were derived by fitting five experimental points, the weighted average value of them was taken as the spectroscopic factor. The average value of these four spectroscopic factors was found to be 0.68 ± 0.14, the uncertainty arose from the error transfer of four spectroscopic factors and their deviation. We neglected the contribution of compound nucleus (CN) process because the influence on the spectroscopic factor is less than 3%. The present result accords with that ($S_{1,3/2} = 0.73$) extracted from the mirror reaction $^8B(d,n)^9C$ [18] within the error bar.

TABLE 1. Optical potential parameters used in DWBA calculations and the corresponding ANCs, where V, W are in MeV, r and a are in fm, the geometrical parameter of single particle bound state was set to be $r_0 = 1.25$ fm and $a = 0.65$ fm.

Set No.	1		2		3		4	
Channel	Entrance*	Exit*	Entrance*	Exit*	Entrance*	Exit†	Entrance*	Exit†
V_r	142.9	41.9	118.0	41.9	118.0	66.37	142.9	66.37
r_{0r}	0.908	1.38	1.00	1.38	1.00	1.14	0.908	1.14
a_r	0.88	0.65	0.94	0.65	0.94	0.57	0.88	0.57
W_s	3.7	10.2	6.87	10.2	6.87	6.85	3.7	6.85
r_{0s}	2.26	1.5	1.98	1.5	1.98	1.14	2.26	1.14
a_s	0.67	0.37	0.59	0.37	0.59	0.5	0.67	0.5
V_{SO}	5.7	4.5	8.5	4.5	8.5	5.5	5.7	5.5
r_{0SO}	0.908	1.35	1.00	1.35	1.00	1.14	0.908	1.14
a_{SO}	0.88	0.33	0.94	0.33	0.94	0.57	0.88	0.57
r_{0c}	1.38	1.33	1.30	1.33	1.30	1.14	1.38	1.14
$S_{1,3/2}$	0.86 ± 0.12		0.68 ± 0.10		0.65 ± 0.10		0.52 ± 0.08	

* The optical potential parameters were taken from Ref. [20]
† The optical potential parameters were taken from Ref. [21]

THE ^{8}Li(n,γ)^{9}Li REACTION RATES

The ^{8}Li$(n,\gamma)^9$Li$_{g.s.}$ cross section was calculated by assuming that the reaction proceeds via direct E1 neutron capture to the ground state of ^{9}Li. At low energies of astrophysical interest, the contribution of d-wave is negligible, the capture reaction is almost totally determined by the s-wave neutron capture process. The cross section for E1 capture of neutron to the ground state of ^{9}Li with the orbital and total angular momenta l_f and j_f is given by

$$\sigma_{(n,\gamma)} = \frac{16\pi}{9}\left(\frac{E_\gamma}{\hbar c}\right)^3 \frac{e_{eff}^2}{k^2} \frac{1}{\hbar v} \frac{(2I_3+1)}{(2I_1+1)(2I_2+1)} S_{l_f j_f} |\int_0^\infty r^2 w_{l_i}(kr) u_{l_f}(r) dr|^2, \quad (2)$$

where E_γ is the γ-ray energy, v the relative velocity between neutron and ^{8}Li, k the incident wave number, $I_{1,2,3}$ are the spins of neutron, ^{8}Li and ^{9}Li, respectively. e_{eff}

$= -eZ/(A+1)$ is the neutron effective charge for the E1 transition in the potential produced by a target nucleus with mass number A and atomic number Z. $w_{l_i}(kr)$ is the distorted radial wave function for the entrance channel, $u_{l_f}(r)$ the radial wave function of the bound state neutron in ^{9}Li. Both wave functions can be calculated with optical potential model.

The optical potential for the neutron scattering on unstable nucleus ^{8}Li is unknown experimentally. We adopted a real folding potential which was calculated using the ^{8}Li density distribution from the measured interaction cross section [22] and an effective nucleon-nucleon interaction DDM3Y [23]. The imaginary part of the potential is very small because of the small flux into other reaction channels and can be neglected in most cases involving neutron capture reaction. The depth of the real potential was scaled to the volume integral of potential per nucleon which was obtained by fitting the experimental cross sections of ^{6,7}Li and ^{12}C neutron capture reaction [24, 25, 26]. One can see that the optical potential changes considerably, while the volume integral of potential per nucleon is a more stable quantity relatively for the 1p-shell nuclei, and the cross section of (n,γ) reaction can be reproduced if the volume integral of potential per nuclei for 1p-shell nuclei is assumed to be a constant. Thus, the cross section of the ^{8}Li$(n,\gamma)^9$Li reaction can be calculated with the Eq. (2) by constraining the optical potentials between ^{8}Li and neutron.

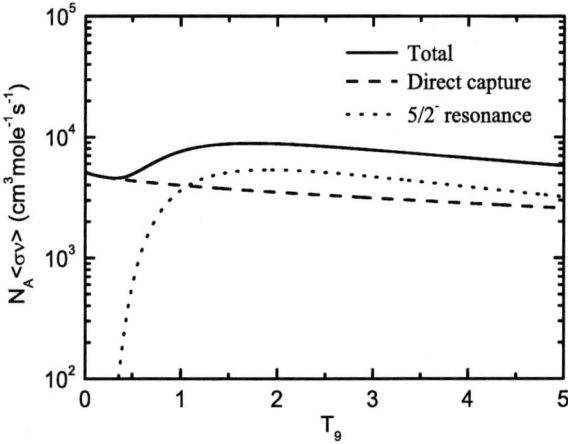

FIGURE 2.   Temperature dependence of the reaction rates for $^8Li(n,\gamma)^9Li$.

Figure 2 demonstrates the temperature dependence of the reaction rate for the direct capture in ^{8}Li$(n,\gamma)^9$Li together with that for the resonant capture at $5/2^-$ (E_x = 4.31 MeV) state in ^{9}Li (deduced by Rauscher et al. [4] with Γ_γ = 0.65 eV and Γ_{tot} = 100 keV), as well as the total reaction rate. Direct capture to the first excited state (E_x = 2.69 MeV) is not included in the calculation since the transition strength is believed to be negligible [10, 11]. It can clearly be seen that the direct capture plays an important role in the ^{8}Li$(n,\gamma)^9$Li reaction, especially in the temperature range of $T_9 < 1$. The reaction rate for the direct capture was found to be $N_A\langle\sigma v\rangle = 3970 \pm 950$ cm^{3}mole^{-1}s^{-1} at $T_9 = 1$, the uncertainty arises from the errors of spectroscopic factor and the volume integral of potential per nucleon. The ^{8}Li$(n,\gamma)^9$Li reaction rates for the direct capture derived from theoretical calculations and experiments are shown in Fig. 3, our result is significantly higher than the upper limit from the most recent Coulomb dissociation experiment [10] and approximately in agreement with the theoretical estimations reported in Refs. [4, 7]. The present result removes a longstanding debate about the absolute cross section of ^{8}Li$(n,\gamma)^9$Li.

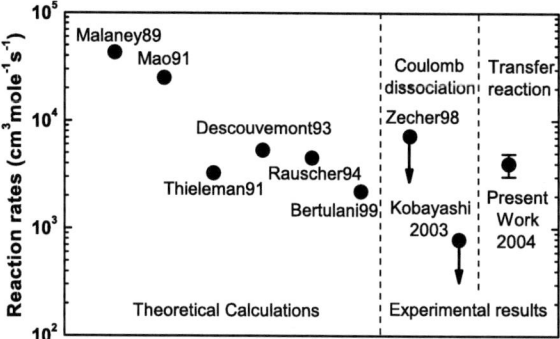

FIGURE 3. The comparison of $^8Li(n,\gamma)^9Li$ direct capture reaction rates derived from theoretical calculations and experiments.

SUMMARY

In summary, we have measured the angular distribution of the ^{8}Li$(d,p)^9$Li$_{\text{g.s.}}$ reaction at $E_{\text{c.m.}} = 7.8$ MeV, through coincidence detection of ^{9}Li and recoil proton. By using the spectroscopic factor deduced from the ^{8}Li$(d,p)^9$Li$_{\text{g.s.}}$ angular distribution, we have successfully derived the ^{8}Li$(n,\gamma)^9$Li$_{\text{g.s.}}$ direct capture cross sections and reaction rates. This work presents the first experimental constraint for the ^{8}Li$(n,\gamma)^9$Li reaction rates of astrophysical relevance.

ACKNOWLEDGMENTS

This work was supported by the Major State Basic Research Development Program under Grant Nos. G200077400 and 2003CB716704, the National Natural Science Foundation of China under Grant No. 10375096.

REFERENCES

1. T. Kajino, and R. N. Boyd, *Astrophys. J.* **359**, 267 (1990).
2. M. Terasawa, K. Sumiyoshi, T. Kajino, G. J. Mathews, and I. Tanihata, *Astrophys. J.* **562**, 470 (2001).
3. H. Miyatake, H. Ishiyama, M.-H. Tanaka, Y. X. Watanabe, N. Yoshikawa, S. C. Jeong, Y. Matsuyama, Y. Fuchi, T. Nomura, T. Hashimoto, T. Ishikawa, K. Nakai, S. K. Das, P. K. Saha, T. Fukuda,

K. Nishio, S. Mitsuoka, H. Ikezoe, S. Ichikawa, M. Matsuda, Y. Mizoi, T. Furukawa, H. Izumi, T. Shimoda, and M. Terasawa, *Nucl. Phys. A* **738**, 401 (2004).
4. T. Rauscher, J. H. Applegate, J. J. Cowan, F.-K. Thielemann, and M. Wiescher, *Astrophys. J.* **429**, 499 (1994).
5. R. A. Malaney, and W. A. Fowler, *Astrophys. J.* **345**, L5 (1989).
6. Z. Q. Mao, and A. E. Champagne, *Nucl. Phys. A* **522**, 568 (1991).
7. K. K-Thieleman, J. H. Cowan, and M. Wiescher, *Nuclei in the Cosmos*, ed. H. Oberhummer et al. (1991), (Berlin:Springer).
8. P. Descouvemont, *Astrophys. J. Lett.* **405**, 518 (1993).
9. C. A. Bertulani, *J. Phys. G* **25**, 1959 (1999).
10. H. Kobayashi, K. Ieki, Á. Horváth, A. Galonsky, N. Carlin, F. Deák, T. Gomi, V. Guimaraes, Y. Higurashi, Y. Iwata, Á. Kiss, J. Kolata, T. Rauscher, H. Schelin, Z. Seres, and R. Warner, *Phys. Rev. C* **67**, 015806 (2003).
11. P. D. Zecher, A. Galonsky, S. J. Gaff, J. J. Kruse, G. Kunde, E. Tryggestad, J. Wang, R. E. Warner, D. J. Morrissey, K. Ieki, Y. Iwata, F. Deak, A. Kiss, Z. Seres, J. J. Kolata, J. von Schwarzenberg, and H. Schelin, *Phys. Rev. C* **57**, 959 (1998).
12. C. Rolfs, *Nucl. Phys. A* **217**, 29 (1973).
13. P. Mohr, *Phys. Rev. C* **67**, 065802 (2003).
14. X. Bai, W. liu, J. Qin, Z. Li, S. Zhou, A. Li, Y. Wang, Y. Cheng, and W. Zhao, *Nucl. Phys. A* **588**, 273c (1995).
15. W. P. Liu, Z. H. Li, X. X. Bai, Y. B. Wang, G. Lian, S. Zeng, S. Q. Yan, B. X. Wang, Z. X. Zhao, T. J. Zhang, H. Q. Tang, B. F. Yang, X. L. Guan, and B. Q. Cui, *Nucl. Instrum. Methods Phys. Res. B* **204**, 62 (2003).
16. Z. H. Li, W. P. Liu, X. X. Bai, B. Guo, G. Lian, S. Q. Yan, B. X. Wang, S. Zeng, Y. Lu, J. Su, Y. S. Chen, K. S. Wu, N. Shu, and T. Kajino, *Phys. Rev. C* **71**, 052801(R) (2005).
17. B. Guo, Z. H. Li, W. P. Liu, X. X. Bai, G. Lian, S. Q. Yan, B. X. Wang, S. Zeng, J. Su, and Y. Lu, *Nucl. Phys. A* **761**, 162 (2005).
18. D. Beaumel, T. Kubo, T. Teranishi, H. Sakurai, S. Fortier, A. Mengoni, N. Aoi, N. Fukuda, M. Hirai, N. Imai, H. Iwasaki, H. Kumagai, H. Laurent, S. M. Lukyanov, J. M. Maison, T. Motobayashi, T. Nakamura, H. Ohnuma, S. Pita, K. Yoneda, and M. Ishihara, *Phys. Lett. B* **514**, 226 (2001).
19. M. H. Macfarlane, and S. C. Pieper, *Argonne National Laboratory Report* (unpublished).
20. C. M. Perey, and F. G. Perey, *Atomic data and nuclear data tables* **17**, 1 (1976).
21. B. A. Watson, P. P. Singh, and R. E. Segel, *Phys. Rev.* **182**, 977 (1969).
22. I. Tanihata, H. Hamagaki, O. Hashimoto, Y. Shida, N. Yoshikawa, K. Sugimoto, O. Yamakawa, T. Kobayashi, and N. Takahashi, *Phys. Rev. Lett.* **55**, 2676 (1985).
23. A. M. Kobos, B. A. Brown, R. Lindsay, and G. R. Satchler, *Nucl. Phys. A* **425**, 205 (1984).
24. T. Ohsaki, Y. Nagai, M. Igashira, T. Shima, H. Kitazawa, K. Takaoka, M. Kinoshita, Y. Nobuhara, A. Tomyo, H. Makii, and K. Mishima, *AIP Conference Proceedings* **529(1)**, 458 (2000).
25. J. C. Blackmon, A. E. Champagne, J. K. Dickens, J. A. Harvey, M. A. Hofstee, S. Kopecky, D. C. Larson, D. C. Powell, S. Raman, and M. S. Smith, *Phys. Rev. C* **54**, 383 (1996).
26. Y. Nagai, M. Igashira, N. Mukai, T. Ohsaki, F. Uesawa, K. Takeda, T. Ando, H. Kitazawa, S. Kubono, and T. Fukuda, *Astrophys. J.* **381**, 444 (1991).

1.3 MOST METAL-DEFICIENT STARS: OBSERVATION and THEORY

Stellar Abundances – First Generation to Solar

John E. Norris

Research School of Astronomy & Astrophysics, The Australian National Observatory, Mount Stromlo Observatory, Cotter Road, Weston Creek, ACT 2611, Australia

Abstract.
The discovery and analysis of metal-poor stars lead to insight into conditions when the Universe and Galaxy were young. We present the rationale for studying such objects (which become progressively rarer at lowest abundance), with a description of their systematic discovery, culminating in the recent analysis of two objects having [Fe/H] < –5.0. We discuss the Metallicity Distribution Function of metal-poor stars and the abundance patterns of several elements, from Li through to the heavy-neutron-capture elements. Relatively few (∼50) stars with [Fe/H] < –3.0 have been analyzed at high spectral resolution and high signal-to-noise. As one proceeds to lowest abundance one finds astounding overabundances of some or all of the CNO group and the lighter elements. This diversity among the most metal-poor stars has yet to be fully understood.

Keywords: Galaxy: abundances – nuclear reactions, nucleosynthesis, abundances – stars: abundances – stars: Population II
PACS: 97.10.Tk, 97.20.Tr, 97.20.Wt, 97.60.Bw, 98.80.Ft

1. INTRODUCTION

The abundances of the chemical elements in stars more metal-poor than the sun hold crucial clues for an understanding of the manner in which they formed and of the conditions that existed at earlier times – including the Big Bang – through the epoch of the formation of the first stars, and for all subsequent generations.
A short list of the rationale for the search for, and analysis of, metal-poor stars includes:

- They encompass the stars closest to the Big Bang. In particular, Li has the potential to constrain conditions in the Big Bang itself.
- The most metal-poor objects were formed at epochs corresponding to redshifts > 4–5, and probe conditions when the first heavy element producing objects formed. The study of these objects is regarded by many as "near-field" cosmology: in particular, objects with, say, [Fe/H] < – 3.5 permit insight into conditions at the earliest times that are not readily afforded by the study of objects at high redshift.
- They constrain our understanding of the nature of the first stars, the early stellar mass function, the explosion of super- and hypernovae, and the manner in which their ejecta were incorporated into subsequent generations of stars.
- They inform our understanding of how the Galaxy formed. Relationships between abundances, kinematics, and ages permit choice between the various paradigms.
- In some stars with [Fe/H] ∼–3.0, the overabundances of the heavy-neutron-capture elements are so large that measurement of Th and U is possible and leads to independent estimates of their ages and hence of the Galaxy.

The present contribution is organized as follows. §2 presents a brief overview of the discovery of metal-poor stars, while §3 discusses their Metallicity Distribution Function (MDF). This is followed in §4 by a discussion of the abundances of several elements as a function of [Fe/H]. Finally, §5 sets the scene for subsequent discussion in this meeting with a brief description of suggestions on the nature of the first objects responsible for the abundance patterns in stars now being discovered with –4.0 < [Fe/H] < –5.5.

Previous useful reviews relevant to this topic include Sandage (1986), Wheeler, Sneden, & Truran (1989), McWilliam (1997), Norris (2004), and Beers & Christlieb (2005).

2. DISCOVERY

For an account of the initial discovery of metal-poor stars by Chamberlain & Aller (1951), and its implications, the reader is referred to Sandage (1986; §4, see in particular his footnote 2). Since that momentous result, ever increasing efforts have been made to discover the most metal-deficient objects, culminating some four decades later in the discovery of two stars with [Fe/H] $\sim$ –5.5 (Christlieb et al. 2002; Frebel et al. 2005). A brief history of progress between these two epochs is shown in Figure 1, which presents the abundance [Fe/H] of the most metal-poor star then known, as a function of epoch. Also shown in the figure is the prediction of Iben (1983) of the minimum abundance one might expect to observe, given that a star formed some 14 Gyr ago with no heavy elements would accrete material from the interstellar medium since that time. It will be interesting to see if the Iben prediction stands the test of future observations!

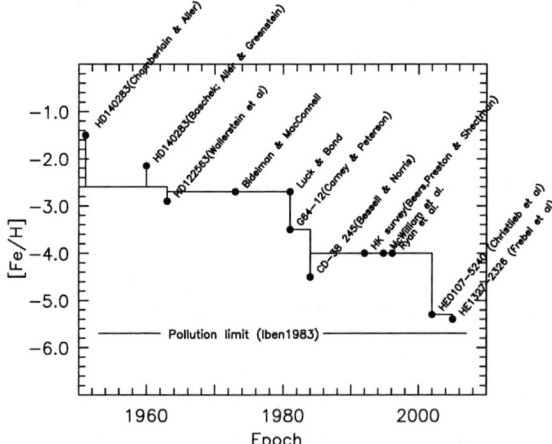

FIGURE 1. The abundance [Fe/H] of the most metal-poor star then known as a function of epoch. The symbols refer to the values published by the authors, and are connected to currently accepted values.

The search for metal-poor objects proceeds in a number of ways: informed serendepity (a euphemism for good luck! e.g. CD–38°245 ([Fe/H] = –4.0) Bessell & Norris 1984), surveys of high-proper-motion stars (Ryan & Norris 1991; Carney et al. 1994), objective prism studies (Beers 1999 and references therein – the HK survey; Christlieb et al. 1999 – the HES). In future one might expect colormetric/spectroscopic surveys such as Sloan SEGUE to play a major role (see e.g. Beers et al. 2006).

3. METALLICITY DISTRIBUTION FUNCTION

The MDF of a population contains important clues to the manner in which it formed. The bimodal distribution of the MDF of the Galactic globular clusters, for example, attests to the existence of distinct halo and disk components. Early work on the MDF of halo field stars suggested that in the regime $-3.5 <$ [Fe/H] < -1.5 the observations were consistent with chemical enrichment in terms of the simple (closed-box) model of Hartwick (1976) (see e.g. Ryan & Norris 1991), but more recent and perhaps more realistic models, such as those of Tsujimoto, Shigeyama, & Yoshii (1999), have been able to provide a somewhat better fit as perhaps might have been expected.

The current situation on the data available for the halo MDF has been recently presented by Beers et al. (2006). To summarize: from the low-resolution spectroscopic HK and HES surveys, some 2750 and 380 objects are now known with [Fe/H] < -2.0 and -3.0, respectively. The reader is referred to that work for details.

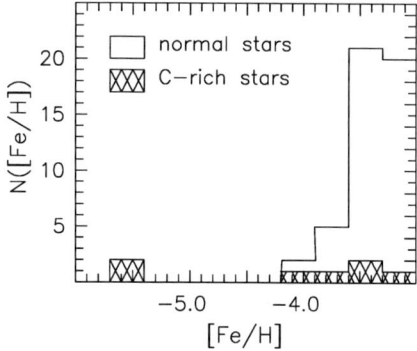

FIGURE 2. Histogram of [Fe/H] for stars with [Fe/H] < -3.0, from high-resolution, high S/N abundance analyses (Carretta et al. 2002, Cayrel et al. 2004, Cohen et al. 2004, Christlieb et al. 2002, Frebel et al. 2005, McWilliam et al. 1995, Plez & Cohen 2005, Norris et al. 1997, 2000, 2001, Ryan et al. 1996) Note that the proportion of carbon-rich objects increases dramatically as one moves to lower abundances.

What has become clear in the past decade, however, is that no currently-proposed model is capable of explaining the results in the regime [Fe/H] < -4.0. There are two points that should be made. First, in spite of extensive searches since the discovery of CD–38°245 in 1984, only two objects with [Fe/H] < -4.2 have been found, where one would have expected some 20–30, according to the simple model of galactic chemical enrichment. Second, both of these objects (HE0107–5480, Christlieb et al. 2002; and HE1327–2326, Frebel et al. 2005) have [Fe/H] ~ -5.5, and as shown in Figure 2, there are no objects, for which high-resolution, high signal-to-noise data are available, currently known in the range $-5.3 <$ [Fe/H] < -4.2. We draw the reader's attention to the fact that these two objects also possess strong chemical peculiarities (in particular large enhancements of the CNO group, which will be discussed below). While the numbers of objects with [Fe/H] < −3.0 for which the required high quality data are currently available are indeed small, they are surely suggestive of interesting non-canonical possibilities. Further work is crucial to address this question.

4. INDIVIDUAL ELEMENTS

4.1 Lithium

The demonstration by Spite & Spite (1982) that the abundance of lithium in metal-poor near main-sequence-turnoff stars is independent of the star's heavy element abundance, and the realisation that the lithium in these objects provides insight into the amount of lithium produced in the Big Bang were fundamental results. Further work by Ryan et al. (1999) and others then led to the interesting claim that as one observes turnoff stars of increasingly lower metal abundance, lithium decreases with decreasing [Fe/H]. It was also argued that, with the exception of a few problematic objects, the regression of A(Li) = (log N(Li)/N(H) + 12.0) on [Fe/H] showed extremely small scatter, ∼0.03 dex, commensurate with the observational uncertainties. This in shown in Figure 3, together with the result for the most metal-poor dwarf/subgiant HE1327–2326. Note in particular that Li is not detected in HE1327–2326, at a level below what might have been expected from any reasonable extrapolation of the earlier results.

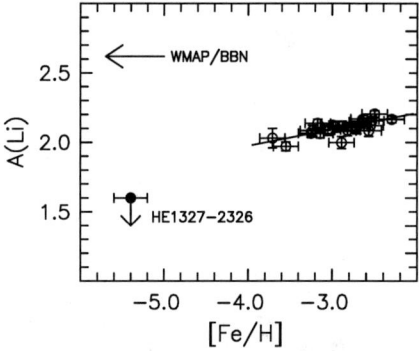

FIGURE 3. A(Li) versus [Fe/H], from Coc et al. (2004), Frebel et al. (2005), Norris et al. (2000), and Ryan et al. (1999).

Ryan et al. reported primordial A(Li)≈ 2.0. Also shown in Figure 3 is the result for the abundance of Li resulting from the WMAP determination of the density of the Universe and the prediction of Big Bang Nucleosynthesis, A(Li) = 2.6. The discrepancy between the stellar observations and the WMAP result is stark (Coc et al. 2004). Possible explanations include: stellar astration of Li (by e.g. rotational mixing and destruction (Pinsonneault et al. 2002) or an erroneous effective temperature scale in the stellar analyses (Meléndez & Ramírez 2004)).

Against this background, the recent work of Asplund et al. (2005) on the abundance of not only ^{7}Li but also ^{6}Li in near main sequence turnoff stars is of particular significance. They confirm the essential result of Ryan et al. (1999), and find no need for a revision of the effective temperature scale. More basically, they detect ^{6}Li in amounts (^{6}Li/^{7}Li ∼0.05) that appear inconsistent with the existing models of rotational mixing and destruction of Li. While detection of ^{6}Li is painstaking work, further observations of this type will be fundamental to a full understanding of this question. The need for ^{7}Li abundances for more dwarfs with [Fe/H] < −4.0 should also be clear.

4.2 Carbon, nitrogen, and oxygen

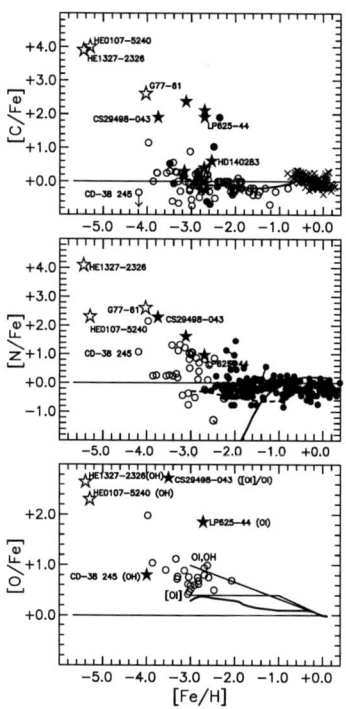

FIGURE 4. Relative C, N, and O abundances. The sources of the data are as in Norris (2004), updated by Cayrel et al. (2004), Frebel et al. (2005, 2006) and Plez & Cohen (2005). For oxygen, schematic representations are shown of recent results based on [O I] λ6300 Å, on the one hand, and the O I infrared triplet and ultraviolet OH lines, on the other. The full lines are from Timmes, Woosley, & Weaver (1995), while the dashed line in the middle panel represents their result with the arbitrary inclusion of convective overshoot.

Abundances for carbon, nitrogen, and oxygen, relative to iron, as a function of [Fe/H] are presented in Figure 4. The large relative overabundances of the CNO group for objects with [Fe/H] < −4.0 suggest that we are witnessing a basically different production mechanism of these elements in the earliest stars, compared with what occurred at higher abundance. We shall return to this in §5. It should be emphasised that this is not an observational selection effect: with the exception of G77–61 (Gass, Wehrse, & Liebert 1988), nothing was known about the CNO abundances in the discovery process of these objects.

4.3 Abundance spreads for the α- and heavier elements

The work of Cayrel et al. (2004) shows that metal-poor stars, with a few important exceptions, show well-defined trends in [X/Fe] as a function of [Fe/H], with only small dispersions, for atomic numbers up to and including the iron-peak. For example, they report σ[Mg/H] = 0.13 for a sample having [Fe/H] < −2.5. Arnone et al. (2005) find an even stronger result: σ[Mg/Fe] = 0.06. These observational results are in strong contrast with the predictions of the stochastic galactic chemical enrichment models at the earliest times of Argast et al. (2002), who predict dispersions of ∼0.4. The implications of this difference are far-reaching, and include the following possibilities: Mg/Fe in supernova ejecta does not depend on stellar mass; there is a very narrow range of stellar masses that produce Mg and Fe at the earliest times; the interstellar medium is well-mixed on very short timescales; and cooling timescales are longer than mixing timescales. The reader should refer to Arnone et al. (2005) for more detail.

The situation beyond the iron peak is quite different. It has been well-established since the work of Ryan et al. (1991) that there exists a large scatter in [X/Fe] as a function of [Fe/H] for the heavy-neutron-capture elements, the degree of which depends on the element under consideration. [Sr/Fe], for example, shows a spread of some 2 dex at lowest abundance. This has led to the suggestion that two processes are needed to produce the heavy-neutron-capture elements in the early phase of the Galaxy. See Aoki et al. (2005) for a comprehensive discussion of this topic.

5. THE MOST METAL-POOR STARS

Figure 5 shows element abundance, [M/H], as a function of atomic number for four of the most metal-poor stars currently known. CD–38°245 ([Fe/H] = −4.0) has an essentially "normal" abundance pattern – that is to say, it is representative of the bulk of material with −4.0 < [Fe/H] < −3.0. The other objects depart significantly from "normality", and show a different, but somewhat similar, pattern: large enhancement of the CNO group, and in two cases significant enhancements of some or all of the light elements Na, Mg, Al, and Si. The search for an explanation of these abnormalities for [Fe/H] < −4.0 is the subject of much current activity.

The question that perhaps begs to be answered, however, is: what is "normal" for [Fe/H] < −4.0. As may be seen in Figure 4, the two objects with [Fe/H] < −4.0 show similar large CNO enhancements. It will be important to learn whether this is "normal" in this metallicity regime. A second question that remains to be answered: is the absence of stars in the range, say, −5.2 < [Fe/H] < −4.2 real, or the result of small numbers?

We conclude with a brief description of various suggestions that have been made to explain these abundances in the most metal-poor objects. While not all of them may be correct or relevant, what now needs to be done is to determine which might provide the ingredients for a fuller understanding of the data. The suggestions include:

- 250–500 $M_\odot$, zero-heavy-element, first generation objects (Woosley & Weaver 1982; Fryer, Woosley, & Heger 2001).
- 25–30 $M_\odot$, zero-heavy-element, first generation supernovae, with mixing and fall-back (e.g. Umeda & Nomoto 2003; Iwamoto et al. 2005).

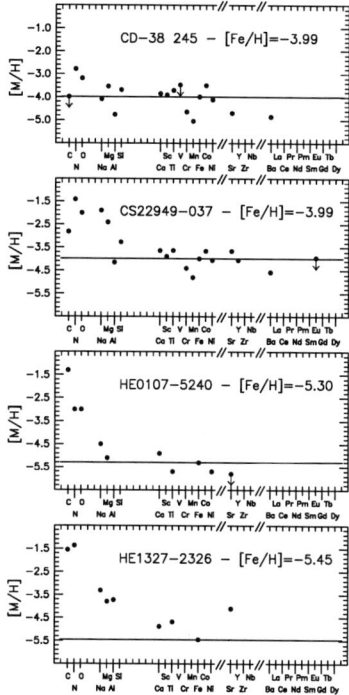

FIGURE 5. Abundances [M/H] as a function of atomic number. The sources of the data are as in Norris (2004), together with Frebel et al. (2005, 2006).

- Canonical evolution of rotating 60 $M_\odot$, $Z = 10^{-5}$ supernovae (Meynet, Ekström, & Maeder 2005).
- Concurrent enrichment by at least two supernovae – one ($\sim$15 $M_\odot$) of which underwent a "normal" explosion, and the other ($\sim$35 $M_\odot$) which experienced "fallback" (Limongi, Chieffi, & Bonifacio 2003).
- Nucleosynthesis and mass transfer in a first generation binary system, with subsequent accretion from the interstellar medium which contained supernova ejecta from the first generation (Suda et al. 2004).

For further discussion of the various possibilities the reader should see the contributions of Aoki, Hirschi, Suda, and Umeda in the present volume.

ACKNOWLEDGMENTS

It is a pleasure to thank several colleagues who have contributed to my understanding of the most metal-poor stars and the enjoyment of their study over the past three decades: W. Aoki, M. Asplund, T. C. Beers, M. S. Bessell, N. Christlieb, A. Frebel, and S. G. Ryan. This work has been supported by Australian Research Council grant DP0342613.

REFERENCES

1. Aoki, W. et al. 2005, ApJ, 632, 611
2. Argast, D., Samland, M., Thielemann, F.-K., & Gerhard, O. E. 2002, A&A, 388, 842
3. Arnone, E., Ryan, S. G., Argast, D., Norris, J. E., & Beers, T. C. 2005, A&A, 430, 507
4. Asplund, M., Lambert, D. L., Nissen, P. E., Primas, F., & Smith, V. V. 2005, arXiv:astro-ph/0510636
5. Beers, T. C. 1999, *The Third Stromlo Symposium: The Galactic Halo*, eds. B. K. Gibson et al. (San Francisco: ASP), 202
6. Beers, T. C. & Christlieb, N. 2005, ARA&A, 43, 531
7. Beers, T. C. et al. 2006, *From Lithium to Uranium: Elemental Tracers of Early Cosmic Evolution, IAU Symposium 228*, eds. V. Hill et al. in press (arXiv:astro-ph/0508423)
8. Bessell, M. S. & Norris, J. 1984, ApJ, 285, 622
9. Carney, B. W., Latham, D. W., Laird, J. B., & Aguilar, L. A. 1994, AJ, 107, 2240
10. Carretta, E., Gratton, R., Cohen, J. G., Beers, T. C., & Christlieb, N. 2002, AJ, 124, 481
11. Cayrel, R. et al. 2004, A&A, 416, 1117
12. Chamberlain, J. W. & Aller, L. H. 1951, ApJ, 114, 52
13. Christlieb, N., Bessell, M. S., Beers, T. C., Gustafsson, B., Korn, A., Barklem, P. S., Karlsson, T., Mizuno-Wiedner, M., & Rossi, S. 2002, Nature, 419, 904
14. Christlieb, N., Wisotzki, L., Reimers, D., Gehren, T., Reetz, J., & Beers, T. C. 1999, *The Third Stromlo Symposium: The Galactic Halo*, eds. B. K. Gibson et al. (San Francisco: ASP), 259
15. Coc, A., Vangioni-Flam, E., Descouvemont, P., Adahchour, A., & Angulo, C. 2004, ApJ, 600, 544
16. Cohen, J. G. et al. 2004, ApJ, 612, 1107
17. Frebel, A. et al. 2005, Nature, 434, 871
18. Frebel, A., Christlieb, N., Norris, J. E., Aoki, W., & Asplund, M. 2006, ApJ, in press (arXiv:astro-ph/0512543)
19. Fryer, C. L., Woosley, S. E., & Heger, A. 2001, ApJ, 550, 372
20. Gass, H., Wehrse, R., & Liebert, J. 1988, A&A, 189, 194
21. Hartwick, F. D. A. 1976, ApJ, 209, 418
22. Iben, I. Jr. 1983, Mem.S.A.It., 54, 321
23. Iwamoto, N., Umeda, H., Tominaga, N., Nomoto, K., & Maeda, K. 2005, Science, 309, 451
24. Limongi, M., Chieffi, A., & Bonifacio, P. 2003, ApJ, 594, L123
25. McWilliam, A. 1997, ARA&A, 35, 503
26. McWilliam, A., Preston, G. W., Sneden, C., & Searle, L. 1995, AJ, 109, 2757
27. Meléndez, J. & Ramírez, I. 2004, ApJ, 615, L33
28. Meynet, G., Ekström, S., & Maeder, A. 2005, *From Lithium to Uranium: Elemental Tracers of Early Cosmic Evolution, IAU Symposium 228*, eds. V. Hill et al. in press (arXiv:astro-ph/0511074)
29. Norris, J. E. 2004, *Origin and Evolution of the Elements*, eds. A. McWilliam & M. Rauch (Cambridge: CUP), 140
30. Norris, J. E., Beers, T. C., & Ryan, S. G. 2000, ApJ, 540, 456
31. Norris, J. E., Ryan, S. G., & Beers, T. C. 1997, ApJ, 489, L169
32. Norris, J. E., Ryan, S. G., & Beers, T. C. 2001, ApJ, 561, 1034
33. Pinsonneault, M. H., Steigman, G., Walker, T. P., & Narayanan, V. K. 2002, ApJ, 574, 398
34. Plez, B. & Cohen, J. G. 2005, A&A, 434, 1117
35. Ryan, S. G. & Norris, J. E. 1991, AJ, 101, 1865
36. Ryan, S. G., Norris, J. E., & Beers, T. C. 1996, ApJ, 471, 254
37. Ryan, S. G., Norris, J. E., & Beers, T. C. 1999, ApJ, 523, 654
38. Ryan, S. G., Norris, J. E., & Bessell, M. S. 1991, AJ, 102, 303
39. Sandage, A. 1986, ARA&A, 24, 421
40. Spite, F., & Spite, M. 1982, A&A, 115, 357
41. Suda, T., Aikawa, M., Machida, M. N., Fujimoto, M. Y., & Iben, I. Jr 2004, ApJ, 611, 476
42. Timmes, F. X., Woosley, S. E., & Weaver, T. A. 1995, ApJS, 98, 617
43. Tsujimoto, T., Shigeyama, T., & Yoshii, Y. 1999, ApJ, 519, L63
44. Umeda, H. & Nomoto, K. 2003, Nature, 422, 871
45. Wheeler, J. C., Sneden, C., & Truran, J. W., Jr. 1989, ARA&A, 27, 279
46. Woosley, S. E. & Weaver, T. A. 1982, *Supernovae: A Survey of Current Research*, eds. M. J. Rees & R. J. Stoneham (Dordrecht: Reidel), 79

An abundance study of the most iron-poor star HE1327-2326 with Subaru/HDS[1]

W. Aoki[*,†], A. Frebel[**], N. Christlieb[‡], J.E. Norris[**], T.C. Beers[§], T. Minezaki[¶,||], P.S. Barklem[††], S. Honda[*], M. Takada-Hidai[‡‡], M. Asplund[**], S.G. Ryan[§§], S. Tsangarides[§§], K. Eriksson[††], A. Steinhauer[¶¶], C. P. Deliyannis[***], K. Nomoto[†††], M.Y. Fujimoto[‡‡‡], H. Ando[*], Y. Yoshii[¶,||] and T. Kajino[*,†]

[*]*National Astronomical Observatory, Mitaka, Tokyo, 181-8588 Japan, aoki.wako@nao.ac.jp*
[†]*Department of Astronomy, Graduate University of Advanced Studies, Mitaka, Tokyo, 181-8588 Japan*
[**]*Research School of Astronomy and Astrophysics, Australian National University, Cotter Road, Weston, ACT 2611, Australia*
[‡]*Hamburger Sternwarte, University of Hamburg, Gojenbergsweg 112, D-21029 Hamburg, Germany*
[§]*Department of Physics and Astronomy and JINA, Michigan State University, East Lansing, MI 48824, USA*
[¶]*Institute of Astronomy, School of Science, University of Tokyo, Mitaka, Tokyo 181-0015, Japan*
[||]*Research Center for the Early Universe, School of Science, University of Tokyo, Bunkyo-ku, Tokyo 113-0033, Japan*
[††]*Uppsala Astronomical Observatory, Box 515, SE-75120 Uppsala, Sweden*
[‡‡]*Liberal Arts Education Center, Tokai University, Hiratsuka-shi, Kanagawa 259-1292, Japan*
[§§]*Department of Physics and Astronomy, Open University, Walton Hall, Milton Keynes MK76AA, UK*
[¶¶]*Department of Astronomy, University of Florida, 211 Bryant Space Science Center, Gainesville, FL 32611-2055, USA*
[***]*Department of Astronomy, Indiana University, 727 East 3rd Street, Swain Hall West 319, Bloomington, IN 47405-7105, USA*
[†††]*Department of Astronomy, School of Science, University of Tokyo, Bunkyo-ku, Tokyo 113-0033, Japan*
[‡‡‡]*Department of Physics, Hokkaido University, Sapporo 060-0810, Japan*

Abstract.
We present an elemental abundance analysis of HE 1327–2326, the most iron-deficient star known, based on a comprehensive investigation of spectra obtained with the Subaru Telescope. HE 1327–2326 is either in its main sequence or subgiant phase of evolution, hence it is essentially unevolved. The chemical abundances of this star have the following properties, which provide new constraints on models of nucleosynthesis processes that occurred in first-generation objects:

(1) The iron abundance (NLTE) is [Fe/H]= -5.45. This value is 0.2 dex lower than that of HE 0107–5240, the previously most iron-poor object known. No object having [Fe/H]= $-5 \sim -4$ is known to date.

(2) This star, as well as HE 0107–5240, exhibits extremely large overabundances of carbon relative to solar ratios ([C/Fe]$\sim +4$).

(3) HE 1327–2326 exhibits remarkable overabundances of the light elements (N, Na, Mg and Al), while HE 0107–5240 shows only relatively small excesses of N and Na.

(4) A large overabundance of Sr is found in HE 1327–2326 as compared to other extremely low metallicity stars.

(5) The Li I 6707 Å line, which is detected in the great majority of metal-poor dwarfs and warm subgiants, is not found in HE 1327–2326. The upper limit on the Li abundance we determine ($\log \varepsilon$(Li) < 1.5) is clearly lower than the expected value from the Spite plateau.

Keywords: Galaxy: abundances – nuclear reactions, nucleosynthesis, abundances – stars: abundances – stars: Population II
PACS: 97.10.Tk, 97.20.Tr, 97.20.Wt, 97.60.Bw, 98.80.Ft

INTRODUCTION

Elemental abundance measurements for extremely metal-poor stars found in our Galaxy have played a unique role in studies of the first stellar generations, since a record of their nucleosynthetic yields is believed to be preserved in the atmospheres of the most metal-deficient stars (see Norris 2006). The large objective-prism surveys for metal-poor stars that have been conducted over the past two decades (e.g., Beers et al. 1992; Christlieb 2003) have provided substantial samples of very metal-poor stars, but no object with [Fe/H]< -4 was identified until quite recently[2].

The discovery of HE 0107–5240 (a giant with [Fe/H] $= -5.3$) in 2001 (Christlieb et al. 2002) has already had a large impact on the study of the first generations of stars and supernovae (SNe), as well as on star formation processes that were operating in the early Universe. In contrast to its low Fe abundance, HE 0107–5240 exhibits extreme overabundances of the light elements carbon, nitrogen, and oxygen relative to solar ratios (Christlieb et al. 2002; Bessell et al. 2004). A number of possibilities have been proposed to account for the distinctive abundance pattern of elements in this object.

Our recent observations discovered HE 1327–2326, an unevolved star having lower iron abundance than HE 0107–5240 (Frebel et al. 2005; Aoki et al. 2006). Here we report a brief summary of the abundance analyses and discuss the implications of its chemical composition for studies of nucleosynthesis in the early Universe.

---

[1] Based on data collected with the Subaru Telescope, which is operated by the National Astronomical Observatory of Japan.
[2] [A/B] = $\log(N_A/N_B) - \log(N_A/N_B)_\odot$, and $\log \varepsilon_A = \log(N_A/N_H) + 12$ for elements A and B.

OBSERVATIONS

HE 1327−2326, as well as HE 0107−5240, was identified as a candidate metal-poor star by the Hamburg/ESO objective-prism survey (Christlieb 2003). The metallicity of this object was estimated to be as low as [Fe/H]~ -4 from follow-up spectroscopy of moderate resolution ($R \sim 2000$). This star became a high-priority target in the ongoing observing program of extremely metal-poor stars with the Subaru High Dispersion Spectrograph (HDS) that started from 2003. High-resolution spectra of HE 1327−2326 were obtained in May and June 2004 covering 3050–6800 Å with a resolving power of $R = 60,000$. The signal-to-noise (S/N) ratios of our final co-added spectrum were 20/1 at 3150 Å, 160/1 at 4000 Å and 170/1 at 6700 Å.

The effective temperature of this object ($T_{\text{eff}} = 6180$ K) was estimated from its optical and near-infrared colors, a value also supported by analyses of Balmer line profiles. The surface gravity was constrained by the proper motion of HE 1327−2326, assuming that the transverse velocity of the star must not be larger than the Galactic escape velocity. This constraint indicates that this star is a subgiant ($\log g \sim 3.7$) or a main-sequence star ($\log g \sim 4.5$). In either case this is an unevolved low-mass star.

The upper panel of Figure 1 shows a comparison of the medium-resolution spectrum of HE 1327−2326 with the solar spectrum. While the solar spectrum is covered by atomic and molecular absorption features, only the Balmer sequence of hydrogen is seen in the spectrum of HE 1327−2326. In the lower panel, the high-resolution spectrum of HE 1327−2326 around 387 nm is compared with that of the Sun. Even in the high-resolution spectrum, only one weak absorption line of Fe is identified in this spectral region, indicating the extremely low iron abundance of this object. On the other hand, weak molecular absorption lines of CH are seen in the spectrum of HE 1327−2326, suggesting a relatively high abundance of carbon in this star.

ABUNDANCE ANALYSES AND RESULTS

The abundance analysis was carried out initially using model atmospheres assuming local thermodynamic equilibrium (LTE). Corrections for non-LTE effects estimated for extremely metal-poor stars (e.g. Asplund 2005) were then applied in order to derive final abundances.

The Fe abundance was determined from four absorption lines in the UV-blue spectral range. The non-LTE corrected value is [Fe/H]$= -5.45$, which is 0.25 dex lower than that of HE 0107−5240 (Christlieb et al. 2004), if similar non-LTE corrections are applied. This Fe abundance is more than 1 dex lower than the Fe abundances of other metal-poor stars.

The carbon abundance was measured from the absorption features of the CH molecule. The derived abundance, [C/Fe]$= +4.0$, is as high as that of HE 0107−5240, and remarkably higher than other extremely metal-poor stars. The results are summarized in Figure 2, where non-LTE corrected abundances are shown for HE 1327−2326 and HE 0107−5240, and are compared with the average abundance patterns of extremely metal-poor stars with [Fe/H]~ -4. The low Fe abundance and the excess of carbon in HE 1327−2326 and HE 0107−5240 suggest that the abundance patterns of

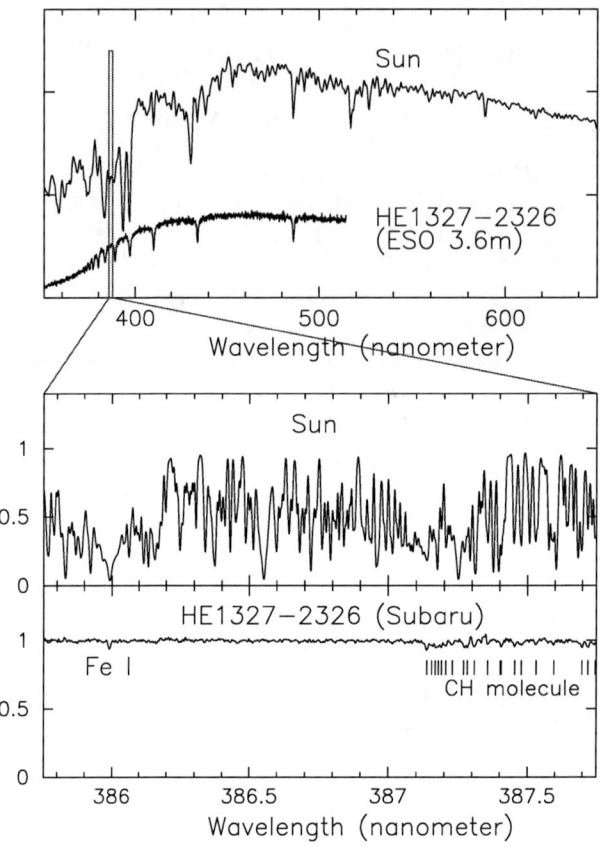

FIGURE 1. The upper panel compares the medium-resolution spectrum of HE 1327−2326 obtained with the ESO 3.6m telescope with the solar spectrum. The spectral energy distribution of HE 1327−2326 is uncalibrated. The prominent absorption features found in HE 1327−2326 are all Balmer lines. The lower panel shows the same comparison for high-resolution spectra over a narrow wavelength range containing the Fe I 386 nm line and the CH molecular lines.

these two stars have a similar origin.

Despite their overall similarity, significant differences in the abundance patterns are found for several of the light elements in these two stars. The nitrogen abundance of HE 1327−2326 was measured from the NH molecular band at 3360 Å. The result, [N/Fe]∼ +4, is higher than that of HE 0107−5240 by about 2 dex. The abundances of Na, Mg, and Al of HE 1327−2326 are also about 1 dex higher than those of HE 0107−5240 (see Figure 2). This result suggests that a large variation is allowed in the yields of the elements from N to Al in the nucleosynthesis processes that preceded the formation of these two stars.

One unexpected result is the detection of Sr, a species located near the first abundance peak of neutron-capture elements. As seen in Figure 2, the Sr abundance of

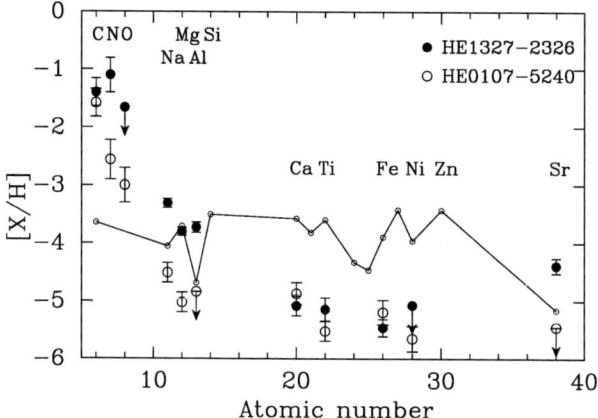

FIGURE 2. Chemical abundance patterns of HE 1327−2326 (filled circles) and HE 0107−5240 (open circles). The line indicates the abundance pattern of the average of four extremely metal-poor stars with [Fe/H]< −3.5 (CD −38°245, CS 22885−096, BS 16467−062, and CS 22172−002: François et al. 2003; Cayrel et al. 2004).

HE 1327−2326 is higher than that of HE 0107−5240 by more than 1 dex. The Sr abundance is, in general, also quite low in other extremely metal-poor stars with [Fe/H]∼ −4. In combination with the non-detection of Ba, an element near the second neutron-capture abundance peak, the observed Sr abundance provides a useful constraint on identifying the progenitor of HE 1327−2326. Frölich et al. (2005), for example, discuss a new process (the ν-p process) in which anti-neutrino interactions with protons in SNe may produce a neutron source capable of synthesizing large amounts of light neutron-capture elements, up to and including Sr.

Another unexpected result is the non-detection of Li in HE 1327−2326. The resonance line of neutral Li at 6708 Å is found in most metal-poor stars with sufficiently high effective temperatures ($T_{eff} > 6000$ K); the Li abundances of such stars produce the so-called Spite plateau at $\log \varepsilon(Li) \sim 2.2$ (see Yong et al. 2006). However, the observed upper limit on the Li abundance of HE 1327−2326 is $\log \varepsilon(Li) \sim 1.5$ (see Norris 2006). This indicates that some as-yet-unidentified mechanism may have depleted the Li in the atmosphere of this object, one that does not operate for most metal-deficient subgiants and dwarfs.

DISCUSSION AND CONCLUDING REMARKS

Even though a low-mass star may have formed from a metal-free cloud, the surface of such a star could be polluted by intersteller matter including metals during its long life. The low iron abundance found in HE 1327−2326 could possibly be interpreted as a result of such surface pollution. The relatively high abundances of carbon and other light elements are, however, not explained by this scenario. One idea is to assume that this object belongs to a binary system, and that mass transfer from the primary star in its

Asymptotic Giant Branch (AGB) phase has produced the elements now observed on the surviving companion. This scenario is discussed in detail by Suda et al. (2006).

Another idea is to explain HE 1327−2326 as an extreme population II star, which formed from clouds of gas including a small amount of metals. In this case, the peculiar abundance pattern found in this object must be provided by a progenitor (or progenitors). Umeda et al. (2006) proposed a so-called "faint supernova" model, which well reproduces the abundance patterns of HE 1327−2326 as well as HE 0107−5240. Another model is discussed by Hirsch & Meynet (2006), who predict that large mass loss, including significant amounts of light elements, may have occurred in massive rotating stars with very low metallicity.

The detailed abundance study of HE 1327−2326 provided a constraints on these various models, and further observing of this object will no doubt prove useful. Our recent observation of this star with ESO/VLT resulted in a determination of its oxygen abundance (Frebel et al. 2006), which will also help in the interpretation of the above suggested models. Further searches for stars with the lowest metallicities, in order to check on the stability of the abundance patterns for such Hyper Metal-Poor stars, are also strongly desired.

REFERENCES

1. Aoki, W., Frebel, A., Christlieb, N. et al. 2006, ApJ, 639, 897
2. Asplund, M. 2005, Ann. Rev. Astron. Astrophys., 43, 481
3. Beers, T.C., Preston, G.W., & Shectman, S.A. 1992, AJ, 103, 1987
4. Bessell, M., Christlieb, N., & Gustafsson, B. 2004, ApJ, 612, L61
5. Cayrel, R., Depagne, E., Spite, M. et al. 2004, A&A, 416, 1117
6. Christlieb, N. 2003, Rev. Mod. Astron., 16, 191
7. Christlieb, N., Bessell, M.S., Beers, T.C. et al. 2002, Nature, 419, 904
8. Christlieb, N., Gustafsson, B., Korn, A. J., Barklem, P. S., Beers, T. C., Bessell, M. S., Karlsson, T., & Mizuno-Wiedner, M. 2004, ApJ, 603, 708
9. François, P., Depagne, E., Hill, V. et al. 2003, A&A, 403, 1105
10. Frebel, A., Aoki, W., Christlieb, N. et al. 2005, Nature, 434, 871
11. Frebel, A., Christlieb, N., Norris, J.E. et al. 2006, ApJ, 638, L17
12. Fröhlich, C. et al. 2005, submitted to Phys. Rev. (Letters) (astro-ph/0511376)
13. Hirsch, R. & Meynet, G. 2006, this volume
14. Norris, J. E. 2006, this volume
15. Suda, T., Nishimira, T., Iwamoto, N., Aikawa, M., Fujimoto, M.Y., & Iben, I. Jr. 2006, this volume
16. Umeda, H., Iwamoto, N., Tominaga, N., Nomoto, K., & Maeda, K. 2006, this volume
17. Yong, D., Carney, B.W., Aoki, W., McWilliam, A., & Schuster, W.L 2006, this volume

Nucleosynthetic Signatures of Pop.III Survivors and the Origin of HE0107-5240 and HE1327-2326

T. Suda[*,†], T. Nishimura[†], N. Iwamoto[**], M. Aikawa[‡], M. Y. Fujimoto[†] and I. Iben Jr.[§]

[*]*Meme Media Laboratory, Hokkaido University*
[†]*Department of Physics, Hokkaido University*
[**]*Nuclear Data Center, Japan Atomic Energy Research Institute*
[‡]*Institut d'Astronomie et d'Astrophysique, Université Libre de Buruxelles*
[§]*Departments of Astronomy and of Physics, University of Illinois*

Abstract. The discoveries of two extremely iron-poor stars with [Fe/H]<-5, HE0107-5240 and HE1327-2326 provided the great opportunities of verifying whether these stars are the survivors of the first generation stars; the very weak lines of detected metals in their atmospheres can be extrinsic source that is the accretion from interstellar gas or from binary companion, rather than intrinsic one. In this work, we explore the possibility that these stars were born in binary systems from the primordial clouds by considering the results of stellar evolution and nucleosynthesis in metal-free models of low to intermediate mass AGB stars. Observed abundance patterns for these 2 stars are in agreement with the results and can be explained by the binary scenario that observed stars disguise their surface abundances by the mass transfer in the binary system. In particular, we first demonstrated the reproduction of Sr without large enhancement of Ba through the neutron capture reactions in the helium flash convective region of AGB models without any iron seeds for s-process. The apparent lack of stars in -5 < [Fe/H] < -4 may suggest the effect of dilution by the surface convection at the red giant branch. If this is true, other Pop.III survivors can be discovered at the main sequence having [Fe/H] ~ -3, whose surface abundances are changed by the mass transfer from evolved companions in binary systems.

Keywords: stellar evolution, first stars, nucleosynthesis
PACS: 97.10.Cv; 97.20.Wt

INTRODUCTION

The stars with [Fe/H] < -5.0 (HE0107-5240 [1] and HE1327-2326 [2]) have great impact on the understandings of our universe. The origins of these stars have become the great cynosure because such low content of iron can be interpreted in terms of the surface pollution from the interstellar matter. We cannot conclude yet whether these stars are the survivors of the first generation stars (Pop.III) or the second generation stars formed from the ejecta of supernovae of the massive first stars. One of the keys to solve the problem may be the anomalous excess of CNO elements in the surface compositions of these stars. The existence of two stars with [Fe/H] < -5, while no stars with -5 < [Fe/H] < -4 may cast a new light on the subject, and on the current status of possible low-mass survivors of Pop.III.

For the most iron-poor carbon rich star HE0107-5240, we have proposed the binary scenario [3] which demands that: (1) the star is born in binary system consisting of the primary star whose mass is in the range $1.5 < M/M_\odot < 3$ and the secondary star of mass $0.8M_\odot$: (2) the primary experiences helium flash driven deep mixing (hereafter He-FDDM) mechanism, triggered by the mixing of hydrogen into the helium convective zone [3, 4] and enriched its envelope with CNO elements, Na, Mg, Al, and possibly s-process elements during the early phase of thermal pulsating AGB: (3) then, the primary undergoes standard third dredge-up episodes and eventually loses its envelope via mass loss to be a now unseen white dwarf: (4) the secondary now evolves to a red giant after accreting gas from the primary. This scenario can reproduce the quantitative abundance pattern of HE0107-5240 and predicts the current orbital period of the binary at ~ 100 years, which is subject to the confirmation by the monitoring of radial velocity in the future. For the Pop.III binary, iron and other elements of scant abundances can be readily explained by the external surface pollution through the encounter with the metal-rich interstellar gas, which is inevitable during their long lives.

In this paper, we advance the binary scenario to the origin of extremely metal-poor carbon rich star with the characteristics of newly discovered dwarf HE1327-2326 taken into account, such as several ten times larger enhancement of Na, Mg, and Al, the lack of Li, and the enhancement of Sr. We present the recent results of model computations to explore the condition that He-FDDM is triggered in extremely metal-poor stars of the relevant mass range. We also compute the progress of neutron capture reactions and resultant nucleosynthesis in the base of helium convective zone induced by the injection of ^{13}C [5] in the wider range of parameters by the nuclear network program [6, see also the paper of T. Nishimura et al. and K. Yamamoto in this symposium] to compare the abundance pattern with observations. Based on all these computations, we discuss the general picture of extremely metal-poor stars in binary systems.

MODELS AND ASSUMPTIONS

Stellar evolution program and adopted input physics are based on those in [1] for the computations of model stars. Model grid is chosen as 0.8, 0.9, 1.0, 1.2, 1.5, 2.0, 2.5, 3.0, 4.0, 5.0, 6.0, and $7.0M_\odot$ in initial mass and $Z = 0$, [Fe/H] = -5, -4, -3, and -2 in initial metallicity, although not all grid points are covered. All models are computed from zero-age main sequence to the early phase of asymptotic giant branch (AGB). For models of metallicity below -2.5, the engulfment of hydrogen by the helium flash convection occurs at the thermally pulsating AGB, which promotes the neutron-capture reactions via $^{12}C(p,\gamma)^{13}N(e^+\nu)^{13}C(\alpha,n)^{16}O$. For simplicity, we do not consider the detailed processes of hydrogen mixing and assume that the mixed protons turn into ^{14}N via CN cycle. To follow the whole process of He-FDDM is beyond the scope of this paper and to be revealed in the future work.

Instead, we followed the changes of the abundances in the helium flash convective region after the ingestion of hydrogen (See Fig.3 of [3] or Fig.2 of [5]), by computing the nuclear network under the assumption of one-zone model of thermal pulse at AGB phase, by means of the recipe of nucleosynthesis for elements of $Z \leq 16$ [6]. For $Z \geq 16$, only neutron capture reactions are followed by given density and temperature in the helium convective shell. In the present computations, the degree of mixing is specified by the amount of ^{13}C relative to ^{12}C by requiring that the mixed protons are converted

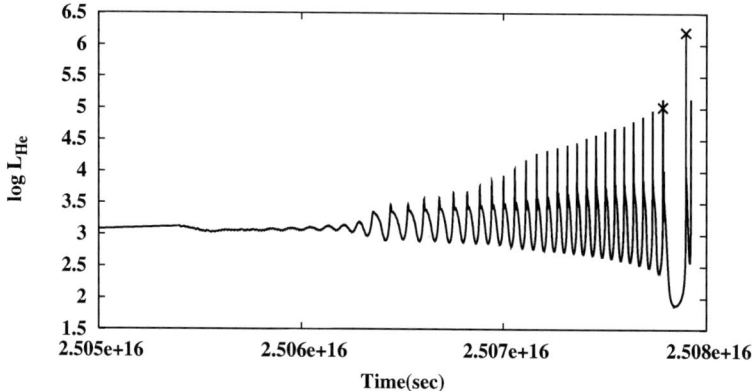

FIGURE 1. Time variation of helium burning luminosity at the early phase of AGB models of $2.0M_\odot$ and Z=0. Crosses denote the point where the contact between the top of the helium flash convection and the bottom of hydrogen burning shell occurs.

into ^{13}C through the partial CN cycle. The onset of ^{13}C mixing into the layer consisting of helium, carbon and oxygen is set at the peak of thermal pulse where the density and the temperature nearly reach its peak. The mixing parameter ranges from 10^{-4} to 0.07 in ^{13}C/^{12}C values and the timescale of mixing and nucleosynthesis are adjusted not to exceed the evolutionary timescale of thermal pulses.

RESULTS

Our computations are almost in good agreement with the privious results of [7]. We confirmed the following results: (1) The evolution of extremely metal-poor models is classified into some groups such as Case I, II, II′, III, IV, and so on as in [7] (2) The models having metallicity below -2.5 undergo He-FDDM at red giant or AGB phase. (3) The border between Case I and II falls between 1.1 and $1.2M_\odot$ for Z=0 and between 1.4 and $1.5M_\odot$ for [Fe/H] = -5 differently from the expectation by [7].

Figure 1 shows the evolutionary results of model with $M = 2.0M_\odot$ and Z=0 at thermally pulsating phase. In this case, we found two events of weak mixing during the thermal pulses, which suggests that neutron capture elements are produced in the helium flash convective zones through the nuclear reaction paths stated before, and that the nuclear products are stored in the stellar interior. These elements should be finally dredged-up to the surface by He-FDDM. We terminated the computation at the onset of third event of hydrogen mixing because of the strong flash to invoke the splitting of convection by the hydrogen burning. In other models, the computations are terminated

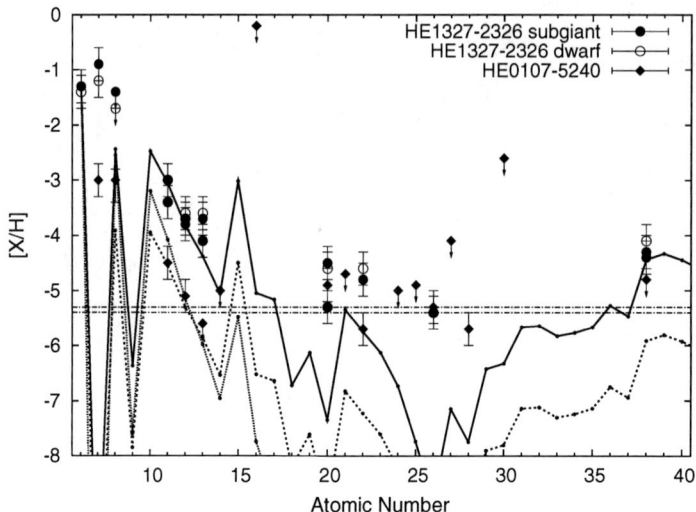

FIGURE 2. Comparisons between observed abundance patterns of HE0107-5240 and HE1327-2326 with theoretical models. Solid, dashed, and dotted line represent the result obtained by the amount of mixing as $^{13}C/^{12}C = 0.01$, the same abundances as the former but multiplied by a factor of 1/30 for elements except for ^{12}C, which assumes the effect of dilution by the third dredge-up by the He-FDDM, and the result by setting $^{13}C/^{12}C = 0.001$, respectively. Horizontal lines are the levels of metallicity for HE0107-5240 and HE1327-2326.

at the onset of the first event of hydrogen mixing, but similar situation should occur for $M < 4.0 M_\odot$.

Based on these stellar models, the abundance pattern of carbon to bismuth in the envelope of primary star is obtained with the nuclear network program. In Figure 2, the observed abundances are compared with the best fit models. Detailed processes of nuclear reactions in the helium convective region is discussed in the paper by Nishimura in the same volume. Our results are consistent with the derived abundances of both HE0107-5240 and HE1327-2326 for elements such as C, O, Na, Mg, Al, and Sr. The parameter to reproduce the abundance pattern of HE0107-5240 is set at $^{13}C/^{12}C = 0.001$ or 30 times smaller value than the case of $^{13}C/^{12}C = 0.01$. The former means the relatively weak contact of helium flash convection with the hydrogen burning shell, though it is enough to split the convection. On the other hand, the latter means that the subsequent third dredge-up dilutes the nuclear products in the envelope after the similar degree of mixing to the case of HE1327-2326. The trend of Na, Mg, and Al abundances seems to favor the latter case. For s-process elements, if the mixing parameter is as much as the order of 0.01, only Sr can be enhanced so that the value of [Sr/Ba] is kept high. The resultant abundance of Ba is comparable to the derived upper limit for HE1327-2326.

The other sources are responsible for the elements that the nucleosynthesis triggered by the hydrogen mixing does not seem to reproduce the observed abundance patterns. The large enhancement of nitrogen for both stars should be complemented directly by the He-FDDM. On the other hand, the origin of Ca, Ti, and iron group elements such

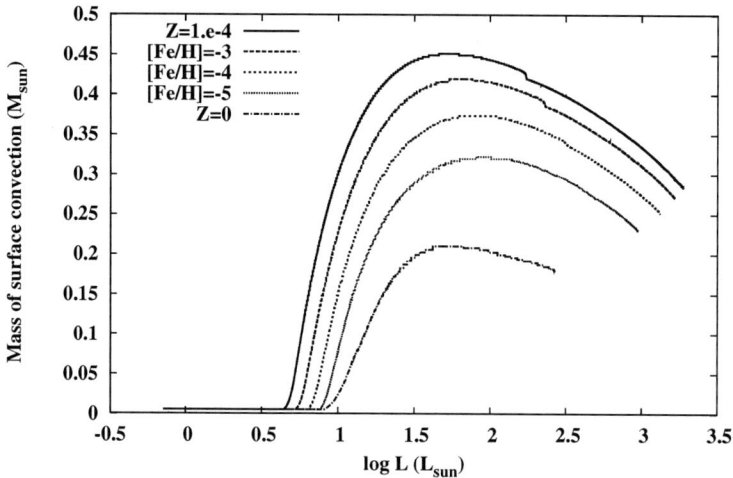

FIGURE 3. Change of the mass of the surface convective zone as a function of total luminosity of model stars. The evolution of $0.8M_\odot$ is followed from zero-age mains sequence to the beginning of core helium flash phase for various initial metallicity.

as Cr, Mn, Fe, Co, and Ni are ascribed to the accretion from interstellar gas of the primordial cloud, polluted by the first supernovae born in the same cloud [3].

In our results, the final abundance of P is quite large. Moreover, Mg isotopic ratio becomes abnormal compared with the models without He-FDDM, i.e., ^{26}Mg is highly enhanced relative to ^{24}Mg. These facts should be the evidence for the occurrence of convective nuclear reactions at AGB phase for extremely metal-poor or Pop.III stars. The follow-up observations of these elements will put the further constraints on the origin of these stars.

CONCLUSIONS AND DISCUSSION

Our binary scenario can explain consistently not only the observed properties of HE0107-5240 and HE1327-2326 but also their differences, in particular, the differences in the abundances of Na, Mg, and Al, which are attributable to the different masses of their primaries; a lower-mass primary of $1.2M_\odot \leq M < 1.5M_\odot$ that experiences the He-FDDM only for HE1327-2326 and a more massive primary of $1.5M_\odot < M \leq 3M_\odot$ that experiences both the He-FDDM and the third dredge-up for HE0107-5240. The abundance patterns of these stars are well reproduced quantitatively by the nucleosynthesis in the helium convective zone of primary stars at the early phase of AGB.

Although the variations of radial velocity of these stars have not yet detected at present, the no detection of Li for HE1327-2326 which still stays at the main sequence strongly supports our scenario. However, the comprehensive simulations of stellar evolution are required to confirm the effect of the internal mixing and dredge-up mechanisms on the structure and evolution of extremely metal-poor stars.

Since the mass of the convective envelope is thought to be small for HE1327-2326, the prediction of binary period will be alternative; the case that the star belongs to long period binary due to the small accreting mass from binary companion, or the case that the surface convective region of observed star is entirely covered by the accreting matter and that the binary period can be relatively arbitrary. According to this speculation, we may found other stars with [Fe/H] < -5 showing no enhancement of carbon because of the large separation to the extent that the mass transfer hardly affect the surface chemical composition of observed stars.

Finally, our scenario with the external pollution of iron group elements gives a reasonable account to why these two stars are found at metallicity [Fe/H] < -5 despite all other stars have [Fe/H] > -4. The lack of known stars in the metallicity range can be interpreted in terms of the dilution effect of polluting layer by the deepening of surface convection by a factor of $\sim$ 100 as the star evolves from the main sequence to the giant branch or to the asymptotic giant branch as shown by Figure 3. One should note that the mass of the surface convection abruptly reaches its maximum in short timescale after settling on the red giant branch and declines with increasing luminosity. The dilution effect implies that Pop. III survivors can be found among main sequence stars of [Fe/H] $\sim$ -3, whose surfaces are polluted by accreting interstellar gas in the primordial clouds.

REFERENCES

1. N. Christlieb, et al., *Nature*, **419**, 904–905 (2002)
2. A. Frebel, W. Aoki, N. Christlieb, et al. 2005, *Nature*, **434**, 871–872 (2005)
3. T. Suda, M. Aikawa, M. N. Machida, M. Y. Fujimoto, & I. Iben, Jr., *The Astrophysical Journal*, **611**, pp.476–493 (2004)
4. M. Y. Fujimoto, I. Iben, Jr., & D. Hollowell, *The Astrophysical Journal*, **349**, pp.580–592 (1990)
5. T. Suda, M. Aikawa, T. Nishimura, & M. Y. Fujimoto, 2005, Proceedings of the conference, *"Origin of Matter and Evolution of Galaxies 2003"*, ed. M. Terasawa et al., *World Scientific*, Singapore, 2005, pp.529-532
6. M. Aikawa, M. Y. Fujimoto, & K. Kato, *The Astrophysical Journal*, **560**, pp.937–956 (2001)
7. M. Y. Fujimoto, Y. Ikeda, & I. Iben, Jr., *The Astrophysical Journal Letters*, **529**, pp.L25–L28 (2000)

The Abundance Pattern and Formation of Extremely Meta-Poor Stars

H. Umeda*, N. Tominaga*, N. Iwamoto[†], K. Nomoto*,** and K. Maeda[‡]

Department of Astronomy, School of Science, University of Tokyo, Bunkyo-ku, Tokyo 113-0033, Japan
[†]*Nuclear Energy Basic Engineering Research Sector Japan Atomic Energy Agency, Tokai, Ibaraki 319-1195, Japan*
**Research Center for the Early Universe, School of Science, University of Tokyo, Bunkyo-ku, Tokyo 113-0033, Japan*
[‡]*Department of Earth Science and Astronomy, College of Arts and Sciences, University of Tokyo, Tokyo 153-8902, Japan*

Abstract. The recent discovery of a hyper metal-poor (HMP: $-5 \lesssim [\text{Fe/H}] \lesssim -4$) star have raised a challenging question if these HMP stars are first generation stars in the Universe. We argue that these HMP stars are the second generation stars being formed from gases which were chemically enriched by the first generation supernovae. The key to this solution is the very unusual abundance patterns of these HMP stars with important similarities and differences. We can reproduce these abundance features with the core-collapse "faint" supernova models which undergo extensive matter mixing and fallback during the explosion (mixing-fallback model). We also show that the abundance patterns of extremely metal-poor (EMP: $-4 \lesssim [\text{Fe/H}] \lesssim -3$) stars are well-reproduced by a 25 $M_\odot$ hypernova mixing-fallback model and those of very metal-poor (VMP: $-3 \lesssim [\text{Fe/H}] \lesssim -2$) stars are well-reproduced by a model integrated by Salpeter's initial mass function over $13 - 50$ $M_\odot$ models.

Keywords: Galaxy: halo — nuclear reactions, nucleosynthesis, abundances — stars: abundances — stars: Population III — supernovae: general
PACS: 26.20.+f, 26.30.+k, 26.50.+x, 97.10.Cv, 97.10.Tk, 97.20.Tr, 97.20.Wt, 97.60.Bw, 98.80.Ft

HE1327–2326 & HE0107–5240

Identifying the first stars in the Universe, i.e., metal-free, Population III (Pop III) stars which were born in a primordial hydrogen-helium gas cloud is one of the important challenges of the current astronomy [1, 2]. Recently two hyper metal-poor (HMP) stars, HE0107–5240 [3] and HE1327–2326 [4], were discovered, whose metallicity Fe/H is smaller than 1/100,000 of the Sun (i.e., [Fe/H] < −5), being more than a factor of 10 smaller than previously known extremely metal-poor (EMP) stars. (Here [A/B] = $\log_{10}(N_A/N_B) - \log_{10}(N_A/N_B)_\odot$, where the subscript $\odot$ refers to the solar value.) This discovery was raised an important question as to whether the observed low mass (~ 0.8 $M_\odot$) HMP stars are actually Pop III stars, or whether these HMP stars are the second generation stars being formed from gases which were chemically enriched by a single first generation supernova (SN), e.g., [5, 6]. This is related to the questions of how the initial mass function depends on the metallicity. Thus identifying the origin of these HMP stars is indispensable to the understanding of the earliest star formation and chemical enrichment history of the Universe.

The elemental abundance patterns of these HMP stars provide a key to answer the

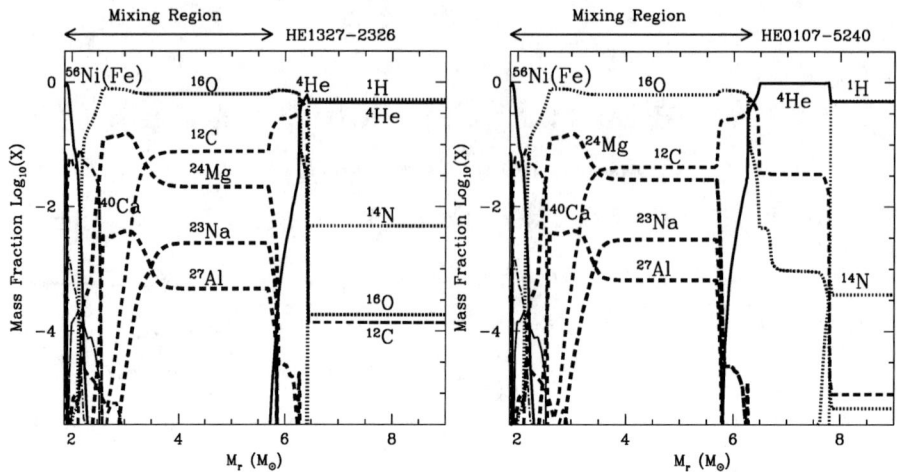

FIGURE 1. (Left):Internal abundance distribution for nuclei (by mass fraction) in the Pop III $25M_\odot$ SN model for the explosion energy of $E_{51} = 0.74$ (i.e., HE1327–2326). The mixing is assumed to take place in the region of $M_r = 1.9 - 5.8M_\odot$. The mass fraction of the ejected materials with respect to the mixed fallback materials is $f = 8.7 \times 10^{-5}$. The ejecta contains $1.0 \times 10^{-5} M_\odot$ ^{56}Ni and 0.20 $M_\odot$ ^{12}C. (Right):Same as the left panel, but for HE0107–5240 ($E_{51} = 0.71$). The mixing is assumed to take place in the region of $M_r = 1.9 - 6.3M_\odot$. The mass fraction of the ejected materials with respect to the mixed fallback materials is $f = 1.2 \times 10^{-4}$. The ejecta contains $1.4 \times 10^{-5} M_\odot$ ^{56}Ni and 0.12 $M_\odot$ ^{12}C.

above questions. The abundance patterns of HE1327–2326 [4] and HE0107–5240 [7, 8] are quite unusual. The striking similarity of [Fe/H] (=−5.4 and −5.2 for HE1327–2326 and HE0107–5240, respectively) and [C/Fe] (∼ +4) suggests that similar chemical enrichment mechanisms operated in forming these HMP stars. However, the N/C and (Na, Mg, Al)/Fe ratios are more than a factor of 10 larger in HE1327–2326. In order for the theoretical models to be viable, these similarities and differences should be explained self-consistently.

Here we show that the above similarities and variations of the HMP stars can be well reproduced in a unified manner by nucleosynthesis in the core-collapse "faint" supernovae (SNe) which undergo mixing-and-fallback. We thus argue that the HMP stars are the second generation low mass stars, whose formation was induced by the first generation (Pop III) SN with efficient cooling of carbon-enriched gases [5, 9].

The similarity of [Fe/H] and [C/Fe] suggests that the progenitor's masses of Pop III SNe were similar for these HMP stars. We therefore choose the Pop III 25 $M_\odot$ models and calculate their evolution and explosion. The abundance distribution after explosive nucleosynthesis is shown in Figures 1 for the kinetic energy E of the ejecta $E_{51} \equiv E/10^{51}$ erg $= 0.74$ and 0.71. In the "faint" SN model, most part of materials that underwent explosive nucleosynthesis are decelerated by the influence of the gravitational pull [10] and will eventually fall back onto the central compact object (the left panel of Figs. 2). Such "fallback" actually takes place in the present modeling if $E_{51} < 0.71$. (For the 50 $M_\odot$ star, the fallback is found to occur for $E_{51} < 2$ because of deeper gravitational potential.) In general, smaller E_{51} leads to a larger amount of fallback (larger $M_{\rm cut}$).

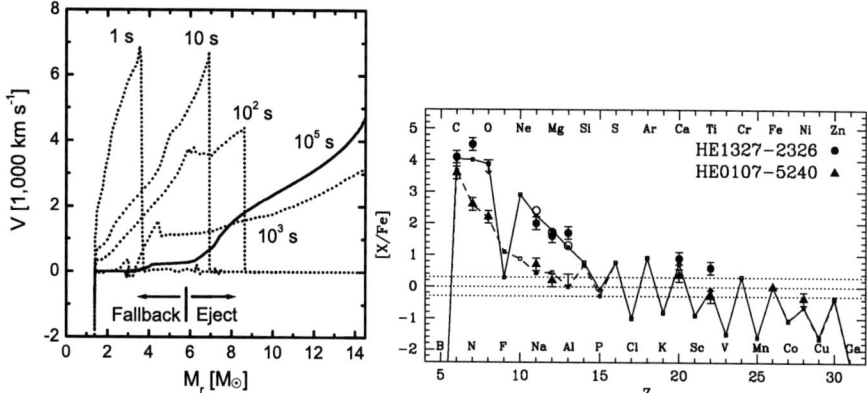

FIGURE 2. (Left):Propagation of the shock wave and the fallback of the model for HE1327–2326. The progenitor is the 25 $M_\odot$ star. As the shock propagates through the H envelope and breaks out of the surface, the materials in the inner region continue to be decelerated and will eventually fallback onto the central remnant. The mass cut (that divides the materials fallen onto the central remnant and ejected outward) is determined by comparing the velocity and the escape velocity at 10^5 seconds after the explosion. (Right):Comparison of elemental abundance ratios observed in HE1327–2326 (filled circles) and HE0107–5240 (filled triangles) with those of our supernova models (connected by the solid line for HE1327–2326 and by the dashed line for HE0107–5240) as a function of atomic number Z. Here the new solar abundances [11] are used. For Na and Al, the importance of accurate non-local thermodynamic equilibrium (LTE) corrections are demonstrated from the comparison with the LTE values indicated by the open circles.

The explosion energies of $E_{51} = 0.74$ and 0.71 lead to the mass cut $M_{\text{cut}} = 5.8 M_\odot$ and $6.3 M_\odot$, respectively. We use the former and the latter models to explain the abundance patterns of HE1327–2326 and HE0107–5240, respectively.

During the explosion, we assume that the SN ejecta undergoes mixing, i.e., materials are first uniformly mixed in the mixing-region extending from $M_r = 1.9 M_\odot$ to the mass cut at $M_r = M_{\text{cut}}$ (where M_r is the mass coordinate), and only a tiny fraction, f, of the mixed material is ejected from the mixing-region together with all materials at $M_r > M_{\text{cut}}$; most materials interior to the mass cut fall back onto the central compact object. Such a mixing-fallback mechanism (which might mimic a jet-like explosion) is required to extract Fe-peak and other heavy elements from the deep fallback region into the ejecta [5, 12].

The right panel of Figures 2 shows the calculated abundance ratios in the SN ejecta models for suitable choice of f which are respectively compared with the observed abundances of the two HMP stars. To reproduce [C/Fe] $\sim +4$ and other abundance ratios of HMP stars in the right panel of Figures 2, the ejected mass of Fe is only $1.0 \times 10^{-5} M_\odot$ for HE1327–2326 and $1.4 \times 10^{-5} M_\odot$ for HE0107–5240 (see [9] for detail). These SNe are much fainter in the radioactive tail than the typical SNe and form massive black holes of $\sim 6 M_\odot$.

The question is what causes the large difference in the amount of Na-Mg-Al between the SNe that produced HE0107–5240 and HE1327–2326. Because very little Na-Mg-Al is ejected from the mixed fallback materials (i.e., $f \sim 10^{-4}$) compared with the materials exterior to the mass cut, the ejected amount of Na-Mg-Al is very sensitive to the location of the mass cut. As indicated in Figures 1, M_{cut} is smaller (i.e., the fallback mass is smaller) in the model for HE1327–2326 ($M_{cut} = 5.8M_\odot$) than HE0107–5240 ($M_{cut} = 6.3M_\odot$), so that a larger amount of Na-Mg-Al is ejected from the SN for HE1327–2326. Since M_{cut} is sensitively determined by the explosion energy, the (Na-Mg-Al)/Fe ratios among the HMP stars are predicted to show significant variations and can be used to constrain E_{51}. Note also that the explosion energies of these SN models with fallback are not necessarily very small (i.e., $E_{51} \sim 0.7$). Further these explosion energies are consistent with those observed in the actual "faint" SNe [13].

Here we note that our previous models [5] tend to underproduce Na compared with the abundances of HE0107–5240. This problem can be solved as follows [9]. Na and Al are mainly produced by C shell-burning, and their production is very sensitive to the treatment of overshooting in the convective C burning shell as well as the ^{12}C abundance left after core He burning, e.g.,[14]. By including overshooting with the overshooting length less than one-fifth of a pressure scale height for whole presupernova evolution, large enough abundances of Na and Al are obtained as seen in Figures 1. Such an overshooting length has been estimated from the comparison with the HR diagrams of many young stellar clusters. After the mixing-and-fallback, the resultant abundance patterns with Na and Al are in reasonable agreement with HE1327–2326 and HE0107–5240 (the right panel of Figs. 2). This enhancement of Na and Al may better explain the small odd-even effect in the elemental abundance patterns observed in EMP stars [15].

The next question is why HE1327–2326 has a much larger N/C ratio than HE0107–5240. N can be produced by the mixing between the He convective shell and the H-rich envelope during the presupernova evolution [17], where C created by the triple-α reaction is burnt into N through the CNO cycle. For the HE1327–2326 model, we assume about 30 times larger diffusion coefficients (i.e., faster mixing) for the H and He convective shells to overcome an inhibiting effect of the mean molecular weight gradient (and also entropy gradient) between H and He layers. Thus, larger amounts of protons are carried into the He convective shell. Then [C/N] ~ 0 is realized as observed in HE1327–2326. Such an enhancement of mixing efficiency has been suggested to take place in the present-day massive stars known as fast rotators, which show various N and He enrichments due to different rotation velocities [18].

Alternatively, if HE1327–2326 is in a binary system and its companion star had experienced the asymptotic giant branch (AGB) phase, only the odd-elements such as N, Na and Al can be efficiently enriched through mass transfer from the companion (AGB) star. However, as shown in the next section, the deficiency of N, Na, Al appear to be a quite general problem in the SN nucleosynthesis. Therefore, the solution by the extra-mixing in the progenitor stars may be better.

Our models offer several predictions for future observations of HMP stars. (1) The metallicity Fe/H of an HMP star is determined by the mass ratios between the ejected Fe M_{Fe} and mixed interstellar H M_{Hmix}, and small M_{Fe} (i.e., small f) is responsible for the small [Fe/H]. Our spherical explosion models predict a continuous distribution of [Fe/H] in metal-poor stars. Thus, if the gap at [Fe/H] ~ -5 to -4 is real, jet-induced mixing might be responsible for constraining the distribution of the f-value. (2) Assuming that C/H needs to be higher than a certain value in order to form low-mass HMP stars, C/Fe would tend to be larger for smaller Fe/H. (3) The (Na-Mg-Al)/Fe ratios in HMP stars

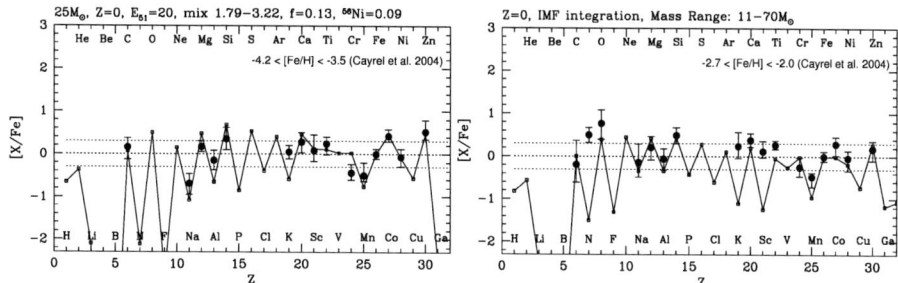

FIGURE 3. (Left):The comparison between the abundance pattern of EMP stars [15] (*filled circles*) and the theoretical individual hypernova ($M_{MS} = 25M_\odot$, $E_{51} = 20$) yields with the mixing-fallback model. Model parameters are shown in the top label. (Right):The comparison between the abundance pattern of VMP stars given by [15] (*filled circles*) and the IMF integrated yield of Pop III SNe from $11M_\odot$ to $70\,M_\odot$ (*solid line*).

would show a continuous distribution because their variations are the result of variation of E. (4) If the large N/Fe is attributable to rotation and if rotation can contribute to enhance E, N/Fe would show a positive correlation with (Na-Mg-Al)/Fe.

EMP & VMP STARS

Next we consider metal richer stars, i.e., EMP and VMP stars [15, 16]. [15] provide abundance patterns of 35 metal-poor stars with small error bars for $-4.2 \lesssim$ [Fe/H] $\lesssim -2.0$. Each EMP star may be formed from the ejecta of a single Pop III SN, although some of them might be a second or later generation stars. The yields of SNe with [Fe/H] $\lesssim -3$ can be approximated by those of Pop III SNe since they are almost same [17, 10]. In this section the theoretical yields are compared with the averaged abundance pattern of four EMP stars, CS 22189-009, CD-38:245, CS 22172-002 and CS 22885-096, which have low metalicities ($-4.2 <$ [Fe/H] < -3.5) and normal [C/Fe] ~ 0. Stars with large [C/Fe] ($\gtrsim 1$), called C-rich EMP stars, are discussed in [5] and [12]. In this section, we do not include overshooting and other extra-mixings. The origin of those stars may be a faint SN and is different from those of [C/Fe] ~ 0 stars.

Comparisons between the yields of a high-energy supernova (hypernova) model and the abundance patten of EMP stars are made in the left panel of Figures 3. The importance of the high-energy explosions are stressed in [12]. Since this model eject the materials synthesized in a deep region of a hypernova and undergo fallback, [(Fe-peak elements)/Fe] and [α/Fe] give good agreements with the observations. Since no extra-mixing is included, (N-Na-Al)/Fe tend to be underproduced.

[15] also provided the abundance patterns of VMP stars whose metallicities ([Fe/H]~ -2.5) are higher than the EMP stars. The observed abundance pattern is represented by the averaged abundance pattern of five stars BD+17:3248, HD 2796, HD 186478, CS 22966-057 and CS 22896-154, which have relatively high metalicities ($-2.7 <$ [Fe/H] < -2.0). Most VMP stars are considered to have the abundance pattern averaged over IMF and metallicity of the progenitors, thus we also compare with IMF-integrated

yields (the right panel of Figs. 3). Since the abundance patterns of [Fe/H] $\lesssim -2.5$ stars are quite similar to Pop III, we can use the Pop III yields for these stars as well [17, 10]. The IMF integration is performed from 11 $M_\odot$ to 70 $M_\odot$ with 7 models, 13, 15, 20, 25, 30, 40, 50 $M_\odot$. Here $M > 20 M_\odot$ models are assumed to be hypernovae and applied the mixing-fallback model so that [O/Fe] = 0.5 (case A in [19]). We assume Salpeter's power-law IMF integrations as follows:

$$\phi(M) = KM^{-2.35} \tag{1}$$

where $\phi(M)dM$ is the number of stars within the mass range of $[M, M+dM]$, and K is a normalization constant. The integration is performed as follow:

$$X(A) = \frac{\int_{11M_\odot}^{70M_\odot} X_M(A) M_{ej} \phi(M) dM}{\int_{11M_\odot}^{70M_\odot} M_{ej} \phi(M) dM} \tag{2}$$

where $X(A)$ is an integrated mass fraction of an element, A, $X_M(A)$ is mass fraction of A in a model whose mass is nearest to M, and M_{ej} is an ejected mass of the nearest model.

Comparing the integrated yields with the Salpeter's IMF to the abundance pattern of VMP stars (the right panel of Figs. 3), most α-elements show reasonable agreements. Its implications and more detailed explanations will be given in [20].

REFERENCES

1. A. Weiss, T. G. Abel, V. Hill, Eds., *The First Stars* (Springer-Verlag, Berlin, 2000).
2. T. Abel, G. L. Bryan, M. L. Norman, *Science* **295**, 93–98 (2002).
3. N. Christlieb et al., *Nature* **419**, 904–906 (2002).
4. A. Frebel et al., *Nature* **434**, 871–873 (2005).
5. H. Umeda, K. Nomoto, *Nature* **422**, 871–873 (2003).
6. R. Schneider, A. Ferrara, R. Salvaterra, K. Omukai, V. Bromm, *Nature* **422**, 869–871 (2003).
7. N. Christlieb et al., *Astrophys. J.* **603**, 708–728 (2004).
8. M. S. Bessell, N. Christlieb, B. Gustafsson, *Astrophys. J.* **612**, L61–L63 (2004).
9. N. Iwamoto, H. Umeda, N. Tominaga, K. Nomoto, & K. Maeda, *Science*, **309**, 451–453 (2005)
10. S. E. Woosley, T. A. Weaver, *Astrophys. J. Suppl. Ser.* **101**, 181–235 (1995).
11. M. Asplund, N. Grevesse, J. Sauval, in *Cosmic abundances as records of stellar evolution and nucleosynthesis*, F. N. Bash, T. G. Barnes, Eds. (Astronomical Society of the Pacific Conf. Ser., San Francisco, 2004), in press (available at http://arxiv.org/abs/astro-ph/0410214).
12. H. Umeda, K. Nomoto, *Astrophys. J.* **565**, 385–404 (2002); **619**, 427–445(2005).
13. M. Turatto, et al. *Astrophys. J.* **498**, L129–L133 (1998).
14. A. Chieffi, M. Limongi, *Astrophys. J.* **577**, 281–294 (2002).
15. R. Cayrel, et al., *Astron. Astrophys.*, **416**, 1117–1138 (2004)
16. S. Honda, et al., *Astrophys. J.*, **607**, 474–498 (2004).
17. H. Umeda, K. Nomoto, T. Nakamura, in *The First Stars*, A. Weiss, T. G. Abel, V. Hill, Eds. (Springer-Verlag, Berlin, 2000), pp. 150-173.
18. A. Heger, N. Langer, S. E. Woosley, *Astrophys. J.* **528**, 368–396 (2000).
19. N. Tominaga, H. Umeda, & K. Nomoto 2006, this volume.
20. N. Tominaga, H. Umeda, & K. Nomoto, in preparation (2006).

Rotating massive stars @ very low Z: high C & N production

Raphael HIRSCHI

Dept. of physics and Astronomy, University of Basel, Klingelbergstr. 82, 4056 Basel, Switzerland

Abstract. Two series of models and their yields are presented in this paper. The first series consists of 20 $M_\odot$ models with varying initial metallicity (solar down to $Z = 10^{-8}$) and rotation ($v_{ini} = 0 - 600\,\mathrm{km\,s^{-1}}$). The second one consists of models with an initial metallicity of $Z = 10^{-8}$, masses between 20 and 85 $M_\odot$ and average rotation velocities at these metallicities ($v_{ini} = 600 - 800\,\mathrm{km\,s^{-1}}$). The most interesting models are the models with $Z = 10^{-8}$ ([Fe/H]~ -6.6). In the course of helium burning, carbon and oxygen are mixed into the hydrogen burning shell. This boosts the importance of the shell and causes a reduction of the size of the CO core. Later in the evolution, the hydrogen shell deepens and produces large amount of primary nitrogen. For the most massive models ($M \gtrsim 60\,M_\odot$), significant mass loss occurs during the red supergiant stage. This mass loss is due to the surface enrichment in CNO elements via rotational and convective mixing.

The yields of the fast rotating 20 $M_\odot$ models can best reproduce (within our study) the observed abundances at the surface of extremely metal poor (EMP) stars. The wind of the massive models can reproduce the CNO abundances of the carbon–rich UMPs, in particular for the most metal poor star known to date, HE1327-2326.

Keywords: Stars: abundances – evolution – rotation – mass loss
PACS: 97.10.Cv

INTRODUCTION

Precise measurements of abundances of extremely metal poor (EMP) stars have recently been obtained by Cayrel et al. [1], Spite et al. [2], Israelian et al. [3], ... These provide new constraints for the stellar evolution models [see 4, 5, 6]. The most striking constraint is the need for primary ^{14}N production in very low metallicity massive stars. Other possible constraints are an upturn of the C/O ratio with a [C/Fe] about constant or slightly decreasing (with increasing metallicity) at very low metallicities, which requires an increase (with increasing metallicity) of oxygen yields below [Fe/H]$\sim$ -3. About one quarter of EMP stars are carbon rich (C-rich EMP, CEMP stars). Ryan et al. [7] propose a classification for these stars. They find two categories: about three quarter are main s-process enriched (Ba-rich) stars and one quarter are enriched with a weak component of s-process (Ba-normal). The two most metal poor stars known to date, HE1327-2326 [8, 9] and HE 0107-5240 [10] are both CEMP stars. These stars are believed to have been enriched by only one to several stars and we can therefore compare our yields to their observed abundances without the filter of a galactic chemical evolution model (GCE). In an attempt to explain the origin of the abundances observed as well as the metallicity trends, I computed pre-supernova evolution models of rotating single stars with metallicities ranging from solar metallicity down to $Z = 10^{-8}$ following the work of Meynet et al. [11].

DESCRIPTION OF THE STELLAR MODELS

The computer model used to calculate the stellar models is described in detail in Hirschi et al. [12]. At low metallicities the mixture of the heavy elements we adopted is the one used to compute the opacity tables for Weiss 95's alpha–enriched composition [13]. The mass loss rates are described and discussed in Meynet et al. [11]. Very little was known about the mass loss of very low metallicity stars with a strong enrichment in CNO elements until recently. Vink and de Koter [14] study the case of WR stars but a crucial case, which has not been studied in detail yet, is the case of red supergiant stars (RSG). As we shall see later, due to rotational and convective mixing, the surface of the star is strongly enriched in CNO elements during the RSG stage. Awaiting for future studies, it is implicitly assumed in this work (as in Meynet et al. [11]) that CNO elements have a significant contribution to opacities and mass loss rates. This assumption is supported by the possible formation of molecular lines in the RSG stage. Therefore the mass loss rates depend on metallicity as $\dot{M} \sim (Z/Z_\odot)^{0.5}$, where Z is the mass fraction of heavy elements at the surface of the star. The evolution of the models was in general followed until core Si–burning and the stellar yields are calculated as in Hirschi et al. [15]. The main characteristics of the models are presented in Table 1. More details about the models are presented in Hirschi [16].

The value of $300\,\mathrm{km\,s^{-1}}$ used for the initial rotation velocity at solar metallicity corresponds to an average velocity of about $220\,\mathrm{km\,s^{-1}}$ on the Main Sequence (MS) which is very close to the average observed value [see for instance 17]. It is unfortunately not possible to measure the rotational velocity of very low metallicity massive stars since they all died a long time ago. Nevertheless, there is indirect evidence that stars with a lower metallicity have a higher rotation velocity. This can be due to the difficulty of evacuating angular momentum during the star formation, which is even more important at lower metallicities [see 18]. Furthermore, a very low metallicity star containing the same angular momentum as a solar metallicity star has a higher surface rotation velocity due to its smaller radius (one quarter of $Z_\odot$ radius for 20 $M_\odot$ stars). In order to compare the models at different metallicities and with different initial masses, the ratio $v_{\mathrm{ini}}/v_{\mathrm{crit}}$ is used (see Table 1). v_{crit} is the critical velocity at which matter becomes gravitationally unbound. $v_{\mathrm{ini}}/v_{\mathrm{crit}}$ increases only as $r^{-1/2}$ for models with the same angular momentum (J) but lower metallicity, whereas the surface rotational velocity increases as r^{-1} ($J \sim vr$). The angular momentum can be compared as well but one has to bear in mind that it varies significantly for models of different initial masses. Finally, $v_{\mathrm{ini}}/v_{\mathrm{crit}}$ is a good indicator for the impact of rotation on mass loss.

In the first series of models, the aim is to scan the parameter space of rotation and metallicity with 20 $M_\odot$ models since a 20 $M_\odot$ star is not far from the average massive star concerning stellar yields. For this series, two initial rotational velocities were used at very low metallicities. The first one is the same as at solar metallicity, $300\,\mathrm{km\,s^{-1}}$. The second v_{ini} is 500 at $Z=10^{-5}$ ([Fe/H]$\sim$-3.6) and $600\,\mathrm{km\,s^{-1}}$ at $Z=10^{-8}$ ([Fe/H]$\sim$-6.6). These second values have ratios of the initial velocity to the break–up velocity, $v_{\mathrm{ini}}/v_{\mathrm{crit}}$, around 0.55, which is only slightly larger than the solar metallicity value (0.44). The 20 $M_\odot$ model at $Z=10^{-8}$ and with $600\,\mathrm{km\,s^{-1}}$ has a total initial angular momentum $J_{\mathrm{tot}} = 3.3\,10^{52}$ erg s which is the same as the solar metallicity 20 $M_\odot$ model

TABLE 1. Initial parameters of the models (columns 1–5): mass, metallicity, rotation velocity [km s^{-1}], total angular momentum [10^{53} erg s] and v_{ini}/v_{crit}. Total lifetime [Myr] and various masses [$M_\odot$] (7–10): final mass, masses of the helium and carbon–oxygen cores and the remnant mass. Total stellar yields (wind + SN) [$M_\odot$] for carbon (11), nitrogen (12) and oxygen (13).

M_{ini}	Z_{ini}	v_{ini}	J_{tot}^{ini}	$\frac{v_{ini}}{v_{crit}}$	τ_{life}	M_{final}	M_α	M_{CO}	M_{rem}	^{12}C	^{14}N	^{16}O
20	2e-02	300	0.36	0.44	11.0	8.763	8.66	6.59	2.57	0.433	4.33e-2	2.57
20	1e-03	000	–	0.00	10.0	19.557	6.58	4.39	2.01	0.373	3.31e-3	1.46
20	1e-03	300	0.34	0.39	11.5	17.190	8.32	6.24	2.48	0.676	3.10e-3	2.70
20	1e-05	000	–	0.00	9.80	19.980	6.24	4.28	1.98	0.370	4.27e-5	1.50
20	1e-05	300	0.27	0.34	11.1	19.930	7.90	5.68	2.34	0.481	1.51e-4	2.37
20	1e-05	500	0.42	0.57	11.6	19.575	7.85	5.91	2.39	0.648	5.31e-4	2.59
20	1e-08	000	–	0.00	8.96	19.999	4.43	4.05	1.92	0.262	8.52e-3	1.20
20	1e-08	300	0.18	0.28	9.98	19.999	6.17	5.18	2.21	0.381	1.20e-4	1.96
20	1e-08	600	0.33	0.55	10.6	19.952	4.83	4.36	2.00	0.823	5.90e-2	1.35
40	1e-08	700	1.15	0.55	5.77	35.795	13.5	12.8	4.04	1.79	1.87e-1	5.94
60	1e-08	800	2.41	0.57	4.55	48.975	25.6	24.0	7.38	3.58	4.14e-2	12.8
85	1e-08	800	4.15	0.53	3.86	19.868	19.9	18.8	5.79	7.89	1.75e+0	12.3

with 300 km s^{-1} ($J_{tot} = 3.6\,10^{52}$ erg s). So a velocity of 600 km s^{-1}, which at first sight seems extremely fast, is probably the average velocity at Z=10^{-8}. In the second series of models, I follow the exploratory work of Meynet et al. [11] and compute models at Z=10^{-8} with initial masses of 40, 60 and 85 $M_\odot$ and initial rotational velocities of 700, 800 and 800 km s^{-1} respectively. Note that, for these models as well, the initial total angular momentum is similar to the one contained in solar metallicity models with rotational velocities of 300 km s^{-1}. Since this is the case, velocities between 600 and 800 km s^{-1} are considered in this work as the average rotational velocities at these very low metallicities.

EVOLUTION OF THE 20 $M_\odot$ MODELS

Mass loss becomes gradually unimportant as the metallicity decreases in the 20 $M_\odot$ models. At solar metallicity, the rotating 20 $M_\odot$ model loses more than half of its mass and at $Z = 10^{-8}$ less than 0.3% (see Table 1). This means that at very low metallicities, the dominant effect of rotation is mixing for the mass range around 20 $M_\odot$. The impact of rotational mixing is best pictured in the Kippenhahn diagram (see Fig. 1). During hydrogen burning and the start of helium burning, mixing increases the core sizes. Mixing of helium above the core suppresses the intermediate convective zones linked to shell H–burning. So far the impact of mixing at $Z = 10^{-8}$ is the same as at higher metallicities. However, after some time in He–burning, the mixing of primary carbon and oxygen into the H–burning shell is important enough to boost significantly the strength of the shell. As a result, the size of the helium burning core becomes and remains smaller than in the non–rotating model. The yield of ^{16}O being closely correlated with the size of the CO core, it is therefore reduced due to the strong mixing. At the same time carbon yields are increased. This produces an upturn of C/O at very low metallicities.

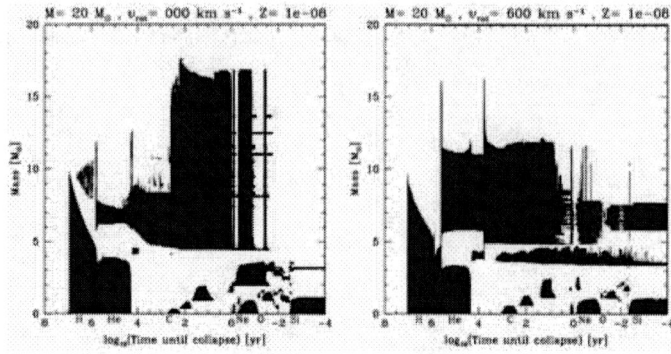

FIGURE 1. Kippenhahn diagrams of $20 M_\odot$ models at $Z = 10^{-8}$ with $v_{ini} = 0\,{\rm km\,s^{-1}}$ (*left*) and $600\,{\rm km\,s^{-1}}$ (*right*).

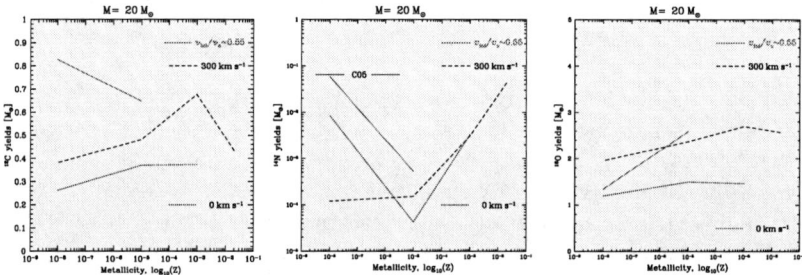

FIGURE 2. Stellar yields of ^{12}C (*left*), ^{14}N (*center*) and ^{16}O (*right*) as a function of the initial metallicity of the models. The solid red, dashed blue and dotted black lines represent respectively the models with $v_{ini}/v_c \sim 0.55$ (v_{ini}=500 km s^{-1} at $Z = 10^{-5}$ and v_{ini}=600 km s^{-1} at $Z = 10^{-8}$), with v_{ini}=300 km s^{-1} and without rotation. For nitrogen, the horizontal mark with C05 in the middle corresponds to the value deduced from the chemical evolution models of Chiappini et al. [4].

Stellar yields of CNO elements

The yields of ^{12}C, ^{14}N and ^{16}O are presented in Fig. 2 and their numerical values are given in Table 1 [see 16, for more details]. The most stringent observational constraint at very low Z is a very high primary ^{14}N production [4, 6], of the order of 0.06 $M_\odot$ per star. In Fig. 2 (*center*), we can see that only the model at $Z = 10^{-8}$ and with v_{ini}=600 km s^{-1} can reach such high values. The bulk of ^{14}N is produced in the convective zone created by shell hydrogen burning (see Fig. 1 *right*). If this convective zone deepens enough to engulf carbon (and oxygen) rich layers, then significant amounts of primary ^{14}N can be produced ($\sim$0.01$M_\odot$). This occurs in both the non–rotating model and the fast rotating model but for different reasons. In the non–rotating model, it occurs due to structure rearrangements similar to the third dredge–up at the end of carbon burning. In the model with v_{ini} =600 km s^{-1} it occurs during shell helium burning because of the strong mixing of carbon and oxygen into the hydrogen shell burning zone.

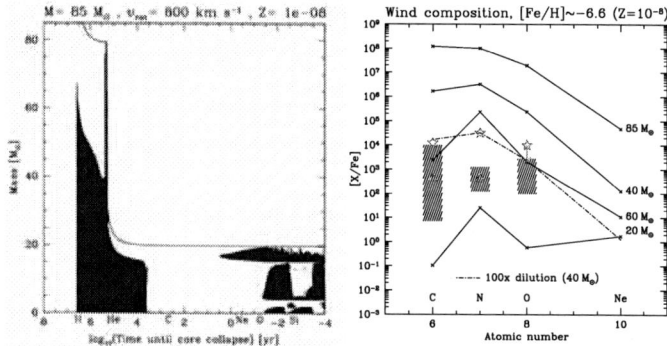

FIGURE 3. *Left*: Kippenhahn diagrams of the $85 M_\odot$ model at $Z = 10^{-8}$ with $v_{ini} = 800\,\mathrm{km\,s^{-1}}$. *Right*: The solid lines represent the chemical composition of the wind material of the of the different models at $Z = 10^{-8}$. The hatched areas correspond to the range of values measured at the surface of giant CEMP stars: HE 0107-5240, [Fe/H]$\simeq$ -5.3 [10]; CS 22949-037, [Fe/H]$\simeq$ -4.0 [19, 20]; CS 29498-043, [Fe/H]$\simeq$ -3.5 [21]. The empty triangles [22]([Fe/H]$\simeq$ −4.0) and stars [8] ([Fe/H]$\simeq$ −5.4, only an upper limit is given for [O/Fe]) correspond to non-evolved CEMP stars.

TABLE 2. Initial mass (1), metallicity (2) and rotation velocity [km s^{-1}] (3) and stellar wind ejected masses [$M_\odot$] for carbon (4), nitrogen (5) and oxygen (6).

M_{ini}	Z_{ini}	v_{ini}	^{12}C	^{14}N	^{16}O
20	1e-08	600	3.44e-12	3.19e-10	6.69e-11
40	1e-08	700	5.34e-03	3.63e-03	2.42e-03
60	1e-08	800	1.80e-05	6.87e-04	5.49e-05
85	1e-08	800	6.34e+00	1.75e+00	3.02e+00

Models with higher initial masses at $Z = 10^{-8}$ also produce large quantities of primary nitrogen. More computations are necessary to see over which metallicity range the large primary production takes place and to see whether the scatter in yields of the models with different masses and metallicities is compatible with the observed scatter.

EVOLUTION OF THE MODELS AT $Z = 10^{-8}$

Contrarily to what was initially expected from very low metallicity stars, mass loss can occur in massive stars [11]. The mass loss occurs in two phases. The first phase is when the star reaches break–up velocities towards the end of the main sequence. Due to this effect stars, even metal free ones, are expected to lose about 10% of their initial masses for an average initial rotation. The second phase in which large mass loss can occur is during the RSG stage. Indeed, stars more massive than about $60 M_\odot$ at $Z = 10^{-8}$ become RSG and dredge–up CNO elements to the surface. This brings the total metallicity of the surface to values within an order of magnitude of solar and triggers large mass loss. The final masses of the models are given in Table 1. The case of the $85 M_\odot$ model is extremely interesting (see Fig. 3 *left*) since it loses more than three quarter of its initial mass. It even becomes a WO star.

Wind composition and CRUMPS stars

In Fig. 3 (*right*), we compare the chemical composition of the wind material with abundances observed in non-evolved carbon rich extremely and ultra [8] metal poor stars. The ejected masses of the wind material are also given in Table 2. It is very interesting to see that the wind material can reproduce the observed abundance in two ways. Either, the wind material is richer than necessary and dilution (by a factor 100 for example for the 40 $M_\odot$ models and HE1327-2326) with the ISM is needed or the wind has the right enrichment (for example the 60 $M_\odot$ and HE1327-2326) and the low mass star could form from pure wind material. The advantage of the pure wind material is that it has a ratio $^{12}C/^{13}C$ around 5 [11] and it can explain Li depletion. With or without dilution, the wind material has the advantage that it brings the initial metallicity of the low mass star above the critical value for its formation [23].

ACKNOWLEDGMENTS

I would like to thank warmly the organisation and Prof. Nomoto and his group for the financial support and the kind hospitality.

REFERENCES

1. R. Cayrel, E. Depagne, M. Spite, et al., *A&A* **416**, 1117–1138 (2004).
2. M. Spite, R. Cayrel, B. Plez, et al., *A&A* **430**, 655–668 (2005).
3. G. Israelian, A. Ecuvillon, R. Rebolo, et al., *A&A* **421**, 649–658 (2004).
4. C. Chiappini, F. Matteucci, and S. K. Ballero, *A&A* **437**, 429–436 (2005).
5. P. François, F. Matteucci, R. Cayrel, et al., *A&A* **421**, 613–621 (2004).
6. N. Prantzos, *astro-ph/0411392*, *NIC8* (2004), arXiv:astro-ph/0411392.
7. S. G. Ryan, W. Aoki, J. E. Norris, and T. C. Beers, *ApJ, in press* (2005), astro-ph/0508475.
8. A. Frebel, W. Aoki, N. Christlieb, et al., *Nature* **434**, 871–873 (2005).
9. W. Aoki, A. Frebel, N. Christlieb, et al., (2005), arXiv:astro-ph/0509206.
10. N. Christlieb, B. Gustafsson, A. J. Korn, et al., *ApJ* **603**, 708–728 (2004).
11. G. Meynet, S. Ekström, and A. Maeder, *A&A accepted* (2005), astro-ph/0510560.
12. R. Hirschi, G. Meynet, and A. Maeder, *A&A* **425**, 649–670 (2004).
13. C. A. Iglesias, and F. J. Rogers, *ApJ* **464**, 943–+ (1996).
14. J. S. Vink, and A. de Koter, (2005), arXiv:astro-ph/0507352.
15. R. Hirschi, G. Meynet, and A. Maeder, *A&A* **433**, 1013–1022 (2005).
16. R. Hirschi, *A&A, in prep* (2006).
17. I. Fukuda, *PASP* **94**, 271–284 (1982).
18. T. Abel, G. L. Bryan, and M. L. Norman, *Science* **295**, 93–98 (2002).
19. J. E. Norris, S. G. Ryan, and T. C. Beers, *ApJ* **561**, 1034–1059 (2001).
20. E. Depagne, V. Hill, M. Spite, et al., *A&A* **390**, 187–198 (2002).
21. W. Aoki, J. E. Norris, S. G. Ryan, et al., *ApJ* **608**, 971–977 (2004).
22. B. Plez, and J. G. Cohen, *A&A* **434**, 1117–1124 (2005).
23. V. Bromm, (2005), arXiv:astro-ph/0509354.

1.4 COSMIC and GALACTIC CHEMICAL EVOLUTION and STRUCTURE FORMATION

The chemical evolution of the Milky Way : from light to heavy elements

Francesca Matteucci

Astronomy Department, Trieste University, Via G.B. Tiepolo, 11, 34124, Trieste, Italy

Abstract. We present results for the chemical evolution of the Milky Way including predictions for elements from Deuterium to Europium. A comparison with the most accurate and recent data allows us to draw important conclusions on stellar nucleosynthesis processes as well as on mechanisms of galaxy formation.

Keywords: Document processing, Class file writing, LaTeX 2_ε
PACS: 43.35.Ei, 78.60.Mq

INTRODUCTION

The Milky Way is the best studied system and a great deal of accurate data such as high resolution spectroscopy relative to the solar neighbourhood have recently accumulated. Detailed chemical evolution models able to follow in detail the evolution of the abundances of several chemical species in the interstellar gas are also available and by means of them we can compare theory with observations and infer many important constraints. In particular, by adopting different set of stellar yields for stars of all masses we can test the stellar nucleosynthesis and the origin of elements. Particularly important are the plots relative to abundance ratios, such as [X/Fe], as functions of stellar metallicity, measured by the quantity [Fe/H]. Recently, Cayrel et al. (2004) have provided a sample of abundances for extremely metal poor stars in the Galactic halo, in particular for stars with metallicity as low as [Fe/H]=-4.0dex. The elements studied are mainly α-elements (O, Mg, Si, Ca, S, Ti), Fe-peak elements and heavier. Abundances for C and N in extremely metal poor halo stars were also measured (Spite et al. 2005; Israelian et al. 2004). All of these data represent an invaluable template for testing chemical evolution models. Here we test the two-infall model (Chiappini et al. 1997) for the chemical evolution of the Milky Way on a large sample of data including the abundances discussed above plus the abundance trends recently derived for light elements such as deuterium. The main ingredients of the chemical evolution model are: i) the initial conditions, ii) the star formation history, iii) the initial mass function (IMF), iv) the stellar nucleosynthesis, v) the rate of gas accretion. In the two-infall model the main assumptions are that the halo and thick disk formed by gas accretion on a relatively short timescale (1-2 Gyr), whereas the thin disk formed by means of an independent accretion episode on longer timescales increasing with galactocentric distances. In particular, the timescale for the formation of the solar neighbourhood is 7-8 Gyr, as imposed by the stellar metallicity distribution. The assumed rate of gas accretion (primordial gas) is:

$$IR = A(R)e^{-t/\tau_H(R)} + B(R)e^{-(t-t_{max})/\tau_D(R)} \tag{1}$$

with the timescale for the first infall episode $\tau_H(R)= 2$ Gyr for the halo, and the timescale for the disk infall $\tau_D(R)$ linearly increasing with R which implies an "Inside-Out Model". The time t_{max} is the time of the maximum infall on the thin disk.

The star formation rate is:

$$SFR = \nu \sigma_{tot}^{k_2} \sigma_{gas}^{k_1} \qquad (2)$$

with $k_1 = 1.5$, $k_2 = 0.5$, and σ_{tot} being the total surface mass density and σ_{gas} the surface gas density. A threshold density for the SFR of $7 M_\odot pc^{-2}$ is also assumed. This threshold regulates the star formation regime producing an oscillatory behaviour for the SFR when the gas density is close to its value (see Figure 1).

The assumed IMF is multi-slope (x_1, x_2, ..) and follows the prescriptions of Scalo (1986).

Element production

The element production in stars can be summarized as follows:

- Low & intermediate mass stars ($0.8 \leq M/M_\odot \leq 8$) produce mainly ^4He, ^{12}C, ^{14}N and heavy s-process elements (Ba, Y, Sr) (in stars with masses from 1 to $3M_\odot$).
- Type II SNe ($M > 8M_\odot$) produce mainly α-elements (O, Ne, Mg, Si, S, Ca), part of Fe and other Fe-peak elements as well as r-process elements (Eu, Ba).
- Type Ia SNe produce mainly Fe-peak elements ($\sim 0.6 - 0.7 M_\odot$ of ^{56}Fe per supernova).
- Novae can possibly be important in producing ^7Li, ^{13}C, ^{15}N, ^{17}O. ^7Li can be produced also in SNII, red giant, asymptotic giant branch stars and cosmic rays.
- Deuterium is only destroyed in stars, whereas ^3He is destroyed but also produced in stars from 1 to $3M_\odot$.

Type Ia SN rates

The type Ia SNe are the most important Fe producers in the universe and represent a fundamental ingredients in galactic chemical evolution models. The most popular scenarios for the progenitors of Type Ia SNe can be summarized as follows:

- **Single degenerate scenario**. This is the classical scenario of Whelan and Iben (1973), namely C-deflagration in a C-O white dwarf reaching the Chandrasekhar mass, M_{Ch}, after accreting material from a companion which can be either a red giant or a Main Sequence star. In this case the maximum mass allowed for both the primary and secondary star in the binary system is $8M_\odot$ and consequently the minimum timescale for the occurrence of type Ia SNe is $t_{SNIa_{min}}$=0.03 Gyr (Greggio & Renzini 1983; Matteucci & Recchi, 2001).
- **Double Degenerate scenario**. This is the merging of two C-O white dwarfs, due to gravitational wave radiation, which explode by C-deflagration when the M_{Ch} is reached (Iben & Tutukov 1984). The minimum timescale for the occurrence of

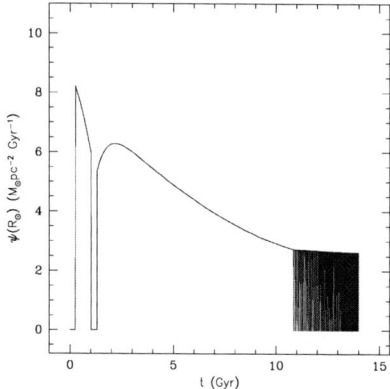

FIGURE 1. The star formation rate as predicted by the two-infall model by Chiappini et al. (1997) for the solar neighbourhood. Note the oscillatory behaviour at late times in the thin disk evolution which is a consequence of the assumed star formation threshold.

such systems is given by the lifetime of a $8M_\odot$, as above, plus the gravitational time delay $t_{SNIa_{min}} = 0.03 + \Delta t_{grav} = 0.03 + 0.15$Gyr (see Tornambè & Matteucci, 1986).
- A more recent model by Hachisu et al. (1999) is based on the single degenerate scenario but with a metallicity effect. In particular, no type Ia system can form if the metallicity of the progenitor stars is lower than [Fe/H] $= -1.0$. This model suggests also that only systems with secondary stars with masses lower than $2.6M_\odot$ can be accepted. This leads to a minimum time for the occurrence of type Ia SNe no shorter than $t_{SNIa_{min}} = 0.33$ Gyr, to which one should add the metallicity delay, namely the time taken by the chemical enrichment process to reach a metallicity of [Fe/H]=-1.

It is important to know the minimum timescale for the occurrence of type Ia SNe as well as the time for the maximun in the type Ia SN rate, which depends not only upon the SN progenitor model but also upon the SFR.

Results

The G-dwarf metallicity distribution. In Figure 2 we present the predicted G-dwarf metallicity distribution compared with the observed distributions. The good fit to the data is obtained by means of a timescale for the formation of the thin disk in the solar vicinity of 7-8 Gyr. A much lower timescale would produce a very bad fit to the observed data since it would move the peak of the distribution toward too low metallicities (see Kotoneva et al. 2000). The stellar metallicity distribution is, in fact, very sensitive to the star formation history which in turn depends strongly on the rate of gas accretion.

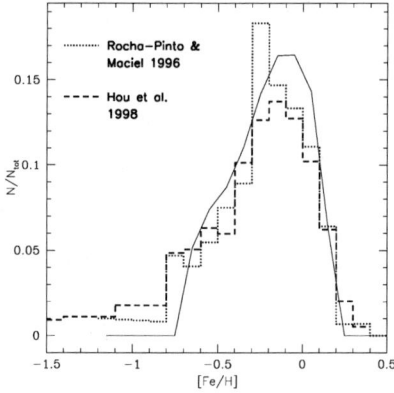

FIGURE 2. The continuous line represents the predictions of the two-infall model with a timescale of disk formation of 8 Gyr. The histograms represent the observational data as indicated in the figure.

Abundance ratios versus iron abundance. In Figures 3 and 4 we show the predictions relative to the [α/Fe] ratios versus [Fe/H] compared with a complete set of data including Cayrel et al.'s (2004) data, as described in François et al. (2004), who tested different sets of yields on the data. Their conclusion was that the yields of massive stars by Woosley & Weaver (1995, WW95) and those of Iwamoto et al. (1999, I99) for type Ia SNe are the best to reproduce the observations altough some variations are required in the yields of some elements such as Mg, Zn and some Fe-peak elements. The good fits shown in Figures 3 and 4 were obtained by including these variations.

A remarkable feature of Figures 3 and 4 is the lack of spread at very low metallicity, suggesting that the halo of the Milky Way was already well mixed at [Fe/H]=-4.0. This should be taken into account in inhomogeneous models of chemical evolution of the halo.

The evolution of light elements. In Figure 5 we present the evolution of deuterium during the lifetime of the Galaxy in the solar vicinity. The predictions are from the models of Chiappini et al. (2002) and Tosi (1988). The observed values at the time of birth of the solar system and in the local interstellar medium (ISM) are also shown. The assumed primordial value for D is the one suggested by the WMAP experiment (Spergel et al. 2003). The large spread shown for the local abundance of D, as measured by FUSE and HST-GHRS probably indicates the effects of D depletion into dust grains. It is important to notice that, in order to reproduce the data, the astration factor for D should have been lower than a factor of 1.5 (see Romano et al. 2003).

Very heavy elements. The s- and r- process elements are generally produced by neutron capture on Fe seed nuclei. The former are formed during the He-burning phase both in low-intermediate and massive stars, whereas the latter occur in explosive events such as Type II SNe. Recently, François et al. (2006 and this conference) have measured the abundances of several very heavy elements (e.g. Ba and Eu) in extremely metal poor

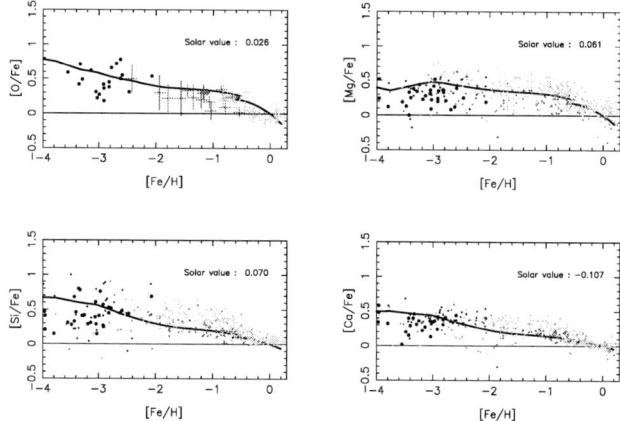

FIGURE 3. Predicted and observed [α/Fe] ratios for a large range of metallicities. The model predictions are from François et al. (2004) and are normalized to the predicted solar abundances. In the top right of each figure we indicate our predicted solar ratios, all in good agreement with the observed ones([α/Fe]=0.0). The black dots represent the data from Cayrel et al. (2004). For the other sources of data see François et al. (2004).

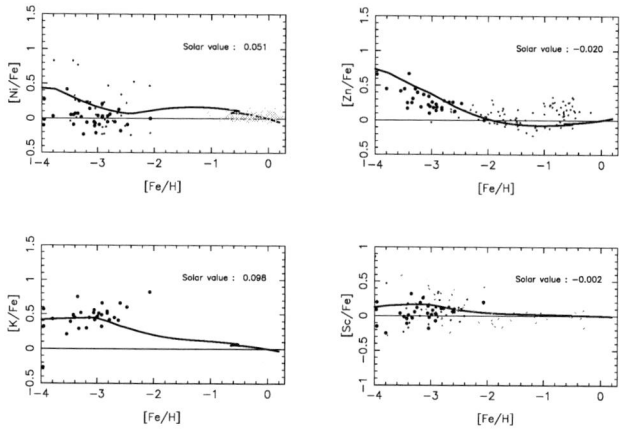

FIGURE 4. The same as in Figure 3 for other chemical species. The predictions are from François et al. (2004). The black dots are the data from Cayrel et al. (2004). For the other sources of data see François et al. (2004).

stars of the Milky Way. Previous work on the subject had shown a large spread in the abundance ratios of these elements to iron, especially at low metallicities. This spread is confirmed although is less than before, and is at variance with the lack of spread observed in the other elements shown before. Apart from this problem, not yet solved, these diagrams can be very useful to put constraints on the nucleosynthetic origin of these elements. In particular, Cescutti et al. (2006) by adopting the two-infall model

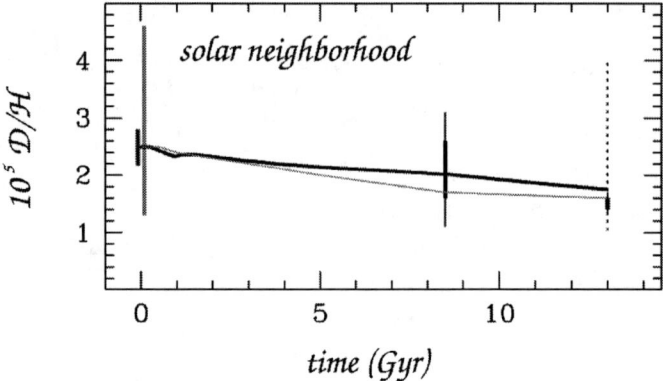

FIGURE 5. The evolution of deuterium in the solar vicinity during the galactic lifetime. The assumed primordial D value is from WMAP (small bar at t=0), the large bar at t=0 represents the primordial D as measured from QSO absorption spectra. The bar at t=8.5 represents the solar system value from Geiss & Gloecker (1988). The D abundance in the local ISM (at t=13 Gyr) is from Linsky (1998, small bar) and from FUSE and HST-GHRS data (large dotted bar, Vidal-Majar et al. 1998; Moos et al. 2002). The thick line represents the predictions of the two-infall model whereas the thin line the predictions of Tosi's (1988) model, as described in Romano et al. (2003).

predicted the evolution of [Ba/Fe] and [Eu/Fe] versus [Fe/H], as shown in figures 6 and 7. They can well fit the average trend but not the spread at very low metallicities since the model assumes instantaneous mixing. In order to fit the Ba evolution, they assumed that Ba is mainly produced as s-process element in low mass stars but that a fraction of Ba is also produced as an r-process element in stars with masses 10-30$M_\odot$. Europium is only an r-process element produced also in the range 10-30$M_\odot$.

CONCLUSIONS

Our main conclusion on the origin of the elements and the mechanism of formation of the Milky Way are:

- The elements ^{12}C and ^{14}N are mainly produced in low and intermediate mass stars ($0.8 \leq M/M_\odot \leq 8$). The amounts of primary and secondary N are still uncertain. Nitrogen should be primary in massive stars.
- The α-elements originate in massive stars: the nucleosynthesis of O is rather well understood and there is agreement between different authors.
- Magnesium is generally underproduced by standard nucleosynthesis models. Starting from the yields of WW95 we suggest that to fit the data the Mg production should be increased in stars with $M \leq 20M_\odot$ and decreased in stars with $M > 20M_\odot$. Silicon should be only slightly increased in stars with $M > 40M_\odot$ (François et al. 2004).

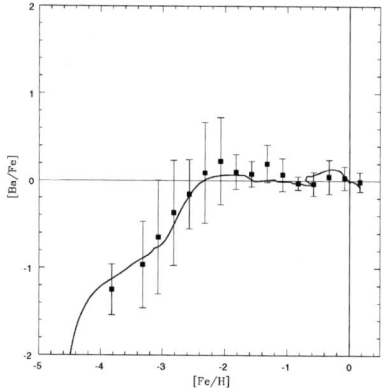

FIGURE 6. The evolution of Barium in the solar vicinity as predicted by the two-infall model (Cescutti et al. 2006). Data are from François et al. (2006)

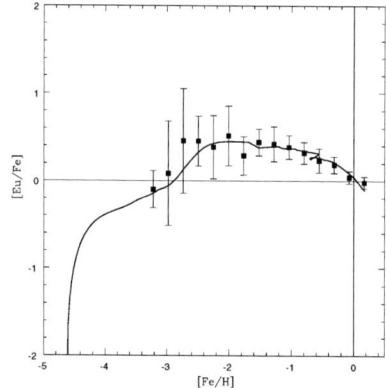

FIGURE 7. The evolution of Europium in tthe solar vicinity (Cescutti et al. 2006). Data are from François et al. (2006)

- Iron originates mostly in type Ia SNe, and the Fe production from massive stars is still uncertain. The metallicity dependent yields of Fe from WW95 overestimate Fe in stars with $M < 30 M_\odot$. It is better to adopt the Fe yields for solar metallicity all over the whole metallicity range. The Fe yields of I99 do not show this problem and they are computed only for solar metallicity (François et al. 2004).
- The Fe-peak elements: Cr, Mn, Ni, Co. Always starting from WW95 we suggest that the production of Cr and Mn should be increased in stars with 10-20 $M_\odot$. Cobalt production should be higher in type Ia SNe and lower in 10-20$M_\odot$, Ni production should be lower in type Ia SNe. The type Ia SN yields adopted for comparison are those of I99 (François et al. 2004).

- In order to reproduce the evolution of Zinc relative to Iron and in particular its overabundance at very low metallicity, new yields from massive stars are required. They should produce more Zn than the standard yields. A good approximation is to use the Zn yields from WW95 for solar metallicity. Moreover, Zn production from type Ia SNe should also be increased (François et al. 2004).
- Barium is mainly an s-process elements produced in the 1-3$M_\odot$ range, but it must have also an r-component from 10-30$M_\odot$ stars. Europium is only an r-process element formed in the stellar mass range 10-30$M_\odot$ (Cescutti et al. 2006).
- Deuterium is only destroyed in stars. A very good agreement with observations is obtained when the primordial abundance $(D/H)_p = 2.5 \cdot 10^{-5}$ derived from WMAP results is adopted, and the astration factor during the galactic lifetime is < 1.5 (Romano et al., 2003).
- In order to fit the G-dwarf metallicity distribution in the solar vicinity, a timescale for the thin disk formation at the solar galactocentric distance of 7-8 Gyr is required. To reproduce the observed abundance gradients along the galactic disk an inside-out formation mechanism with the timescale for disk formation increasing with galactocentric distance should be assumed (Matteucci & François, 1989; Chiappini et al. 2001) .

REFERENCES

- Cayrel, R., Depagne, E., Spite, M. et al. 2004, A&A, 416, 117
- Cescutti, G., Matteucci, F., François, P. 2006, A&A, in press
- Chiappini, C., Matteucci, F., Gratton, R. 1997, ApJ, 477, 765
- Chiappini, C., Renda, A., Matteucci, F. 2002, A&A, 395, 789
- Chiappini, C., Matteucci, F., Romano, D. 2001, ApJ, 554, 1044
- François, P., Matteucci, F., Cayrel, R., Spite, M., Spite, F., Chiappini, C. 2004, A&A, 421, 613
- François, P. et al. 2006, in preparation
- Geiss, J., Gloecker, G. 1998, Space Sci. Rev., 84, 239
- Greggio, L., Renzini, A. 1983, A&A, 118, 217
- Kotoneva, E., Flynn, C. Chiappini, C., Matteucci, F. 2000, MNRAS, 336, 879
- Hou, J.L., Prantzos, N., Boissier, S. 2000, A&A, 363, 921
- Hachisu, I., Kato, M., Nomoto, K. 1999, ApJ, 522, 487
- Iben, I.Jr., Tutukov, A. 1984, ApJ, 284, 719
- Israelian, G., Shchukina, N., Rebolo, R., Basri, G., González-Hernandez, J.I., Kajino, T. 2004, A&A, 419, 1095
- Iwamoto, K., Brachwitz, F., Nomoto, K. 1999, ApJS, 125, 439 (I99)
- Linsky,J.L. 1998, Space Sci. Rev., 84, 285
- Matteucci, F., François, P. 1989, A&A, 239, 885
- Matteucci, F., Recchi, S. 2001, ApJ, 558, 351
- Moos, H.W. et al. 2002, ApJS, 140, 3
- Rocha-Pinto, H.J., Maciel, W. 1996, MNRAS, 279, 447
- Romano, D., Tosi, M., Matteucci, F., Chiappini, C. 2003, A&A, 346, 295
- Scalo, J.M. 1986, Fund. Cosmic Phys., 11, 1
- Spergel, D.N., et al. 2003, ApJS, 148, 175
- Spite, M., Cayrel, R., Plez, B., Hill, V., Spite, F., Depagne, E., François, P. et al. 2005, A&A, 430, 655
- Tornambé, A., Matteucci, F. 1986, MNRAS, 223, 69
- Tosi, M. 1988, A&A, 197, 33
- Vidal-Majar, A. et al. 1998, A&A, 338, 694
- Woosley, S.E., Weaver, T.A. 1995, ApJS, 101, 181 (WW95)

Chemical evolution in the Milky Way Disk

Birgitta Nordström

Niels Bohr Institute, Juliane Maries Vej 30, DK-2100 Copenhagen, Denmark
birgitta@astro.ku.dk

Abstract. Classical models of galactic evolution predict a smooth rise in heavy-element abundance (metallicity) with time. We test this prediction with a new, large and unbiased sample of long-lived stars in the solar neighbourhood and find that several of the key tests fail to support the classical predictions. In agreement with earlier studies, our observed metallicity distribution function is deficient in low-mass metal-poor stars from the generation that produced the heavy elements seen in the Sun and younger stars. In contrast to some earlier studies, we find no clear rise in overall metallicity with time in the Solar neighbourhood; we also find that the galactic disk has experienced kinematic heating throughout its life, and identify groups of stars that may be traces of dwarf galaxies that have merged with the Milky Way.

Keywords: Composition and chemical evolution, Galactic disk, Galactic evolution, Solar neighbourhood, Stellar age, Dynamical evolution
PACS: 98.35.Ac, 98.35.Bd, 98.35.Pr, 97.20.Jg, 98.35.Df, 98.35.-a, 98.35.Hj, 98.62.Ai, 98.65.Fz, 98.10.+z, 98.62.Bj, 97.10.Wn, 97.10.Cv

THE MILKY WAY AS A PROTOTYPE

"The formation and evolution of galaxies is one of the great outstanding problems in astrophysics." (Freeman and Bland-Hawthorn 2002). From the point of view of this meeting, the key function of galaxies is to turn a fraction of the primordial 'soup' of hydrogen and helium into the heavier elements that populate the rest of the periodic system. The nuclear processes by which stars in galaxies synthesize heavier elements are by now elementary textbook material; but our understanding of even the main features – let alone the details – of how this actually occurs throughout a galaxy is still very incomplete.

Spiral galaxies are an important part of the visible Universe, with our own Milky Way as a prototype, and we are ideally positioned for observing the most important component of the Milky Way and other spiral galaxies – the disk. In the Solar neighbourhood we can study stars from the entire history of the disk and obtain data on chemical abundances, ages, and kinematics to a level of accuracy and detail that are impossible elsewhere. The stars in the solar neighbourhood therefore provide a fundamental benchmark for all theoretical models of the chemical and dynamical evolution of galaxy disks. Our large survey of 14,000 F- and G dwarf stars (Nordström et al. 2004) was designed to provide a new, superior basis for our understanding of the evolution of the Milky Way disk.

THE NEW DATA

The key feature of our study is the definition of a large, unbiased, and complete sample of nearby F- and G-type dwarf stars. Thus, our data (Nordström et al. 2004) include new metallicities, ages, kinematics, and Galactic orbits for 16,682 such stars.

Our 63,000 new, accurate radial-velocity observations are combined with Hipparcos (ESA 1997) distances and angular motions to provide complete 3D space motions and galactic orbits for over 14,200 stars; they also allow detection of most binary stars for which the derived astrophysical parameters would be incorrect. Metallicities and temperatures are determined from Strömgren *uvby* photometry (Olsen 1994). In addition, we have determined accurate ages for the stars, a parameter missing in many other studies and crucial comparing with galaxy evolution models. The new technique to determine ages is described in Jørgensen and Lindegren (2005).

TESTING GALACTIC EVOLUTION MODELS

As galaxies evolve, new generations of stars continue to produce heavy elements, adding to the overall metal content of the galaxy. How fast this happened and in which proportions the elements formed is something that in principle can be tested by direct observation. For truly incisive tests of the models we need detailed chemical abundances, ages, and galactic orbits for large, homogeneous samples of long-lived stars. That can be done by combining our new overall metallicities with the detailed, accurate elemental abundances for a sub-sample of 189 stars by Edvardsson et al. (1993). That paper shed much new light on the nucleosynthesis history of the galactic disk, but suffered from selection effects, which our sample was designed to overcome.

Here we illustrate the model tests with two diagrams showing the observed metallicities and space velocities as functions of age for the stars in our sample.

The Age-Metallicity Relation

The age-metallicity relation for disk stars is often used to test classical models for the chemical enrichment of the galaxy. Earlier studies had found a clear age-metallicity relation as predicted by evolution models (Twarog 1980), but this was questioned by several later studies (e.g. Feltzing et al. 2001). However, most previous results were based on samples that were biased against stars with certain combinations of age and metallicity (notably old, metal-rich stars), which our new data set is not.

Figure 1 shows *all* single F and G stars within 40 pc and with well-determined ages. There is clearly no significant rise of overall iron abundance ([Fe/H]) with time, but and the scatter in [Fe/H] at all ages greatly exceeds the observational error of ~0.1 dex. Limiting the sample to the very best-determined ages leads to the same conclusions. The challenge is to understand why this is so, when after all nucleosynthesis continues to take place inside the stars, and how the solar neighbourhood has *actually* evolved.

The key features of our age-metallicity relation for field stars agree well with recent studies of open clusters (e.g. Friel et al. 2002), which show the same constant mean metallicity and large scatter at all ages as we find. The metallicity-redshift relation for distant QSO absorption line systems provides an interesting analogy (Pettini 1999).

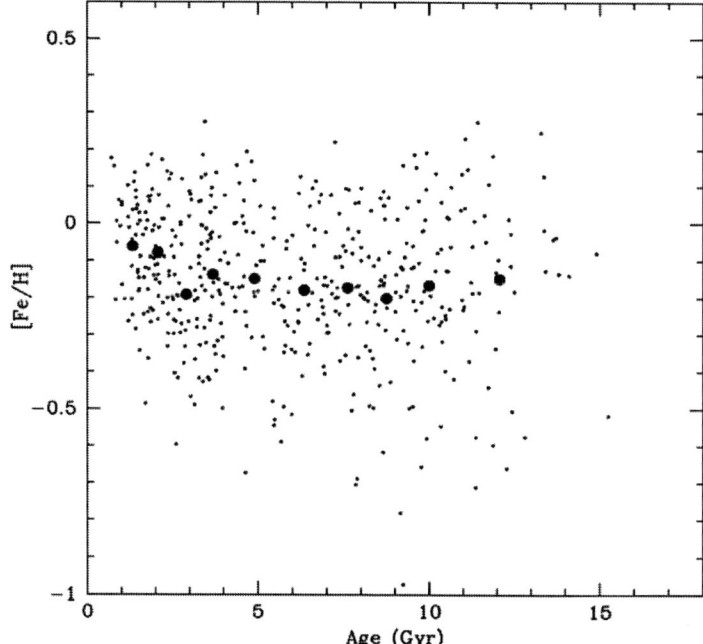

FIGURE 1. Ages and iron abundances ([Fe/H]) for 462 single F- and G stars within a distance of 40 pc from the Sun. The large dots show mean ages and metallicities in 10 bins with equal numbers of stars. Figure from Nordström et al. (2004)

Dynamical heating and signs of past accretion

In order to understand the age-metallicity diagram, the dynamical evolution must also be taken into account. This can also be done from our data, which include complete space velocities and galactic orbits for all the stars.

An important test of the dynamical evolution of the disk is the relation between age and space velocity for stars in the Solar neighbourhood. For this, we have computed the 3D velocity components U (radial) and V (tangential) in the galactic plane, as well as W, perpendicular to the plane.

The velocity dispersions as a function of time reflect the gradual and continuing dynamical heating of the galactic disk. The classic Oort relation predicts that σ_V/σ_U should be constant and equal to 0.5 for a flat galactic rotation curve. Mühlbauer & Dehnen (2003) show, however, that accounting for the true velocity dispersions and non-axisymmetric components of the disk potential may result in large variations.

The smaller exponents for the in-disk heating (σ_U, σ_V) compared to the out-of-the-disk heating (σ_W) seen in Fig. 1 give further constraints on models designed to explain the observed kinematic heating of the disk. Several mechanisms have been proposed to explain the heating: Molecular clouds and black holes in the disk, and on a larger scale, transient spiral arms (De Simone et al. 2004) or the Galactic bar (Fux 2001).

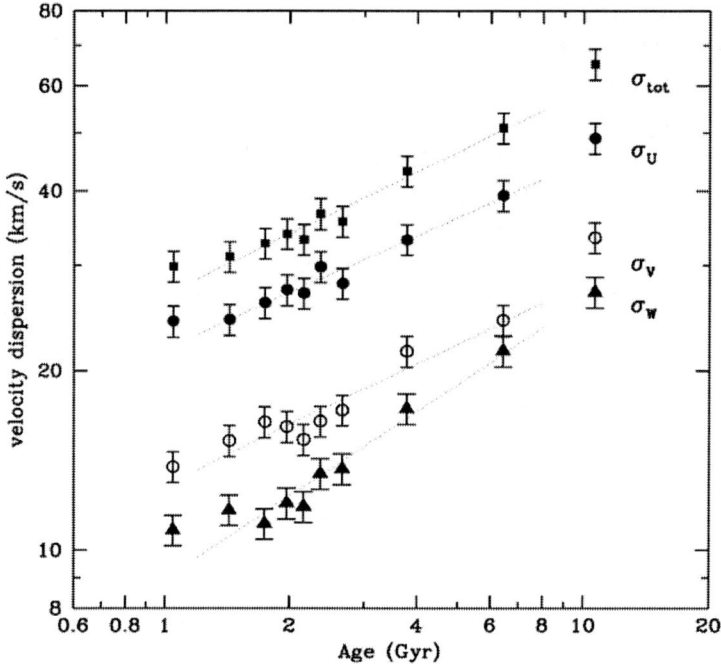

FIGURE 2. Velocity dispersions as functions of age for the subsample of single stars with age errors below 25%. σ_{tot}, σ_V, σ_U, σ_W denote the dispersion in the total, U, V, and W velocity components computed in 10 bins with equal numbers of stars. The lines show fitted power laws; see Nordström et al. (2004) for full details. Figure from Nordström et al. (2004)

Massive black holes are ruled out by other observational constraints, and their heating index of 0.5 is also too high. Mergers of satellite galaxies have certainly contributed, but major mergers should result in episodic heating of the disk, possibly resulting in the creation of the thick disk. Stable spiral arm patterns increase the random motions of stars within the plane, but not perpendicular to it. Molecular clouds heat stars in all three directions, but in isolation they are inefficient heaters with a total exponent of only 0.21 and a vertical exponent of 0.26 (see simulations by Hänninen & Flynn (2002). Further simulations including realistic descriptions of all heating components are needed, and indeed recent work on the effects of stochastic, transient spiral wave structures by De Simone et al. (2004) seems able to produce exponents in the observed range.

A search for signatures of past accretion events in the Milky Way in the new observational sample is discussed by Helmi et al. (2006). Signs of past accretion are detected by using numerical simulations of invariant dynamical parameters to characterise the orbital properties of debris from any disrupted satellites. The results have important implications on galaxy formation and structure formation as they give strong evidence for a hierarchical formation scenario for the Milky Way disk.

CONCLUSIONS

We find that standard evolution models for the evolution of the element content of the Galactic disk fail several classical tests. The observed lack of an overall increase in average metal abundance with time shows interesting analogies with the metallicity-redshift for distant QSO absorption line systems (Pettini 1999), which may help to identify the mechanisms that control the heavy-element production locally.

We find that the galactic disk has been heated continuously throughout its lifetime. Several mechanisms to explain this have been proposed, but most do not seem to be sufficiently efficient to explain the observations.

Finally, the identification of substantial amounts of debris in the Galactic disk whose origin can be traced back to more than one satellite galaxy (Helmi et al. 2006) gives evidence in favour of a hierarchical formation scenario for the Milky Way disk.

ACKNOWLEDGMENTS

I thank the many colleagues who collaborated in completing the Geneva–Copenhagen Survey. The Danish and Swedish Research Councils, the Carlsberg Foundation, the Royal Physiographic Society in Lund, the European Southern Observatory, the Nordic Academy for Advanced Studies (NordForsk), and the Smithsonian Institution provided vital financial support, which is gratefully acknowledged.

REFERENCES

1. B. Edvardsson, J. Andersen, B. Gustafsson, D. L. Lambert, P. E. Nissen, J. Tomkin, *Astron. & Astrophys.*, 275, 101-152 (1993)
2. ESA, *The Hipparcos and Tycho Catalogues*, ESA-SP 1200 (1997)
3. S. Feltzing, J., Holmberg, and J. R. Hurley, *Astron. & Astrophys*, 377, 911-914 (2001)
4. K. C. Freeman, and J. Bland-Hawthorn, *Ann. Rev. Astron & Astrophys*, 40, 487- 537 (2002)
5. E. D. Friel, K. A. Janes, M. Tavarez, et al. 2002, *Astron. Journ.*, 124, pp. 2693-2720 (2002)
6. R. Fux, *Astron. & Astrophys.*, 373, pp. 511- 535 (2001)
7. A. Helmi, J. F., Navarro, B. Nordström, J. Holmberg, M. G. Abadi, M. Steinmetz, 2005, *astro-ph/050540*, (*Mon. Not. R. Astr. Soc.,* Jan. 2006)
8. J. Hänninen, and C. Flynn, *Mon. Not. R. Astr. Soc.*, 337, pp. 731-742 (2002)
9. B.R. Jørgensen, *Astron. & Astrophys.,* 363, pp. 947-957 (2000)
10. B.R. Jørgensen, L. Lindegren, *Astron. & Astrophys.,* 436, pp. 127-143 (2005)
11. G. Mühlbauer, and W. Dehnen, *Astron. & Astrophys.*, 401, pp. 975-984 (2003)
12. B. Nordström, M. Mayor, J. Andersen, J. Holmberg, F. Pont, B. R. Jørgensen, E. H. Olsen, S. Udry, and Mowlavi *Astron. & Astrophys.*, 418, pp. 989-1089 (2004)
13. E.H., Olsen, *Astron. & Astrophys Suppl Ser.*, 106, pp. 257-266 (1994)
14. M. Pettini, *Chemical Evolution from Zero to High Redshift*, in Proc. ESO Workshop edited by J. R. Walsh and M. R. Rosa (Berlin: Springer), 1999, p. 233
15. R. De Simone, X. Wu, S. Tremaine, *Mon. Not. Roy. Astron. Soc.*, 350, pp. 627-643 (2004)
16. B. A. Twarog, *Astrophys. Journ.,* 242, pp. 242-259 (1980)

Galactic Chemical Evolution with Heavy Metals Produced by the First Generation Stars

Yuhri Ishimaru*, Shinya Wanajo[†], Wako Aoki**, Sean G. Ryan[‡] and Nikos Prantzos[§]

Academic Support Center, Kogakuin University, 2665-1, Nakanomachi, Hachioji, Tokyo 192-0015, Japan
[†] *Research Center for the Early Universe, Graduate School of Science, University of Tokyo, 7-3-1 Hongo, Bunkyo-ku, Tokyo 113-8654, Japan*
**National Astronomical Observatory, 2-21-1 Osawa, Mitaka, Tokyo 181-8588, Japan*
[‡] *Centre for Astrophysics Research, STRI, University of Hertfordshire, College Lane, Hatfield, AL10 9AB, UK*
[§] *Institut d'Astrophysique de Paris, 98 bis, Boulevard Arago, 75014, Paris, France.*

Abstract. Observed large scatters in abundance ratios of neutron-capture elements relative to iron in metal-poor stars may suggest incomplete mixing of the interstellar medium at the beginning of the Galaxy. On the other hand, recent studies of metal-poor stars show considerable small dispersions for abundance ratios of C-Zn. We discuss whether such variations of scatters in abundance ratios can be explained by a consistent chemical evolution model. We also attempt to constrain the origins of *r*-process elements, comparing predictions by an inhomogeneous chemical evolution model with new observational results with *Subaru* HDS.

Keywords: nucleosynthesis, stars: abundances, Galaxy: evolution, Galaxy: halo
PACS: 26.30.+k; 97.10.Tk; 97.20.Tr; 97.20.Wt; 98.35.Bd; 98.35.Gi

INTRODUCTION

Metal-poor stars record enrichment history of the Galaxy at the early epoch. Abundance analysis of these stars reveals large star-to-star scatters in *r*-process elements [1, 2]. This may be interpreted as a result of incomplete mixing of the interstellar medium (ISM) at the beginning of the Galaxy. On the other hand, recent studies show considerable small dispersions for abundance ratios of C-Zn of extremely metal-poor stars [3]. However, if we assume well-mixed ISM to account for this observation, it causes difficulty to understand the large scatters in *r*-process elements. We discuss about a consistent scenario to explain distributions of abundance ratios of lighter to heavier elements, using an inhomogeneous chemical evolution model.

If metal-poor stars contain products of a single or a few supernovae (SNe), huge dispersions in abundances of *r*-process elements possibly imply that their yields are highly dependent on SN progenitor mass. However, no consensus about the origins of *r*-process elements has been achieved, although a few scenarios show some promise [4, 5]. In particular, observed enhancement of Sr comparing to Ba in metal-poor stars suggests the presence of the 'weak' *r*-process which produce mainly lighter *r*-elements. Comparing observations with predictions by inhomogeneous chemical evolution models, we discuss the enrichment of Sr, Pd, Eu, and Ba, and attempt to constrain the origin of *r*-process.

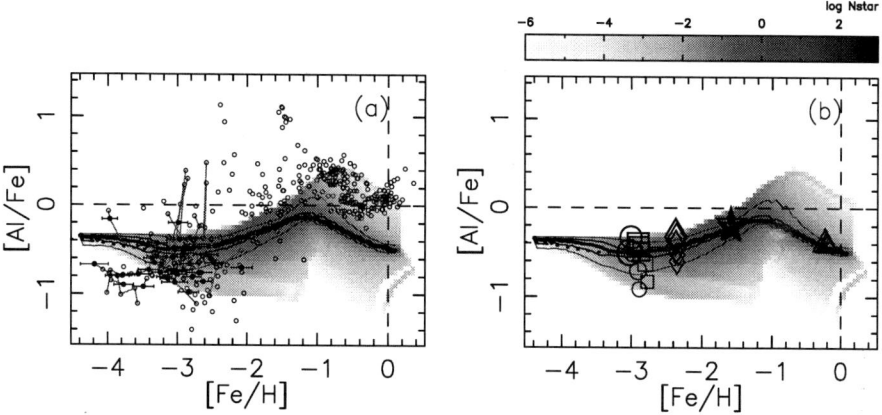

FIGURE 1. [Al/Fe] as a function of [Fe/H]. Gray-scale indicates distribution of stellar fraction predicted by a model using the data of supernova yield by Chieffi & Limongi (2004). The average stellar distributions are indicated by thick-solid lines with the 50% (solid lines) and 90% confidence intervals (thin-solid lines). (a) Comparison of the prediction with observational data of Cayrel et al. (2004) denoted by small filled circles and other data (small open circles). Stars formed through supernova from progenitors of zero metal (circles), $5 \times 10^{-5} Z_\odot$ (squares), $5 \times 10^{-3} Z_\odot$ (diamonds), $5 \times 10^{-2} Z_\odot$ (stars), and $0.3 Z_\odot$ (triangles) are indicated in (b) The size of symbols increase with progenitor masses from $13 M_\odot$ to $35 M_\odot$.

INHOMOGENEOUS CHEMICAL EVOLUTION OF O–ZN

In our inhomogeneous chemical evolution model, star formation is assumed to be induced through individual SNe, and chemical compositions of new stars are given by the mass average of the remnant of the SN which induce their formation and the surrounding ISM. For comparison, we take three sets of SN yields; Nomoto et al. (1997)[6], Woosley & Weaver (1995)[7], and Chieffi & Limongi (2004)[8]. In particular, the latter two yields take into account dependency of yields on stellar metallicity. Using this model with various SN yields, we calculate distributions of relative abundance ratios of [O–Zn/Fe] vs. [Fe/H], and discuss implications of observational scatters in abundance ratios of metal-poor stars.

Figure 1 shows an example of [Al/Fe] as a function of [Fe/H]. Although the nucleosynthesis of Al still contains significant uncertainties and the predicted distribution (gray-scale) is not completely consistent with observations (Fig. 1a), the observational data shows typical trend of metal-poor stars and is useful for our purpose. Extremely metal-poor stars of [Fe/H]< -3 show a mild correlation with metallicity, with very small scatters. Typically, the scatter is relatively large at [Fe/H]~ -3 but decrease again towards higher metallicity. Fig. 1b suggests that distribution of stars produced through SNe of zero-metal progenitors (circles) is concentrated at [Fe/H]~ -3 with large dispersions. It is because chemical compositions of these stars are less contaminated by products by other SNe and only reflect the yields of zero-metal stars, those for iron take similar values irrespective of progenitor mass, while abundance ratios relative to iron depends on stellar mass (cf. the caption of Fig. 1: the size of symbols indicates difference of stellar mass).

Chemical evolution model also shows that stars of [Fe/H] < −3 are also formed from gas containing products of zero-metal or extremely metal-poor stars but much more efficiently diluted by the ISM. Therefore, the mass dependency of yields of metal-poor stars is elongated by dilution in this region, and a trend of [X/Fe] vs. [Fe/H] with small scatter appears. On the other hand, we can also see from Fig. 1b that a trend with metallicity for stars of −3 <[Fe/H] < −1 come from dependency of SN yields on progenitor metallicity. The decrease of [Al/Fe] with metallicity at [Fe/H]> −1 is caused by significant production of iron by type Ia SNe.

In conclusion, it is basically possible to reproduce the observed variations in scatters and trends in various elements, although up to now, no SN models can completely explain the trends and scatters in chemical abundances of all elements. Therefore, observational data must give strong constraints on future works of SN models.

BIMODAL R-PROCESS

In our previous study, observed wide spread of Eu in metal-poor stars are shown to be well-reproduced by inhomogeneous enrichment scenario. In particular, sub-solar values of [Eu/Fe] in stars of [Fe/H]∼ −3 can restrict the site of r-process as SNe of low-mass end stars such as $8-10M_\odot$ [9]. Distribution of abundance ratio [Ba/Fe] also supports this result.

However, recent comprehensive spectroscopic analyses of extremely metal-poor stars have further shown that lighter neutron-capture elements ($Z < 56$) significantly deviate from the scaled solar r-process pattern that fits to the heavier elements [10, 11, 12]. In particular, relative abundance of Sr exhibits wide variations not shared by Ba[1, 2], suggesting that Sr at least does not come from a universal process. Although Sr is known to be produced also by weak s-process in massive stars, weak s-process yields at [Fe/H]∼ −3 cannot be enough high to exhibit the large enhancement of Sr, since weak s-process requires a secondary element ^{22}Ne as neutron source. The second origin of Sr in metal-poor stars should be in primary source by massive stars. Thus we suggest bimodal r-processes for Sr. While only specific SNe produce heavier elements, e.g., Ba and Eu, by main r-process, lighter elements such as Sr may be produced by residual SNe as 'weak' r-process.

Assuming the site of weak r-process as SNe of $20-30M_\odot$ stars, we examine scatters in [Sr/Ba] of metal-poor stars by an inhomogeneous chemical evolution model. As shown in fig. 2, this scenario can well explain the observational distribution of [Sr/Ba], when weak r-process produces ∼ 60% of Sr but only ∼ 1% of Ba in metal-poor stars [13].

ORIGIN OF PALLADIUM

Lighter r-process elements possibly have a distinct origin from heavier elements. However, the location of the boundary in atomic number (or in mass number) between the two r-processes has been unresolved. Intermediate mass elements between Sr and Ba must provide clues to understand the nucleosynthesis of weak r-process. Thus, we sug-

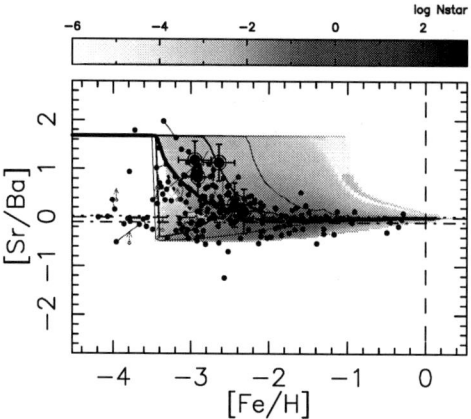

FIGURE 2. [Sr/Ba] as a function of [Fe/H]. Gray-scale indicates predicted distribution of stellar fraction, using the Fe yield by Nomoto et al. (1997). Solid lines are the same with Fig. 1. Weak *r*-process fraction for Sr and Ba are assumed as 60% and 1%, respectively. The observational data of this study are given by large circles, with other data (small circles).

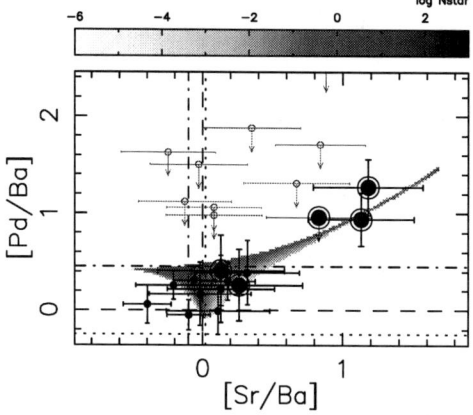

FIGURE 3. Same with Fig. 2 but for [Pd/Ba] vs. [Sr/Ba]. Weak *r*-process fraction for Pd is assumed to be 10%.

gest to use Pd, which is located between *typical* light (e.g., Sr, Y, and Zr) and heavy (e.g., Ba, Eu, and Th) *r*-process elements, as diagnostics of the two *r*-processes.

We estimated Pd abundances for 5 metal-poor stars, using *Subaru* HDS. Figure 3 show abundance ratios [Pd/Ba] as a function of [Sr/Ba]. By definition, [Sr/Ba] should increase with the fractional contribution of weak *r*-process to the stellar abundances. Thus, if Pd originates from weak *r* like Sr, i.e., its weak *r* fraction is ∼ 60%, [Pd/Ba] must show a correlation with the slope of unity to [Sr/Ba]. If Pd comes from main *r*-process like Ba, [Pd/Ba] must be constant irrespective of [Sr/Ba]. Our data show a mild correlation with the slope less than unity, suggesting that the weak *r*-process fraction for Pd takes intermediate value between those of Sr and Ba. This trend is well explained by a model, assuming the weak *r*-process fraction for Pd as $\sim 10\%$. Therefore, this result possibly implies that the weak *r*-process fraction is large at lighter elements, e.g., Sr, but decreases with atomic mass towards heavier elements such as Ba and Eu.

Recent observations result using *Subaru* shows that the abundance pattern of a metal-poor star HD 122563 is consistent with the above discussions: its abundance pattern gradually decreases with atomic number from Sr towards Yb[14]. This star may be the first evidence for 'weak' *r*-process nucleosynthesis.

CONCLUSIONS

It is shown that variations of scatters in observed trends of abudance ratios of C–Zn with metallicity can be understood by inhomogeneous chemical evolution scenario. The maximum dispersion at [Fe/H]~ -3 comes from stellar mass-dependencies of yields of zero-metal or extremely metal-poor stars. Stars formed from the highly diluted gas containing products of zero-metal stars cause the trend with small scatters shown in stars of [Fe/H]< -3.

Origins of *r*-process elements are constrained by comparing predictions from an inhomogeneous chemical evolution model with chemical compositions of extremely metal-poor stars obtained by *Subaru* HDS. Sub-solar values of [Eu/Fe] in three metal-poor stars imply the site of main *r*-process to be low-mass end of the SN mass range. Enhancements of Sr comparing to heavier elements suggest existence of two distinct sources for *r*-process. New data of Pd abundance suggest that 'weak' *r*-process yield is large at around Sr, but decreases with atomic mass towards heavier elements such as Ba and Eu.

REFERENCES

1. Ryan, S. G., Norris, J. E., & Beers, T. C. 1996, ApJ, 471, 254
2. Honda, S., Aoki, W., Kajino, T., et al., 2004, ApJ, 607, 474
3. Cayrel, R., et al., 2004, A&A, 416, 1117
4. Woosley, S. E., Wilson, J. R., Mathews, G. J., et al. 1994, ApJ, 433, 229
5. Wanajo, S., Tamamura, M., Itoh, N., et al. 2003, ApJ, 593, 968
6. Nomoto, K., Hashimoto, M., Tsujimoto, T., et al. 1997, Nucl. Phys. A, 616, 79
7. Woosley, S. E., Weaver, T. A., 1995, ApJS, 101, 181
8. Chieffi, A., Limongi, M., 2004, ApJ, 608, 405
9. Ishimaru, Y., Wanajo, S., Aoki, W., & Ryan, S. G., 2004, ApJ, 600, L47
10. Hill, V., Plez, B., Cayrel, R., et al. 2002, A&A, 387, 560
11. Cowan, J. J., Sneden, C., Burles, S.,et al. 2002, ApJ, 572, 861
12. Sneden, C., Cowan, J. J., Lawler, J. E., et al. 2003, ApJ, 591, 936
13. Ishimaru, Y. & Wanajo, S. 2000, in *First Stars*, ed. A. Weiss, T. Abel,& V. Hill (Springer), p.189
14. Honda, S., Aoki, W., Ishimaru, Y., Wanajo, S., Ryan, S. G., 2006, ApJ in press.

1.5 EXPLOSIVE NUCLEOSYNTHESIS in SUPERNOVAE

Nucleosynthesis in Core Collapse Supernovae

Marco Limongi[*] and Alessandro Chieffi[¶]

[*]*INAF – Osservatorio Astronomico di Roma, Via Frascati 33, I-00040, Monteporzio Catone, Roma, Italy. Email: marco@oa-roma.inaf.it*
[¶]*INAF – Istituto di Astrofisica Spaziale e Fisica Cosmica, Via Fosso del Cavaliere, I-00133, Roma, Ital. Email: achieffi@rm.iasf.cnr.it*

Abstract. We present the basic properties of the yields of our latest set of presupernova evolution and explosive nucleosynthesis of massive stars in the range between 11 and 120 $M_\odot$ having solar and zero metallicity.

Keywords: nucleosynthesis, abundances – stars:evolution – supernovae: general
PACS: 97.10.Cv, 97.10.Tk, 97.20.Wt, 97.60.-s, 97.60.Bw

INTRODUCTION

Massive stars, those massive enough to explode as supernovae, play a key role in many fields of astrophysics. They are crucial in determining the evolution of the galaxies because: 1) they light up regions of stellar birth and hence induce star formation; 2) they are responsible for the production of most of the elements (among which those necessary to life); 3) they may induce mixing of the interstellar medium through stellar winds and radiation; 4) they leave, as remnant, exotic objects like neutron stars and black holes. Massive Population III Stars could play an important role in Cosmology because they contribute to 1) the reionization of the universe ad $z>5$, 2) the production of massive black holes that could have been the progenitors of active galactic nuclei, 3) the pregalactic metal enrichment. Finally Massive Stars play an important role in the field of γ-ray astrophysics because 1) they are responsible for the production of some long-lived γ-ray emitter nuclei as ^{26}Al, ^{56}Co, ^{57}Co, ^{44}Ti and ^{60}Fe, and 2) they are likely connected to the Gamma Ray Bursts. As a consequence, the understanding of these stars, i.e., their presupernova evolution, their explosion as supernovae and, especially, their nucleosynthesis, is crucial for the interpretation of many astrophysical objects.

In this paper we present the basic properties of the yields of our latest presupernova evolution and explosion of massive stars, in the range between 11 and 120 $M_\odot$ having an initial solar and zero metallicity. A more detailed discussion of these models will be presented in a forthcoming paper.

HYDROSTATIC AND HYDRODYNAMIC CODES

The results presented in this paper are based on a new set of presupernova models and explosions of solar and zero metallicity stars in the mass range between 11 and 120 $M_\odot$, covering therefore the full range of masses that are expected to give rise to Type II/Ib/Ic supernovae as well as those contributing to the Wolf-Rayet populations. All these models, that will be presented shortly in a forthcoming paper [1], have been computed by means of the more recent version (5.050218) of the FRANEC (Frascati Raphson Newton Evolutionary Code), whose main differences with the respect to the previous versions [2] are the following: 1) the time dependent mixing is taken into account by means of a classical diffusion equation [3]; 2) the equation for the convecitive mixing and the ones describing the chemical evolution of the matter due to the nuclear burning are coupled together and solved simultaneously; 3) a moderate amount of overshooting (0.2 H_p) is assumed during core H burning; 4) mass loss is taken into account following the prescription of [4,5] for the blue supergiant phase (T_{eff}>12000 K), [6] for the red supergiant phase (T_{eff}<12000 K) and [7] during the Wolf-Rayet phase; 5) updated cross sections have been adopted whenever possible (see electronic references table in [1]).

The explosion is simulated by means of a piston of initial velocity v_0 located at ~1 $M_\odot$ in the presupernova model and moving along a balistic trajectory under the gravitational field of the compact remnant. The developement and the evolution of the shock wave that forms, is followed by means of a 1D PPM lagrangian hydro code [8]. The explosive nucleosynthesis is computed by using the same nuclear network adopted in the hydrostatic evolution. For each model, several hydro calculations have been performed by iterating on v_0 in order to obtain a given amount of ^{56}Ni ejected and a corresponding final kinetic energy at the infinity. Since, at present, there is no self consistent model for the explosion of a core collapse supernova, the relation between the initial mass and the remnant mass is essentially unknown. However, observations seem to indicate that stars with mass $M_{MS} \leq 25$ $M_\odot$ form neutron stars producing ~ 0.1 $M_\odot$ of ^{56}Ni while stars with mass $M_{MS} \geq 25$ $M_\odot$ form black hole producing either ~ 10^{-3} $M_\odot$ or ~ 0.1 $M_\odot$ of ^{56}Ni depending on many factors among which rotation, stellar wind, magnetic fields, metallicity and binarity [9]. Therefore, guided by the observations, we choose two initial mass-remnant mass relations: the first one (*trend*) in which we assume that stars with mass $M_{MS} \leq 25$ $M_\odot$ produce 0.1 $M_\odot$ of ^{56}Ni while stars more massive than this limit produce 10^{-3} $M_\odot$ of ^{56}Ni, the second one (*flat*) in which we assume that all the core collapse supernovae are assumed to eject 0.1 $M_\odot$ of ^{56}Ni, independently of the initial mass.

THE SOLAR METALLICITY MODELS

The first set of models discussed in the present paper is the one computed with initial solar metallicity [10] and including masses in the range between 11 and 120 $M_\odot$, namely, 11, 12, 13, 14, 15, 16, 17, 20, 25, 30, 35, 40, 60, 80 and 120 $M_\odot$. These models will be presented in more detail in a forthcoming paper. The first result worth to be mentioned is reported in Fig. 1, where the element production factors (PFs) averaged over a Salpeter Initial Mass Function ($dm/dn=km^{-2.35}$) are shown for the two chosen initial mass-remnant mass relations.

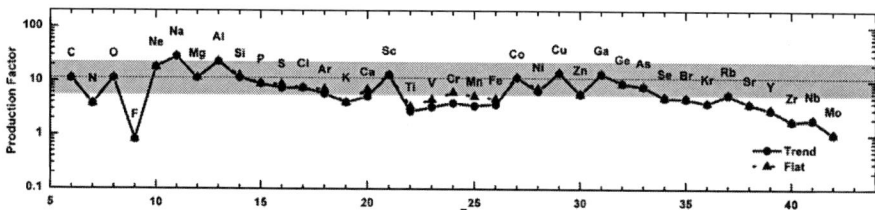

FIGURE 1. Element production factors averaged over a Salpeter Initial Mass Function for solar metallicty stars in the mass range 11- 120 $M_\odot$: the filled circles connected by the solid line refer to the *trend* case (see text); the filled triangles connected by the dotted line refer to the *flat* case. The horizontal dashed line refers to the production factor of O. All the nuclei whose production factors falls within a factor of 2 of the oxygen one (shaded area) are produced in scaled solar proportions.

In the reasonable assumption that the average metallicity grows continuously and slowly compared to the evolutionary timescales of the stars contributing to the global enrichment of the gas, it would be desirable that a generation of solar metallicity stars provides yields in roughly solar proportions or, in other words, that the PFs of the various isotopes remain essentially flat. Since oxygen is produced mainly by core collapse supernovae and it is also the most abundant element produced by these stars we use its production factor as the one that better represents the overall increase of the average metallicity and to verify whether or not the other nuclei follow its behavior. In particular, we assume that all the elements whose production factor falls within a factor of 2, taken as a suitable warning threshold, of the oxygen one are compatible with a flat distribution relative to oxygen, while those outside this compatibility range may potentially constitute a problem. The things worth noting in Fig. 1 are the following: (1) the only elements that vary substantially between the cases *trend* and *flat* are the iron peak ones, i.e., Ti, V, Cr, Mn and Fe; (2) the majority of the elements are produced in scaled solar proportions relative to O: the exceptions are N, F, K, the iron peak elements (Ti, V, Cr, Mn, Fe) and s-process elements heavier than Br. N and s-process elements above Br are under abundant, as expected, because these elements are mainly produced by intermediate mass stars that are not included in the mass interval analyzed in this paper. The underproduction of both F and K is mainly due to the lack, in these calculations, of the neutrino induced reactions [11]. The iron peak elements significantly depend on the adopted mass cut, i.e., the initial mass-remnant mass relation. Indeed, a changing from the *trend* to the *flat* case lead to an increase of the yields of these elements, pushing them to a closer scaled solar distribution and hence leaving less room for the SNIa contribution because of the larger amount of Fe (^{56}Ni) produced.

Figure 2 shows the element production factors averaged over a Salpeter IMF in the *trend* case, with two different upper mass limits, i.e., M_{top}=35 $M_\odot$ and M_{top}=120 $M_\odot$. Interestingly, Fig. 2 shows that the PFs of all the elements from N to Ca are almost independent on the upper mass limit, i.e., the PFs of all these elements are quite similar in all the models with initial masses in the range 35-120 $M_\odot$. The PFs of the iron peak elements increase by reducing M_{top}, as expected, because no iron is produced in stars more massive than 25 $M_\odot$ in the *trend* case.

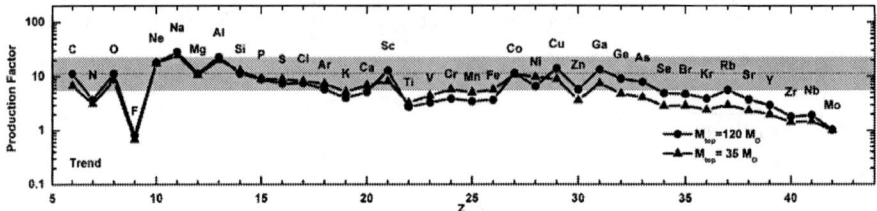

FIGURE 2. Element production factors averaged over a Salpeter Initial Mass Function for solar metallicty models in the *trend* case with two choices of the upper mass limit M_{top} of the IMF: the filled circles connected by the solid line refer to the (standard) case in which M_{top}=120 $M_\odot$; the filled triangles connected by the dotted line refer to the case in which M_{top}=35 $M_\odot$. The horizontal dashed line refers to the production factor of O in the standard case.

The PFs of C and of the s-process elements decrease by reducing M_{top} as a consequence of the substantial production of these elements in stars more massive than 35 $M_\odot$, hence the inclusion of stars in the mass range 35-120 $M_\odot$ changes the relative scaling of C and of the s-process relative to all the other elements (in particular O).

THE ZERO METALLICITY MODELS

The set of zero metallicity models includes masses in the range between 13 and 80 $M_\odot$, namely, 13, 15, 20, 25, 30, 35, 50 and 80 $M_\odot$. At variance with the solar metallicity case, these models have been computed without mass loss and without overshooting and will be presented in more details in a forthcoming paper. The first interesting result to be mentioned is reported in Fig. 3 that shows the element production factors for all the computed models in the *trend* case. A large spread of the PFs of all the elements with Z<14 is clearly shown in Fig. 3. In particular a large primary N production is obtained in stars in the mass range between 25 and 35 $M_\odot$; such a primary production is connected, in these stars, with the ingestion of protons by the He convective shell that penetrates into the overlying H rich layer. Indeed, the protons ingested by the He convective shell activate the CNO cycle at high temperatures, typical of He burning, leading to a substantial production of N. Such a partial mixing between the He convective shell and the overlying H rich mantle is a rather common feature in stellar models of initial zero metallicity [11, 12, 13] because of the low entropy barrier present at the H-He interface in these stars. At variance with the lightest elements, the intermediate mass elements ($14 \leq Z \leq 21$) show PFs almost independent on the initial mass.

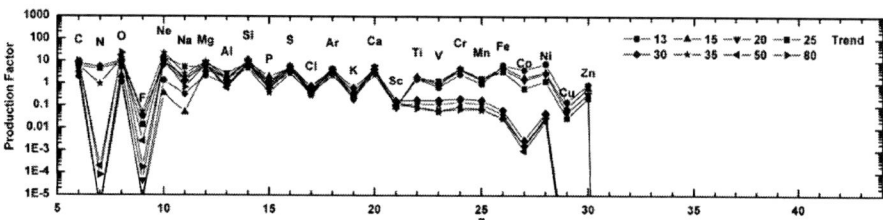

FIGURE 3. Production factors of all the elements obtained in the *trend* case for zero metallicity models: the symbols refer to the eight computed models as shown in the legend in the upper right corner.

The PFs of the iron peak elements, as expected, are larger for models producing more iron, i.e., stars below 25 $M_\odot$. As a consequence, such a behavior is strongly dependent on the initial mass-remnant mass relation. A last feature worth to be mentioned is that there is a cutoff in the PFs at the level of Zn, i.e., no elements heavier than Zn are produced in zero metallicity massive stars. This has the obvious consequence that the observed abundances of elements above Zn in very metal poor stars must be attributed to stars (or, in general, to processes) outside the range presently discussed [15].

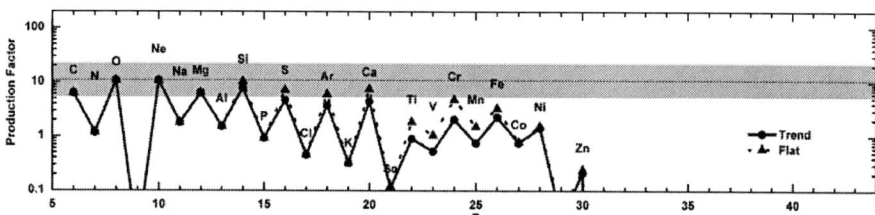

FIGURE 4. Element production factors averaged over a Salpeter Initial Mass Function for zero metallicity models: the filled circles connected by the solid line refer to the *trend* case (see text); the filled triangles connected by the dotted line refer to the *flat* case. The horizontal dashed line refers to the production factor of O. All the nuclei whose production factors falls within a factor of 2 of the oxygen one (shaded area) are produced in scaled solar proportions.

Although the primordial Initial Mass Function is still presently unknown, it is interesting to integrate the element yields of all the models over a Salpeter IMF. Figure 4 shows the element PFs averaged over a Salpeter IMF in the *trend* and *flat* cases. The first thing worth to be noted is that, as for the solar metallicity models, the PFs of all the elements are essentially independent on the initial mass-remnant mass relation, except those of the iron peak elements. The second interesting feature is the well known *odd-even effect*, i.e., the large difference between the PFs of the odd (from

N to Sc) and the even nuclei (from C to Ca). In particular, the elements C, Ne, Mg and Si are produced in almost scaled solar proportions relative to O while S, Ar and Ca are deficient by a factor of ~ 2. The odd elements, from N to Sc, are underproduced by a factor of 10 to 100 relative to O. The iron peak elements are deficient by a factor of about 10, although their production factors depend on the initial mass-remnant mass relation, i.e., they increase by changing from the *trend* to the *flat* case. As already shown in Fig. 3, no production of elements beyond Zn is obtained in these models.

ACKNOWLEDGMENTS

Marco Limongi warmly thanks Ken'ichi Nomoto and the Organizing Committee for their financial support necessary to attend the OMEG05 meeting. Marco Limongi and Tatiana Grilli also thank Ken'ichi Nomoto, Keiichi Maeda, Nozomu Tominaga and Masaomi Tanaka for their very kind hospitality and support during their visit in Tokyo.

REFERENCES

1. M. Limongi and A. Chieffi, *in preparation*
2. M. Limongi and A. Chieffi, *ApJ* **592**, 404 (2003).
3. T.A. Weaver, G.B. Zimmerman and S.E. Woosley, *ApJ* **225**, 1021 (1978).
4. J.S. Vink, A. de Koter and H.J.G.L.M. Lamers, *A&A* **362**, 295 (2000).
5. J.S. Vink, A. de Koter and H.J.G.L.M. Lamers, *A&A* **369**, 574 (2001).
6. C. de Jager, H. Nieuwenhuijzen and K.A. van der Hucht, *A&AS* **72**, 529 (1988).
7. T. Nugis and H.J.G.L.M. Lamers, *A&A* **360**, 227 (2000).
8. P. Colella and P.R. Woodward, *Journal of Computational Physics* **54**, 174 (1984).
9. K. Nomoto, N. Tominaga, H. Umeda, K. Maeda, T. Ohkubo, J. Deng and P.A. Mazzali, in "Hypernovae, Black-Hole forming Supernovae, and First Stars", in *The Fate of The Most Massive Stars*, edited by R.M. Humphreys and K. Z. Stanek, ASP Conference Series 332, 2005, pp. 374-384
10. E. Andres and N. Grevesse, *Geochim. Cosmochim. Acta* **53**, 197 (1989).
11. S.E. Woosley and T.A. Weaver, *ApJS* **101**, 181 (1995).
12. A. Chieffi, I. Dominguez, M. Limongi and O. Straniero, *ApJ* **554**, 1159 (2001).
13. M.Y. Fujimoto, I. Jr. Iben and D. Hollowell, *ApJ* **349**, 580 (1990).
14. M. Limongi, A. Chieffi and P. Bonifacio, *ApJL* **594**, L123 (2003)
15. P.S. Barklem, N. Christlieb, T.C. Beers, V. Hill, M.S. Bessell, J. Holmberg, B. Marsteller, S. Rossi, Zickgraf, S.J. and D. Reimers, *A&A* **439**, 129 (2004)

Light Element Production in Type Ic Supernovae

Toshikazu Shigeyama[*], Ko Nakamura[*,†], Shinya Wanajo[*] and Susumu Inoue[**]

[*]*Research Center for the Early Universe, Graduate School of Science, University of Tokyo, Bunkyo-ku, Tokyo, 113-0033, Japan*
[†]*Department of Astronomy, Graduate School of Science, University of Tokyo, Bunkyo-ku, Tokyo, 113-0033, Japan*
[**]*National Astronomical Observatory of Japan, Mitaka, Tokyo 181-8588, Japan*

Abstract. We discuss production of light elements (Li, Be) in energetic type Ic supernovae (SNe Ic) and how newly synthesized light elements are transferred to stars of the next generation. We have pointed out that spallation reactions involving N and He become important in an explosion of a rotating metal-poor star if the progenitor still keeps a fraction of the He layer at SN explosion. In this kind of explosions, ^{6}Li is produced by a fusion reaction He+He. Simultaneously, most of Be is produced through the spallation reaction He+N. This scenario suggests that there must be intrinsic scattering in abundance of Li isotopic ratios among metal-poor stars because the mass range of stars concerned here is limited to $\gtrsim 40\,M_\odot$ and light element yields should be sensitive to the degree of rotation.

Keywords: Supernovae, Light elements, Population II stars
PACS: 97.20.Tr, 97.60.Bw, 26.30.+k, 26.40.+r

INTRODUCTION

Supernovae showing neither H lines or He lines are classified as Type Ic (SN Ic). A SN Ic is thought to be an explosion of a massive star that lost its He layers as well as H-rich envelope before the gravitational core collapse. The size of the progenitor star at explosion is comparable to or less than the solar radius. Thus higher pressure at the shock breakout is expected to accelerate the surface layers to higher energies than the other types of core collapse SNe in which the progenitors have more extended envelopes. Light element production in SNe Ic was first discussed using a hydrodynamical model for a SN Ic, SN 1994I in M51 [1]. This SN is thought to be an explosion of a C+O star that originating from a $\sim 13-15 M_\odot$ main sequence star by losing its H-rich and He envelopes [2]. The original hydrodynamical models in these studies do not have sufficiently accelerated ejecta with energies above the threshold for spallation reactions probably due to coarse zoning in the surface layers and relatively low explosion energies.

The discovery of an energetic SN Ic, SN 1998bw associated with a γ-ray burst GRB 980425 in 1998 has promoted further studies on this acceleration mechanism [3, 4]. In particular, the energy distribution of the surface layers was formulated [5] as a function of parameters to characterize the explosion as

$$\frac{M(>\varepsilon)}{M_{\rm ej}} = 2.2 \times 10^{-4} \left(\frac{E_{51}}{M_{\rm ej}}\right)^{3.67} \left(\frac{\varepsilon}{10\,{\rm MeV}}\right)^{-3.67}, \quad (1)$$

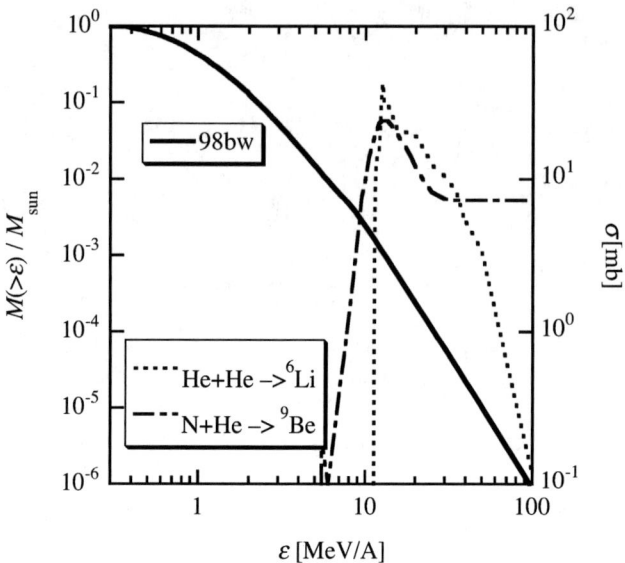

FIGURE 1. The mass of ejecta with the energy per nucleon greater than ε (Solid curve) as a function of ε in a SN 1998bw model [7]. Together shown are the cross sections for He+He→^{6}Li (dashed curve) and N+He→^{9}Be (dash-dotted curve) reactions.

based on a self-similar solution for shock breakout [6] and the corresponding numerical calculations [3]. Here $M(>\varepsilon)$ denotes the mass with energy per nucleon greater than ε, M_{ej} the mass of ejecta in units of the solar mass, and E_{51} is the explosion energy in units of 10^{51} ergs. Using this formula, they showed that a considerable amount of C+O layers have energies above the threshold for light element production and estimated yields for light elements from SNe Ic including such an energetic event as SN 1998bw. However their yields are turned out to be overestimated partly due to very limited applicability of the above energy distribution [7].

Nakamura & Shigeyama (2004) calculated an explosion of a C+O star with the ejecta mass of $15 M_\odot$ and explosion energy of 3×10^{52} ergs determined for SN 1998bw [8] by using a special-relativistic hydrodynamical code. They obtained the energy distribution of the ejecta shown in Figure 1. A substantial amount of ejecta have energies exceeding the threshold for light element productions through C, O+H, He→Li, Be, B. They also calculated explosions for well observed SNe Ic with $E_{51} \sim 1$ and showed that some fraction of the surface layer is accelerated to enough energies for light element production in contrast to the previous work [1].

CIRCUMSTELLAR MATTER

These previous studies on light element production from SNe Ic were concerned with interactions of the accelerated ejecta with the interstellar medium and apparently ignored interactions with the circumstellar matter (CSM) that the progenitor has supplied via wind before explosion.

Energy Loss

The CSM around a progenitor of a SN Ic is so thick that ejecta composed of He with the energy ~ 10 MeV/A will lose most of its energy inside the CSM. If the energy is lost by ionizing He and/or by Coulomb scattering off electrons in the CSM depending on the ionization states. If the CSM is fully ionized the ratio of the stopping range R due to Coulomb scattering to the column depth σ of the CSM is given by the formula

$$\frac{R}{\sigma} \sim 0.072 \left(\frac{\varepsilon}{10\,\text{MeV/A}}\right)^2 \left(\frac{\dot{M}}{10^{-6} M_\odot/\text{yr}}\right)^{-1} \left(\frac{v_w}{1,000\,\text{km s}^{-1}}\right) \left(\frac{r}{R_\odot}\right), \quad (2)$$

where the CSM is a steady stellar wind with the velocity v_w and the mass loss rate $\dot{M}$ starting at the distance R from the center of the star. The ejecta with $\varepsilon = 10$ MeV/A is expanding at $v \sim 42,000$ km s^{-1}. This formula clearly shows that the ejecta responsible for light element production lose their energies inside the CSM. Even when the CSM is composed of neutral matter the amount of energy lost through electron scattering is reduced to $\sim 60\%$ of the previous case. Thus the ejecta loses most of the energies inside the CSM irrespective of the ionization states. Therefore it is expected that light elements are enhanced in a region with the scale of 100 pc at most after one SN Ic event and inherited by stars of the next generation while the previous studies assumed that light elements produced by an SN Ic are distributed over 10 kpc scales. This local enhancement leads to efficient enrichment of light elements in metal-poor stars that inherited the abundance patterns of SNe Ic ejecta [9] and it also results in various abundance ratios of light elements to heavy elements in metal-poor stars.

Chemical Compositions

The CSM interacting with the accelerated ejecta is mainly composed of He because the ejecta lose most of their energies in the inner part of the CSM where the density is higher than the outer layers. Thus reactions with He dominate production of light elements.

EXTREMELY METAL-POOR MASSIVE STARS

An SN Ic is thought to be an explosion of a massive star that has lost its He layers. A star needs to have heavy elements in the outer layer to undergo intense mass loss. Therefore SNe Ic may not be able to contribute to nucleosynthesis in early evolutionary stages of a galaxy, where the matter is deficient in heavy elements.

Recent theoretical calculations [10, 11], however, have shown that some extremely metal-poor massive stars lose their He layers in addition to the H-rich envelopes. Thus it is likely that SNe Ic occurred even in the early evolutionary stages of the Milky Way.

Rotation-induced Mixing

Meynet et al (2005) calculated evolution of rotating massive stars and showed that rotation induces mixing of heavy element products (C) of He burning into the H-rich envelope and protons into He layers. As a result of the mixing, the H-rich envelope is enriched with C at the level of [C/H]~ -1 and N is more enhanced due to the CNO cycle.

They have shown that incorporation of a metallicity dependent mass loss rate with enhanced metallicities in the outer layers causes intense mass loss even in some extremely metal-poor stars with $Z = 10^{-8}$. For example, a star with initial mass of $60 M_\odot$ loses its He layer and will explode as an SN Ic. If a star still keeps a fraction of its He layer at explosion, then N as well as He will be accelerated to energies above the threshold for light element production.

LOCAL ENHANCEMENT OF LIGHT ELEMENTS

From the above consideration, SNe Ic in the early evolutionary stages of a galaxy could enrich the CSMs with light elements. Such CSMs were swept up by blast waves of SNe Ic and mixed with the interstellar matter also swept up by the same blast waves.

Light Elements in Population II Stars

According to the supernova-induced star formation scenario [9], stars were formed from the swept up matter. Thus in a metal-deficient environment, the abundance patterns of these stars are determined by individual SNe Ic and not affected by other preceding SNe. As was pointed out in the preceding section, the important reaction for the production of ^{6}Li in Population II stars is

$$\text{He} + \text{He} \rightarrow \, ^6\text{Li}, \tag{3}$$

in place of

$$\text{C, O} + \text{He, H} \rightarrow \, ^6\text{Li, Be, B}. \tag{4}$$

This fusion reaction has a threshold at $\varepsilon \sim 6$ MeV/A and a high cross section of ~ 40 mb at the peak (dotted curve in Fig. 1).

For ^{9}Be, the spallation reaction

$$\text{N} + \text{He} \rightarrow \, ^9\text{Be} \tag{5}$$

dominates the production (see dash-dotted curve in Fig. 1). This reaction has a threshold at ~ 6 MeV/A and the peak value is ~ 24 mb whereas the reaction O+He→^{9}Be has somewhat higher threshold at ~ 8 MeV/A and a lower peak value of ~ 8 mb.

Recent observations of ^{6}Li in extremely metal-poor stars [12] revealed that some stars exhibit high ^{6}Li/H ratios that cannot be reconciled with the bigbang nucleosynthesis. Thus the observed ^{6}Li needs to be produced by other processes. SNe Ic described above are a potential candidate.

At the metal-poor end, the star LP 815-43 with [Fe/H]$= -2.74$ and [O/H]$= -2.2$ is observed to have ^{6}Li/H$\sim 7.9 \times 10^{-14}$ [12]. At the same time, ^{9}Be is enhanced in this star as compared with other stars with similar metallicities [13, 14]. Furthermore, an observation [15] has suggested that this star is N rich. There are a few more stars with ^{6}Li detections that show Be and N enhancement. As an origin of these elements, explosions of metal-poor massive stars that have lost their H-rich envelope and left a fraction of their He layers with enhanced N are likely to match observed abundance features at first sight.

Yields of these elements from the process discussed here are calculated and compared with recent observations for extremely metal-poor stars in Nakamura et al. (2005). They concluded that the explosion of a star with mass of $40 M_\odot$ can account for the observed ^{6}Li/O and Be/O ratios if it keeps its He layer with mass of $\sim 0.01 - 0.1 M_\odot$ and the mass fraction of N is ~ 0.01.

The metallicity of stars formed from the matter swept up by this SN Ic can be estimated as follows. The mass M_O of O in the ejecta is $M_O \sim 10 M_\odot$ [8, 7] and the mass M_H of H in the swept up matter $M_H \sim 1.4 \times 10^6 M_\odot$ [9]. From these values, [O/H]~ -2.7 is obtained. Because of the uncertainty of the mass of Fe ejected from any core collapse SNe, we can not make a reliable estimate for [Fe/H]. This [O/H] is marginally consistent with the observed value of [O/H] for the star LP 815-43 given the inherent dispersion of metallicities in the swept up matter [17]. Furthermore, a significantly aspherical explosion is expected from a rotating massive star considered here while the adopted explosion model is spherically symmetric. Light element production in aspherical SNe Ic and the subsequent evolution as a SN remnant needs to be explored.

ACKNOWLEDGMENTS

This work has been partially supported by the grant in aid (16540213) of the Ministry of Education, Science, Culture, and Sports in Japan.

REFERENCES

1. Fields, B. D., Cassé, M., Vangioni-Flam, E., & Nomoto, K. 1996, ApJ, 462, 276
2. Iwamoto, K., Nomoto, K., Höflich, P., Yamaoka, H., Kumagai, S., & Shigeyama, T. 1994, ApJ, 437, L115
3. Matzner, C. D., & McKee, C. F. 1999, ApJ, 510, 379
4. Tan, J. C., Matzner, C. D., & McKee, C. F. 2001, ApJ, 551, 946
5. Fields, B. D., Daigne, F., Cassé, M., & Vangioni-Flam, E. 2002, ApJ, 581, 389
6. Sakurai, A. 1960, Commun.Pure Appl.Math.,13,353
7. Nakamura, K. & Shigeyama, T. 2004, ApJ, 610, 888

8. Iwamoto, K., et al. 1998, Nature, 395, 672
9. Shigeyama, T., & Tsujimoto, T. 1998, ApJ, 507, L135
10. Meynet, G., & Maeder, A. 2002, A & Ap, 390, 561
11. Meynet, G., Ekstrom, S., & Maeder, A. 2005, preprint (astro-ph/0510560)
12. Asplund, M., Lambert, D., L., Nissen, P. E., Primas, F. & Smith, V. V. 2005, astro-ph/0510636
13. Primas, F., Molaro, P., Bonifacio, P., & Hill, V. 2000a, A & Ap, 362, 666
14. Primas, F., Asplund, M., Nissen, P. E., & Hill, V. 2000b, A & Ap, 364, L42
15. Israelian, G., Ecuvillon, A., Rebolo, R., García-López, R., Bonifacio, P., & Molaro, P. 2004, A & Ap, 421, 649
16. Nakamura, K., Inoue, S., Wanajo, S., & Shigeyama, T. 2005, this proceedings
17. Nakasato, N., & Shigeyama, T. 2000, ApJ, 541, L59

Nucleosynthesis and Emission Processes in Aspherical Supernovae

Keiichi Maeda

Department of Earth Science and Astronomy, Graduate School of Arts and Science, University of Tokyo, Meguro-ku, Tokyo 153-8902, Japan: maeda@esa.c.u-tokyo.ac.jp

Abstract. Features in nucleosynthesis and emission processes in jet-like aspherical hypernova explosions are presented. The aspherical model yields large (Co, Zn)/Fe and small (Mn, Cr)/Fe as are consistent with abundance patterns in metal-poor halo stars, indicating important contribution of hypernovae in the early Galactic chemical evolution. The same model also yields large amount of ^{44}Ti. As for emission features, this model is found to reproduce successfully the optical light curve and spectra of hypernova SN 1998bw. The viewing angle is close to the polar direction. The same model explains a peculiar [OI] 6300Å profile observed in SN 2003jd, only if the viewing angle is different from that for SN 1998bw. These analyses support the validity of the aspherical models, therefore the use of the models as a reference model for hypernova nucleosynthesis. In addition, theoretical prediction is presented for high energy emissions from the decays of ^{56}Ni and ^{56}Co.

Keywords: supernovae – radiative transfer – nuclear reactions, nucleosynthesis, abundances
PACS: 26.30.+k, 26.50.+x, 97.60.Bw

INTRODUCTION

Lately, much emphasis has been placed on a possible role of asymmetry in core-collapse supernovae, motivating a number of multi-dimensional explosion simulations (e.g., see [1, 2] for recent reviews). However, there have been surprisingly few studies on nucleosynthesis and emission processes in asymmetric supernovae, despite the importance as a tool to test the validity of explosion models. Especially interesting in this aspect is a hypernova, a very energetic supernova with the explosion energy $E_{51} \equiv E/10^{51}$ ergs $\gtrsim 10$ [3]. A prototypical hypernova SN 1998bw was discovered in association with a Gamma-Ray Burst (GRB980425) [4]. Two other Gamma-Ray Bursts are reported to be associated with hypernovae similar to SN 1998bw (see [3] for a review). In this paper, we present results from our 2D/3D computations of aspherical (jet-like) supernovae, especially focusing on hypernovae. The paper includes (1) nucleosynthetic features, comparing them with abundance patterns of metal-poor stars, (2) optical emission features, comparing them with SNe 1998bw and 2003jd, and (3) high energy emission features.

NUCLEOSYNTHESIS IN ASPHERICAL SUPERNOVAE

If supernova explosions are not spherically symmetric, the issue of nucleosynthesis should be revisited and reconsidered. In this paper, we focus on nucleosynthesis in bipolar (jet-like) explosions. Nucleosynthesis in bipolar supernova explosions was computed by [5, 6, 7, 8]. In [5, 6], instantaneous aspherical energy injection is assumed (i.e., as-

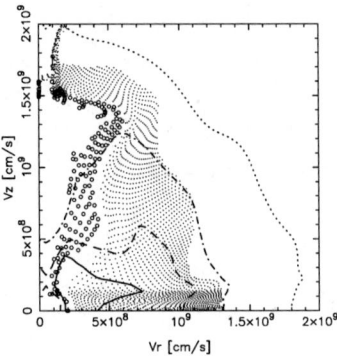

FIGURE 1. Distribution of ^{56}Ni (which decays into Fe: points) and ^{16}O (dots) in 2D velocity space. Note that the ejecta is already at a homologous expansion phase so that the radius of each point is proportional to the radial velocity. Also shown is the contours of densities. From [6].

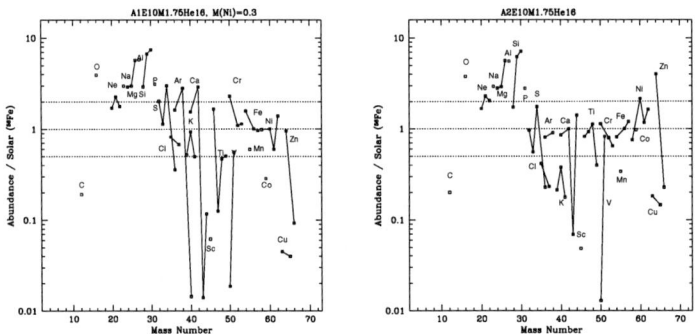

FIGURE 2. Isotopic yields of spherical (left) and aspherical (right) explosions [8]. In both cases, the progenitor is a $16M_\odot$ He core of a $40M_\odot$ star, and the explosion energy is $E_{51} \equiv E/10^{51}$ erg $= 10$. The mass of ^{56}Ni is $0.3M_\odot$ (spherical) and $0.24M_\odot$ (aspherical), and these values give [O/Fe] ~ 0.5 being consistent with observed in extremely metal poor stars (e.g., [11]).

suming that the energy injection timescale is much shorter than the time scale of the jet propagation), while in [7] the energy input is continuous with the time scale comparable to the jet propagation time scale (for a few seconds). In [8] the time scale of the energy injection is varied as a parameter and its effect on nucleosynthesis was investigated.

The different time scale of the energy injection gives quantitatively different behavior in nucleosynthesis. Noticeably, it is found that the ejected mass of ^{56}Ni and other iron peak elements sensitively depend on the time scale of the explosion. For a given total energy input, the longer time scale leads to the smaller amount of these isotopes and elements. See [7, 8, 9, 10] for detailed discussion. Qualitative behavior, on the other hand, is similar, especially as long as ratios of different isotopes are considered. We mainly discuss the results of instantaneous energy injection models in what follows.

Figure 1 shows spatial distribution of nucleosynthesis products in a jet-like explosion

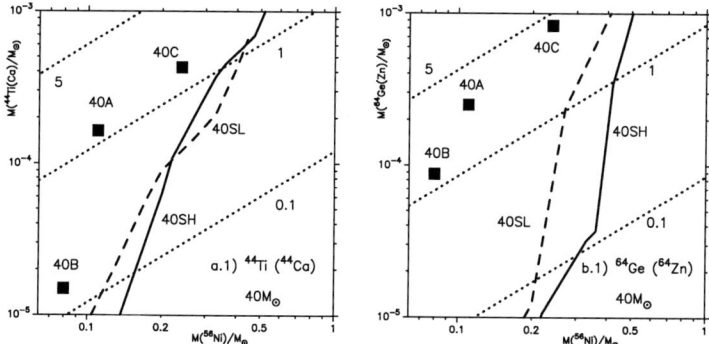

FIGURE 3. Ejected masses of ^{44}Ti (left) and ^{64}Ge (right) as a function of $M(^{56}\text{Ni})$ for some bipolar models (40A, B, and C: filled squares) and for spherical models (40SH, SL: lines). The dotted lines show the ratio (^{44}Ca, ^{64}Zn)/^{56}Fe relative to the solar value. The models are from [7].

model. In this model, the energy injected by the central collapsing core is assumed to be larger toward the z-axis than in the r-plane, with the initial velocity 8 times larger in the former. ^{56}Ni is synthesized preferentially along the z-axis, where the shock is stronger, while a lot of unburned materials, mainly O, are left at low velocity in the r-plane.

The aspherical models also yield the overall abundance patterns different from spherical models (Fig. 2). [(Zn, Co)/Fe] ([X/Y] $\equiv \log(X/Y) - \log(X/Y)_\odot$) are enhanced, while [(Mn, Cr)/Fe] are suppressed. The reason is that the aspherical (jet-like) models eject preferentially higher temperature materials through the strong shock at the z-axis. Note that these trends are similar to ones observed in extremely metal poor halo stars ([11], also see references in [7]), therefore suggesting important roles of hypernovae in the early Galactic chemical evolution (assuming that our models represent hypernovae very well – in the next section we show this is the case). The enhancement of the α-rich freezeout products is seen in Figure 3. For a given mass of ^{56}Ni, the aspherical models (denoted as 40A, B, C) eject larger amount of ^{44}Ti and ^{64}Zn. This mechanism could be important to solve problems related to the production of ^{44}Ti [12].

SIGNATURE OF ASPHERICITY IN HYPERNOVAE

A Prototypical Hypernova SN 1998bw

Figure 4 shows that the bolometric light curve [13] of SN 1998bw is in good agreement with our aspherical models viewed from the direction close to the pole (z), not with any spherical models. Specifically, for the ejecta mass $\sim 10 M_\odot$, the aspherical model with $E_{51} \sim 20$ yields the best match to the observation, especially in early phases up to ~ 2 months. The same model also yields the late time light (~ 1 year), as well as photospherical velocities, being well consistent with the observations [14].

The same model also explains late phase nebular spectra at ~ 1 year (Figure 5) [15]. A sharply peaked [OI] 6300, 6363Å doublet and a broad feature around 5200Å (blend

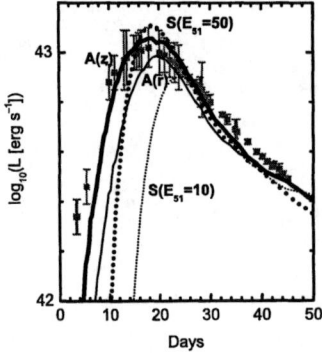
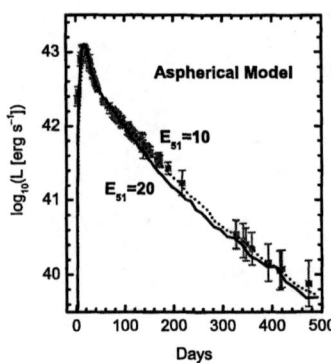

FIGURE 4. Optical light curves for the aspherical models of [6] with the ejecta mass $M_{ej} = 10.4 M_\odot$, as compared with the bolometric light curve of a prototypical hypernova SN 1998bw [13]. The left panel shows early phase light curves. The model with the kinetic energy of the expansion $E_{51} = 20$ is shown for the viewing direction along the z-axis (A(z)) and for the r-plane (A(r)). Also shown are the spherical models (S) with $E_{51} = 50$ and 10. The right panel shows late phase behavior. The aspherical models with $E_{51} = 20$ and 10 are shown. From [14].

of FeII] in our model) are reproduced by the aspherical model, if viewed from the polar direction – the same expected from the light curve!

From these modeling, we conclude that the aspherical model with the ejecta mass $\sim 10 M_\odot$ and $E_{51} \sim 20$ are in good agreement with all the optical observations of SN 1998bw. The viewing angle is determined to be close to the polar axis, which strengthens the case of the association with the Gamma-Ray Burst GRB980425.

SN 2003jd: SN 1998bw-lke hypernova but viewed off-axis?

The conclusion in the previous section, i.e., the aspherical model for SN 1998bw, immediately implies that there must be similar events, but viewed at different directions. Alternatively, if we find this, it will give a strong confirmation of the aspherical model.

What is the noticeable characteristics for such off-axis events? First of all, we do not expect to see a Gamma Ray Burst because of the relativistic beaming effect. Next, we expect the early phase behavior similar to SN 1998bw (Figure 4). Finally, the nebular spectra will be very atypical. It will give a double-peaked [OI] 6300, 6363Å doublet since we see a disk-like and expanding oxygen distribution [6] (Figure 5). These is a supernova having all these features – SN 2003jd. Figure 5 (right) shows its nebular spectra taken by Subaru and Keck telescopes. They clearly show the double-peaked [OI] profile. Indeed, the observation matches the theoretical line profile perfectly [6, 16]. From these arguments, we suggest that SN 2003jd is a hypernova, possibly associated with a GRB, similar to SN 1998bw, but viewed from the equatorial direction. In this interpretation, it is different from SN 1998bw only in the viewing angle.

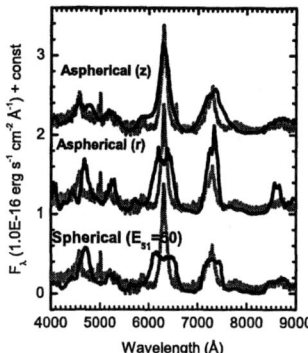 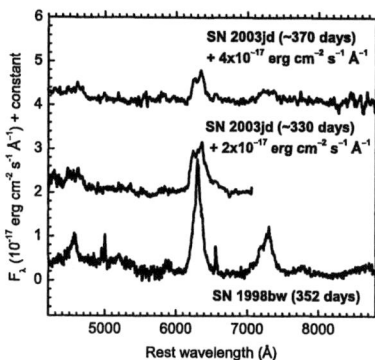

FIGURE 5. (Left) Synthetic spectra at 350 days after the explosion as compared with the spectrum of SN 1998bw at 337 days after the B maximum (gray curves; from [13]). Models are shown for an aspherical model with $E_{51} \sim 20$ viewed at the z-axis (top), the one viewed at the r-plane (middle), and a spherical model with $E_{51} \sim 50$ (bottom). From [15]. (Right) Observed spectra of SN 2003jd [16].

Expected High Energy Emission

The conclusion in the previous sections is derived by modeling optical emission. High energy radiation from decays ^{56}Ni $\rightarrow$ ^{56}Co $\rightarrow$ ^{56}Fe should also reflect the asphericity in the ejecta. Since distribution of ^{56}Ni is very sensitive to asphericity at the explosion (see Fig. 1), it will provide an ideal tool to test the explosion dynamics.

Unfortunately, these has been no detection of either γ-ray or hard X-ray emission from the ^{56}Ni & ^{56}Co decays in core collapse supernovae, except very nearby SN 1987A. However, in light of suggested future improvement of the sensitivities of detectors in this wavelength range by an order of 1 or 2 [17], it should be important to present theoretical expectation taking into account deviation from spherical symmetry.

[18] examined details of the γ and hard X-ray emission from aspherical hypernovae (for SN 1998bw). Figure 6 shows the theoretical prediction. In sum, they found the followings. (1) The aspherical model yields large line-to-continuum ratio and large cut-off energy, irrespective of the viewing angle. (2) Line profiles are very sensitive to both the asphericity and the viewing angle. (3) The luminosity depends the viewing angle weekly. Combination of these (future) observations will be a strong tool to distinguish models and observer's direction. See [18] for more details, including estimate of detectability by the current (*INTEGRAL*) and future instruments.

CONCLUDING REMARKS

We show in the present paper that (1) all the optical observations of SN 1998bw, a prototypical hypernova, can be understood in the context of the aspherical model, viewed basically on-axis, (2) SN 2003jd fits well into the same context, with only viewing angle

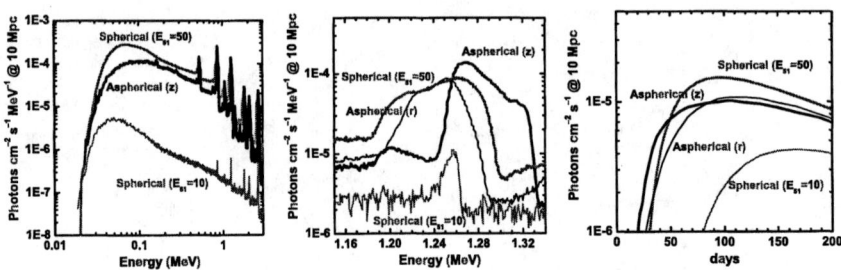

FIGURE 6. γ-ray and hard X-ray signatures of aspherical/spherical hypernova explosions [18]. The aspherical model with $E_{51} = 20$ is shown for the overall spectrum (left), 1,238 keV ^{56}Co line profiles (middle), and for the light curve of the same 1,238 keV line (right). The effect of viewing angle (either along the z-axis or the r-plane) is especially evident in the line profile. From [18]

different from SN 1998bw, and (3) further support of the model will be obtained by future γ and hard X-ray observations of similar events.

ACKNOWLEDGMENTS

The author is supported through the JSPS (Japanese Society for the Promotion of Science) Research Fellowship for Young Scientists.

REFERENCES

1. A. Mezzacappa, et al., in this volume
2. S. Yamada, et al., in this volume
3. K. Nomoto, K. Maeda, N. Tominaga, T. Ohkubo, J. Deng, & P.A. Mazzali, *Ap&SS* **298**, 81 – 86 (2005).
4. T.J. Galama, et al., *Nature* **395**, 670
5. S. Nagataki, *ApJS* **127**, 141 – 157 (2000).
6. K. Maeda, T. Nakamura, K. Nomoto, P.A. Mazzali, F. Patat, & I. Hachisu, *ApJ* **565**, 405 – 412 (2002).
7. K. Maeda, & K. Nomoto, *ApJ* **598**, 1163 – 1200 (2003).
8. K. Maeda, PhD Thesis, University of Tokyo (2004).
9. S. Nagataki, A. Mizuta, S. Yamada, H. Takabe, & K. Sato, *ApJ* **596**, 401 – 413 (2003).
10. S. Nagataki, A. Mizuta, & K. Sato, *ApJ* submitted (astro-ph/0601111) (2006).
11. A. McWilliam, et al., *AJ* **109**, 2757 (1996).
12. L.-S. The, et al., *A&A* in press (astro-ph/0601039) (2006).
13. F. Patat, et al., *ApJ* **555**, 900 – 917 (2001).
14. K. Maeda, P.A. Mazzali, & K. Nomoto, *ApJ* submitted (astro-ph/0511389) (2006).
15. K. Maeda, K. Nomoto, P.A. Mazzali, & J. Deng, *ApJ* **640**, in press (astro-ph/0508373) (2006).
16. P.A. Mazzali, et al., *Science* **308**, 1284
17. T. Takahashi, T. Kamae, & K. Makishima, "Future Hard X-ray and Gamma-ray Observations," in *New Century of X-Ray Astronomy*, edited by H.Inoue & H.Kunieda, ASP Conf. Series 251, Astronomical Society of the Pacific, San Francisco, 2001, pp. 210–213
18. K. Maeda, *ApJ* submitted (astro-ph/0511480) (2006).

1.6 WEAK INTERACTION and NEUTRINO PHYSICS

Results from KamLAND

Kunio Inoue

Research Center for Neutrino Science, Tohoku University
Aramaki Aoba, Aoba, Sendai, Miyagi 980-8578, Japan

Abstract. Earth is our most familiar astronomical object. However, its properties are not very well known, because its interior is optically invisible. One of the most important parameters in understanding Earth is heat generation. A large part of this heat is generated by the decay of radioactive elements, accompanying neutrino emissions, in the earth. KamLAND has provided precise determination of neutrino oscillation parameters and revealed how neutrinos travel observing neutrinos from distant nuclear power reactors. Consequently, KamLAND has made neutrinos new tool to see through astronomical objects that are opaque. Success in the first observation of geologically-produced neutrinos with KamLAND is a break-through for observational geophysics and is the start of "Neutrino Geophysics".

Keywords: neutrino, geo-neutrino, reactor, KamLAND
PACS: 14.60.Pq, 26.65.+t, 28.50.HW

KAMLAND

KamLAND[1] is a monolithic liquid scintillator detector containing 1200 m³ liquid scintillator (1,2,4-trimethylbenzene 20%, dodecane 80% and PPO 1.52g/l as a fluor) and is covered by 2700 m.w.e. mountain rock under Mt. Ikenoyama. The light output of the LS is $\sim$ 8000 photons/MeV and these photons are monitored by 1879 PMTs at 34% photo-coverage, providing $\sim$ 500 p.e./MeV. The liquid scintillator was purified by the water extraction technique and the achieved impurity level is $3.5 \pm 0.5 \times 10^{-18}$ g/g for Uranium and $5.2 \pm 0.8 \times 10^{-17}$ g/g for Thorium. Thanks to the high light output and achieved low impurity level, the primary trigger threshold is set at 0.7 MeV (visible). Thus, the KamLAND has a big advantage to observe anti-electron-neutrinos with the delayed coincidence method to the inverse beta decay reaction, $\bar{v}_e + p \rightarrow e^+ + n$. ^{210}Pb, a daughter nucleus of ^{222}Rn, is about 50 mBq/m³. It is a major background below 1.5 MeV (^{210}Bi) as singles and alpha decay of ^{210}Po is the main source of $^{13}C(\alpha,n)$ background that mimics antineutrino delayed coincidence signal. Future low energy solar neutrino observation via $v_e e$ elastic scattering requires further reduction of remaining ^{210}Po and ^{85}Kr background. And current primary interests are a precise determination of neutrino oscillation parameters with reactor antineutrinos and an investigation of interior of the earth with geologically produced antineutrinos.

REACTOR NEUTRINOS

The first discovery of neutrinos by Reines has been performed with the Savannah river nuclear reactor. Since then, nuclear reactors have been intensively used for studying neutrino properties. Through the development and improvements of various reactor experiments, neutrino production at nuclear reactors and observation of antineutrinos with inverse beta decay reaction on proton have become one of the most established method.

Neutrinos from nuclear power reactors are more than 99.999% pure anti-electron-neutrino at $E_v > 1.8$ MeV. Only 4 fissile nuclei (^{235}U, ^{238}U, ^{239}Pu and ^{241}Pu) dominate the neutrino production and similar energy release from those fissile nuclei (^{235}U:201.7, ^{238}U:205.0, ^{239}Pu:210.0, ^{241}Pu:212.4 MeV [2]) makes strong correlation between thermal power output and neutrino flux. Neutrino spectra from their fission, except ^{238}U have been experimentally obtained [3] at ILL through a procedure, (i) measurement of beta spectrum with spectrometer, (ii) fitting with 30 hypothetical beta spectra and (iii) conversion of beta spectra to neutrino spectra. Since ^{238}U doesn't go to fission with thermal neutrons, its fission spectra has been theoretically calculated [4] considering 744 traces of fission products. Contribution of 4 fissile nuclei changes in time as fuel burns, but its evolution is tracable from initial composition. Thus, neutrino flux is calculable at a few % level from initial enrichment and history of thermal power output. The uncertainty is usually dominated by an accuracy of cooling-water flow-meters.

Neutron decay is an inverse reaction of the neutrino detection reaction and the cross section of the inverse beta decay is related with the neutron life time through the formula [5],

$$\sigma_{\text{tot}}^{(0)} = \frac{2\pi^2 m_e^5}{f_{\text{p.s.}}^R \tau_n} E_e^{(0)} p_e^{(0)}. \tag{1}$$

Recent precise measurement of neutron lifetime with ultra cold neutrons, $\tau_n = 885.7 \pm 0.8$ sec, greatly improved the precision of the inverse beta decay cross section. Applying order $1/M$ corrections, its precision at relevant energies for reactor neutrino observation (< 10 MeV) is better than $\sim 0.2\%$ [6].

Overall interaction rate and also neutrino spectra models have been experimentally examined at a good accuracy [7, 8]. Thanks to thorough previous experiments, current reactor experiments can predict expected spectrum at a few % level without reference detectors close to reactor cores. The latest long baseline experiment, KamLAND, observes neutrinos from country-wide many reactor cores. Its successful observation without near detectors became possible thanks to the knowledge from previous efforts for understanding reactor neutrinos.

Neutrino oscillation causes deficit or spectral distortion of reactor neutrinos under the formula,

$$P_{\bar{e}\bar{e}} \simeq 1 - \sin^2 2\theta_{13} \sin^2\left(\frac{\Delta m_{31}^2 L}{4E}\right) - \cos^4 \theta_{13} \sin^2 2\theta_{12} \sin^2\left(\frac{\Delta m_{21}^2 L}{4E}\right). \tag{2}$$

Relevant distances to search for oscillation effects (and to measure oscillation parameters) are, a km for 1-3 deficit (θ_{13}), a few km for 1-3 spectral distortion (Δm_{31}^2), several tens km for 1-2 deficit (θ_{12}) and a few hundred km for 1-2 spectral distortion (Δm_{21}^2).

Commercial reactors distributed at 130-220 km distance from KamLAND generate 7% of world total reactor power (70 GW$_{\text{thermal}}$ over 1.1 TW$_{\text{thermal}}$) and neutrinos from those reactors contribute to 80% of neutrino flux at the KamLAND site. This situation provides the effective baseline of about 180 km for neutrino oscillation study and gives superior sensitivity to determine Δm_{21}^2.

GEO-NEUTRINOS

Heat production in the earth controls earth formation and evolution, and is a driving force of mantle and outer core convection. Thus, it is important for understanding the structure of the earth and also various phenomena close to us, such as earthquake, volcano, eruption and terrestrial magnetism. Total heat released from the surface of the earth was once calculated to be 44.2 ± 1 TW [9]. The estimation is based on data from $\sim 25,000$ bore-holes at more than 20,000 locations around the world. Despite the small quoted error, a re-evaluation of the same data provided much smaller value, 31 ± 1 TW [10], telling difficulty of extrapolation of the data to the whole surface. Estimation of the heat sources requires an earth model such as the bulk silicate earth (BSE) model, but the model itself requires heat budget model during the earth formation and evolution. Compromising global knowledge, the BSE model, using analysis of the CI carbonaceous chondrite meteorite as an origin of the earth, predicts radiogenic heat generation of 19 TW (about 8 TW U, 8 TW Th and 3 TW K).

This major heat source may be experimentally determined by an observation of anti-neutrinos emitted in their decays. Neutrino emission is directly connected to their heat generation with the formula.

$$^{238}U \rightarrow\, ^{206}Pb + 8\,^4He + 6e^- + 6\bar{\nu}_e + 51.7\,\text{MeV} \tag{3}$$

$$^{232}Th \rightarrow\, ^{208}Pb + 6\,^4He + 4e^- + 4\bar{\nu}_e + 42.7\,\text{MeV} \tag{4}$$

$$^{40}K \rightarrow\, ^{40}Ca + e^- + \bar{\nu}_e + 1.311\,\text{MeV}\, (89.28\%) \tag{5}$$

$$^{40}K + e^- \rightarrow\, ^{40}Ar + \nu_e + 1.505\,\text{MeV}\, (10.72\%) \tag{6}$$

In 1953, Gamov has already pointed the possibility of these neutrinos as a background of reactor neutrino observation in a letter to Reines. Possibility of using neutrinos to study Earth science was first suggested by Marx [11], Markov [12] and Eder [13] in 1960's, and then revived several times by several authors. However, experimental observation has not been performed for a long time due to the severe requirements to very low background condition (10^{-15} g/g U and Th) and very big target volume (order of metric-kton). KamLAND meets the both requirements and is the first detector which has an opportunity to practically access the radiogenic heat generation in the earth with neutrinos. Endpoint energies of anti-neutrinos from above decays are 3.27 MeV (^{238}U), 2.25 MeV (^{238}Th) and 1.31 MeV (^{40}K). Only neutrinos from U/Th exceed the reaction threshold of the inverse beta decay, 1.806 MeV. And an observation of ^{40}K is a future interest with a different experimental technique.

Translation of neutrino flux to the heat generation is affected by inhomogeneous distribution of radioactive nuclei. Large nuclei, U/Th, are thought to be educed from dense material during earth formation and evolution and concentration of them ranges orders of magnitude at various layers of the earth. A model is used for estimating geo-neutrino flux at the experimental site. CC/OC stand for continental and oceanic crust and CS and OS stand for their sediments in Table.1. Roughly, half are in the thin crust and rest are in the mantle. Thorium to uranium ratio is fairly stable as ~ 3.9 despite large variation of their absolute concentrations. For a practical estimation, local geology must be also considered [14, 15]. And the most important thing to use neutrinos as a probe for studying interior of the earth is to understand how neutrinos travel. Determination of neutrino oscillation parameters with reactor and solar neutrino experiments is the starting point of the applied neutrino physics.

TABLE 1. A reference model[15].

layer type	U [ppm]	Th [ppm]	layer type	U [ppm]	Th [ppm]
upper CC	2.8	10.7	OS	1.7	6.9
middle CC	1.5	6.1	OC	0.10	0.22
lower CC	0.2	1.2	mantle	0.012	0.048
CS	2.8	10.7	core	0	0

KAMLAND RESULTS

Reactor neutrino results

Reactor neutrino analysis uses events with visible energies more than 2.6 MeV ($E_{\text{vis}} \simeq E_{\bar{\nu}_e} - 0.8$ MeV) for avoiding uncertainty of geo-neutrino contribution ($E_{\bar{\nu}_e} < 3.27$ MeV). The first results [16] were deduced from 162 ton-year exposure from March 4th to October 6th of 2002. Observed number of events was 54 where expected signal is 86.8 ± 5.6 and expected background is 2.8 ± 1.7, resulting 99.95% C.L. significance of neutrino disappearance. This evidence of reactor neutrino disappearance has solved the long-standing "solar neutrino problem" in a process of elimination, combining with the solar neutrino data. Since then, KamLAND has improved analysis tools, fiducial volume enlargement by factor 1.33 (5.5 m radius) with more uniform energy scale and less vertex bias, relaxed coincidence criteria gaining factor 1.15 better detection efficiency (89.8%), factor 3.55 longer run-time (515.1 days). Considering factor 0.77 lower reactor operation, total statistical improvement is factor 4.2. So far achieved systematic errors and background summary for the improved analysis [17] are shown in Table 2 and 3.

TABLE 2. Estimated systematic errors (%).

Fiducial volume	4.7	Reactor power	2.1
Energy threshold	2.3	Fuel composition	1.0
Cut efficiency	1.6	$\bar{\nu}_e$ spectra	2.5
Live time	0.06	Cross-section	0.2
Total			6.5

TABLE 3. Background summary (events).

Accidental	2.69 ± 0.02
Spallation	4.8 ± 0.9
Fast neutron	< 0.9
$^{13}C(\alpha,n)$	10.3 ± 7.1
Total	17.8 ± 7.3

The 2nd results with 766.3 ton-year exposure (from March 9th, 2002 to January 11th, 2004) accumulated 258 events where 365.2 ± 23.7 signals and 17.8 ± 7.3 backgrounds are expected. Improved statistics strengthened the significance of neutrino disappearance to 99.9998% C.L. and made the spectral distortion analysis more powerful. Observed

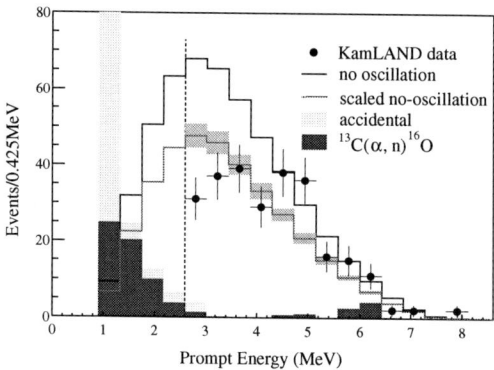

FIGURE 1. Observed reactor neutrino spectrum is shown (plots) together with estimated backgrounds (red and green solid histograms for accidental and $^{13}C(\alpha,n)^{16}O$, respectively), expected no-oscillation neutrino spectrum (black histogram) and scaled no-oscillation neutrino spectrum (blue histogram with error band).

spectrum doesn't agree with a scaled no-oscillation spectrum as shown in Fig.1. Its significance of deviation from no oscillation was tested by χ^2 method dividing the data into arbitrarily chosen 20 equal expected-rate bins. The obtained significance was 99.6% ($\chi^2/\text{dof} = 37.3/19$). If one choose smaller number of bins to discriminate low frequency distortion such as neutrino oscillation, the significance further increases. Combining the rate and shape information, no-oscillation is excluded at 99.999995% C.L.. Survival probability of neutrino is expressed with L/E in models, neutrino oscillation, decay and decoherence. To illustrate relation between survival probability and distance, L_0/E_ν distribution is plotted in Fig.2 together with the best-fit models. Here, different distances to various reactors are represented by effective distance, $L_0 = 180$ km. KamLAND data show clear oscillatory behavior and matches with neutrino oscillation, but monotonically decreasing decay and decoherence models can not explain the behavior.

Clear oscillation pattern also provides better determination of oscillation parameters, especially Δm_{21}^2. Combining with solar neutrino data, allowed region of oscillation parameters is obtained as shown in Fig.3. Mass difference is mainly determined by KamLAND as $\Delta m^2 = 7.9^{+0.6}_{-0.5} \times 10^{-5} \text{eV}^2$ and mixing angle is by solar data, $\tan^2\theta = 0.40^{+0.10}_{-0.07}$. KamLAND has a plan to improve fiducial volume estimation by deploying an all-volume calibration device, by enlarging fiducial volume and by eliminating $^{13}C(\alpha,n)$ background with further purification of the liquid scintillator. Assuming 3% rate error and 1% scale error with these improvements and 3 kton-year data accumulation, one can draw a sensitivity plot of KamLAND alone as solid (black) lines in Fig.3. Future KamLAND has capability to reject full mixing, to improve Δm^2 by factor ~ 2 and to determine mixing angle at comparable precision with the current solar data.

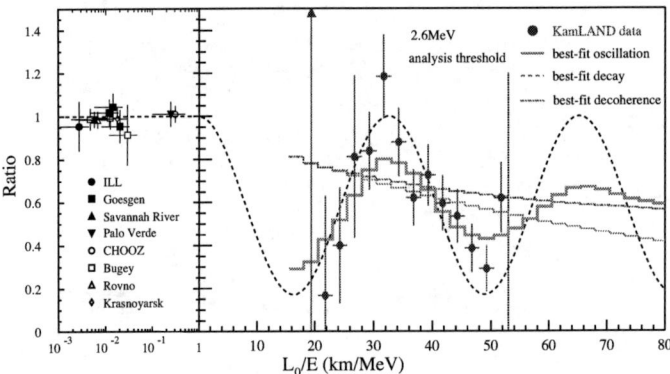

FIGURE 2. The ratio of observed signal rate to no-oscillation expectation is plotted as a function of $L_0/E_{\bar{\nu}_e}$. KamLAND data is divided according to observed energies and the constant L_0 is chosen at the effective distance, 180 km. Dashed curve shows an ideal oscillation pattern as if one reactor existed and histograms are with real reactor distribution for the best fit oscillation parameters $(\tan^2\theta, \Delta m^2) = (0.46, 7.9 \times 10^{-5} \text{eV}^2)$.

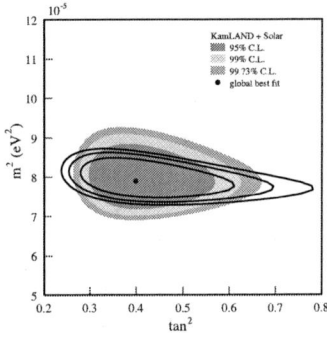

FIGURE 3. Contour shown with filled region is the allowed region from the combined analysis of solar and KamLAND reactor neutrino results at 95, 99 and 99.73% confidence level. Contour with black lines is an estimation for future sensitivity of KamLAND alone with 3 kton-year data accumulation at the same confidence level interval.

Geo-neutrino results

Precise determination of neutrino oscillation parameter with reactor and solar neutrinos made possible to use neutrinos as a probe to investigate astronomical objects. And the reactor neutrino contribution below 2.6 MeV (now the background for geo-neutrino observation) can be estimated at a good accuracy.

KamLAND has made the first experimental investigation of geologically-produced antineutrinos [18] with data acquired from March 9th, 2002 through October 30th, 2004, total exposure of $(4.87 \pm 0.24) \times 10^{31}$ target proton years. Energy window for the geo-neutrino analysis is 0.9 to 2.6 MeV in visible energy. In order to improve more difficult background situation at lower energy, tighter selection criteria are used as shown in Table.4 and overall efficiency is reduced from 0.898 (reactor analysis) to 0.687 ± 0.007 (geo-neutrino analysis).

TABLE 4. Selection criteria for geo-neutrino analysis.

selection	geo-neutrino analysis	reactor neutrino analysis
prompt energy window	$0.9 < E_p < 2.6$ MeV	$2.6 < E_p < 8.5$ MeV
delayed energy window	$1.8 < E_d < 2.6$ MeV	$1.8 < E_d < 2.6$ MeV
prompt signal fiducial volume	$\|\mathbf{r}_p\| < 5$ m	$\|\mathbf{r}_p\| < 5.5$ m
delayed signal fiducial volume	$\|\mathbf{r}_d\| < 5$ m	$\|\mathbf{r}_d\| < 5.5$ m
delayed signal cylinder cut	$r_d^{xy} > 1.2$ m	-
timing correlation	$0.5 < \Delta T < 500 \mu s$	$0.5 < \Delta T < 1000 \mu s$
vertex correlation	$\|\mathbf{r}_p - \mathbf{r}_d\| < 1.0$ m	$\|\mathbf{r}_p - \mathbf{r}_d\| < 2.0$ m

Background estimation for geo-neutrino analysis is summarized in Table.5. The largest background is from reactor neutrino and its estimation error is coming from the uncertainty of the neutrino oscillation parameters. In this low energy region, contribution from long-lived nuclei such as ^{106}Ru ($T_{1/2}=372$ days), ^{144}Ce (285 days) and ^{90}Sr (28.6 years) [19] in reactor cores is not negligible. Neutrino flux from those nuclei doesn't have strong correlation with reactor power output and long term average power is used, instead, to estimate their contributions. Another serious background is $^{13}C(\alpha,n)^{16}O_{g.s.}$. The reaction is 99.7% dominated by ^{210}Po α decay. Neutron energy from the reaction with ^{210}Po α ranges up to 7.3 MeV but appears below 2.6 MeV visible energy due to quenching effect of the liquid scintillator. Estimation of the background requires various knowledge, number of ^{210}Po decay, $^{13}C(\alpha,n)$ cross section, branching ratio, angular distribution and quenching effect. Although new cross section measurement[20] improved the uncertainty of the total cross section from 20% to 5%, published KamLAND result uses old measurements and thus the estimated error is quite large. Validity of the estimation has been tested with "Pulse shape discrimination" and event rate below the positron annihilation energy where neutrino signal appears only as resolution tail.

TABLE 5. Background summary for geo-neutrino analysis.

reactor neutrino	80.4 ± 7.2	$^{13}C(\alpha,n)$	42 ± 11
accidental	2.38 ± 0.01	reactor long-lived	1.9 ± 0.2
spallation	0.30 ± 0.05		
		Total background	127 ± 13

By subtracting estimated background from total observed number of events, 152, we obtain an excess of 25^{+19}_{-18} events consistent with the prediction of 19 events from a reference geological model [15]. The excess corresponds to interaction rate of $5.1^{+3.9}_{-3.6} \times 10^{-31}$ $\bar{\nu}_e$/target-proton/year.

Uranium and Thorium contribution can be individually determined using spectrum information in an ideal case. But with the limited statistics shown in Fig.4, there is a sensitivity only on the total number of U/Th events. By fixing Th/U ratio to 3.9 from geophysics knowledge and performing likelihood rate+shape analysis, the best-fit number of observed geo-neutrino events was obtained as 28 events and 90% confidence interval was obtained as 4.5-54.2 events. Since the significance of the geo-neutrino detection is still only about 95% level, 99% C.L. upper limit of geo-neutrino flux was

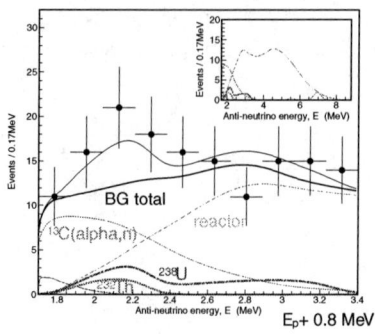

FIGURE 4. Observed spectrum below 3.4 MeV ($E_{vis} < 2.6$ MeV) is shown with estimated backgrounds and expected geo-neutrino signal. Thin solid (black) line is the total expected spectrum and thick solid (black) line is excluding geo-neutrino signal. Inset shows expected spectrum for the extended energy region covering entire reactor neutrino spectrum.

calculated ($\phi < 1.62 \times 10^7\ \bar{\nu}_e/\text{cm}^2/\text{sec}$). This flux upper limit is still 3.8 times larger than the model expectation and statistics is yet poor to discriminate geophysical models. But this is the first step on "Neutrino geophysics" and KamLAND first established the method to observe geo-neutrinos.

Future purification of liquid scintillator will improve statistical significance a lot at KamLAND. And proposed experiments far from any nuclear reactors and on the oceanic plate will have statistical advantages. Adding such experiments and performing global measurement, substantial improvements on geophysics will be brought by neutrino experiments.

FUTURE SOLAR NEUTRINO OBSERVATION

Another application of neutrinos is an investigation of the nuclear fusion reaction in the Sun. Original motivation of the solar neutrino observation by Davis was a determination of dominant fusion process in the Sun between the pp-chain and the CNO-cycle. The pp-chain predicts smaller event rate, but too small observed number of events has brought the solar neutrino problem. After KamLAND and solar neutrino experiments, the solar neutrino problem has been solved and we can come back to Davis's original motivation. So far, individually observed solar neutrinos are only ^{8}B neutrinos. Its branching ratio is very small (0.02%) and is not very adequate to verify the standard solar model. More abundant neutrinos have lower energies and realtime observation at low energy region is indispensable to proceed. On the other hand, contribution of the CNO-cycle will be important to study stellar evolution because its contribution increases as the Sun glows in the main sequence.

KamLAND has a plan to upgrade its purification system in 2006. New distillation and Nitrogen purge system are supposed to eliminate obstacles to observing low energy solar neutrinos, such as ^{210}Pb, ^{40}K, ^{222}Rn, ^{85}Kr. After the successful purification, KamLAND is going to measure ^{7}Be solar neutrino (branching ratio 14%) flux. Low energy ^{8}B neutrinos down to ~4 MeV is also in the scope. At the energy region, spectral distortion due to the MSW effect is expected.

Spallation background, ^{11}C, lies at CNO and pep neutrino region and they are not removed by the purification. The ^{11}C production by muon-spallation accompanies neutron at 95% probability. Efficient tagging of the neutron and precise measurement of neutron association probability will allow us to subtract ^{11}C background in the CNO and pep energy region. A crude estimation for 3 year run and 4 m radius fiducial volume gives ~3% accuracy for CNO+pep neutrino flux measurement or several % for CNO neutrino flux measurement assuming the SSM pep neutrino flux. Upgrade of electronics is also considered to maximize the neutron tagging efficiency.

SUMMARY

KamLAND reactor results provided spectral distortion at more than 99.6% CL and excluded no oscillation at 99.999995% CL. The L/E plot shows clear oscillatory behavior and it resulted in precise measurement of oscillation parameters. Especially, mass squared difference has been determined at several % accuracy, $\Delta m^2 = 7.9^{+0.6}_{-0.5} \times 10^{-5} \text{eV}^2$. Anti-electron-neutrino has now become a good probe and the first geo-neutrino investigation pioneered "Neutrino geophysics". Future experiments will provide substantial improvements on geophysics. KamLAND 2nd phase is aiming at observing ^{7}Be, CNO+pep and low energy ^{8}B (important for testing the MSW effect) solar neutrinos, individually.

REFERENCES

1. K.Inoue, *New J. Phys.* **6**(2004)147.
2. M.F.James, *J. Nucl. Energy* **23** (1969) 517.
3. K.Schreckenbach et al., it Phys.Lett. B**160** (1985) 325. A.A.Hahn et al., *Phys. Lett.* B**218** (1989) 365.
4. P.Vogel et al., *Phys. Rev.* C**24** (1981) 1543.
5. P.Vogel and J.F.Beacom, *Phys. Rev.* D**60** (1999) 053003.
6. A.Kurylov et al., *Phys. Rev.* C**67** (2003) 035502.
7. Y Declais et al., *Phys. Lett.* B**338** (1994) 383.
8. B.Achkar et al., *Phys. Lett.* B**374** (1996) 243.
9. H.N .Pollack et al., *Rev. Geophys.* **31**(1993)267.
10. A.M.Hofmeister et al., *Techtonophysics* **395**(2005)159.
11. G.Marx, N.Menyard, Mitteilungen der Sternwarte, Budapest, 48(1960). G.Marx, Geophysics by neutrinos, *Czech. J. Phys.* B**19**(1969)1471.
12. M.A.Markov, NEUTRINO 1964, NAUKA, in Russian (1964).
13. G.Eder, Terrestrial neutrinos, *Nucl. Phys.* **78**(1966)657.
14. F.Mantovani, L.Carmignani, G.Fiorentini, M.Lissia, *Phys. Rev.* D**69**(2004)013001.
15. S.Enomoto, E.Ohtani, K.Inoue, A.Suzuki, hep-ph/0508049.
16. KamLAND collaboration, *Phys. Rev. Lett.* **90** (2003) 021802.
17. KamLAND collaboration, *Phys. Rev. Lett.* **94** (2005) 081801.
18. KamLAND collaboration, *Nature* Vol.**436** Num.7050 pp499-503.
19. V.I.Kopeikin, L.A.Mikaelyan, V.V.Sinev, *Phys. of Atomic Nuclei* **64**(2001)849.
20. S.Harissopulos et al., *Phys. Rev.* C**72** (2005) 062801.

Neutrino Magnetic Moment

A.B. Balantekin

University of Wisconsin, Physics Department, Madison, WI 53706 USA[1]

Abstract. Current experimental and observational limits on the neutrino magnetic moment are reviewed. Implications of the recent results from the solar and reactor neutrino experiments for the value of the neutrino magnetic moment are discussed. It is shown that spin-flavor precession in the Sun is suppressed.

Keywords: Neutrino magnetic moment, spin-flavor precession, solar antineutrinos
PACS: 14.60.Pq, 13.40.Em, 26.65.+t, 96.15.Gh

INTRODUCTION

A minimal extension of the Standard Model (with non-zero neutrino masses) yields a neutrino magnetic moment of [1]

$$\mu_v = \frac{3eG_F m_v}{8\pi^2 \sqrt{2}} = \frac{3G_F m_e m_v}{4\pi^2 \sqrt{2}} \mu_B \qquad (1)$$

where $\mu_B = e/2m_e$ is the Bohr magneton. Note that the neutrino magnetic moment is proportional to the neutrino mass as required by the symmetry principles. Since the recent solar, atmospheric and reactor neutrino experiments indicate the existence of non-zero neutrino masses, we also know that neutrino has a magnetic moment. Using the neutrino parameters deduced from analyses of those experiments [2] we get $\mu_v \geq (4 \times 10^{-20}) \mu_B$. Larger values of magnetic moments are possible in extensions of the Standard Model, as indicated by the inequality sign in this value. If the magnetic moment is generated by physics at scale Λ we can write

$$\mu_v \sim \frac{e\mathcal{G}}{\Lambda}, \qquad (2)$$

where $\mathcal{G}$ represents the combination of the coupling constants and appropriate 2π factors. If we remove the external photon from the diagrams leading to Eq. (2) we get a contribution to the mass of the order

$$\delta m_v \sim \mathcal{G} \Lambda. \qquad (3)$$

These equations imply that

$$\delta m_v \sim \frac{\Lambda^2}{m_e} \left(\frac{\mu_v}{\mu_B} \right). \qquad (4)$$

---

[1] E-Mail: baha@physics.wisc.edu

If one assumes that the scale Λ is not significantly higher than the electroweak scale, current neutrino mass limits imply $|\mu_\nu| \leq 10^{-14} \mu_B$ for Dirac neutrinos [3]. It is however possible to introduce models where the magnetic moment and mass do not come from the same number of loops, and relax this bound (see e.g. Ref. [4]).

It is well-established that the neutrinos mix and the discussion above illustrates that magnetic moment is properly defined in the mass basis [5]. In this basis Dirac neutrinos can have both diagonal and off-diagonal moments, whereas Majorana neutrinos can only have transition moments. More specifically, if the magnetic moment operator is designated by $\boldsymbol{\mu}$, then $\boldsymbol{\mu} = \boldsymbol{\mu}^\dagger$ for Dirac neutrinos, and $\boldsymbol{\mu}^T = -\boldsymbol{\mu}$ for Majorana neutrinos.

LIMITS ON THE NEUTRINO MAGNETIC MOMENT

There are a number of possible physical processes involving a neutrino with a magnetic moment. Among these are the $\nu - e$ scattering, spin-flavor precession in an external magnetic field, plasmon decay, and the neutrino decay. For the first process, using the magnetic moment operator $\boldsymbol{\mu}$, the total cross section at an experiment where the final neutrino is not observed can be written in the Born approximation as

$$\sigma \sim \sum_i |\langle \nu_i | \boldsymbol{\mu} | \nu_e \rangle|^2, \qquad (5)$$

where $|\nu_i\rangle, i = 1, 2, 3$ represent the mass eigenstates and we assumed that electron neutrinos are used. Since the neutrino mixing matrix in $|\nu_e\rangle = \sum_i U_{ei} |\nu_i\rangle$ is unitary, Eq. (5) takes the form

$$\sigma \sim \langle \nu_e | \boldsymbol{\mu}^\dagger \boldsymbol{\mu} | \nu_e \rangle. \qquad (6)$$

Detecting a neutrino magnetic moment then implies detecting them in mass eigenstates. Consequently the measured magnetic moment of the neutrino, in principle, depends on the distance from its source [5]:

$$\mu_e^2 = \sum_i |\sum_j U_{ej} \mu_{ij} \exp(-iE_j L)|^2. \qquad (7)$$

The differential scattering cross section for electron neutrinos or antineutrinos on electrons is given by [6]

$$\begin{aligned} \frac{d\sigma}{dT} &= \frac{G_F^2 m_e}{2\pi} \left[(g_V + g_A)^2 + (g_V - g_A)^2 \left(1 - \frac{T}{E_\nu}\right)^2 + (g_A^2 - g_V^2) \frac{m_e T}{E_\nu^2} \right] \\ &\quad + \frac{\pi \alpha^2 \mu_\nu^2}{m_e^2} \left[\frac{1}{T} - \frac{1}{E_\nu} \right], \end{aligned} \qquad (8)$$

where T is the electron recoil kinetic energy, $g_V = 2\sin^2\theta_W + 1/2$, $g_A = +1/2(-1/2)$ for electron neutrinos (antineutrinos), and the neutrino magnetic moment is expressed in units of μ_B. The first line in Eq. (8) is the standard electroweak contribution and the second line represents the contribution of the neutrino magnetic moment. Clearly the magnetic moment contribution is dominant at low recoil energies. The magnetic moment cross section will exceed the standard electroweak cross-section for recoil energies

$$\frac{T}{m_e} < \frac{\pi^2 \alpha^2}{(G_F m_e^2)^2} \mu_\nu^2, \qquad (9)$$

i.e. the lower the smallest measurable recoil energy is, the smaller values of the magnetic moment can be probed. To perform such an experiment either solar or reactor neutrinos have been used. SuperKamiokande collaboration looked for distortions in the energy spectrum of solar neutrinos scattered off the electrons in their detector. No clear signal was observed. Combined with the other solar neutrino and KamLAND experiments a limit of $\mu_\nu \leq 1.1 \times 10^{-10}$ μ_B at 90% C.L. was obtained [7]. The MUNU collaboration, using reactor neutrinos, recently obtained a slightly better limit of $\mu_\nu \leq 9 \times 10^{-11}$ μ_B at 90% C.L. [8]. Another possibility for doing such experiments is to utilize low-energy beta beams [9]. A detailed study of neutrino-electron scattering using low-energy beta-beams in general is given in Ref. [10] and limits on the neutrino magnetic moment in particular are given in Ref. [11]. The latter work finds that a tritium source may yield a better bound.

Neutrinos change helicity in magnetic moment scattering. This fact has been used to put limits from astrophysics and cosmology on neutrino magnetic moment. If μ_ν is sufficiently large, then the proto-neutron star formed in a core-collapse supernova can cool faster since the right-handed components are sterile. It was found that $\mu_\nu \geq 10^{-12} \mu_B$ would be inconsistent with the observed cooling time of SN1987a [12]. In the Early Universe the existence of right-handed Dirac neutrinos that may be produced in magnetic scattering increase the number of effective degrees of freedom altering neutrino counting through the big-bang nucleosynthesis yields [13]. (Similar limits do not apply to Majorana neutrinos since antineutrino states are already counted).

The tightest astrophysical bound on neutrino magnetic moment comes from the red-giant stars. A large enough magnetic moment implies enhanced plasmon decay rate, $\gamma^* \to \nu\nu$, inside the star. Since the neutrinos freely escape the stellar environment this process in turn cools a red giant star faster, delaying helium ignition. Existing observations of globular cluster stars lack any evidence of this effect, yielding a limit of $\mu_\nu \leq 3 \times 10^{-12} \mu_B$ [14].

The discussion above shows the neutrino magnetic moment is presently known to be in the range

$$(9 \times 10^{-11}) \mu_B \geq \mu_\nu \geq (4 \times 10^{-20}) \mu_B. \qquad (10)$$

The large width of this range represents possible physics beyond the standard model which can be explored using the neutrino magnetic moment measurements.

IMPLICATIONS OF NEUTRINO SPIN-FLAVOR PRECESSION

If neutrinos have magnetic moments, large magnetic fields that exist in astrophysical environments may give rise to an additional spin-flavor precession coupled to the usual matter-enhanced neutrino oscillations [15, 16]. Spin-flavor precession changes the helicity of the neutrinos, and if the neutrinos are of Majorana type, this yields a solar antineutrino flux [17, 18, 19]. Since the electron antineutrino yields a very distinctive two-neutron signal on charged-current deuteron break-up, Sudbury Neutrino Observatory (SNO) measurements were able to put a limit of $\Phi_{\bar{\nu}_e} \leq 3.4 \times 10^4 \text{cm}^{-2}\text{s}^{-1}$ at 90% C.L. [20]. This corresponds to less than 0.8% of the standard solar model ^8B flux. The KamLAND, experiment, being directly sensitive to antineutrino scattering in their scintillator liquid, provides a slightly better bound of $\Phi_{\bar{\nu}_e} \leq 3.7 \times 10^2 \text{cm}^{-2}\text{s}^{-1}$ at 90% C.L. [21]. This is less than 2.8×10^{-4} of the Standard Solar Model ^8B ν_e flux.

A complete analysis of the spin-flavor precession scenario in the Sun requires detailed knowledge of the solar magnetic fields. Unfortunately information about solar magnetic fields is rather incomplete. If the magnetic field is greater than 10^8 G, magnetic pressure becomes the same order of magnitude as the matter pressure obviating the Standard Solar Model. For the neutrino masses and the mixing angles deduced from the solar and reactor neutrino experiments, both the spin-flavor and the MSW resonances are very close together in the inner radiative zone. It was shown that magnetic fields greater than $\sim 10^7$ G, localized at about $0.2 R_\odot$, would cause the sound speed profile to be at variance with the helioseismic observations [22].

The MSW resonance takes place in the Sun where the condition

$$\sqrt{2} G_F N_e = \frac{\delta m^2}{2 E_\nu} \cos 2\theta \tag{11}$$

is satisfied. The spin-flavor precession resonance takes place before the solar neutrinos reach the MSW resonance point. It is where

$$\frac{G_F}{\sqrt{2}} (2 N_e - N_n) = \frac{\delta m^2}{2 E_\nu} \cos 2\theta \tag{12}$$

for the Dirac neutrinos and where

$$\sqrt{2} G_F (N_e - N_n) = \frac{\delta m^2}{2 E_\nu} \cos 2\theta \tag{13}$$

for the Majorana neutrinos. In these equations N_e and N_n are the electron and neutron densities, respectively. For the observed neutrino parameters these resonances significantly overlap, hence previous approaches treating them as isolated resonances [19] are not applicable [23, 24]. However, there is a limit which may provide an analytical insight into the problem of overlapping resonances. In the limit $N_n = 0$, clearly the resonances of Eqs. (11), (12), and (13) are at the same location. The electron neutrino survival probability is [25]

$$P(\nu_e \to \nu_e) = \frac{1}{2} - \frac{1}{2} \cos 2\theta \left(1 - 2 P_{\text{hop}}\right), \tag{14}$$

where P_{hop} is the hopping probability between matter eigenstates. It can be shown that [23], in the limit $N_n = 0$, the hopping probability is given by

$$P_{hop}(\mu B \neq 0) = P_{hop}(\mu B = 0) \times$$

$$\exp\left\{ \frac{i}{\pi} \int_{r_0}^{r_0^*} dr \frac{\delta m^2}{2E} \left[\frac{(\mu B)^2}{\sqrt{\zeta^2(r) - 2\zeta(r) \cos 2\theta_\nu + 1}} \right] \right\}, \tag{15}$$

where we used the semiclassical treatment of the matter-enhanced neutrino oscillations to calculate the hopping probability [26]. In Eq. (15), $\zeta(r) = \frac{2\sqrt{2}G_F N_e(r)}{\delta m^2/E}$, and r_0^* and r_0 are the turning points (zeros) of the integrand. Matter-enhanced neutrino oscillations in the Sun are primarily adiabatic [2], hence the hopping probability is very small to begin with. Eq. (15) implies that the reduction factor of the hopping probability,

$$\exp\left[-\frac{\pi}{\alpha}\frac{(\mu B)^2 2E}{\delta m^2}\right], \tag{16}$$

is also very small. For a 10^5 G magnetic field, a magnetic moment of $10^{-12}\mu_B$, and $E \sim 10$ MeV, this hopping reduction factor is $\sim 10^{-3}$. An exact numerical calculation, using the neutrino mixing parameters $\delta m^2 = 8 \times 10^{-5}$ eV^2, $\tan^2\theta = 0.4$, and relatively large values of $\mu_\nu = 10^{-11}\mu_B$ and $B = 10^5$ G, finds that the electron neutrino survival probabilities calculated with the MSW resonance only ($B=0$) and calculated with both resonances ($B \neq 0$) differ by less than 10^{-5} [23]. For reasonable values of the solar magnetic fields the effect of the neutrino magnetic moment on the solar neutrino flux is minuscule. It should be noted however that large fluctuations of the magnetic fields could impact spin-flavor precession [27].

ACKNOWLEDGMENTS

This work was supported in part by the U.S. National Science Foundation Grant No. PHY-0244384, and in part by the University of Wisconsin Research Committee with funds granted by the Wisconsin Alumni Research Foundation.

REFERENCES

1. W. J. Marciano and A. I. Sanda, *Phys. Lett. B* **67**, 303–305 (1977); B. W. Lee and R. E. Shrock, *Phys. Rev. D* **16**, 1444–1473 (1977).
2. See e.g. A. B. Balantekin and H. Yuksel, *J. Phys. G* **29**, 665–682 (2003) [arXiv:hep-ph/0301072]; *Phys. Rev. D* **68**, 113002 (2003) [arXiv:hep-ph/0309079] and references therein.
3. N. F. Bell, V. Cirigliano, M. J. Ramsey-Musolf, P. Vogel and M. B. Wise, *Phys. Rev. Lett.* **95**, 151802 (2005) [arXiv:hep-ph/0504134].
4. H. Georgi and L. Randall, *Phys. Lett. B* **244**, 196–202 (1990).
5. J. F. Beacom and P. Vogel, *Phys. Rev. Lett.* **83**, 5222–5225 (1999) [arXiv:hep-ph/9907383].
6. P. Vogel and J. Engel, *Phys. Rev. D* **39**, 3378–3383 (1989).
7. D. W. Liu et al. [Super-Kamiokande Collaboration], *Phys. Rev. Lett.* **93**, 021802 (2004) [arXiv:hep-ex/0402015].
8. Z. Daraktchieva et al. [MUNU Collaboration], *Phys. Lett. B* **615**, 153–159 (2005) [arXiv:hep-ex/0502037].
9. C. Volpe, *J. Phys. G* **30**, L1–L6 (2004) [arXiv:hep-ph/0303222].
10. A. B. Balantekin, J. H. de Jesus and C. Volpe, arXiv:hep-ph/0512310.
11. G. C. McLaughlin and C. Volpe, *Phys. Lett. B* **591**, 229–234 (2004) [arXiv:hep-ph/0312156].
12. J. M. Lattimer and J. Cooperstein, *Phys. Rev. Lett.* **61**, 23–26 (1988); R. Barbieri and R. N. Mohapatra, *Phys. Rev. Lett.* **61**, 27–30 (1988).
13. J. A. Morgan, *Phys. Lett. B* **102**, 247–250 (1981).
14. G. G. Raffelt, *Phys. Rev. Lett.* **64**, 2856–2858 (1990).
15. C. S. Lim and W. J. Marciano, *Phys. Rev. D* **37**, 1368–1373 (1988); E. K. Akhmedov, *Phys. Lett. B* **213**, 64–68 (1988).

16. A. B. Balantekin, P. J. Hatchell and F. Loreti, *Phys. Rev. D* **41**, 3583–3593 (1990).
17. R. S. Raghavan, A. B. Balantekin, F. Loreti, A. J. Baltz, S. Pakvasa and J. T. Pantaleone, *Phys. Rev. D* **44**, 3786–3790 (1991).
18. E. K. Akhmedov, *Phys. Lett. B* **255**, 84–88 (1991).
19. A. B. Balantekin and F. Loreti, *Phys. Rev. D* **45**, 1059–1065 (1992).
20. B. Aharmim *et al.* [SNO Collaboration], *Phys. Rev. D* **70**, 093014 (2004) [arXiv:hep-ex/0407029].
21. K. Eguchi *et al.* [KamLAND Collaboration], *Phys. Rev. Lett.* **92**, 071301 (2004) [arXiv:hep-ex/0310047].
22. S. Couvidat, S. Turck-Chieze and A. G. Kosovichev, *Astrophys. J.* **599**, 1434–1448 (2003).
23. A. B. Balantekin and C. Volpe, *Phys. Rev. D* **72**, 033008 (2005) [arXiv:hep-ph/0411148].
24. A. Friedland, arXiv:hep-ph/0505165.
25. See e.g. A. B. Balantekin, *Phys. Rept.* **315**, 123–135 (1999) [arXiv:hep-ph/9808281].
26. A. B. Balantekin and J. F. Beacom, *Phys. Rev. D* **54**, 6323–6337 (1996) [arXiv:hep-ph/9606353]; A. B. Balantekin, S. H. Fricke and P. J. Hatchell, *Phys. Rev. D* **38**, 935–943 (1988).
27. F. N. Loreti and A. B. Balantekin, *Phys. Rev. D* **50**, 4762–4770 (1994) [arXiv:nucl-th/9406003].

The Effect of Neutrino Oscillations on Supernova Light Element Synthesis

Takashi Yoshida*, Toshitaka Kajino[†,**], Hidekazu Yokomakura[‡], Keiichi Kimura[‡], Akira Takamura[§] and Dieter H. Hartmann[¶]

*Astronomical Institute, Graduate School of Science, Tohoku University, Miyagi 980-8578, Japan
[†]National Astronomical Observatory of Japan, and The Graduate University for Advanced Studies, Tokyo 181-8588, Japan
[**]Department of Astronomy, Graduate School of Science, University of Tokyo, Tokyo 113-0033, Japan
[‡]Department of Physics, Graduate School of Science, Nagoya University, Aichi 464-8602, Japan
[§]Department of Mathematics, Toyota National College of Technology, Aichi 471-8525, Japan
[¶]Department of Physics and Astronomy, Clemson University, South Carolina 29634, USA

Abstract. We investigate light element synthesis through the ν-process during supernova explosions considering neutrino oscillations and investigate the dependence of ^{7}Li and ^{11}B yields on neutrino oscillation parameters mass hierarchy and θ_{13}. The adopted supernova explosion model for explosive nucleosynthesis corresponds to SN 1987A. The ^{7}Li and ^{11}B yields increase by about factors of 1.9 and 1.3 in the case of normal mass hierarchy and adiabatic 13-mixing resonance compared with the case without neutrino oscillations. In the case of inverted mass hierarchy or nonadiabatic 13-mixing resonance, the increase in ^{7}Li and ^{11}B yields is much smaller. Astronomical observations of ^{7}Li/^{11}B ratio in stars formed in regions strongly affected by prior generations of supernovae would constrain mass hierarchy and the range of θ_{13}.

Keywords: supernovae, nucleosynthesis, neutrinos
PACS: 26.30.+k, 14.60.Pq, 97.60.Bw, 25.30.Pt, 13.15.+g

INTRODUCTION

At the final stage of massive star evolution, stellar core collapses, a proto-neutron star forms, and the surrounding material is exploded as a supernova. The gravitational energy is released from the forming proto-neutron star by a huge number of neutrinos, $N_\nu \sim 10^{58}$. The neutrinos interact with nuclei in the surrounding material and some elements are newly produced despite small reaction cross sections. This is called the ν-process (e.g., [1]). In light elements, ^{7}Li and ^{11}B are mainly produced through the ν-process [1, 2, 3]. We have shown that the ^{7}Li and ^{11}B yields strongly depend on neutrino temperature, i.e., neutrino energy spectra [3].

Recent neutrino experiments have confirmed that neutrinos have finite masses and that their flavors are changed during their propagation [4, 5, 6, 7]. Most of neutrino oscillation parameters have been determined by these experiments. However, mass hierarchy has not been clarified and only upper limit of the mixing angle θ_{13} has been determined. Supernova neutrinos also change their flavors in propagating from a proto-neutron star to the surface of the star. The mixing probabilities of supernova neutrinos have been investigated [8, 9] and it has been found that neutrino oscillations change their energy spectra in the surrounding material. Thus, it is expected that the production of ^{7}Li and

^{11}B are affected by neutrino oscillations. Their yields will depend on mass hierarchy and the mixing angle θ_{13}.

In this study, we investigate the effect of neutrino oscillations on light element synthesis through the ν-process in a supernova explosion. We investigate the dependence of ^{7}Li and ^{11}B yields on mass hierarchy and the mixing angle θ_{13}.

SUPERNOVA EXPLOSION MODEL

In order to calculate the ν-process in supernova explosions, we use a supernova neutrino model. The neutrino luminosity is assumed to decay exponentially with time. The total neutrino energy is set to be 3×10^{53} ergs and the decay time of neutrino luminosity is set to be 3 s (after [1]). The neutrino luminosity is equally divided among neutrino flavors. The neutrino energy spectra at the neutrino sphere are assumed to obey Fermi Distribution with zero chemical potential. The neutrino temperatures of ν_e, $\bar{\nu}_e$, and $\nu_{\mu,\tau}$ and $\bar{\nu}_{\mu,\tau}$ are set to be 3.2 MeV, 5 MeV, and 6 MeV, respectively [2, 3].

Recent neutrino experiments such as SuperKamiokande [5], SNO [6], KamLAND [7], and CHOOZ [4] have determined precisely most of neutrino oscillation parameters. We adopted neutrino oscillation parameter values of Large Mixing Angle (LMA) solutions. Mass-squared differences are set to be $\Delta m_{31}^2 = \pm 2.4 \times 10^{-3}$ eV^2 and $\Delta m_{21}^2 = 7.9 \times 10^{-5}$ eV^2 where positive and negative values of $\Delta^2 m_{31}$ correspond to normal and inverted mass hierarchies. Mixing angles are set to be $\sin^2 2\theta_{12} = 0.816$ and $\sin^2 2\theta_{23} = 1$. We investigate the effect of mass hierarchy and $\sin^2 2\theta_{13}$ in the range of $\sin^2 2\theta_{13} \leq 0.1$.

We used the supernova explosion model in [2, 3]. The presupernova structure is adopted from a 16.2 $M_\odot$ star corresponding to the progenitor of SN 1987A [10]. We calculated the supernova explosion using a spherically symmetrical hydrodynamic code [11]. We set the explosion energy and the mass cut to be 1×10^{51} ergs and 1.61 $M_\odot$.

We numerically calculated detailed nucleosynthesis during the supernova explosion by postprocessing. The nuclear reaction network consists of 291 species of nuclei [2]. The reaction rates of the ν-process are adopted from [12]. We have already shown in [2] that the ν-process reactions from ^{4}He and ^{12}C are important to produce ^{7}Li and ^{11}B. In order to investigate the effect of neutrino oscillations on ^{7}Li and ^{11}B production, we analytically approximated the cross sections of charged-current ν-process reactions of ^{4}He and ^{12}C in the form of $\sigma_\nu = a(\varepsilon_\nu - \varepsilon_{th})^b$, where ε_{th} is threshold of a ν-process reaction. The coefficients of the functions were determined to fit the corresponding reaction rates averaged by Fermi distribution in [12].

NEUTRINO MIXING PROBABILITIES

We show the mixing probabilities of neutrinos passing through the presupernova in the case of LMA, normal mass hierarchy, and $\sin^2 2\theta_{13} = 1 \times 10^{-2}$ in Fig. 1. There are two resonances of neutrino oscillations; the resonance due to Δm_{31}^2, denoted by H-resonance, is located in the O/C layer and the resonance due to Δm_{21}^2, denoted by L-resonance, is located in the He layer. Since the H-resonance is adiabatic and the mass hierarchy is normal, large flavor mixing of neutrinos occurs in the O/C layer.

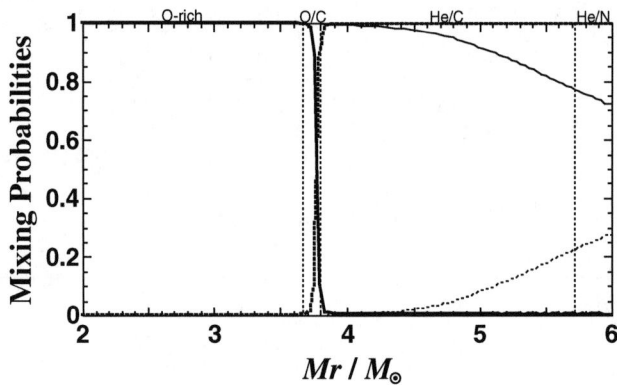

FIGURE 1. Mixing probabilities of supernova neutrinos in the case of normal mass hierarchy, $\sin^2 2\theta_{13} = 1 \times 10^{-2}$, and the neutrino energy of 50 MeV. Thick solid line and thick dashed line indicate the mixing probabilities of $\nu_e \to \nu_e$ and $\bar{\nu}_e \to \bar{\nu}_e$, respectively. Thin solid line shows the sum of the mixing probabilities of $\nu_\mu \to \nu_e$ and $\nu_\tau \to \nu_e$. Thin dashed line corresponds to antineutrino case.

The mixing probability from ν_e to ν_e changes from unity to close to zero in the O/C layer, i.e., all ν_e emitted from the proto-neutron star change to other flavors. At the same time, the sum of the mixing probabilities from ν_μ to ν_e and from ν_τ to ν_e changes from zero to unity. So, if the numbers of ν_e, ν_μ, and ν_τ are the same at the neutrino sphere, the number of ν_e in the He layer is almost the same as at the neutrino sphere and all ν_e are converted from ν_μ and ν_τ. On the other hand, the mixing probability from $\bar{\nu}_e$ to $\bar{\nu}_e$ changes from 1 to 0.72 gradually in the He layer. The sum of the mixing probabilities from $\bar{\nu}_\mu$ to $\bar{\nu}_e$ and from $\bar{\nu}_\tau$ to $\bar{\nu}_e$ also changes gradually from 0 to 0.28. Since there are no resonances of antineutrinos, flavor change of antineutrinos is smaller than that of neutrinos. In the O-rich layers, flavor change probabilities are almost zero; the neutrino mixing does not occur practically.

Since the mixing probabilities change in the O/C and He layers, the neutrino spectra also change there. Especially, the spectrum of ν_e in the He layer becomes close to that of $\nu_{\mu,\tau}$ at the neutrino sphere. This change affects the contribution of charged-current ν-process reactions and ^{7}Li and ^{11}B yields.

^{7}LI AND ^{11}B PRODUCTION WITH NEUTRINO OSCILLATIONS

Mass fraction distributions of ^{7}Li and ^{11}B in the case of normal mass hierarchy and $\sin^2 2\theta_{13} = 1 \times 10^{-2}$ are shown in Fig. 2. The mass fractions of ^{7}Be and ^{11}C with neutrino oscillations are clearly larger than those without oscillations in the He layer. On the other hand, the mass fractions of ^{7}Li and ^{11}B with neutrino oscillations are slightly larger even in the He layer. In inner O-rich layers, the mass fractions of these species are not affected by neutrino oscillations. The effect of neutrino oscillations on the mass fractions of these elements depend on their production processes.

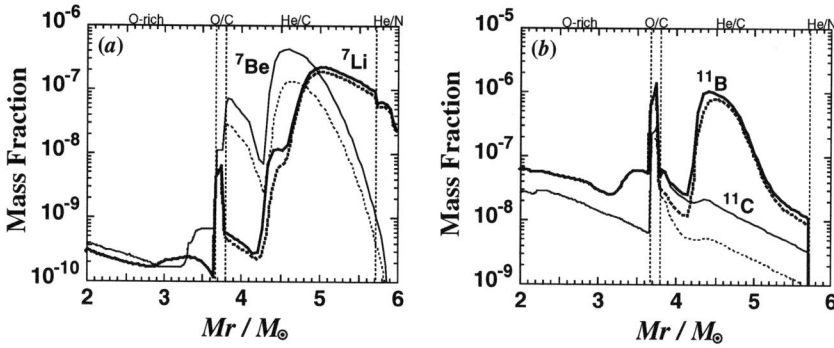

FIGURE 2. Mass fraction distributions of ^{7}Li (a) and ^{11}B (b) in the case of normal mass hierarchy and $\sin^2 2\theta_{13} = 1 \times 10^{-2}$. The mass fractions of ^{7}Li (thick lines) and ^{7}Be (thin lines) in panel (a) and those of ^{11}B (thick lines) and ^{11}C (thin lines) in panel (b) are shown separately. Solid lines and dashed lines correspond to the mass fractions with neutrino oscillations and without neutrino oscillations, respectively.

In the He layer, ^{7}Be is mainly produced through ^{4}He($\nu,\nu' n$)^{3}He(α,γ)^{7}Li; ^{4}He($\nu,\nu' n$)^{3}He is an important ν-process reaction. A charged-current ν-process reaction ^{4}He($\nu_e,e^- p$)^{3}He also produces ^{3}He but the contribution of this reaction is small. This is because the temperature of ν_e is smaller than $\nu_{\mu,\tau}$. When neutrino oscillations are taken into account, however, almost all ν_e in the He layer are converted from $\nu_{\mu,\tau}$. The ν_e temperature becomes close to that of $\nu_{\mu,\tau}$ at the neutrino sphere. This increase in the ν_e temperature raises the reaction rate of ^{4}He($\nu_e,e^- p$)^{3}He. Thus, the mass fraction of ^{7}Be increases by neutrino oscillations in the He layer.

The increase in the mass fraction of ^{11}C in the He layer is due to a similar reason. Most of ^{11}C is produced through ^{12}C($\nu,\nu' n$)^{11}C. The corresponding charged-current ν-process reaction is ^{12}C($\nu_e,e^- p$)^{11}C. The increase in the ν_e temperature due to the neutrino oscillations raises the reaction rate of ^{12}C($\nu_e,e^- p$)^{11}C.

In the case of ^{7}Li, main production process is ^{4}He($\nu,\nu' p$)^{3}H(α,γ)^{7}Li. The corresponding charged-current reaction is ^{4}He($\bar{\nu}_e,e^+ n$)^{3}H. As mentioned above, there are no resonances for antineutrinos. So, the average $\bar{\nu}_e$ energy scarcely increases and, therefore, the reaction rate of ^{4}He($\bar{\nu}_e,e^+ n$)^{3}H scarcely does. The increase in the mass fraction of ^{7}Li due to ^{4}He($\bar{\nu}_e,e^+ n$)^{3}H is seen in the range of $M_r \geq 4.8 M_\odot$. In the inner range of the He layer ^{7}Li is also produced through ^{7}Be(n,p)^{7}Li. The increase in the mass fraction of ^{7}Li in there is partly due to the increase in the rate of ^{4}He($\nu_e,e^- p$)^{3}He.

The production process of ^{11}B is similar to that of ^{7}Li. Most of ^{11}B is produced in the region inside the ^{7}Li abundant region. In this region ^{11}B is mainly produced through ^{7}Li(α,γ)^{11}B during explosive nucleosynthesis. Therefore, the increase in the ^{11}B production by neutrino oscillations is similar to that of ^{7}Li. A small fraction of ^{11}B is produced through ^{12}C($\nu,\nu' p$)^{11}B. The corresponding charged-current reaction is ^{12}C($\bar{\nu}_e,e^+ n$)^{11}B but its contribution is small even including the neutrino oscillations.

We show the relation of ^{7}Li and ^{11}B yield ratios to the mixing angle $\sin^2 2\theta_{13}$ in Fig. 3. The yield ratio means the ratio of the yield taking account of neutrino oscillations to that

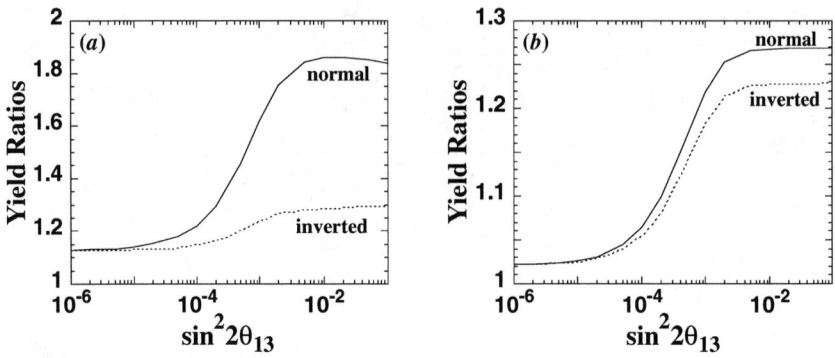

FIGURE 3. The relation of yield ratios of ^{7}Li (*a*) and ^{11}B (*b*) to $\sin^2 2\theta_{13}$. Solid lines and dashed lines indicate normal mass hierarchy and inverted mass hierarchy, respectively.

without neutrino oscillations. The ^{7}Li and ^{11}B yields without neutrino oscillations are $2.36 \times 10^{-7} M_\odot$ and $5.63 \times 10^{-7} M_\odot$, respectively. Both ^{7}Li and ^{11}B yield ratios depend on mass hierarchy and $\sin^2 2\theta_{13}$.

The ^{7}Li yield ratio is about 1.1 in the range of $\sin^2 2\theta_{13} \lesssim 2 \times 10^{-5}$ independent of mass hierarchy. Since the H-resonance is nonadiabatic, there is no clear difference in mixing probabilities between normal and inverted mass hierarchies. Slight increase in the yield is due to the increase in the production in outer region of the He layer. In the range of $2 \times 10^{-5} \lesssim \sin^2 2\theta_{13} \lesssim 2 \times 10^{-3}$, the ^{7}Li yield increases with $\sin^2 2\theta_{13}$. This is because the H-resonance changes from nonadiabatic to adiabatic with increasing $\sin^2 2\theta_{13}$. The difference of the yield ratios by mass hierarchies is also seen. In the range of $\sin^2 2\theta_{13} \gtrsim 2 \times 10^{-3}$ the ^{7}Li yield ratio is about 1.9 in the normal mass hierarchy case and is about 1.3 in the inverted mass hierarchy case; it scarcely depends on $\sin^2 2\theta_{13}$ but depends on mass hierarchy. In this $\sin^2 2\theta_{13}$ range, H-resonance is adiabatic. In the normal mass hierarchy case, the energy spectrum of ν_e in the He layer is close to Fermi distribution of $T_{\nu_e} = 6$ MeV. This change raises the reaction rate of ^{4}He$(\nu_e, e^- p)^3$He and the yield of ^{7}Be. In the inverted mass hierarchy case, the increase in the average energy of $\bar{\nu}_e$ is not so large because the temperature of $\bar{\nu}_e$ is 5 MeV even at the surface of the neutron star. Therefore, the increase in ^{7}Li yield is small.

The dependence of the ^{11}B yield ratio is similar to that of ^{7}Li but the change is smaller and the difference of the yield ratios by mass hierarchy is also smaller. The ^{11}B yield ratio is about 1.02 in the range of $\sin^2 2\theta_{13} \lesssim 2 \times 10^{-5}$. It is about 1.27 in the normal mass hierarchy case and is about 1.23 in the inverted mass hierarchy case in the range of $\sin^2 2\theta_{13} \gtrsim 2 \times 10^{-3}$. Since ^{11}B is mainly produced by way of ^{7}Li, small increase in the reaction rate of ^{4}He$(\bar{\nu}_e, e^+ n)^3$H by neutrino oscillations brings about small increase in the production of ^{11}B. Additional contribution from ^{4}He$(\nu_e, e^- p)^3$He by way of ^{7}Be raises the yield ratio in the normal mass hierarchy case.

We have shown that the ^{7}Li and ^{11}B yields depend on mass hierarchy and $\sin^2 2\theta_{13}$. However, the evaluated yields also depend on the energy spectra of neutrinos emitted from a proto-neutron star [3] and the energy spectra have uncertainties. In order to reduce the dependence on the neutrino energy spectra, we consider the ^{7}Li/^{11}B abundance ratio. Since both ^{7}Li and ^{11}B are mainly produced through the ν-process, their yields are roughly proportional to the ν-process reaction rates. Therefore, the dependence of ^{7}Li/^{11}B on the ν-process reaction rates would be cancelled out. When we do not consider neutrino oscillations, the ^{7}Li/^{11}B abundance ratio is 0.59. In the case of normal mass hierarchy and adiabatic H-resonance, it is 0.88. Thus, the increase in ^{7}Li/^{11}B will be seen if neutrino oscillations with normal mass hierarchy and adiabatic H-resonance occur. Observations of ^{7}Li/^{11}B in supernova remnants or stars formed in regions strongly affected by prior generations of supernovae may constrain neutrino oscillation parameters such as mass hierarchy and $\sin^2 2\theta_{13}$. Details are discussed in [13, 14].

SUMMARY

We investigated the ^{7}Li and ^{11}B production in a supernova taking account of neutrino oscillations. Neutrino oscillations in the supernova raise the reaction rates of charged-current ν-process reactions such as ^{4}He$(\nu_e, e^- p)^3$He and ^{4}He$(\bar{\nu}_e, e^+ n)^3$H. This increase enhances the production of ^{7}Li and ^{11}B. In the case of normal mass hierarchy and $\sin^2 2\theta_{13} \gtrsim 2 \times 10^{-3}$ corresponding to adiabatic H-resonance, the ^{7}Li yield is larger by about a factor of 1.9 than that without neutrino oscillations. The ^{11}B yield is also larger but the increase is smaller than the ^{7}Li case. The ^{7}Li/^{11}B ratio would not be affected by uncertainties by supernova neutrino spectra, so that the observations of ^{7}Li/^{11}B may provide more strict constraints to mass hierarchy and the mixing angle θ_{13}.

This work has been supported in part by the Ministry of Education, Culture, Sports, Science and Technology, Grants-in-Aid for Young Scientist (B) (17740130) and Scientific Research (17540275), for Specially Promoted Research (13002001), and by Mitsubishi Foundation.

REFERENCES

1. S. E. Woosley, D. H. Hartmann, R. D. Hoffman, and W. C. Haxton, *ApJ* **356**, 272–301 (1990).
2. T. Yoshida, M. Terasawa, T. Kajino, and K. Sumiyoshi, *ApJ* **600**, 204–213 (2004).
3. T. Yoshida, T. Kajino, and D. H. Hartmann, *Phys. Rev. Lett.* **94**, 231101 (2005).
4. M. Apollonio, et al., *Eur. Phys. J. C* **27**, 331–374 (2003).
5. Super-Kamiokande Collaboration, Y. Ashie, et al., *Phys. Rev. Lett.* **93**, 101801 (2004).
6. SNO Collaboration, S. N. Ahmed, et al., *Phys. Rev. Lett.* **92**, 181301 (2004).
7. KamLAND Collaboration, T. Araki, et al., *Phys. Rev. Lett.* **94**, 081801 (2005).
8. A. S. Dighe, and A. Y. Smirnov, *Phys. Rev. D* **62**, 033007 (2000).
9. K. Takahashi, M. Watanabe, K. Sato, and T. Totani, *Phys. Rev. D* **64**, 093004 (2001).
10. T. Shigeyama, and K. Nomoto, *ApJ* **360**, 242–256 (1990).
11. T. Shigeyama, K. Nomoto, H. Yamaoka, and F.-K. Thielemann, *ApJ* **386**, L13–L16 (1992).
12. R. D. Hoffman, and S. E. Woosley (1992), URL http://www-phys.llnl.gob/Research/RRSN/nu_csbr/neu_rate.html.
13. T. Yoshida, T. Kajino, H. Yokomakura, K. Kimura, A. Takamura, and D. H. Hartmann, *Phys. Rev. Lett.* **96**, 091101 (2006).
14. T. Yoshida, T. Kajino, H. Yokomakura, K. Kimura, A. Takamura, and D. H. Hartmann, *Submitted to ApJ* (2006).

Supernova Detection Via a Network of Neutral Current Spherical TPC's

J.D. Vergados* and Y. Giomataris[†]

*University of Ioannina, Ioannina, Gr 451 10, Greece and RCNP, Osaka University, Ibaraki,Japan
[†]CEA, Saclay, DAPNIA, Gif-sur-Yvette, Cedex,France

Abstract. The coherent contribution of all neutrons in neutrino nucleus scattering due to the neutral current offers a realistic prospect of detecting supernova neutrinos. For a typical supernova at 10 kpc, about 1000 events are expected using a spherical gaseous detector of radius 4 m and employing Xe gas at a pressure of 10 Atm. We propose a world wide network of several such simple, stable and low cost supernova detectors with a running time of a few centuries.

Keywords: spherical gaseous TPC, neutral current neutrino detectors, supernova neutrinos, neutrino cross section, nuclear recoils, supersymmetry, R-parity violating SUSY
PACS: 13.15.+g, 14.60Lm, 14.60Bq, 23.40.-s, 95.55.Vj, 12.15.-y

INTRODUCTION.

In a typical supernova an energy of about 10^{53} ergs is released in the form of neutrinos [1],[2]. These neutrinos are emitted within an interval of about 10 s after the explosion and they travel to Earth undistorted, except that, on their way to Earth, they may oscillate into other flavors. Thus for traditional detectors relying on the charged current interactions the precise event rate may depend critically on the specific properties of the neutrinos. The time integrated spectra in the case of charged current detectors, like the SNO experiment, depend on the neutrino oscillations [3]. An additional problem is the fact that the charged current cross sections depend on the details of the structure of the nuclei involved. With neutral current detectors one exploits the fact that the vector component of the current can lead to coherence, i.e. an additive contribution of all neutrons in the nucleus (the proton component is tiny). Furthermore the deduced neutrino fluxes do not depend on the neutrino oscillation parameters (e.g. the mixing angles). Even in our case, however, the obtained rates depend on the assumed characteristic temperature for each flavor.

Recently it has become feasible to detect neutrinos by measuring the recoiling nucleus and employing gaseous detectors with much lower threshold energies. Thus one is able to explore the advantages offered by the neutral current interaction, exploring ideas put forward more than a decade ago [4]. A description of the NOSTOS project and details of the spherical TPC detector are given in [5]. We have built a spherical prototype 1.3 m in diameter which is described in [6]. The outer vessel is made of pure Cu (6 mm thick) allowing to sustain pressures up to 5 bar. The inner detector is just a small sphere, 10 mm in diameter and developments are currently under way to build a spherical TPC detector using new technologies. First tests were performed by filling the volume with argon mixtures and are quite promising. High gains are easily obtained and the signal to noise is large enough for sub-keV threshold. The whole system looks stable and robust made of stainless steel as a proportional counter located at the center of curvature of the TPC. Furthermore this interaction, through its vector component, can lead to coherence, i.e. an additive contribution of all neutrons in the nucleus (the vector contribution of the protons is tiny).

In this paper we will derive the amplitude for the differential neutrino nucleus coherent cross section. Then we will estimate the expected number of events for all the noble gas targets. We will show that these results can be exploited by a network of small and relatively cheap spherical TPC detectors placed in various parts of the world (for a description of the apparatus see our earlier work [5]). The operation of such devices as a network will minimize the background problems. There is no need to go underground, but one may have to go sufficiently deep underwater to balance the high pressure of the gas target. Other types of detectors have also been proposed [7],[8].

Large gaseous volumes are easily obtained by employing long drift technology (i.e TPC) that can provide massive targets by increasing the gas pressure. Combined with an adequate amplifying structure and low energy thresholds, a three-dimensional reconstruction of the recoiling particle, electron or nucleus, can be obtained. The use of new micropattern detectors and especially the novel Micromegas [9] provide excellent spatial and time accuracy that is a precious tool for pattern recognition and background rejection [10],[11]. The virtue of using such large gaseous volumes and the new high precision microstrip gaseous detectors has been recently discussed in a dedicated workshop [12] and their relevance for low energy neutrino physics and dark matter detection has been widely recognized. Such low-background low-energy threshold systems are actually successfully used in the CAST [13], the solar axion experiment, and are under development for several low energy neutrino or dark matter projects [5],[14].

STANDARD AND NON STANDARD WEAK INTERACTION

The standard neutral current left handed weak interaction can be cast in the form:

$$\mathcal{L}_q = -\frac{G_F}{\sqrt{2}} \left[\bar{v}_\alpha \gamma^\mu (1-\gamma^5) v_\alpha \right] \left[\bar{q}\gamma_\mu (g_V(q) - g_A(q)\gamma^5) q \right] \quad (1)$$

(diagonal in flavor space). At the nucleon level we get:

$$\mathcal{L}_N = -\frac{G_F}{\sqrt{2}} \left[\bar{v}_\alpha \gamma^\mu (1-\gamma^5) v_\alpha \right] \left[\bar{N}\gamma_\mu (g_V(N) - g_A(N)\gamma^5) N \right] \quad (2)$$

with

$$g_V(p) = \frac{1}{2} - 2\sin^2\theta_W \simeq 0.04 \ , \ g_A(p) = 1.27\frac{1}{2} \ ; \ g_V(n) = -\frac{1}{2} \ , \ g_A(n) = -\frac{1.27}{2} \quad (3)$$

Beyond the standard level one has further interactions which need not be diagonal in flavor space:

$$g_V(q) - g_A(q)\gamma^5 \rightarrow \left(g_V^{SM}(q) - g_A^{SM}(q)\gamma^5 \right) \delta_{\alpha\beta} + \left(\lambda^{qL}\delta_{\alpha\beta} + \varepsilon_{\alpha\beta}^{qL} \right)(1-\gamma^5)$$

$$\left[\bar{v}_\alpha \gamma^\mu (1-\gamma^5) v_\alpha \right] \rightarrow \left[\bar{v}_\alpha \gamma^\mu (1-\gamma^5) v_\beta \right] \quad (4)$$

Furthermore at the nucleon level

$$g_V(N) - g_A(N)\gamma^5 \to \left(g_V^{SM}(N) - g_A^{SM}(N)\gamma^5\right)\delta_{\alpha\beta} + \left(\lambda^{NL}\delta_{\alpha\beta} + \varepsilon_{\alpha\beta}^{NL}\right)(1 - 1.27\gamma^5)$$
$$\left[\bar{v}_\alpha\gamma^\mu(1-\gamma^5)v_\alpha\right] \to \left[\bar{v}_\alpha\gamma^\mu(1-\gamma^5)v_\beta\right] \quad (5)$$

with

$$\lambda^{pL} = 2\lambda^{uL} + \lambda^{dL}, \quad \lambda^{nL} = \lambda^{dL} + 2\lambda^{dL}, \quad \varepsilon_{\alpha\beta}^{pL} = 2\varepsilon_{\alpha\beta}^{uL} + \varepsilon_{\alpha\beta}^{dL}, \quad \varepsilon_{\alpha\beta}^{nL} = \varepsilon_{\alpha\beta}^{uL} + 2\varepsilon_{\alpha\beta}^{dL} \quad (6)$$

In the above expressions λ^{qL} can arise, e.g., from radiative corrections, see e.g. PDG [15] and $\varepsilon_{\alpha\beta}^{qL}$ from R-parity violating interactions in supersymmetric models [16]-[17]. Indeed since R-parity conservation has no robust theoretical motivation one may accept an extended framework of the MSSM with R-parity non-conservation MSSM. In this case the superpotential W acquires additional R-parity violating terms:

$$W_{\cancel{R}_p} = \lambda_{ijk}L_iL_jE_k^c + \lambda'_{ijk}L_iQ_jD_k^c + \lambda''_{ijk}U_i^cD_j^cD_k^c + \mu_jL_jH_u \quad (7)$$

Of interest to us here is the $\lambda'_{ijk}L_iQ_jD_k^c$ involving first generation quarks and s-quarks, i.e the term $\lambda'_{\alpha 11}L_\alpha Q_1 D_1^c$. From this term in four component notation we get the contribution

$$\lambda'_{\alpha 11}\left(\bar{d}_R^c v_{\alpha L} - \bar{u}_R^c \alpha_L\right)\tilde{d}^c, \quad \alpha = e, \mu, \tau$$

where $v_{\alpha L} = \frac{1}{2}(1-\gamma_5)v_\alpha$ etc. Thus

- The first term at tree level yields the interaction

$$-\frac{\lambda'_{\alpha 11}\lambda'_{\beta 11}}{m_{\tilde{d}_L}^2}\bar{v}_{\alpha L}d_R^c \bar{d}_R^c v_{\beta L} \to \frac{1}{2}\frac{\lambda'_{\alpha 11}\lambda'_{\beta 11}}{m_{\tilde{d}_L}^2}\bar{v}_{\alpha L}\gamma^\mu v_{\beta L}\bar{d}_R^c\gamma_\mu d_R^c \quad (8)$$

The previous equation can be cast in the form:

$$\mathcal{L}_d = -\frac{G_F}{\sqrt{2}}\varepsilon_{\alpha\beta}^d\left[\bar{v}_\alpha\gamma^\mu(1-\gamma^5)v_\beta\right]\left[\bar{d}\gamma_\mu(1-\gamma^5)d\right]; \varepsilon_{\alpha\beta}^d = \lambda'_{\alpha 11}\lambda'_{\beta 11}\frac{m_W^2}{m_{\tilde{d}_L}^2} \quad (9)$$

There is no such term associated with the u quark, $\varepsilon_{\alpha\beta}^u = 0$.

- Proceeding in an analogous fashion the collaborative effect of the first and second term, for $\alpha, \beta = e, \mu, \tau$, yields the charged current contribution:

$$\mathcal{L}_{du} = \frac{G_F}{\sqrt{2}}\varepsilon_{\alpha\beta}^d\left[\bar{\alpha}\gamma^\mu(1-\gamma^5)v_\beta\right]\left[\bar{u}\gamma_\mu(1-\gamma^5)d\right] \quad (10)$$

- Finally the second term, for $\alpha, \beta = e, \mu, \tau$, leads to a neutral current contribution of the charged leptons:

$$\mathcal{L}_u = \frac{G_F}{\sqrt{2}}\varepsilon_{\alpha\beta}^d\left[\bar{\alpha}\gamma^\mu(1-\gamma^5)\beta\right]\left[\bar{u}\gamma_\mu(1-\gamma^5)u\right] \quad (11)$$

The above non standard flavor changing neutral current interaction have been found to play an important role in the in the infall stage of a stellar collapse [18]. Furthermore precise measurements involving the neutral current neutrino-nucleus interactions may yield valuable information about the non standard interactions [19]. They are not, however, going to be further considered in this work.

COHERENT NEUTRINO NUCLEUS SCATTERING

The cross section for elastic neutrino nucleon scattering has extensively been studied [1],[20].

From the expressions of the previous section we see that the coherent contribution [21] may come from the neutrons and is expected to be proportional to the square of the neutron number. The neutrino-nucleus coherent cross section takes the form:

$$\left(\frac{d\sigma}{dT_A}\right)_{weak} = \frac{G_F^2 A m_N}{2\pi}(N^2/4)F_{coh}(A,T_A,E_\nu),$$

$$F_{coh}(A,T_A,E_\nu) = \left(1 + \frac{A-1}{A}\frac{T_A}{E_\nu}\right) + (1 - \frac{T_A}{E_\nu})^2$$

$$\left(1 - \frac{A-1}{A}\frac{T_A}{m_N}\frac{1}{E_\nu/T_A - 1}\right) - \frac{A m_N T_A}{E_\nu^2} \quad (12)$$

SUPERNOVA NEUTRINOS

The number of neutrino events for a given detector depends on the neutrino spectrum and the distance of the source. We will consider a typical case of a source which is about 10 kpc, I.e. $D = 3.1 \times 10^{22}$ cm (of the order of the radius of the galaxy) with an energy output of 3×10^{53} ergs with a duration of about 10 s. Furthermore we will assume for simplicity that each neutrino flavor is characterized by a Fermi-Dirac like distribution times its characteristic cross section and we will not consider here the more realistic distributions, which have recently become available [22]. This is adequate for our purposes. Thus:

$$\frac{dN}{dE_\nu} = \sigma(E_\nu)\frac{E_\nu^2}{1+\exp(E_\nu/T)} = \frac{\Lambda}{JT}\frac{x^4}{1+e^x}, \quad x = \frac{E_\nu}{T} \quad (13)$$

with $J = \frac{31\pi^6}{252}$, Λ a constant and T the temperature of the emitted neutrino flavor. Each flavor is characterized by its own temperature as follows:

$$T = 8 \text{ MeV for } \nu_\mu, \nu_\tau, \tilde{\nu}_\mu, \tilde{\nu}_\tau \text{ and } T = 5 \ (3.5) \text{ MeV for } \tilde{\nu}_e \ (\nu_e)$$

The constant Λ is determined by the requirement that the distribution yields the total energy of each neutrino species.

$$U_\nu = \frac{\Lambda T}{J}\int_0^\infty dx \frac{x^5}{1+e^x} \Rightarrow \Lambda = \frac{U_\nu}{T}$$

We will further assume that $U_\nu = 0.5 \times 10^{53}$ ergs per neutrino flavor. Thus one finds:

$$\Lambda = 0.89 \times 10^{58} \; (\nu_e), \; 0.63 \times 10^{58} \; (\tilde{\nu}_e), 0.39 \times 10^{58} \; \text{(all other flavors)}$$

The differential event rate (with respect to the recoil energy) is proportional to the quantity:

$$\frac{dR}{dT_A} = \frac{\lambda(T)}{J} \int_0^\infty dx F_{coh}(A, T_A, xT) \frac{x^4}{1+e^x} \tag{14}$$

with $\lambda(T) = (0.89, 0.63, 0.39)$ for $\nu_e, \tilde{\nu}_e$ and all other flavors respectively. This is shown in Figs. 1 and 1. The total number of expected events for each neutrino species can be

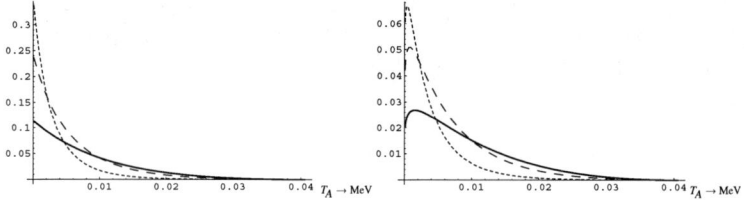

FIGURE 1. The differential event rate as a function of the recoil energy T_A, in arbitrary units, for Xe. On the left we show the results without quenching, while on the right the quenching factor is included. We notice that the effect of quenching is more prevalent at low energies. The short dash, long dash and continuous curve correspond to $\nu_e, \tilde{\nu}_e$ and all other flavors respectively.

cast in the form:

$$\text{No of events} = \tilde{C}_V(T) h(A, T, (T_A)_{th}), \; h(A, T, (T_A)_{th}) = \frac{F_{fold}(A, T, (T_A)_{th})}{F_{fold}(40, T, (T_A)_{th})} \tag{15}$$

with

$$F_{fold}(A, T, (T_A)_{th}) = \frac{A}{J} \int_{(T_A)_{th}}^{(T_A)_{max}} \frac{dT_A}{1 MeV} \times \int_0^\infty dx F_{coh}(A, T_A, xT) \frac{x^4}{1+e^x} \tag{16}$$

and

$$\tilde{C}_V(T) = \frac{G_F^2 m_N 1 MeV}{2\pi} \frac{N^2}{4} \Lambda(T) \frac{1}{4\pi D^2} \frac{PV}{kT_0} \tag{17}$$

Where k is Boltzmann's constant, P the pressure, V the volume, and T_0 the temperature of the gas. Summing over all the neutrino species we can write:

$$\text{No of events} = C_V \frac{K(A, (T_A)_{th})}{K(40, (T_A)_{th})} Qu(A) \tag{18}$$

with

$$C_V = 153 \left(\frac{N}{22}\right)^2 \frac{U_\nu}{0.5 \times 10^{53} ergs} \left(\frac{10 kpc}{D}\right)^2 \frac{P}{10 Atm} \left[\frac{R}{4m}\right]^3 \frac{300}{T_0} \tag{19}$$

$K(A, (T_A)_{th})$ is the rate at a given threshold energy divided by that at zero threshold. It depends on the threshold energy, the assumed quenching factor and the nuclear mass number. It is unity at $(T_A)_{th} = 0$. From the above equation we find that, ignoring quenching, the following expected number of events:

$$1.25, \ 31.6, \ 153, \ 614, \ 1880 \text{ for He, Ne, Ar, Kr and Xe} \tag{20}$$

respectively. For other possible targets the rates can be found by the above formulas or interpolation.

The quantity $Qu(A)$ is the quenching factor [23]-[24], assuming a threshold energy $(T_A)_{th} = 100$eV. The parameter $Qu(A)$ takes the values:

$$0.49, \ 0.38, \ 0.35, \ 0.31, \ 0.29 \text{ for He, Ne, Ar, Kr and Xe} \tag{21}$$

respectively. The effect of quenching is larger in the case of heavy targets, since the average energy of the recoiling nucleus is smaller. Thus the number of expected events for Xe assuming a threshold energy of 100 eV is reduced to about 560.

The effect of quenching is exhibited in Fig 2 for the two interesting targets Ar and Xe. We should mention that it is of paramount importance to experimentally measure the

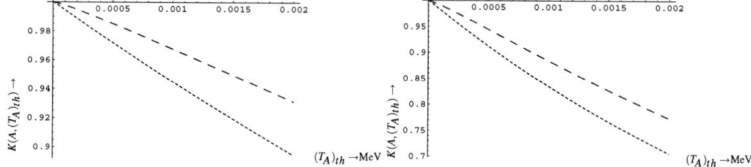

FIGURE 2. The function $K(A, (T_A)_{th})$ versus $(T_A)_{th}$ for the target Ar on the left and Xe on the right. The short and long dash correspond to no quenching and quenching factor respectively. One sees that the effect of quenching is less pronounced at higher thresholds. The differences appear small, since we present here only the ratio of the rates to that at zero threshold. The effect of quenching at some specific threshold energy is not shown here. For a threshold energy of 100 eV the rates are quenched by factors of 3 and 3.5 for Ar and Xe respectively (see Eq. (21)).

quenching factor. The above estimates were based on the assumption of a pure gas. Such an effect will lead to an increase in the quenching factor and needs be measured.

CONCLUSIONS

In the present study it has been shown that it is quite simple to detect typical supernova neutrinos in our galaxy, provided that such a supernova explosion takes place (one explosion every 30 years is estimated [25]). The idea is to employ a small size spherical TPC detector filled with a high pressure noble gas. An enhancement of the neutral current component is achieved via the coherent effect of all neutrons in the target. Thus employing, e.g., Xe at 10 Atm, with a feasible threshold energy of about 100 eV in the detection the recoiling nuclei, one expects between 600 and 1900 events, depending on the quenching factor. We believe that networks of such dedicated detectors, made out of simple, robust and cheap technology, can be simply managed by an international scientific consortium and operated by students. This network comprises a system, which can be maintained for several decades (or even centuries). This is is a key point towards being able to observe few galactic supernova explosions.

acknowledgments: One of the authors (JDV) is indebted for support and hospitality to the OMEG05 organizing committee during the OMEG05 conference and to Professor Hiroshi Toki during a visit to RCNP.

REFERENCES

1. J.F. Beacom, W.M. Farr and P. Vogel, Phys. Rev D **66** (2002) 033001;hep-ph/0205220
2. J.R. Wilson and R.W.Mayle, Phys. Rept. **227** (1993) 97.
   M. Herant, W. Benz, W.R. Hix, C.L. Fryer and S.A. Golgate, Astrophys. J. **435** (1994) 339.
   M. Rampp and H.T. Janka, Astrophys. J. **539** (2000) L33.
   A. Mezzacappa, M. Liebendorfer, O.E. Messer, W.R. Hix, F.K. Thielemann and S.W. Bruenn, Phys. Rev. Lett. **86** (2001) 1935.
   C.L. Fryer and A. Heger Astrophys. J. **541** (2000) 1033.
   G.G. Raffelt, Nuc. Phys. Proc. Suppl. **110** (2002) 254;hep-ph/0201099;
   R. Tomas, M. Kachellriess, G.G. Raffelt, A.Dighe, A-T Janka and L. Schreck, JCAP **0409** (2004) 015;
   R. Tomas, D. Semikoz, G.G. Raffelt, M. Kachellriess and A.S. Dighe, Phys. Rev. D **68** (2002) 093013;
   M.T. Keil, G.G. Raffelt, A-T Janka, Astrophys. J. **590** (2003) 971;
   J.F. Beacom, R.N. Boyd and A. Mezzacappa, Phys. Rev. D **63** (2001) 073011.
   M.K. Sharp, J.F. Beacom J.A. Formaggio, Phys. Rev. D **66** (2002) 013012; hep-ph/0205035.
   A. Burrows, J. Hayes and B.A. Fryxell, Astrophys. J. **450** (1995) 830.
3. K. Takahashi, K. Sato, A. Burrows, T. A. Thompson, Phys.Rev. D **68** (2003) 113009; hep-ph/0306056
4. A. Burrows, D. Klein and R. Gandhi, Phys. Rev. D **45** (1992) 3361.
5. Y. Giomataris and J.D. Vergados, Nucl. Instr. Meth. A **530** (2004) 330.
6. The NOSTOS experiment and new trends in rare event detection, I. Giomataris et al, hep-ex/0502033, submitted to the SIENA2004 International Conference (2005).
7. P. Barbeau, J.I. Collar, J. Miyamoto and I. Shipsey, IEEE Trans. Nucl. Sci. **50** (2003) 1285.
8. C. Hagmann and A. Bernstein, IEEE Trans. Nucl. Sci. **51** (2004) 2151.
9. I. Giomataris et al., Nucl. Instr. Meth. A **376** (1996) 29
10. J.I. Collar and Y. Giomataris, Nucl. Inst. Meth. **471** (2001) 254
11. P. Gorodetzky et al., Nucl.Phys.Proc.Suppl. **138** (2005) 56
12. Second workshop on large TPC for low energy rare event detection, 20-21 December 2004, Paris, France.
13. C.E. Aalseth et al., Nucl.Phys.Proc.Suppl. **110** (2002) 85.
14. I. Giomataris et al., hep-ex/0502033.
    T. Patzak et al., Nucl. Instr. Meth. A **434** (1999) 358.
    B. Ahmed *et al*, Astropart. Phys. **19** (2003) 691.
15. S. Eidelman **et al** [Particle Data Group], Phys. Lett. B **592** (2004) 1.
16. M. Hirsch, M.A. Diaz, W. Porod, C. Romao and J,W.F. Valle, Phys. Rev. D **62** (2000) 11308-1.
17. O. Haug, Amand Faessler, J.D. Vergados and S. Kovalenko, Nuc. Phys. B565 (2000) 38; hep-ph/9909318.
18. P.S. Amanic, G.M. Fuller and B. Gristein, Astropart. Phys. **24** (2005) 160.
19. J. Barranco, O.G. Miranda and T.I. Rashba, Probing new physics with coherent neutrino scattering off nuclei, MPP-2005-85; hep-ph/0508299.
20. P. Vogel and J. Engel, *Phys. Rev. D* **39** (1989) 3378.
21. E.A. Paschos and A. Kartavtsev, hep-ph/0309148.
22. T. Totani, K. Sato, H.E. Dalhed and J.R. Wilson, Astrophys.J. 496 (1998) 216.
    R. Buras, Hans-Thomas Janka (1), M. Th. Keil , G. G. Raffelt and M. Rampp, Astrophys.J. 587 (2003) 320.
23. E. Simon *et al*, Nucl. Instr. Meth. A 507 (2003) 643; astro-ph/0212491.
24. J. Lidhart *et al*, Mat. Phys. Medd. Dan. Vid. Selsk. 33 (10) (1963) 1 Y. Giomataris, P. Rebourgeard, J.P. Robert, Georges Charpak, Nucl. Instrum. Meth. A **376** (1996) 29
25. K. Scholberg, Nucl. Phys. Proc. Suppl. 91 (2000) 331; hep-ex/0008044.

DM search by studying X-rays following WIMP nuclear interactions

H. Ejiri*, Ch.C. Moustakidis[†] and J.D. Vergados**

*INT, University of Washington, Seattle, WA 98195, USA;
JASRI-Spring8, Mikazuki-cho, Hyogo, 679-5198, Japan
[†]Department of Theoretical Physics, Aristotle University of Thessaloniki,
54124 Thessaloniki, Greece
**University of Ioannina, Ioannina, GR 45110, Greece

Abstract. Weakly interacting massive particles (WIMPs) are shown to be studied by measuring X-rays following inner-shell ionizations associated with WIMP nuclear interactions. Inner-shell electrons are well ionized by WIMP nuclear interactions. K shell ionization probabilities are of the order of 0.20 ∼ 0.25 in case of 100 ∼ 300 GeV WIMPs interacting with medium heavy nuclei. Thus K X-rays following K shell ionizations can be used to search for WIMPs. Exclusive studies of K X-rays in coincidence with nuclear recoils provide a good opportunity for high sensitivity measurements of WIMPs in the $\sigma_p \approx 10^{-6}$ pb region. Several detector options for the exclusive X-ray studies are discussed.

Keywords: Dark matter, WIMPs, LSP, X-rays, inner-shell ionization
PACS: 95.35+d, 12.60.Jv.

INTRODUCTION

It is of great interest to search for dark matter (DM) from the point of both cosmology and particle physics. The present report is based mainly on the recent works of possible X-rays following inner-shell ionizations associated with DM nuclear interactions [1, 2, 3]. A part of the present work has been presented at the recent Erice school on neutrinos in 2005 [4].

Recent experimental observations [5, 6, 7, 8, 9] have shown that the universe consists of dark matter with $\Omega_{DM} \approx 30\%$, dark energy with $\Omega_{DE} \approx 65\%$ and baryonic matter with $\Omega_{BM} \approx 5\%$, with Ω_{XY} being the fraction of the mass.

DM is considered to be mainly Weakly Interaction Massive Particles (WIMPs) such as the Lightest Super-symmetric Particles (LSP). Search for WIMPs is very interesting also for new particle physics beyond the standard theory.

Interaction of WIMPs is quite weak, and their velocity is very slow. Accordingly, WIMP signals are extremely rare in rate and are very low in energy. Thus experimental studies of WIMPs require high sensitivity low-BG detectors and/or novel methods in order to select the WIMP signals from all kinds of BG ones. Experimental searches for cold dark matter have been made by attempting to observe the elastic coherent scattering off nuclei [10, 11, 12, 13], since the cross section gain is increased by the factor A^2 with A being the mass number.

The recoil-energy spectrum of the elastic scattering, however, shows a monotonic pattern, decreasing with the recoil energy. This is just the same behavior as that of the background. Thus it is very hard to separate and identify the DM-recoil component from the background ones in the spectrum.

So far several novel methods have been suggested to search for WIMPs, and some of them have been tried experimentally. Observation of seasonal variation of low-energy spectra have been claimed to be due to the cold DM with $M \approx 30 - 100$ GeV [10]. It is noted that the seasonal variation of BGs from Rn and others might contribute to the spectrum modulation, depending on the Rn content [14]. Recent experiments such as EDELWEISS [11, 12] and CDM [13] almost exclude the cold DM with scalar type interaction in the mass region claimed in [10].

It has been shown [15, 16] that γ-rays following inelastic scattering off nuclei provide a sensitive way to study WIMPs since the γ energy is as large as 10 - 100 keV and is monochromatic. The fact that no quenching effect occurs on the electrons/γ-rays, in contrast to the nuclear recoil in solid detectors, makes it easier to measure the γ-rays. Inelastic scattering in ^{127}I and other odd-A nuclei can be used to search for spin-coupled DM candidates.

In the previous paper [1, 2], we have shown that the measurement of ionized electrons via WIMP nucleus interactions can be a good and realistic way for direct detection of WIMPs. Inner-shell excitation probability is shown to be quite sizable [2].

The present paper aims at showing that X-rays from inner-shell excitation following WIMP nuclear interaction can be used to search for medium and heavy WIMPs.

Unique features of using such X-rays are as follow.
(1) K X-rays from medium heavy nuclei give sharp peaks in the energy region of 20-80 keV, which is well above the detector noise and BG levels.
(2) They are used to study WIMPs with both spin-independent and spin dependent interactions.
(3) Exclusive studies of X-rays in coincidence with nuclear recoils reduce BG rates.

Thus such X-rays provide a good opportunity for high sensitivity studies of WIMPs beyond $M_{LSP} \sim 100$ GeV with the sensitivity of the order of $10^{-6} \sim 10^{-7}$ pb proton cross section. Inner-shell ionizations are described in section 2, and production rates for X-rays in section 3. Detectors for such X-rays are discussed in section 4, and concluding remarks are given in section 5.

INNER-SHELL IONIZATION BY WIMP NUCLEAR INTERACTION

Ionization rates of atomic electrons in WIMP nuclear interactions have been discussed in the previous papers [1, 2]. Using the same notations there, the ratio of the $n\ell$-shell electron ionization rate to the nuclear recoil rate is given as [2]

$$Z \frac{\sigma_{n\ell}}{\sigma_r} = Z p_{n\ell} \int |\phi_{n\ell} \sqrt{2m_e T}|^2 f(T, \varepsilon_{n\ell}) m_e \sqrt{2m_e T} \, dT, \tag{1}$$

$$f(T, \varepsilon_{n\ell}) = \frac{\int N v^2 e^{-v^2/v_0^2} \sinh(2v/v_0) dv}{\int D v^2 e^{-v^2/v_0^2} \sinh(2v/v_0) dv}, \tag{2}$$

where $p_{n\ell}$ is the probability of one electron in the $n\ell$ shell, and $[\sigma_{n\ell}/\sigma_r]$ is the ratio normalized to one electron per atom.

TABLE 1. Inner-shell ionization ratios for WIMP nuclear interactions in ^{131}Xe. The ratios mormalized to one electro per atom are given in the 3 - 5 columns. $[X/Y]_L$, $[X/Y]_M$ and $[X/Y]_H$ stand for the cross section ratios $[X/Y]$ for light (L), medium (M) and heavy (H) WIMP's with M=30, 100, and 300 GeV, respectively.

$n\ell$	$-\varepsilon_{n\ell}$ keV	$[\sigma_{n\ell}/\sigma_r]_L$	$[\sigma_{n\ell}/\sigma_r]_M$	$[\sigma_{n\ell}/\sigma_r]_H$	$[Z\sigma_{n\ell}/\sigma_r]_L$	$[Z\sigma_{n\ell}/\sigma_r]_M$	$[Z\sigma_{n\ell}/\sigma_r]_H$
1s	34.56	0.0006	0.0041	0.0047	0.034	0.221	0.255
2s	5.45	0.0224	0.0271	0.0271	1.211	1.461	1.463
2p	4.89	0.0703	0.0834	0.0836	3.796	4.506	4.513
3p	0.96	0.1017	0.1050	0.1050	5.492	5.670	5.670
3d	0.68	0.1734	0.1775	0.1775	9.364	9.585	9.585

The inner-shell ionization rates are evaluated for WIMPs interacting with ^{131}Xe isotopes. The K and L shell excitation rates with respect to the nuclear recoil rates, together with the binding energies, are shown for light, medium and heavy WIMPs with the masses of 30, 100 and 300 GeV in Table 1.

It is very important to note that the K shell ionization probability is as large as 0.2 $\sim$ 0.25 for medium heavy WIMPs with $M = 100 \sim 300$ GeV. The number of the K shell (1s shell) holes increases as the WIMP mass increases. It is quite sizable and almost constant for the WIMP mass ≥ 100 GeV. This is due to the increase of the nuclear recoil velocity. The number of the L shell (2s and 2p) holes is as large as 5$\sim$7 for the WIMP's with $M = 30\sim300$ GeV. Thus L X-rays can be used for light WIMPs if one can separate L X-ray signals from nuclear recoil ones.

X-RAYS FROM INNER-SHELL EXCITATIONS

The $n\ell$ X-ray production rate is simply obtained by using the X-ray branching ratio as

$$\frac{\sigma_{n\ell}(X)}{\sigma_r} = b_{n\ell}\frac{\sigma_{n\ell}}{\sigma_r}, \qquad (3)$$

where $\sigma_{n\ell}(X)$ is the sum of the X-ray rates for X-rays filling the $n\ell$ shell and $b_{n\ell}$ is the fluorescence ratio. The Auger electron branching ratio is simply given by 1-$b_{n\ell}$. Here we assumed that the inner-shell electron holes are filled by outer-shell electrons in the same atom via X-ray emission or the Auger effect. Non-radiative electron transfer to the inner-shell from neighboring atoms is considered to be small, since the nuclear recoil velocity is much smaller than the K-shell electron velocity.

The K shell fluorescence ratio for Xe is 0.89 in case of one K-hole in the atom. Then the K X-ray rates with respect to the recoil rate are 0.03, 0.20, and 0.23 for WIMPs with 30, 100 and 300 GeV, respectively. The X-ray rate increases as the WIMP mass increases.

The K_{ij} X-ray ratio is evaluated as

$$\frac{\sigma_K(K_{ij})}{\sigma_r} = \frac{\sigma_{1s}}{\sigma_r} b_{1s} B(K_{ij}), \qquad (4)$$

where $B(K_{ij})$ is the K-ij X-ray branch [17, 18]. The K X-ray rates are evaluated for the K shell holes given in the Table 1 by using the K-ij X-ray branch for one K-hole in the atom. The K X-ray rates, together with the K X-ray energies for ^{131}Xe isotope, are shown for light, medium and heavy WIMPs in Table 2.

TABLE 2. K X-ray rates and energies in WIMPs nuclear interactions with ^{131}Xe. $[Z\sigma_K/\sigma_r]_L, [Z\sigma_K/\sigma_r]_M$ and $[Z\sigma_K/\sigma_r]_H$ are the ratios for light (30 GeV), medium (100 GeV) and heavy (300 GeV) WIMP's, respectively.

K X-ray	$E_K(K_{ij})$ keV	$B_K(K_{ij})$	$[Z\sigma_K(K_{ij})/\sigma_r]_L$	$[Z\sigma_K(K_{ij})/\sigma_r]_M$	$[Z\sigma_K(K_{ij})/\sigma_r]_H$
$K_{\alpha 2}$	29.5	0.284	0.0086	0.0560	0.0645
$K_{\alpha 1}$	29.8	0.527	0.0160	0.1036	0.1196
$K_{\beta 1}$	33.6	0.154	0.0047	0.0303	0.0350
$K_{\beta 2}$	34.4	0.034	0.0010	0.0067	0.0077

K X-rays are followed by L X-rays, and L X-rays are by M X-rays and so on. Then the energy sum of these X-rays is just the K shell binding energy. These X-rays are converted to electrons via the photo-electric effect in detectors.

Then the sum of the photo-electron energies and the sum of the Auger electron energies are given by the binding energy ($-\varepsilon_{n\ell} \geq 0$). Thus, one may expect the $n\ell$ excitation signal with the rate of $\sigma_{n\ell}/\sigma_r$ and the electron energy of $-\varepsilon_{n\ell}$.

EXPERIMENTS FOR X-RAY STUDIES

X-rays associated with WIMP nuclear interactions are always accompanied by nuclear recoils. Therefore inclusive measurements of X-rays and ionization electrons together with the nuclear recoils modify the continuum recoil spectrum by the additional contributions from the ionization electrons and the X-rays.

Exclusive measurements of the X-rays in coincidence with recoil nuclei are very powerful to select WIMP signals and to reduce BG ones. In order to carry out exclusive experiments, one need to separate the X-ray signal from the recoil one. The separation is possible experimentally by utilizing the large difference between the flight path lengths for the X-ray and the recoil nucleus.

K and L X-rays in case of medium-heavy nuclei such as Xe isotopes have energies of 30 and 5 keV, respectively. The K and L X-rays penetrate through matter of about 100 and 5 mg/cm^2, respectively before depositing their energies via the photo-electric effect. Thus the identification of the X-rays and other charged particles can clearly be made by using good position-resolution detectors.

One option of detectors for exclusive studies of the X-rays following WIMPs scattering off nuclei is a TPC with Xe gas. The trajectory analysis makes it possible to identify the WIMP nuclear interaction point with the recoil and ionization electrons and the X-ray interaction point with the photo-electron track.

A hybrid detector, which consists of Xe gas ionization chambers for nuclear recoils and plastic scintillation-fibers for K X-rays, is an alternative way for exclusive studies of X-rays and nuclear recoils. A detector, which consists of many modules of thin semiconductors/scintillators with thickness of the order of 100 mg/cm^2 is very attractive and realistic for exclusive WIMP experiments.

Recently a highly-segmented NaI scintillator array has been developed at Tokushima group [19]. It consists of 16 layers of thin NaI plates, each with 50 mm long 50 mm wide and 500 μm thick. It is possible to built an NaI detector segmented into thin NaI plates, each with thickness of the order of the ^{127}I K X-ray range. Then nuclear recoils are measured in one layer of NaI in coincidence with the K X-rays in an adjacent layer. Here one may expect the similar K X-ray rate from ^{127}I with $Z = 53$ as the rate from ^{131}Xe given above.

A detector made of multi-layer NaI plates with a thin Si detector plate interleaved between them, is capable to select the WIMP signals and reject most BG electron ones. The nuclear recoil is detected in one NaI plate, while the K X-ray flies through the Si plate and is detected in the next NaI plate. Thus the WIMP signal with the nuclear recoil accompanied by the X-ray is characterized by a recoil signal from one NaI, one discrete X-ray signal from the next NaI and no signal from the Si plate between them. On the other hand most BG electrons by signals from the Si plate as well. In fact research and development works for WIMP detectors with multi-layer thin scintillator plates are under progress at Osaka University.

In case of exclusive measurements, where the BG rate is much reduced, detectors of an order of 10 kg can be used to search for DM in the region of $10^{-6} \sim 10^{-7}$ pb proton cross section. Acturely the signal rate with a 50 kg NaI detector is around 10 per year for 100 GeV WIMP's at 3 10^{-7} pb proton cross section. Here we assumed the density of 0.3 GeV and the velocity of 270 km /ses and the X-ray detection efficiency of 0.2.

CONCLUDING REMARKS

Direct searches for WIMPs are of great interest for cosmology and particle physics. Experimental searches for WIMPs, however, are very hard because of extremely rare event rates and very low-energy signals. Several groups use ultra-low BG detectors and/or novel methods to identify WIMP signals.

The present paper reports one novel way to search for WIMPs by measuring hard X-rays following inner-shell excitations, which are produced in WIMP nuclear interactions. K X-ray production probabilities are of the order of $0.2 \sim 0.25$ for $100 \sim 300$ GeV WIMPs interacting with Xe isotopes. The flight path of the K X-ray with around 30 keV is of the order of 100 mg/cm^2, and detectors with position resolution of the order of the flight path are feasible. Accordingly, exclusive measurements of the had X-rays following inner-shell excitations in coincidence with nuclear recoils are realistic.

Several options of detectors for exclusive studies of such X-rays are under consideration. They are a large TPC, a hybrid detector of TPC and thin scintillator, and a multi-layer detector of thin semi-conductors/scintillators. Such exclusive experiments are free of most backgrounds from natural and cosmogenic radioactive impurities in detector components.

Then remaining backgrounds in exclusive WIMP experiments are those due to cosmogenic neutrons scattered off target nuclei, resulting in inner-shell electron excitations and X-ray emissions. Incoming and scattered neutrons, however, are reduced down to less than 0.01 by means of active shields such as plastic or liquid scintillators since neutrons are strongly interacting particles in contrast to the weakly interacting WIMPs.

In short, X-rays are quite promising to search for medium and heavy WIMPs, and L X-rays are used to search for light WIMPs as well as for medium and heavy WIMPs. Thus it is quite realistic to study WIMPs/LSP in the $10^{-6} \sim 10^{-7}$ pb region.

ACKNOWLEDGMENTS

The authors are grateful to Prof. K. Fushimi and M. Nomachi for valuable discussions. The first author (H.E.) thanks the Institute for Nuclear Theory (INT) at the University of Washington and the Department of Energy for the support during his stay at INT. The second author (Ch.C.M.) acknowledges support by the Greek State Grants Foundation (IKY) under contract (515/2005). Finally the third author (J.D.V.) is indebted to the Greek Scholarship Foundation (IKYDA) for partial support, the Humboldt foundation for the Research Award, Professor H. Toki for his hospitality at RCNP and to Professor A. Faessler for his hospitality in Tuebingen.

REFERENCES

1. J. D. Vergados, and H. Ejiri, 2005 *Phys. Lett.* 2005, B 606, pp.313-322 ; arXiv: hep-ph/041151.
2. Ch. C. Moustakidis, J.D. Vergados and H. Ejiri, *Nucl. Phys.* 2005, B 727, pp.406-420.
3. H. Ejiri, Ch. C. Moustakidis, and J.D. Vergados, *Phys. Lett.* 2006, in press; arXiv: hep-ph/0510042.
4. H. Ejiri, *Proc. Erice School/Workshop on Neutrinos Sept. 2005*; *Progress in Particle and Nuclear Physics.* 2006, 57, pp.153 - 161.
5. M.G. Santos et al., *Phys. Rev. Lett.* 2002, 88, pp.241302-1-5, and references therein.
6. P.D. Mauskopf et al., *Astrophys. J.* 536, 2002, pp.L59-62.
7. N.W. Halverson et al., *Astrophys. J.* 2002, 568, pp.38-45.
8. D.N. Spergel et al., *Astrophys. J. Suppl.* 2003, 148, pp.175-226.
9. M. Tegmark et al., *Phys. Rev. D.* 2004, 69, pp.103501-1-26.
10. R. Bernabei et al., *Phys. Lett.* 1998, B 424, pp.195-201; *Phys. Lett.* 1999, B 450, pp.448-455.
11. A. Benoit et al. EDELWEISS collaboration, *Phys. lett.* 2002, B 545, pp.43-49.
12. V. Sanglar et al. EDELWEISS collaboration, arXiv:astro-ph/0306233.
13. D.S. Akerb et al. CDMS collaboration, *Phys. Rev.* 2003, D 68, pp.082002 (1-4), and arXiv: astro-ph/0405033.
14. K. Fushimi, et al., *Astroparticle Physics* 1999, 12, pp.185 - 192.
15. H. Ejiri, K. Fushimi and H. Ohsumi, *Phys. Lett.* 1993, B 317, pp.14 - 18.
16. J.D. Vergados, P. Quentin and D. Strottmann, Int. J. Mod. Phys. E. 2005, 14, pp.751; arXiv: hep-ph/0310365.
17. S.I. Salem, S.L. Panossian and R.A. Krause, *Atomic and Nuclear Data Table*. 1974, 14, pp.91-109.
18. W. Bambynek et al., *Rev. Mod. Phys.* 1972, 44, pp.716-813.
19. K. Fushimi et al, private communication 2005.

# 1.7 SUPERNOVAE, NEUTRON STARS and HIGH DENSITY MATTER

Constraints on the Dense Matter Equation of State from Observations

James M. Lattimer

Dept. of Physics & Astronomy, Stony Brook University, Stony Brook, NY 11794-3800, USA

Abstract. Neutron stars are laboratories for dense matter physics. New observations of neutron stars from radio pulsars, X-ray binaries, quasi-periodic oscillators, X-ray bursters and thermally-emitting isolated neutron stars are setting bounds to masses, radii, rotation rates, radiation radii, redshifts, moments of inertia, temperatures and ages. Radio pulsar data suggest a broader range of masses than expected. Radiation radii measurements from isolated neutron stars and from X-ray sources in globular clusters are also becoming available. But the most precise radius measurement may come from possible moment of inertia measurement from relativistic binary pulsars. The largest pulsar rotation rates set upper bounds to the ratio R^3/M, and quasi-periodic oscillations, if associated with the innermost stable orbit, set upper limits to both M and R. Observations of cooling neutron stars up to a million years old shed light on their internal compositions, including their superfluid properties, by constraining the neutrino emission rates.

Keywords: Neutron Stars, Equation of State, Pulsars
PACS: 26.20.+f,26.60.+c97.60.Gb,,97.60.Jd

INTRODUCTION

Neutron stars can be considered as having 5 important regions:

- The **atmosphere**, the cm-thick region where observed X-ray and optical thermal radiation is produced.
- The **envelope**, an insulating layer meters in thickness whose composition influences the observed temperatures of thermal emissions.
- The **crust**, extending inwards to densities of order 1/3 the nuclear saturation density, $n_s \simeq 0.16$ fm^{-3} or $\rho_s \simeq 3 \times 10^{14}$ g cm^{-3}, is composed primarily of nuclei surrounded (above densities of 4×10^{11} g cm^{-3}) by dripped neutrons. At the highest densities in the crust, nuclei become deformed, possibly through a sequence of shapes including spaghetti-like rods and lasagna-like plates. The crust terminates in a phase transition to uniform nucleonic matter, which forms the bulk of the
- **outer core**.
- The **inner core**, if it exists, contains more exotic components in addition to nucleonic matter, primarily including strangeness-containing particles such as hyperons, Bose (kaon or pion condensates), and/or deconfined quarks.

In addition, nucleons in the crust and nucleons and hyperons in the core of the star could become superfluid or superconducting. Transitions to these states could be important for the specific heat and for transport properties such as the neutrino opacity and emissivity.

This talk will focus on the structural properties and cooling of neutron stars, as determined from observations. Other important aspects of neutron stars, such as the pulsar mechanism and the origin of neutron stars, are not discussed.

NEUTRON STAR MASSES

Observations of neutron star masses will limit the extent of softening from exotic components that could populate the inner core. These exotic materials invariably result in a reduction of the neutron star maximum mass. A realistic upper limit to the neutron star maximum mass was obtained by Ref. [1] who assumed that above a fiducial density ρ_f the equation of state is causal (i.e., $dP/d\rho \leq c^2$):

$$M \leq 4.2\sqrt{\rho_s/\rho_f}\, M_\odot. \tag{1}$$

For example, the high-density limit of deconfined quarks has $dP/d\rho = c^2/3$. Using this as the upper limit above ρ_f, the coefficient 4.2 in the above is reduced by $\sqrt{3}$ [2].

Secondly, a measured neutron star mass automatically sets an upper limit to the density inside *any* neutron star[3]. For a given equation of state, the mass increases with the central density. The central density of the maximum mass star is the largest possible density. On the other hand, a further consequence of the analyses of Refs. [1, 2] is that the radius of a neutron star is limited to $R_{lim} \geq 2.94 GM/c^2$. The smallest radius is obtained when the mass reaches the maximum mass. If the matter in a neutron star was incompressible, one would then find that

$$\rho_c \leq \frac{3M}{4\pi R_{lim}^3} \simeq 5.9 \times 10^{15} \left(\frac{M}{M_\odot}\right)^{-2} \text{g cm}^{-3}, \tag{2}$$

which means an observed mass would set an upper limit to the central density. In actuality, matter is not incompressible, but the upper limit still scales with M^{-2} (but the coefficient is 5/2 times larger)[3].

New measurements from radio binary pulsars include objects with masses apparently in excess of 1.6 $M_\odot$, as summarized in Fig. 1. For the two binaries Terzan 5 I and J, one of the neutron stars has a mass in excess of 1.68 $M_\odot$ to 95% confidence [4], while PSR J0751+1807 has an estimated mass in excess of 1.6 $M_\odot$ to the same confidence [5].

Masses can also be estimated for another handful of binaries which contain an accreting neutron star emitting X-rays. Some of these systems seem to contain large masses, but the estimated errors are also large. The system of Vela X-1 is noteworthy because its lower mass limit (1.6 to 1.7$M_\odot$) is at least mildly constrained by geometry[6]

RADII

As opposed to mass measurements, which constrain the equation of state at the highest densities, a measurement of the radius of a neutron star is a more direct measure of the properties of matter close to the nuclear saturation density. It has been demonstrated [7]

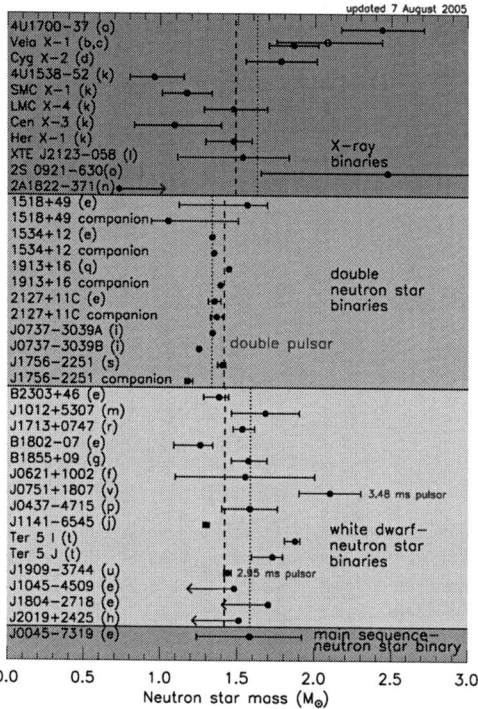

FIGURE 1. Neutron star mass measurements. The upper (green) region contains masses inferred from neutron stars in accreting X-ray binaries, while the other 3 regions show mass measurements from pulsar timing. Error bars are 68% confidence in all cases. Dashed and solid lines indicate inverse error-weighted means and simple averages of measured masses within each category.

that the radius of a $1 - 1.5\,M_\odot$ neutron star is rather tightly correlated with the pressure of matter between 1 and 2 times n_s, such that $R \propto P^{1/4}$. This is true for both equations of state that have maximum masses in excess of $2\,M_\odot$, as well as for those which a large degree of softening above n_s and maximum masses closer to $1.5\,M_\odot$. The simplest way of understanding this correlation is to make use of the fact that in the vicinity of n_s, most equations of state obey a polytropic relation $P = Kn^{1+1/n}$ where K is a constant and $n \sim 1$. In Newtonian gravity, a polytrope scales like

$$R \propto K^{n/(3-n)} M^{(1-n)/(3-n)} \sim K^{1/2} M^0 \tag{3}$$

where $n = 1$ was used. In other words, one could expect that moderate mass neutron stars would have radii dependent solely upon K and not M. An example of a traditional mass-radius diagram is reproduced in Fig. 2, and EOS's that are not too soft show the mass-independence of the radius. However, the above equation suggests that $R \propto K^{1/2}$, or that $R \propto P_*^{1/2} \rho_*^{-1}$, where the $*$ subscript indicates fiducial values. Ref. [7] showed that the phenomenological exponent of 1/4 for P_* results from general relativity.

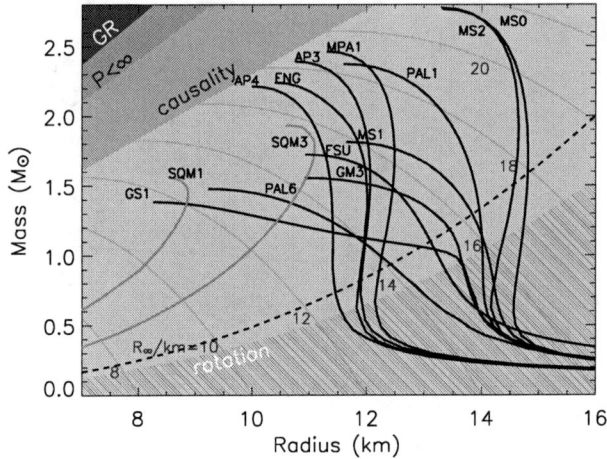

FIGURE 2. Mass-radius diagram for representative neutron star equations of state. Black (green) lines (for labels, see [7]) are normal (strange quark matter) EOS's. Shaded regions at upper left are excluded, respectively, by the event horizon ($R > 2GM/c^2$), finite pressure ($R > (9/4)GM/c^2$), and causality ($R > 2.94GM/c^2$). Fixed values of radiation radii $R_\infty = R/\sqrt{1-2GM/c^2}$ are indicated. The lower shaded region shows the limit imposed by rotation by the 642 Hz pulsar PSR1937+21; the accompanying dashed line corresponds to a statistical limit of 730 Hz[8].

The significance of this result is that a radius measurement rather directly relates to the pressure. Although the pressure of matter inside nuclei near n_s is well understood (by definition, the pressure of symmetric nuclear matter at the saturation density is zero), neutron star matter represents an extreme extrapolation to neutron richness (proton fractions near 0.05). As a result, the pressure of neutron star matter at n_s in various theories is uncertain to about a factor of 6. This uncertainty is directly related to a lack of knowledge of the density dependence of the so-called nuclear symmetry energy E_{sym}, or the isospin dependence of the nuclear interaction. To a good approximation, E_{sym} represents the difference, at a given density, of the energies of symmetric matter and pure neutron matter. For densities near n_s, one finds $P \propto dE_{sym}/dn$.

It is interesting that the density dependence of the nuclear force can also be traced through the neutron skin thickness of neutron-rich nuclei [9, 10].

Spin Frequency

Any object has a limiting spin frequency for which mass will be shed from their equators. In the Newtonian Roche approximation[11],

$$P_{min} = 2\pi \left(\frac{3}{2}\right)^{3/2} \left(\frac{R^3}{GM}\right)^{1/2} \simeq 1.0 \left(\frac{R}{10\text{ km}}\right)^{3/2} \left(\frac{M_\odot}{M}\right)^{1/2} \text{ ms}. \quad (4)$$

It has been found that in full general relativity, that a similar formula holds nearly independently of the underlying EOS, if M and R refer to the *non-rotating* mass and radius, by simply changing the 1.0 to 0.96 ± 0.03. The region excluded region in the $M-R$ plane by the 642 Hz pulsar PSR1937+21 is indicated in Fig. 2; the accompanying dashed line corresponds to a statistical limit of 730 Hz[8].

Radiation Radius

Assuming that it is thermal, the observed flux F and the inferred temperature T_{eff} of X-ray and possibly optical radiation from a neutron star's surface can be utilized to find another relation between M and R. If the neutron star was Newtonian, and a blackbody, Kirchoff's laws would imply that

$$R = \frac{d}{T_{eff}^2}\sqrt{\frac{F}{\sigma}}. \tag{5}$$

However, the temperature and the flux are both redshifted, so if F and T_{eff} refer to values at the Earth, then R must be replaced by $R_\infty = R(1 - 2GM/Rc^2)^{1/2}$. In addition, the existence of the neutron star atmosphere modifies the above relation[12]. Most significantly, an atmosphere has the effect of making the temperature appear non-uniform, as if the X-rays were produced in a high-temperature, small area and the optical radiation in a low-temperature, large area. In the few cases in which both optical and X-radiation from the same star is seen, the optical flux is 3–5 times larger than that which would be inferred by a naive extrapolation of the X-ray blackbody into the optical wavelengths.

These atmospheric effects can be understood by a naive two-temperature blackbody model. The optical radiation is on the Rayleigh-Jeans tail of the emission, and thus proportional to R^2T while the X-ray flux, proportional to R^2T^4, dominates the total flux. We can write

$$L = L_O + L_X \simeq L_X = 4\pi R_X^2 \sigma T_X^4, \quad L_O \propto 4\pi R_X^2 T_X + 4\pi R_O^2 T_O = 4\pi R_X^2 T_X f, \tag{6}$$

where X and O refer to the X-ray and optical, respectively, the radii refer to the radiation radii of the two components, and $f \sim 3-5$ is the optical flux enhancement referred to above. Then, one finds

$$R^2 = R_X^2 + R_O^2 = R_X^2[1 + \frac{T_X}{T_O}(f-1)]. \tag{7}$$

With typical values $T_X/T_O \sim 5, f \sim 5$, one finds $R \simeq 4.5R_X$. In other words, the radiation radius inferred from X-rays alone underpredicts the neutron star radius by a large factor.

Atmospheric effects are heavily dependent upon the assumed composition and magnetic fields. Radii determined by this technique are directly proportional to the assumed distance. Unfortunately, all of these factors are generally uncertain. The best chances will exist in cases for which the distance to the neutron star is well-constrained, and for which the atmosphere of the star is well-understood.

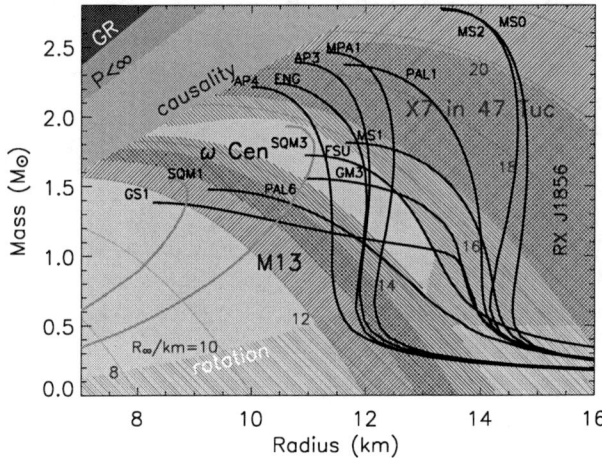

FIGURE 3. Mass-radius diagram for representative neutron star equations of state showing permitted regions for thermally-emitting neutron stars. Explanation of other regions is detailed in Fig. 2.

One of the best-studied, nearby, thermally emitting neutron stars is RX J1856-3754[13, 14]. It's distance can be estimated by parallax. Assuming a heavy-element, non-magnetic atmosphere then implies that $R_\infty > 16.5$ km[14, 15]. Other relatively close, isolated neutron stars (whose emission is not heavily contaminated by a pulsar's magnetospheric emission) have much less well-determined estimates for R_∞.

However, the discovery of many thermally emitting neutron stars in globular clusters is an intriguing development. These stars have experienced recent episodes of accretion from companion stars, which heated the otherwise cold stars, and simultaneously, it is believed, effectively quenched surface magnetic fields. The accreting material is dominated by H, so the atmospheric properties of these stars are well understood. Although these stars are at considerable distances, the fact that they lie within globular clusters allow (or will allow) their distances to be accurately determined, to the few percent level. Recently examined candidates include stars in the globular clusters ω Cen[16], M13[17] and 47 Tuc[18]. Estimated masses and radii for these candidates, as well as for RXJ 1856-3754, are displayed in Fig. 3.

Redshifts, Glitches and QPO's

The unambiguous identification of spectral lines from the surface of a neutron star would provide a different combination of mass and radius through the redshift:

$$z = (1 - 2GM/Rc^2)^{1/2} - 1. \tag{8}$$

Possible lines detected from active X-ray bursters XTE J1814-338 ($z < 0.38$), 4U1820-30 ($0.20 < z < 0.30$) and EXO 0748-676[19] ($z \simeq 0.35$) are claimed but remain controversial. They remain consistent with a wide number of equations of state.

Somewhat more model-dependent are limits to mass and radius inferred from glitches from the Vela pulsar[20]. These glitches, abrupt interruptions of the normally clock-like pulsar signals, are thought to represent global transers of angular momentum within the star. The leading theory is that this occurs from weak coupling between the n superfluid in the neutron star crust and the rest of the star. If true, the magnitude of the glitches and their observed stochastic occurences imply that the crustal contribution to the total moment of inertia of the star is greater than about 1.4%. Ref. [20] showed that, in general, the ratio of crustal to total moment of inertia is determined by the quantity $P_t R^4 M^{-2}$ where the only dependence on the EOS is the transition pressure P_t at the boundary between the outer core and the crust around $1/3\ n_s$, whose apparent upper limit is $P_t < 0.65$ MeV fm^{-3}. This translates into a lower limit for R for a given M.

Several accreting neutron stars display the phenomenon of quasi-periodic oscillations, discovered by Fourier analysis of their X-ray emissions. Besides lower frequencies of 200-400 Hz associated with the star's spin, two higher frequencies are also observed. As an accretion episode continues, these higher frequencies evolve to slightly greater frequencies with reduced coherence. Many models of these QPO's suggest that one or the other of these upper frequencies corresponds to the orbital frequency of the inner edge of the star's accretion disc. If the neutron star is non-rotating, the last stable orbit occurs at a radius of $6GM/c^2$, and the orbital frequency there is $(c^2/6)^{3/2}/GM$, its maximum possible value. If the highest frequency observed from a source does correspond to the orbital frequency of the last stable orbit, two relations can be immediately deduced:

$$R_A > 5GM/c^2, \qquad R_A > R, \qquad (9)$$

where R_A is the radius of the inner edge of the accretion disc. Limits to M and R of the neutron star so derived have a slight dependence on the star's rotational frequency because of general relativity[21].

CONCLUSIONS

Space does not permit discussion of additional limits based upon measurements of the moment of inertia of a neutron star or the cooling histories of neutron stars. Moments of inertia, possibly determined from timing of relativistic binaries containing two neutron stars through spin-orbit coupling[22, 23], are particularly intriguing because, dimensionally, $I \propto MR^2$, so this would provide a tighter constraint on R because M is very well known. This possibility is extensively discussed in Ref. [24].

The cooling of neutron stars adds another dimension of possibilities. For most of their observed lives, neutron stars cool primarily through neutrino emission. An important question is whether this cooling occurs through the so-called direct Urca process

$$n \to p + e^- + \bar{\nu}_e, \qquad p \to n + e^+ + \nu_e \qquad (10)$$

or via the slower modified Urca process mediated by "bystander" nucleons. If the abundance of protons and electrons is too small (proton/baryon ratio less than about 1/9), energy and momentum conservation cannot be achieved without these bystander nucleons. Sufficiently precise age and temperature estimates for neutron stars can determine if cooling is thus relatively slow or rapid. A complication is that even if the direct Urca process is possible, cooling may be quenched by nucleon superfluidity, which also prevents momentum conservation from being possible. Most observations so far avaiable are consitent with slow cooling or fairly large superfluid critical temperatures. Nevertheless, some evidence exists that in a few cases, more rapid cooling is occuring. The situation is summarized in Ref. [25].

ACKNOWLEDGMENTS

I would like to thank the organizers of OMEG05 for the invitation to discuss neutron stars. This work is supported in part by the U.S. Department of Energy under grant DE-AC02-ER40317. Much of this work has been done in collaboration with M. Prakash.

REFERENCES

1. C.E. Rhoades & R. Ruffini, *Phys. Rev. Lett.* **32** (1974) 324.
2. J.M. Lattimer, M. Prakash, D. Masak & A. Yahil, *Ap. J.* **355** (1990) 241.
3. J.M. Lattimer & M. Prakash, *Phys. Rev. Lett.* **94** (2005) 111101.
4. S.M. Ransom, J.W.T. Hessels, I.H. Stairs, P.C.C. Freire, F. Camilo, V.M. Kaspi & D.L. Kaplan, *Science* **307** (2005) 892.
5. D.J. Nice, E.M. Splaver, I.H. Stairs, O. Löhmer, A. Jessner, M. Kramer & J.M. Cordes, *Ap. J.* **634** (2005) 1242.
6. H. Quaintrell et al., *Ast. & Ap.* **401** (2003) 303.
7. J.M. Lattimer & M. Prakash, *Ap. J.* **550** (2001) 426.
8. D. Chakrabarty, E.H. Morgan, M.P. Muno, D.K. Galloway, R. Wijnands, M. van der Klis & C.B. Markwardt, *Nature* **424** (2003) 42.
9. C.J. Horowitz & J. Piekarewicz, *Phys. Rev.* **C64** (2001) 062802.
10. A.W. Steiner, M. Prakash, J.M. Lattimer & P.J. Ellis, *Phys. Rep.* **411** (2005) 325.
11. S.L. Shapiro & S.A. Teukolsky, *Black Holes, White Dwarfs, and Neutron Stars* (Wiley: New York, 1983).
12. R. Romani, *Ap. J.* **313** (1987) 718.
13. F.M. Walter, S.J. Wolk & R. Neuhauser, *Nature* **379** (1996) 233.
14. F.M. Walter & J.M. Lattimer, *Ap. J. Lett* **576** (2002) L145.
15. T.M. Braje & R.W. Romani, *Ap. J.* **580** (2002) 1043.
16. B. Gendre, D. Barret & A. N. Webb, *A & A* **400** (2003) 512.
17. B. Gendre, D. Barret & N. Webb, *A & A* **403** (2003) L11.
18. G.B. Rybicki, C.O. Heinke, R. Narayan & J.E. Grindlay, *Ap. J*, in press; astro-ph/0506563 (2005).
19. J. Cottam, F. Paerels & M. Mendez, *Nature* **420** (2002) 51.
20. B. Link, R.I. Epstein & J.M. Lattimer, *Phys. Rev. Lett.* **83** (1999) 3362.
21. M.C. Miller, F.K. Lamb & D. Psaltis, *Ap. J.* **508** (1998) 791.
22. B.M. Barker & R.F. O'Connell, *Phys. Rev.* **D12** (1975) 329.
23. T. Damour & G. Schaefer, *Nuovo Cimento* **101B** (1988) 127.
24. J.M. Lattimer & B.F. Schutz, *Ap. J.* **629** (2005) 979.
25. D. Page, J.M. Lattimer, M. Prakash & A.W. Steiner, *Ap. J. Supp.* **155** (2004) 623.

Recent Developments in Neutron Star Thermal Evolution Theories and Observation

Sachiko Tsuruta

Physics Department, Montana State University, Bozeman, Montana 59717 USA

Abstract. Recent years have seen some significant progress in theoretical studies of physics of dense matter. Combined with the observational data now available from the successful launch of *Chandra* and *XMM/Newton* X-ray space missions as well as various lower-energy band observations, these developments now offer the hope for distinguishing various competing neutron star thermal evolution models. For instance, the latest theoretical and observational developments may already exclude both nucleon and kaon direct Urca cooling. In this way we can now have a realistic hope for determining various important properties, such as the composition, superfluidity, the equation of state and stellar radius. These developments should help us obtain deeper insight into the properties of dense matter.

Keywords: Neutron Stars, Thermal evolution, Dense Matter
PACS: Geophysics, Astronomy, and Astrophysics; Stars; Neutron Stars

INTRODUCTION

The launch of the *Einstein* Observatory gave the first hope for detecting thermal radiation directly from the surface of neutron stars (NSs). However, the temperatures obtained by the *Einstein* were only the upper limits[1]. *ROSAT* offered the first confirmed detections (not just upper limits) for such surface thermal radiation from at least three cooling neutron stars, PSR 0656+14, PSR 0630+18 (Geminga) and PSR 1055-52[2]. Recently the prospect for measuring the surface temperature of isolated NSs, as well as obtaining better upper limits, has increased significantly, thanks to the superior X-ray data from *Chandra* and *XMM/Newton*, as well as the data in the lower energy bands from optical-UV telescopes such as *Hubble Space Telescope*. Consequently, the number of possible surface temperature detections has already increased to at least seven[3]. Very recently *Chandra* offered an important upper limit to PSR J0205+6449 in 3C58[4]. At the same time, more careful and detailed theoretical investigation of various input microphysics has been in progress[5]-[9]. The current paper is meant as a progress report on these recent developments. Specifically, we try to demonstrate that distinguishing among various competing NS cooling theories has started to become possible, by careful comparison of improved theories with new observations[3][10][11].

NEUTRON STAR COOLING THEORIES

The first detailed cooling calculations[12] showed that isolated NSs can be warm enough to be observable as X-ray sources for about a million years. After a supernova explosion a newly formed NS first cools via various neutrino emission mechanisms before the surface photon radiation takes over. Among the important factors which seriously affect the nature of NS cooling are: neutrino emission processes, superfluidity of constituent particles, composition, mass, and the equation of state (EOS)[3]. In this paper, for convenience, the conventional, slower neutrino cooling mechanisms, such as the modified Urca, plasmon neutrino and bremsstrahlung processes, will be called 'standard cooling'. On the other hand, the more 'exotic' extremely fast cooling processes, such as the direct Urca processes involving nucleons, hyperons, pions, kaons, and quarks, will be called 'nonstandard' processes[3].

The composition of NS interior is predominantly neutrons with only a small fraction of protons, electrons and muons when the interior density is not high (the central density $\rho^c <\sim 10^{15}$ gm/cm^3). For higher densities more 'exotic' particles, such as hyperons, pions, kaons and quarks, may dominate the central core. Therefore, when the star is less massive and hence less dense, we have a neutron star with the interior consisting predominantly of neutrons (no 'exotic' particles) and it will cool with the slower, 'standard' neutrino processes. On the other hand when ρ^c exceeds the transition density to the exotic matter ρ_{tr}, the transition from nucleons to 'exotic' particles takes place. Therefore, more massive stars, whose ρ^c exceeds ρ_{tr}, possess a central core consisting of the exotic particles. In that case, the nonstandard fast cooling takes over. Note, however, that if the proton fraction in the neutron matter is exceptionally high, i.e., $>\sim 15\%$, very fast cooling can take place in a NS without any exotic particles, through the nucleon direct Urca process. This can happen for a certain type of EOS models which allow such high proton concentration above a certain critical density[13]. In order to include this option, in the following discussion we will call any fast nonstandard process, an 'exotic process', rather than 'a process involving exotic particles'. The observational data suggest that there are at least two classes of NSs, the hotter and cooler. The most natural explanation is that the hotter stars are less massive and cool by slower cooling processes, while the cooler ones are more massive and cool by one of the fast nonstandard processes[3][10].

As the central collapsed star cools after a supernova explosion and the interior temperature falls below the superfluid critical temperature, T^{cr}, some constituent particles become superfluid. That causes suppression of both specific heat (and hence the internal energy) and all neutrino processes involving the superfluid particles. The net effect is that in the case of fast nonstandard cooling, the star cools more slowly and hence the surface temperature and luminosity will be higher at a given age during the neutrino cooling era, due to the suppression of neutrino emissivity. Therefore, nonstandard fast cooling will be no longer so fast if the superfluid energy gap, which is proportional to T^{cr}, is significant. In fact, if the gap is large enough, nonstandard cooling is fully suppressed and the cooling curve becomes essentially the same as the standard cooling curve. Therefore, depending on the size of the energy gap, a nonstandard cooling curve can lie anywhere between the standard curve and the nonstandard one without superfluid suppression.

In addition to various neutrino cooling mechanisms conventionally adopted in earlier calculations, recently the 'Cooper pair neutrino emission'[14][15] was 'rediscovered' to be also important under certain circumstances. This process takes place when the participating particles become superfluid, and the net effect is to enhance, for some superfluid models, the neutrino emission right after the superfluidity sets in[15][16].

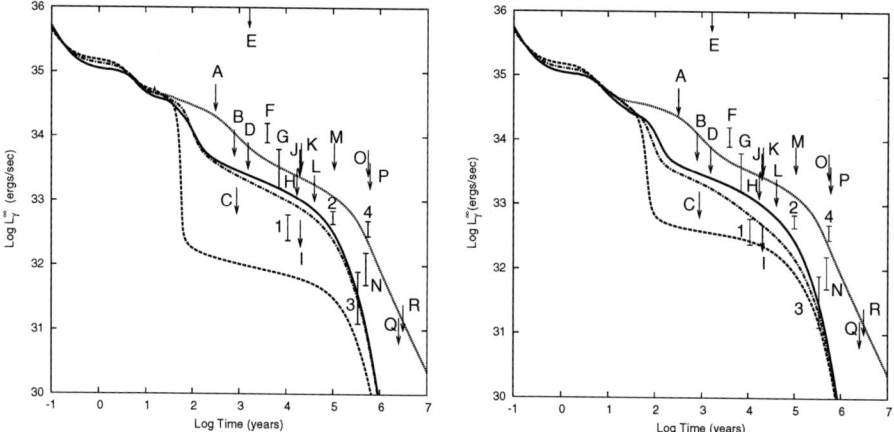

FIGURE 1. Thermal evolution curves with the newest observational data. In Fig. 1a (left panel) the dotted and solid curves refer to the standard cooling of M = 1.4M$_\odot$ neutron stars with and without heating, respectively, while the dot-dashed and dashed curves are for hyperon cooling of 1.6 and 1.8M$_\odot$ stars, respectively. In Fig. 1b (the right panel) the solid, dot-dashed and dashed curves refer to pion cooling of 1.4, 1.6 and 1.8M$_\odot$ stars, respectively. In the same figure the dotted curve refers to thermal evolution of a 1.4M$_\odot$ pion star with heating. The vertical bars refer to temperature detection data with error bars, while the downward arrows refer to the upper limits. The more accurate detection data are sown with numbers, for (1) the Vela pulsar, (2) PSR 0656+14, (3) Geminga, and (4) PSR 1055-52. The rest of the data shown are more rough estimates. Some of more interesting among these are shown with letters, as (A) Cas A point source, (B) the Crab pulsar, (C) PSR J0205+6449 in 3C58, (F) RX J0822-4300, (G) 1E1207.4-5209, (I) PSR 1046-58, (N) RX J1856-3754, and (R) PSR 1929+10. The complete list of all these data sources is found in T06a,b[10][11]. Sources RX J0002+62 and RX J0720.4-3125 are not shown because currently there are still too large uncertainties including the age estimate. See the text for further details.

MOST RECENT THERMAL EVOLUTION MODELS

Most recently we calculated NS thermal evolution[1] adopting the most up-to-date microphysical input and a fully general relativistic, 'exact' evolutionary code (i.e., without making isothermal approximations). This code was originally constructed by Nomoto and Tsuruta (1987)[17] which has been continuously up-dated. Our input neutrino emissivity consists of all possible mechanisms, including Cooper pair emission. See Tsuruta et al. 2006a,b, hereafter referred to as T06a,b[10][11] for the details. In the models presented here, we consider thermal evolution of neutron stars which possess a central core consisting of hyperons and pion condensates at high densities, which we conveniently call hyperon and pion stars. The results are summarized in Figure 1, where surface photon luminosity which corresponds to surface temperature (both to be observed at infinity), is shown as a function of age.

Fig. 1a (the left panel) shows thermal evolution of neutron (lower mass) and hyperon (higher mass) stars. The critical transition density from neutron to hyperon matter, ρ_{tr}^Y, is

---

[1] We adopt the expression 'thermal evolution' when we include not only cooling but also heating.

set at $4\rho_0$, which is estimated by nuclear theories[7]. (Here $\rho_0 = 2.8 \times 10^{14}$ gm/cm^3 is the nuclear density.) For $\rho < \rho_{tr}^Y$ we adopt the TNI6 EOS recently constructed for neutron matter by Takatsuka et al 2006[7], while for $\rho > \rho_{tr}^Y$ it becomes TNI6U, the same EOS but for hyperon matter[7]. This EOS is medium in stiffness[2] and it is very similar to the FP model adopted earlier, e.g. in Umeda et al. 1994[18] and Umeda, Tsuruta and Nomoto 1995[19]. As the superfluid model for hyperons we adopt the Ehime Model[7] and as the neutron superfluid model the OPEG-B Model[6], both recently constructed. The Cooper pair neutrino emissivity derived by Yakovlev, Levenfish, and Shibanov (1999)[14] is adopted for both neutrons and protons. For our heating calculations we adopt the vortex creep heating model with the heating parameter $K = 10^{37}$ ergs m$^{-3/2}$ s^2, which is maximum in strength according to theoretical estimates[19][20], and magnetic field $B = 10^{12}$ Gauss, reasonable for ordinary pulsars[21]. The other input parameters are the same as in Tsuruta 1998[21].

In Fig. 1a the dotted and solid curves refer to thermal evolution of a 1.4M$_\odot$ NS with and without heating, respectively. Since for these stars the central density $\rho^c < \rho_{tr}^Y$, they consist predominantly of neutrons and they cool by the slower 'standard' processes. The dot-dashed and dashed curves present cooling of 1.6M$_\odot$ and 1.8M$_\odot$ hyperon stars. For the TNI6U EOS adopted, we find that $\rho^c = \rho_{tr}^Y$ for a 1.5M$_\odot$ star. Therefore, our stars with mass larger than $\sim$ 1.5M$_\odot$ contain a hyperon core and hence the predominant cooling mechanism is the nonstandard hyperon direct Urca process. However, the 1.6M$_\odot$ star (dot-dashed) does not cool much faster than less massive neutron stars because the superfluid suppression is very large. The 1.8M$_\odot$ star (dashed) cools faster because the superfluid gap decreases significantly for this larger mass (and hence denser) star[3]. See T06a[10] for further details.

Fig. 1b (the right panel) shows thermal evolution of neutron stars with a pion core. The EOS adopted is 'TNI3P Model', which is a modified version of TNI3 EOS for neutron matter recently constructed[6]. This EOS is somewhat stiffer than medium. It was modified by T06b[11], to include pions for densities exceeding ρ_{tr}^π, the critical density for transition to pions. The pion transition density ρ_{tr}^π is set to be $2\rho_0$, significantly lower than that for hyperons, adopting the results of recent careful theoretical studies[8]. As the superfluid model for the pion-condensed phase we adopt the result from the most recent calculations by Tamagaki and Takatsuka (2006)[8] which indicates that the gaps for pion condensates are significantly larger for a significant range of densities above $2\rho_0$, although we assume that they decrease at higher densities. Other microphysical input is the same as in Fig. 1a.

In Fig. 1b the dotted and solid curves refer to thermal evolution of 1.4M$_\odot$ pion stars with and without heating, respectively. Since the transition density is low for pions we find that the central density of a 1.4M$_\odot$ star already exceeds the pion transition density even for this relatively stiff EOS chosen, and hence its core already consists of pion condensates. However, for these stars the gap, and hence T^{cr}, is so large that the

---

[2] Often an EOS is referred to as being 'stiff' when the consequent stellar model is more extended and less dense, while it is referred to as being 'soft' if it is more compact and denser for a given mass.
[3] Note that after the superfluidity sets in the gap first increases, reaches a peak and then decrease to zero as density increases.

superfluid suppression is essentially complete, which means the curves lie essentially in the same positions as the standard cooling. The dot-dashed and dashed curves present cooling of 1.6$M_\odot$ and 1.8$M_\odot$ pion stars, respectively. For the higher central density of these more massive stars the superfluid gap decreases and hence these stars cool faster. However, for these stars the gap is still large enough to keep the superfluid suppression significant. See T06b for further details.

COMPARISON WITH OBSERVATION

In Fig. 1 thermal evolution curves are compared with the latest observational data. We may note that the data suggest the existence of at least two classes of sources, hotter stars (e.g., (F) RX J0822-4300, (G) 1E1207.4-5209, (2) PSR 0656+14, (N) J1856-4754, and (4) PSR 1055-52), and cooler stars (e.g., (C) PSR J0205+6449 in 3C58, (1) the Vela pulsar and (I) PSR 1046-58). The hotter sources are consistent with thermal evolution of less massive stars such as 1.4$M_\odot$ stars. For PSR 1055-52(4) the age uncertainty is relatively large, but still it will probably require at least moderate heating. Source (F) is slightly above the dotted curves, but for this source the distance is quite uncertain and if it is closer away that will bring the data point down. Also if the star has magnetic envelopes with light elements such as Hydrogen cooling will be slower and that will bring the curves somewhat higher[22]. Note that for this source the best fit to the spectral data requires a Hydrogen atmosphere[10].

Comparison of cooler star data with pion and hyperon curves confirms the earlier conclusion[3][21][23] that nonstandard cooling of more massive stars is required for these cooler data. In this case we find that significant superfluid suppression is required, at least for (1) the Vela pulsar detection data. The age uncertainty should not affect this conclusion, especially for younger cooler sources such as the pulsars in (C) 3C58 and (1) Vela, because the slope of the curves in these younger years is relatively flat. Also the uncertainty for the age estimate from pulsar spins is small for younger pulsars, at most a factor of 2 or so. Note that for the pulsar in 3C58 the difference between the age estimates coming from the SNR data and pulsar data is very small[10].

Recently a possibility for very cold NSs in at least four supernova remnants (SNRs) was reported by Kaplan et al. 2004[24]. These authors note that if there is a NS in these SNRs the upper limits to their surface luminosity should be very low[4]. We do not place these upper limits in our Fig. 1 because their data are given as L_x(0.5 - 10 keV), the X-ray luminosity within the limited window between 0.5 - 2 Kev. That should be significantly lower than L_{bol}, the total bolometric luminosity over all wavelengths, which should be the one to be compared with theoretical luminosity in the cooling curves. For instance, $L_{bol} \sim 80\, L_x$(0.5 - 10 keV) for PSR 0656+14 (the former, $\sim 8 \times 10^{32}$ ergs/s, vs the latter, $\sim 10^{31}$ ergs/s). However, even so, we note that these upper limits are safely below the standard cooling curves, and hence a nonstandard fast cooling scenario is required.

---

[4] these authors pointed out that although some of these SNRs may contain no compact collapsed objects or the compact remnants may be black holes, it is quite unlikely that none of them contains a NS.

DISCUSSION

Until recently it was thought that at least for binary pulsars observations offered stringent constraints on the mass of a NS, to be very close to $1.4M_\odot$[21]. If this evidence extends to isolated NSs also, then the EOS should be severely constrained because it has to be such that the mass of the star whose central density is very close to the transition density (where the nonstandard process sets in) should be very close to $1.4M_\odot$[21]. Very recent observational data, however, suggest that the mass range should be much broader, $\sim 1 - 2M_\odot$[9][25]. If so, that still should give some useful constraint on the EOS. For instance, a very soft EOS, such as the BPS Model[21], should be excluded because for this EOS the maximum mass is only $\sim 1.5M_\odot$, and hence stars with mass larger than this cannot be explained. That is why we chose medium to stiff EOSs for our models. We chose a stiffer EOS for pion stars because the transition to pions takes place at lower densities.

The qualitative behavior of all nonstandard scenarios is similar if their transition density is the same[21]. However, here we try to demonstrate that it is still possible to offer comprehensive assessment of at least which options are more likely while which are less likely. First of all, we note that all of the nonstandard mechanisms without suitable suppression are too fast for all the detection data. Significant suppression of neutrino emissivity due to superfluidity is required, to be consistent with cooler stars such as the Vela pulsar. However, Takatsuka and Tamagaki (1997), hereafter TT97[26] already showed, through careful microphysical calculations, that for neutron matter with such high proton concentration as to permit the nucleon direct Urca, the superfluid critical temperature T^{cr} should be extremely low, $\sim$ several x 10^7 K, not only for neutrons but also for protons. Here we emphasize that this conclusion does not depend on the nuclear models adopted for the calculations. On the other hand, most of the the observed NSs, which are to be compared with cooling curves, are hotter (the core temperature being typically $\sim 10^8$ K to several times 10^8 K). That means *the core particles are not yet in the superfluid state* in these observed NSs. Conclusion is that *a star cooling with nucleon direct Urca would be too cold* to be consistent with these detection data. The same argument applies to kaon cooling also[27]. Further details are found in [3][10][11][23].

As to the hyperon cooling scinario, we find that that will be a viable option if recently constructed hyperon superfluid gap models (which include the Ehime Model adopted here)[7] are valid. Recently the Gifu-Kyoto nuclear experimental group[28] reported that the superfluid gap for hyperons would be much smaller. If so, hyperon cooling also would be in trouble. However, since then their experiment has not been confirmed by follow-up experiments. We find that the pion cooling option is still valid.

A few other groups have calculated neutron star cooling. Due to lack of space here we comment on only the recent major work by Yakovlev et al 2004, hereafter referred to as Y04[16]. Although these authors sometimes adopted simplified 'toy models' with the isothermal and other various approximations, their results and ours generally agree, at least qualitatively, when similar input is applied. There are, however, some serious differences. For instance,

(i) In an effort to bring up the standard cooling curves to explain a hot pulsar PSR 1055, these authors conclude that neutron superfluidity must be so weak as to be negligible. However, this conclusion contradicts with the results of serious theoretical studies of neutron superfluidity, which find that neutron superfluidity could not be so small for normal neutron matter where proton concentration is small[6]. On the other hand, we have shown (see Fig. 1) that this apparent discrepancy disappears even with models with significant neutron superfluidity when heating is included. Also, it may be pointed out that the age uncertainty is rather large for this pulsar since it is older[10]. Therefore, the unrealistic assumption of negligible neutron superfluidity for normal neutron matter is not required.

(ii) To explain cooler stars, these authors chose the nucleon direct Urca process, which they called Durca, as their nonstandard cooling option. Also, to explain both hot PSR 1055 and cooler Vela pulsar Y04 require their models to possess large proton superfluidity and yet negligible neutron superfluidity. However, TT97[26] already showed that that is impossible for models where the Durca option works. Specifically, for Durca to operate proton concentration must be significant. In such a case TT97 showed that both neutron and proton superfluidity must be very small - it cannot be that one is very high while the other negligibly small. In other words the models by Y04 are unphysical. Also, for such models where Durca can operate it is impossible for proton superfluidity to be strong enough to offer sufficient suppression to explain the Vela pulsar. Here we emphasize that their models are based on studies of normal neutron matter where proton concentration is small, while that argument breaks down for special models with high proton concentration which will allow nucleon Durca to operate.

SUMMARY AND CONCLUDING REMARKS

We have shown that the most up-to-date observed temperature data are consistent with the current thermal evolution theories of isolated NSs if less massive stars are warmer while more massive stars are cooler. The comparison of theory with observation, especially with the low temperature upper limit for the pulsar in 3C58, shows that fast nonstandard cooling is required for cooler stars. The need for nonstandard cooling is further strengthened by the recent report by Kaplan et al.[24] for the very low upper limits to neutron stars possibly present in some of four SNRs.

Among various nonstandard cooling scenarios, both nucleon Durca and kaon cooling may be excluded. The major reason is that for nucleon Durca to be operative, high proton concentration is required, which weakens superfluidity of both protons and neutrons. Then nucleon Durca will be too fast to be consistent with e.g., the Vela pulsar data. Similar argument applies to kaon cooling. Hyperon cooling may be in trouble if the hyperon superfluid gap should be so small as reported by recent nuclear experiments, although this report is yet to be confirmed by follow-up experiments. On the other hand, pion cooling is still consistent with both observation and theory. The conclusion is that *the presence of 'exotic' particles, possibly pion condensates, will be required within a very dense star.* If the need for larger mass stars most recently reported for binary pulsars is confirmed, the very soft EOSs should be excluded.

The capability of constraining the composition of NS interior matter purely through observation alone will be limited, due to often very large uncertainties, mainly for stellar distance and age. Therefore, it will be very important to *exhaust all theoretical resources.* Theoretical uncertainties are also very large, especially in the supranuclear density regime. However, here we emphasize that we should still be able to set *acceptable ranges* of theoretical feasibility, at least to separate models more-likely from those less-likely.

ACKNOWLEDGMENTS

We acknowledge with special thanks the contributions by our collaborators, M.A. Teter, J. Sadino, J. Thiel, T. Takatsuka, R. Tamagaki, W. Candler, K. Nomoto, H. Umeda, A. Liebmann, and K. Fukumura. Thanks are due to K. Nomoto, T. Tatsumi and others in their groups for their hospitality and help during our visits to Tokyo University and Kyoto University, and the participants of Kyoto workshops for valuable discussions. Our work for this paper has been supported in part by NASA grants NAG5-3159, NAG5-12079, AR3-4004A, and G02-3097X.

REFERENCES

1. K. Nomoto, and S. Tsuruta, *Ap. J*, **305**, L19 (1986).
2. W. Becker, Ph.D. Thesis, University of Munich, Munich (1995).
3. S. Tsuruta, in *Young Neutron Stars and Their Environments*, edited by F. Camilo et al., IAU Symposium Proceedings 218, Astronomical Society of the Pacific, San Francisco, 2004, pp. 21–28.
4. P. O. Slane, D. J. Helfand, and S. S. Murray, *Ap. J*, **571**, L45 (2002).
5. e.g., T. Takatsuka, and R. Tamagaki, *Prog. Theor. Phys.*, **105**, 179 (2001), and references therein.
6. e.g., T. Takatsuka, and R. Tamagaki, *Prog. Theor. Phys.*, **112**, 37–72 (2004), and references therein.
7. T. Takatsuka, S. Nishizaki, Y. Yamamoto, and R. Tamagaki, *Prog. Theor. Phys.*, in press (2006).
8. R. Tamagaki, and T. Takatsuka, *Prog. Theor. Phys.*, in press (2006), and references therein.
9. J. M. Lattimer, this volume (2006).
10. S. Tsuruta, J. Sadino, T. Takatsuka, M. A. Teter, B. Chandler, R. Tamagaki, K. Nomoto, H. Umeda, and A. Liebmann, in preparation (2006a) (T06a).
11. S. Tsuruta, M. A. Teter, J. Thiel, T. Tamagaki, T. Takatsuka, W. Chandler, J. Sadino, K. Nomoto, H. Umeda, and K. Fukumura, in preparation (2006b) (T06b).
12. S. Tsuruta, and A. G. W. Cameron, *Canad. J. Phys.*, **44**, 1863 (1966).
13. J. M. Lattimer, C. J. Pethick, M. Prakash, and P. Haensel, *Phys. Rev. Letters*, **27**, 2701 (1991).
14. D. G. Yakovlev, K. P. Levenfish, and Yu. A. Shibanov, *Physics-Uspekhi*, **42**, 737 (1999).
15. E. G. Flowers, M. Ruderman, and P. G. Sutherland, *Ap. J*, **205**, 541 (1976).
16. D. G. Yakovlev, O. Y. Gnedin, A. D. Kaminger, K. P. Levenfish, and A. Y. Potenkin, *Adv. Sp. Res.*, **33**, 523 (2004) (Y04), and references therein.
17. K. Nomoto, and S. Tsuruta, *Ap. J*, **312**, 711 (1987).
18. H. Umeda, K. Nomoto, S. Tsuruta, T. Muto, and T. Tatsumi, *Ap. J*, **431**, 309–320 (1994).
19. H. Umeda, S. Tsuruta, and K. Nomoto, *Ap. J*, **433**, 256 (1995).
20. M. A. Alpar, K. S. Cheng, D. Pines, and J. Shaham, *Ap. J*, **346**, 823 (1989).
21. S. Tsuruta, *Physics Reports*, **292**, 1–130 (1998).
22. A. Y. Potenkin, D. G. Yakovlev, G. Chabrier, and O. Y. Gnedin, Astroph/0305256 (2003).
23. S. Tsuruta, M. A. Teter, T. Takatsuka, T. Tatsumi, and R. Tamagaki, *Ap. J*, **571**, L143 (2002).
24. D. L. Kaplan, S. R. Kulkarnoi, D. A. Frail, B. M. Gaensler, P. O. Slane, and E. V. Gotthelf, in *Young Neutron Stars and Their Environments*, edited by F. Camilo et al., IAU Sympmposium Proceedings 218, Astronomical Society of the Pacific, San Fransisco, 2004, pp. 123–126, and reference therin.
25. D. J. Nice, and E. M. Splaver, in *Young Neutron Stars and Their Environments*, edited by F. Camilo et al., IAU Sympmposium Proceedings 218, Astronomical Society of the Pacific, San Fransisco, 2004, pp. 49–52.
26. T. Takatsuka, and R. Tamagaki, *Prog. Theor. Phys.*, **97**, 345 (1997) (TT97).
27. T. Takatsuka, and R. Tamagaki, *Prog. Theor. Phys.*, **94**, 457 (1995).
28. H. Takahashi, et al., *Phys. Rev. Letters*, **87**, 21 (2001).

Microscopic origin of the magnetic field in compact stars

Toshitaka Tatsumi

Department of Physics, Kyoto University, Kyoto 606-8502, Japan

Abstract. A magnetic aspect of quark matter is studied by the Fermi liquid theory. The magnetic susceptibility is derived with the one-gluon-exchange interaction, and the critical Fermi momentum for spontaneous spin polarization is found to be 1.4fm^{-1}. A scenario about the origin of magnetic field in compact stars is presented by using this result.

Keywords: ferromagnetism, quark matter, Fermi liquid theory, magnetars
PACS: 04.40.Dg;07.55.Db;12.38.Bx;26.60.+c;97.60.Jd

INTRODUCTION

Nowadays the phase diagram of QCD in the density-temperature plane has been explored by many people: the high-temperature region is relevant for relativistic heavy-ion collisions or cosmological phase transition in early universe, while the high-density region for compact stars. In the high-density and low-temperature region many exciting phenomena have been expected due to the large and sharp Fermi surface, such as color superconductivity, chiral density waves or ferromagnetism. Here we are concentrated in the magnetic aspect of quark matter: we are interested in the spin degree of freedom [1]. If magnetism is realized in quark matter, it should be interesting theoretically, but have important implications for the magnetic properties of compact stars.

Recent discoveries of magnetars, compact stars with huge magnetic field of $O(10^{15}G)$, seem to enforce us to reconsider the origin of the magnetic field in compact stars. They have firstly observed by the $P - \dot{P}$ curve, and some cyclotron absorption lines have been recently observed [2]. A naive working hypothesis of conservation of magnetic flux during their evolution from the main-sequence progenitors cannot be applied to magnetars, because the resultant radius is too small for $O(10^{15}G)$. The dynamo mechanism may work in compact stars, but it might look to be unnatural to produce such a huge magnetic field. The typical energy scale is given by the interaction energy with this magnetic field, which amounts to O(MeV) for electrons, and O(keV) to O(MeV) for nucleons or quarks. On the other hand, considering the atomic energy scale or the strong-interaction energy scale, we can say that $O(10^{15}G)$ is very large for electrons, but not so large for nucleons or quarks. Hence it would be interesting to consider a microscopic origin of magnetic field as an alternative: if ferromagnetism or spontaneous spin polarization occurs inside compact stars, it may be a possible candidate [1] . For

---

[1] In a recent paper Makishima also suggested a hadronic origin from the observation of X-ray binaries [3].

nuclear matter, there have been done many calculations with different nuclear forces and different methods since the first discoveries of pulsars in early seventies, but all of them have given negative results so far [4]. In the following we discuss the magnetic aspect of quark matter and consider a possibility of spontaneous spin polarization.

RELATIVISTIC FERROMAGNETISM

We have considered the possibility of ferromagnetism in quark matter interacting with the one-gluon-exchange (OGE) interaction [5] or with an effective interaction [6], and suggested that quark matter has a potentiality to be spontaneously polarized. To understand the magnetic properties of quark matter more realistically, especially near the critical point, some non-perturbative consideration about the instability of the Fermi surface is indispensable. Recently there are some studies about the effective interaction near the Fermi surface [7, 8], using the idea of the renormalization group [9]. Here we apply the Fermi liquid theory to derive the magnetic susceptibility and discuss the spontaneous spin polarization, considering quarks as quasiparticles [10, 11, 12].

Fermi liquid theory

In the Fermi liquid theory the total energy is given as a functional of the distribution function. As is already shown in ref. [13], the spin degree of freedom is specified by the three vector ζ in the rest frame. Then the quasi-particle energy and the effective interaction near the Fermi surface can be written as,

$$\varepsilon(\mathbf{k}\zeta ci) = \frac{\delta E}{\delta n(\mathbf{k}\zeta ci)}, \quad f_{\mathbf{k}\zeta ci,\mathbf{q}\zeta'dj} = \frac{\delta \varepsilon(\mathbf{k}\zeta ci)}{\delta n(\mathbf{q}\zeta'dj)}, \tag{1}$$

where the subscripts $c(d)$ and $i(j)$ denotes the color and flavor degrees of freedom. The Landau Fermi liquid interaction $f_{\mathbf{k}\zeta ci,\mathbf{q}\zeta'dj}$ is related to the forward scattering amplitude for two quarks on the Fermi surface. [2] In QCD the interaction is flavor independent, $f_{\mathbf{k}\zeta ci,\mathbf{q}\zeta'dj} = \delta_{ij} f_{\mathbf{k}\zeta c,\mathbf{q}\zeta'd}$. Since there is no direct interaction due to color neutrality, the Fock exchange interaction gives the leading contribution in the weak coupling limit, i.e. the color symmetric OGE interaction can be written as

$$f^S_{\mathbf{k}\zeta,\mathbf{q}\zeta'} \equiv \frac{1}{N_c^2} \sum_{c,d} f_{\mathbf{k}\zeta c,\mathbf{q}\zeta'd} = \frac{m}{E_k}\frac{m}{E_q} M_{\mathbf{k}\zeta,\mathbf{q}\zeta'}, \tag{2}$$

with the Lorentz invariant matrix element,

$$M_{\mathbf{k}\zeta,\mathbf{q}\zeta'} = g^2 \frac{N_c^2-1}{4N_c^2 m^2}\left[2m^2 - k\cdot q - m^2 a\cdot b\right]\frac{1}{(k-q)^2}, \tag{3}$$

---

[2] Note that it is also only the non-relevant interaction at the Fermi surface in the context of the renormalization group approach [9].

where we used the Feynman gauge for the gluon propagator. The term including the inner product $a \cdot b$ represents the spin dependence. The spin vector a^μ is explicitly given as a function of ζ and momentum [13]. There are many possible forms about a^μ, but we here use the simplest one,

$$a^0 = \frac{\mathbf{k} \cdot \zeta}{m}, \mathbf{a} = \zeta + \frac{\mathbf{k}(\zeta \cdot \mathbf{k})}{m(E_k + m)}. \tag{4}$$

From the invariance of the properties of quark matter under the Lorentz transformation, the Fermi velocity can be written [10] as

$$v_F^{-1} = \left(\frac{\partial k}{\partial \varepsilon(\mathbf{k}\zeta)}\right)_{k_F} = \frac{\mu}{k_F}\left(1 + \frac{1}{3}F_1^S\right) \tag{5}$$

with the spin-symmetric Landau parameter F_1^S defined by

$$F_1^S = N(0)f_1^S, f_1^S = -\frac{3}{4}\frac{g^2(N_c^2-1)}{4N_c^2\mu^2}\frac{m^2}{k_F^2}\int_{-1}^{1}du u\frac{1}{1-u}, \tag{6}$$

for OGE. Here $N(0)$ is the density of states at the Fermi surface, $N(0) = 2N_c k_F^2/2\pi^2 (\partial k/\partial \varepsilon(\mathbf{k}\zeta)_{k_F})$. Note that f_1^S clearly shows log divergence reflecting the gauge interaction. When we take into account the higher-order corrections for the gluon propagator, the electric propagator is screened by the Debye mass, while the magnetic one receives only the Landau damping. This fact exhibits the non Fermi liquid nature of quark matter. However, we shall see that the magnetic susceptibility becomes finite even in this case.

Applying the weak magnetic field to quark matter, we consider the energy change (see Fig. 1). Using the Gordon identity, the QED interaction Lagrangian can be recast as

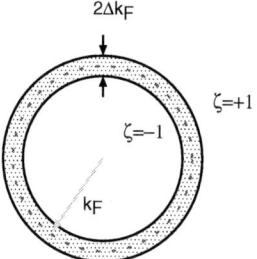

FIGURE 1. Modification of the Fermi surface in the presence of the weak magnetic field.

$$\int d^4 x \mathscr{L}_{\text{ext}}^{\text{QED}} = \sum_f \mu_q^f \int d^4 x \bar{\psi}_f [-i\mathbf{r} \times \nabla + \sigma] \times \mathbf{B}\psi_f \tag{7}$$

with the magnetic moment, $\mu_q^f = e_q^f/2m$. Since the orbital angular momentum gives null contribution on average, we hereafter only consider the spin contribution. In the following we consider only one flavor without loss of generality. For the energy to be minimum (chemical equilibrium),

$$\varepsilon(k_F + \Delta k_F, \zeta = +1) = \varepsilon(k_F - \Delta k_F, \zeta = -1), \qquad (8)$$

i.e.,

$$-\frac{g_D \mu_q B}{2} + \left(\frac{d\varepsilon}{dk}\right)_{k_F} \Delta k_F + (\bar{f}_{++} - \bar{f}_{+-})\Delta N$$
$$= \frac{g_D \mu_q B}{2} - \left(\frac{d\varepsilon}{dk}\right)_{k_F} \Delta k_F + (\bar{f}_{-+} - \bar{f}_{--})\Delta N, \qquad (9)$$

where $\Delta N = N_c k_F^2 \Delta k_F / 2\pi^2$, and g_D is the gyromagnetic ratio [13],

$$g_D = 2 \int \frac{d\Omega_k}{4\pi} \left(a_z - \frac{k_F}{\mu} \cos\theta a_0\right), \qquad (10)$$

which is reduced to be 2 in the non-relativistic limit, $m \gg k_F$. The angle-averaged Fermi liquid interactions $\bar{f}_{\zeta\zeta'}$ are given as

$$\bar{f}_{++} = \bar{f}_{--} = \frac{(N_c^2 - 1)g^2}{4N_c^2 \mu^2} \left[\frac{1}{2} - \frac{m^2}{k_F^2} - \frac{1}{3}\frac{m(\mu - m)}{k_F^2}\right] + \frac{f_1^S}{3}$$

$$\bar{f}_{+-} = \bar{f}_{-+} = \frac{(N_c^2 - 1)g^2}{4N_c^2 \mu^2} \left[\frac{1}{2} + \frac{1}{3}\frac{m(\mu - m)}{k_F^2}\right], \qquad (11)$$

by the use of the standard spin configuration (4). The latter is reduced to null in the non-relativistic limit, which implies there is no interaction between quarks with different spins [11]. From Eqs. (9) and (11) we find the spin susceptibility,

$$\chi_{\text{spin}} \equiv g_D \mu_q \Delta N / V B$$
$$= \chi_{\text{free}} \left[1 - \frac{4\alpha_c}{3\mu\pi} \frac{m(2\mu + m)}{3k_F}\right]^{-1} \qquad (12)$$

for $N_c = 3$, with the corresponding one without any interaction, $\chi_{\text{free}} = g_D^2 \mu_q^2 \mu k_F / 4\pi^2$. Note that log divergence is included in the Fermi liquid interaction (11), but it is canceled by the one coming from the Fermi velocity (5). Fig. 2 shows the ratio of spin susceptibility as a function of the Fermi momentum. It diverges around $k_F = 1.4\text{fm}^{-1}$, which is a signal of the spontaneous magnetization.

It would be interesting to compare the above result with the previous one given by the perturbative calculation with OGE [5], where we can see the *weakly first-order* phase transition at a certain low density, and that the critical density is similar to that given by the Fermi liquid theory. Thus two calculations are consistent with each other in the weak coupling limit.

Here we have discussed only the lowest order contribution, but a recent paper has also suggested the phase transition at low densities by including the higher-order effects [14].

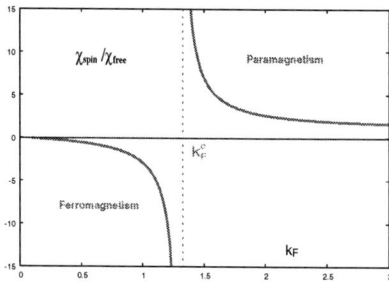

FIGURE 2. Spin susceptibility as a function of the Fermi momentum.

ASTROPHYSICAL IMPLICATIONS

There are two possibilities about the existence of quark matter in compact stars: one is in the quark stars, which are composed of low density strange matter, and the other is in the core region of neutron stars, which are called hybrid stars. Consider magnetars as quark stars or hybrid stars. Then we can easily estimate their magnetic field near the surface, if ferromagnetism is realized in quark matter. The maximum strength of the magnetic field on the surface $r = R$ is estimated by

$$B_{\max} = \frac{8\pi}{3}\left(\frac{r_Q}{R}\right)^3 \mu_Q n_Q, \qquad (13)$$

where r_Q is the radius of quark lump with density n_Q and μ_Q the single quark magnetic moment; it amounts to $O(10^{15-17})$G for $n_Q = 0.1\text{fm}^{-3}$, which might be enough for magnetars. It should be interesting to observe the braking indexes about magnetars; if their values are near three, the dipole radiation is a good picture and we may say that the above scenario looks more realistic.

There may be left another interesting problem about hierarchy of magnetic field in compact stars (Table 1). Unfortunately, the idea of ferromagnetism may not be sufficient for explaining it, and we need to consider the global magnetic structure and some dynamical mechanisms, e.g. formation of magnetic domain or existence of metamagnetism, besides it.

TABLE 1. Hierarchy of magnetic field in compact stars.

	millisecond pulsars	usual radio pulsars	magnetars
Magnetic field [G]	10^9	10^{12}	10^{15}
Period [sec]	10^{-2}	10^0	10^1
Age [year]	10^9	10^6	10^3

We might also consider a scenario about the cosmological magnetic field in the galaxies and extra galaxies. It is well known that magnetic fields are present in all galaxies and galaxy clusters, which are characterized by the strength, $10^{-7} - 10^{-5}$G, with the spatial scale, ≤ 1Mpc [15]. The origin of such magnetic fields is still unknown, but the first magnetic fields may have been created in the early universe. If magnetized quark lumps are generated during the QCD phase transition, they can give the seed fields.

CONCLUDING REMARKS

Magnetic properties of quark matter have been discussed by applying the Fermi liquid theory, which gives one of the non-perturbative tool to analyze them. Spin couples with motion in the relativistic theories, and we must extend the Fermi liquid interaction accordingly. We have seen that the magnetic susceptibility can be given by applying a weak magnetic field and considering the energy change in the tiny region near the Fermi surface. The Landau parameters may be log divergent for the gauge interaction, but the magnetic susceptibility is given to be finite by the cancellation. The critical point is the found to be the same order with the one given by the perturbative evaluation, which may support the relevance of the Fermi liquid theory in this problem. We may extend the present analysis by including higher order diagrams within the Fermi liquid theory or using the renormalization group.

If ferromagnetism is realized in quark matter, we may consider various scenarios: it might give a microscopic origin of the magnetic field in compact stars, and a seed of the primordial magnetic field during the cosmological QCD phase transition. It may also predict the production of small magnets composed of strange matter during the relativistic heavy-ion collisions.

This work is supported by the Japanese Grant-in-Aid for Scientific Research Fund of the Ministry of Education, Culture, Sports, Science and Technology (13640282,16540246).

REFERENCES

1. For a recent review, T. Tatsumi, E. Nakano and K. Nawa, hep-ph/0506002.
2. P.M. Woods and C.J. Thompson, astro-ph/0406133.
3. K. Makishima, Prog. Theor. Phys. Suppl. **151** (2003) 54.
4. For recent results, S. Fantoni et al., Phys. Rev. Lett. **87** (2001) 181101; I. Vidana and I. Bombaci, Phys. Rev. **C66** (2002) 045801.
5. T. Tatsumi, Phys. Lett. **B489** (2000) 280.
6. E. Nakano, T. Maruyama and T. Tatsumi, Phys. Rev. **D68** (2003) 105001; T. Tatsumi, T. Maruyama and E. Nakano, Prog. Theor. Phys. Suppl. **153** (2004) 190; hep-ph/0312351.
7. D.K. Hong, Phys. Lett. **B473**(2000) 118; Nucl. Phys. **B582** (2000) 451.
8. S. Hands, Phys. Rev.**D69** (2004) 014020.
9. J. Polchinski, hep-th/9210046.
   R. Shankar, Rev. Mod. Phys. **66** (1994) 129.
10. G. Baym and S.A. Chin, Nucl. Phys. **A262** (1976) 527.
11. C. Herring, "Exchange Interactions among Itinerant Electrons: Magnetism IV", Academic Press, NY,1966.
    K. Yoshida, "Theory of magnetism", Springer, Berlin, 1998.
12. J. Negele and H. Orland, "Quantum Many-Particle Systems", Addison-Wesley Pub.,1988.
13. T. Maruyama and T. Tatsumi, Nucl. Phys. **A693** (2001) 710.
14. A. Niegawa, Prog. Theor. Phys. **113** (2005) 581.
15. L.M. Widow, Rev. Mod. Phys. **74** (2002) 775.

1.8 SUPERNOVA EXPLOSION MECHANISM

Core Collapse Supernovae: Modeling Requirements and Surprises

Anthony Mezzacappa[*], John M. Blondin[†], O.E. Bronson Messer[**] and Stephen W. Bruenn[‡]

[*]*Physics Division, Oak Ridge National Laboratory, Oak Ridge, TN 37831*
[†]*Department of Physics, North Carolina State University, Raleigh, NC 27695-8202*
[**]*National Center for Computational Sciences, Oak Ridge National Laboratory, Oak Ridge, TN 37831*
[‡]*Department of Physics, Florida Atlantic University, Boca Raton, FL 33431*

Abstract. Past modeling efforts have illuminated that core collapse supernovae may be neutrino driven, MHD driven, or both, but uncertainties in the current models prevent us from being able to answer even this most basic question. Certain, however, is the need for multifrequency, and ultimately multifrequency *and* multiangle, neutrino transport. Moreover, terms in the neutrino transport equations that describe "observer corrections," such as angular aberration and frequency shift, are critical and cannot be neglected, and for massless neutrinos, global conservation of both lepton number and lab-frame specific neutrino energy must be maintained. Recent simulations in three dimensions of the stationary accretion shock instability (SASI) have also clearly demonstrated that two-dimensional models, constrained by axisymmetry, are limited and that three dimensional simulations will yield many surprises, some of them quite remarkable. Two areas that have not received as much attention in the past, neutrino mixing and magnetic fields, must be explored and may yield many additional surprises. The recent discovery that neutrino–neutrino interactions may result in deep neutrino mixing for *both* neutrinos and antineutrinos, across the energy spectrum, and possibly maximal mixing, must be explored in the context of detailed numerical simulations, and parameterized studies of core collapse supernovae with magnetic fields have produced a variety of results, depending on initial magnetic field configurations and strength, that now beg for detailed neutrino radiation magnetohydrodynamics simulations with multifrequency neutrino transport, especially in three dimensions, to determine which possibilities are realized in Nature.

Keywords: core collapse supernovae
PACS: 97.60.Bw

NEUTRINO TRANSPORT

Core collapse supernovae are initiated by the collapse of the iron cores of massive stars at the ends of their lives. The collapse proceeds to super-nuclear densities. The inner core becomes incompressible under these extremes, bounces, and, acting like a piston, launches a shock wave into the outer stellar core. This shock wave will ultimately propagate through the stellar layers beyond the core and disrupt the star in a core collapse supernova explosion. However, the shock stalls in the outer core, losing energy to nuclear dissociation and electron neutrino losses, and exactly how the shock is revived is unknown. This is the central question in core collapse supernova theory. (For a more complete review, the reader is referred to [1].)

The stalled supernova shock is thought to be revived, at least in part, by the charged-current absorption of electron neutrinos and antineutrinos that emerge from the proto-neutron star, a fraction of which are absorbed by protons and neutrons behind the shock. This is known as the delayed shock mechanism, originally proposed by Wilson and Bethe [2, 3]. Between the neutrinosphere and the shock, the material both heats and cools by electron neutrino and antineutrino emission and absorption. The neutrino heating and cooling have different radial profiles. Consequently, this region splits into a net cooling region and a net heating region, separated by a "gain" radius at which heating and cooling balance. We refer to the region between the gain radius and the shock as the "gain region."

The neutrino heating in the gain region can be written as

$$\dot{\varepsilon} = \frac{X_n}{\lambda_0^a} \frac{L_{\nu_e}}{4\pi r^2} <E_{\nu_e}^2><\frac{1}{F}> + \frac{X_p}{\lambda_0^a} \frac{L_{\bar{\nu}_e}}{4\pi r^2} <E_{\bar{\nu}_e}^2><\frac{1}{F}>. \quad (1)$$

The first (second) term corresponds to the absorption of electron neutrinos (antineutrinos). It depends linearly on the electron neutrino luminosity, L_{ν_e}, and inverse flux factor, $1/F$, which is a measure of the isotropy of the neutrino distribution (F is the ratio of the specific neutrino flux to the specific neutrino energy), and quadratically on the electron neutrino rms energy, $\sqrt{<E_{\nu_e}^2>}$. In addition to the dependence on the three key neutrino quantities in the heating rate above, the revival of the stalled supernova shock depends on a complex interplay of neutrino heating, mass accretion through the shock, and mass accretion through the gain radius [4]. It is mass accretion through the shock and gain radii that determines the amount of mass in the gain region, the former being a source of mass in the gain region and the latter being a sink. Moreover, the mass accretion through the gain radius serves to both sustain the neutrino luminosities and to undermine the pressure in the gain region, simultaneously—*i.e.*, it serves a supporting and a detrimental role.

All three quantities in the neutrino heating rate must be computed accurately [4, 5, 6, 7, 8, 9], which requires that we solve the Boltzmann neutrino transport equations, but given the quadratic dependence on the neutrino spectrum it is imperative that, at the very least, the spectrum be computed accurately. This requires the use of multi-neutrino energy (a.k.a. multifrequency or "multigroup") transport. The most compelling argument for the use of multifrequency neutrino transport can be made, of course, by simply taking stock of the results from all past core collapse supernova simulations. With the exception of Wilson's spherically symmetric models that invoke the doubly diffusive neutron finger instability in the proto-neutron star to boost the neutrino luminosities, no simulation to date performed with complete (*i.e.*, including the advection terms and all $O(v/c)$ observer corrections) multifrequency neutrino transport has yielded an explosion. This is a sobering fact [8, 10, 11, 12, 13, 14, 15, 16]. [1] And this list now includes both one- and two-dimensional simulations. Moreover, without neutron fingers, whose existence is a matter of current debate [18, 19], Wilson does not obtain explosions [20].

---

[1] A recent marginal exception to this trend was found by [17], where a *weak* explosion of a less-massive 11 $M_\odot$ progenitor was reported in the context of a two-dimensional model.

In our attempt to simulate core collapse supernova explosions we are presented with a number of underlying technical challenges that will ultimately dictate the degree to which we are confident that our simulation outcomes represent reality at all. Among these is the challenge of maintaining conservation of lepton number (for massless neutrinos) and energy in any given supernova model. The ultimate source of energy in a core collapse supernova is the gravitational binding energy of the remnant neutron star. The energy released in the form of neutrinos during the $\sim 1/3$ s in which we believe the shock is revived is $\sim [(2 \times 5) + 2] \times 10^{52} \times 1/3 \sim 4 \times 10^{52}$ erg. We have used the three-flavor neutrino luminosities from fully general relativistic spherically symmetric models [21] to compute this. This energy is ~ 40 times larger than the $\sim 10^{51}$ erg associated with the explosion. Consequently, total energy must be conserved over the course of a simulation to better than one part in 10^{2-3}. A typical simulation will be carried out over $\sim 10^5$ time steps, which requires that energy be conserved systematically to better than one part in 10^{7-8} per time step. This is a severe requirement. How can this be achieved?

We must first define the energy that *is* conserved. For the radiation field—modulo energy losses and gains owing to the neutrino interactions with the matter—the *lab frame* specific radiation energy is globally conserved. In one approach, we may begin with the comoving frame specific radiation energy and flux. In the continuum limit, the lab frame specific radiation energy conservation is guaranteed by a cancellation of terms in the equations for the *comoving frame* specific radiation energy *and* flux when these equations are added together to give the evolution (conservation) equation for the lab frame specific radiation energy (the lab frame specific radiation energy, by the Lorentz transformation, is a sum of the comoving frame specific radiation energy and flux). In the $O(v/c)$ limit, for example, the cancelling terms arise from *both* the $O(1)$ and $O(v/c)$ ("observer corrections": angular aberration and frequency shift) terms in the original Boltzmann equation from which the comoving frame specific energy and flux equations arise (these are the first two moments of the Boltzmann equation). We must in turn ensure that such cancellations occur in the discrete limit to ensure that energy is conserved in our simulations. This requires that we construct the discrete representation of the terms in the Boltzmann equation from which the cancelling terms arise with great care—*i.e.*, the discrete representation of the $O(1)$ and $O(v/c)$ terms in the Boltzmann equation are not independent. This has been achieved in the spherically symmetric case [22, 23, 24]. The proliferation of $O[(v/c^2)]$ terms in the lab frame conservation equation as we move from one- to three-dimensional models, when we begin with the $O(v/c)$ comoving frame moment equations, has led us to modify this procedure somewhat, to begin instead with a fully relativistic approach [25, 26].

Moreover, the observer corrections are critical in the evolution of the comoving frame neutrino distributions and in the dynamics of stellar collapse [22, 27, 28, 29, 30]. They cannot be excluded from any realistic model, as Figures 1 and 2 show. In these figures, we compare the entropy and velocity profiles from two simulations of Newtonian gravity, $O(v/c)$, spherically symmetric stellar collapse (performed with the AGILE-BOLTZTRAN code [31]). One simulation includes the observer corrections in the Boltzmann equation, the other does not. When the observer corrections are neglected, the inner homologous core mass at bounce is reduced by ~ 0.2 $M_\odot$, and the entropy throughout the inner core is significantly reduced [32]. The observer corrections are responsible for moving the neutrinos to higher energies (shorter mean free paths) as

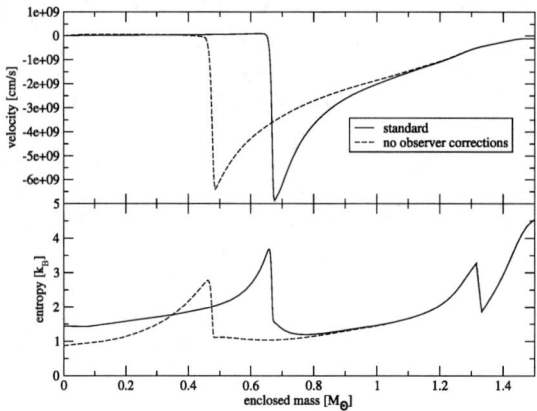

FIGURE 1. The entropy and velocity profiles as a function of enclosed mass at bounce for spherically symmetric models with and without the $O(v/c)$ observer corrections [32]. The differences between the two cases are striking.

the core is compressed [22, 27]. If we neglect these terms, the neutrino distributions in energy will not be correct and will be biased toward lower energies. The net result is increased electron capture and core cooling. Of course, an artificial reduction of this size in the homologous core mass cannot be tolerated. The shock loses $\sim 10^{51}$ erg for every 0.1 $M_\odot$ of iron it dissociates as it propagates through the outer core. In addition, without the $O(v/c)$ terms, energy and lepton number will not be conserved. The work done by the core in compressing the neutrinos is not reflected in the neutrino distributions. Figure 2 illustrates the extent to which the conservation breaks down. Prior to bounce, we have lost nearly 3×10^{51} erg. These results clearly demonstrate the impact these terms have not only on the neutrino distributions, but on the dynamics of stellar collapse and the conservation of total energy and lepton number. They cannot be ignored, and they must be included with great care.

A NEW INSTABILITY IN TWO DIMENSIONS VERSUS THREE

A fundamentally new hydrodynamic instability has been discovered recently [33] that may play a significant role in the core collapse supernova mechanism and that may be key to observables such as neutron star kicks, spectropolarization, and pulsar spin. The postbounce stellar core flow is best characterized as an accretion flow through a quasi-stationary shock. It has been shown in two- and now three-dimensional hydrodynamics studies constructed to reflect the conditions during the postbounce shock reheating epoch that nonspherical perturbations of the accretion shock lead to the development of a "stationary accretion shock instability (SASI)" [33, 34, 35, 36]. Recent studies [37, 38, 39] confirm the existence of the SASI instability in two-dimensional models

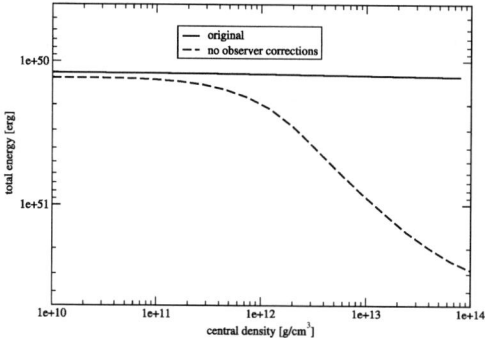

FIGURE 2. The total energy versus central density for the same models shown in Figure 1. The model without the observer corrections has clearly lost an energy $\sim 3 \times 10^{51}$ erg prior to stellar core bounce.

that include radial-ray neutrino transport. The potential ramifications of the SASI for the supernova mechanism and observables were first elaborated by Blondin, Mezzacappa, and DeMarino [33]. Two-dimensional simulations of the SASI lead to bipolar explosions and to a self-similarity in the flow at late times, with an aspect ratio consistent with the supernova spectropolarimetry data [40]. However, in the two-dimensional case the imposition of artificial reflecting boundary conditions to maintain axisymmetry sustains the $l = 1$ mode. Three-dimensional simulations yield a far more complex outcome [36]. In the linear regime, the SASI begins as an $l = 1$ instability in both the two- and three-dimensional cases [41]. However, in the three-dimensional case in the nonlinear regime, the $l = 1$ mode gives rise to an $m = 1$ mode. In addition, the $m = 1$ mode gives rise to a second shock beneath and orthogonal to the supernova shock, forming a shock triple point. In turn, this triple point effectively channels flow inward and is capable of imparting significant angular momentum to the proto-neutron star [36]. Beginning with spherically symmetric initial conditions, the final spin period in the case shown here is estimated to be ~ 35 ms, within the range of observed spins of young pulsars. The development of the $l = 1$ and $m = 1$ SASI modes in the three-dimensional case are shown in Figures 3 and 4. Taken together, these studies clearly indicate that two- and three-dimensional models will not only be quantitatively different, they will be qualitatively different. Moreover, they also make clear that connections between the nature of the progenitor and the final outcome of stellar collapse, and between model predictions and observations, are far more complex than we had anticipated in the past. For example, we have seen that bipolar explosions are possible without rotation [33] and that significant angular momentum may exist in the stellar core after collapse even beginning with nonrotating stellar cores at the onset of collapse [36].

MAGNETIC FIELDS

Arguably, the fundamental question here is whether the magnetic fields will organize into large-scale configurations that will help drive and collimate outflows from the

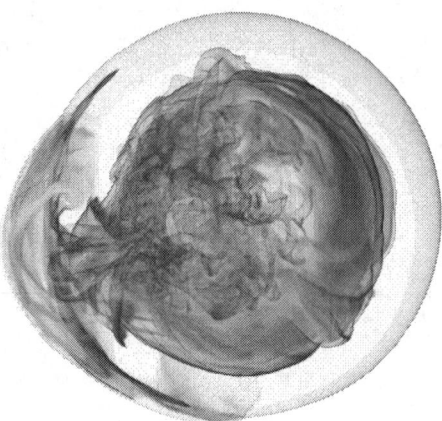

FIGURE 3. Shown here is the development of the stationary accretion shock instability (SASI) in three dimensions. At this stage in the nonlinear evolution, the $l = 1$ mode is dominant, as evidenced by the unipolar outflow.

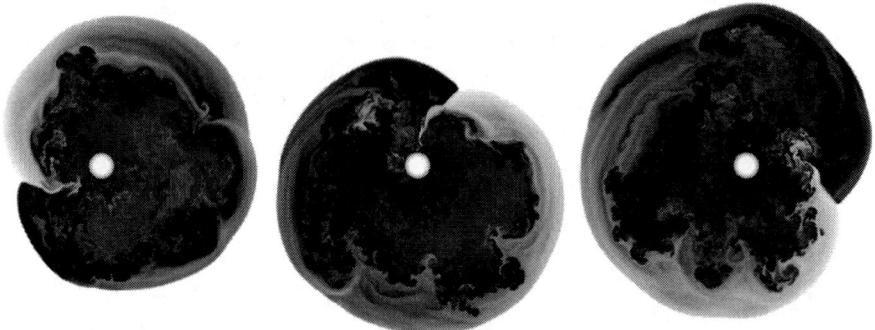

FIGURE 4. At this stage in the nonlinear evolution in three dimensions, the $l = 1$ SASI mode gives rise to an $m = 1$ spiral mode. Shown here is a temporal sequence of two-dimensional equatorial slices of the three-dimensional simulation data. The spiral flow is evident. Most important is the formation of an internal shock below the supernova shock and quasi-orthogonal to it, forming a shock triple point. This shock triple point funnels flow and angular momentum inward.

stellar core. The pioneering simulations of LeBlanc and Wilson [42] and Symbalisty [43] were the first to explore the evolution of stellar core magnetic fields during core collapse and their impact on the explosion mechanism. These simulations exhibited the development of a magnetic bubble deep in the core owing to the dramatic increase in magnetic pressure close to the rotation axis as core field lines are dragged inward and compressed. This magnetic bubble led to buoyant, bipolar outflows that culminated in bipolar explosions [the LeBlanc–Wilson (LW) jet].

More recently, this idea has been extended [44]. Owing to stellar core differential rotation, an initially poloidal field threading the stellar core could be wound up into a potentially significant toroidal field. Moreover, the field may be wound up quickly before it has a chance to expand vertically, later expanding in a spring-like fashion along the rotation axis, evolving into an open helix. In this way, material in the core could be driven outward along the rotation axis in a "spring and fling" manner, the latter arising because the material is accelerated along the rotating field lines by centrifugal forces. The so called hoop stresses would serve to collimate the flow.

The most advanced simulations of stellar core collapse to include magnetic fields were performed more than twenty years ago. Symbalisty [43] concluded that inordinately large rotation and magnetic fields strengths were required for an explosion to develop through the original Leblanc–Wilson mechanism. The magnetic fields in the stellar core can be amplified in the way suggested by Leblanc and Wilson or through wrapping, as described above, but the growth of field strength through other mechanisms must be considered.

Magnetic fields during stellar core collapse may be amplified in one of four ways: (1) Through collapse, as in the Leblanc–Wilson scenario. (2) By wrapping. (3) Through a dynamo, combining the action of fluid instabilities, such as convection, and rotation. (4) Through shear [the magnetorotational instability (MRI) [45]]. In the case of the MRI, the field amplification occurs through the stretching of the field lines in a strongly differentially rotating core. Akiyama et al. [46] were the first to propose that the MRI could be important in the core collapse supernova context. They argued that, with sufficient differential rotation, the MRI could amplify the magnetic field strengths in the stellar core exponentially quickly, rather than linearly through, for example, wrapping. The magnetic fields could then act as a conduit, channeling rotational energy into outflows [44].

In short, magnetic fields may play an important role in supernova dynamics. Even initially small magnetic fields in the stellar core may be amplified quickly after core bounce, through a variety of mechanisms, to participate in the supernova dynamics. A recent suite of simulations by Takiwaki et al. [47], Yamada and Sawai [48], and Sawai, Kotake, and Yamada [49] have further illuminated the landscape. These groups found uniformly that rotation with initially strong magnetic fields led to collimated explosions. However, with regard to initially weak magnetic fields and, in particular, the role of the MRI, the conclusions were mixed. The latter case (initially weak fields) is the case most relevant to core collapse supernovae, and the mixed outcomes in these parameterized numerical studies beg for more detailed studies with multifrequency neutrino transport (multigroup flux-limited diffusion or something more sophisticated) and, especially, three spatial dimensions.

NEUTRINO MASS AND MIXING

It is now an experimental fact that neutrinos have mass and, therefore, mix in flavor. Observations of Solar and atmospheric neutrinos, and experiments at LSND, indicate there may be as many as three independent values of the difference in the square of the neutrino masses (δm^2), and four mixing angles, which would require three active and at least one sterile neutrino. Although we await confirmation of the LSND findings, the data already strongly suggest that neutrino mixing should be included in core collapse supernova models. Neutrino mixing may significantly affect one or more of the following: the supernova mechanism, supernova nucleosynthesis, and terrestrial supernova neutrino detection.

As we discussed earlier, all three active neutrino flavors are involved in core collapse supernova dynamics. Electron, muon, and tau neutrinos and their antineutrinos are produced primarily through thermal emission, nucleon–nucleon bremsstrahlung, and neutral-current neutrino–neutrino annihilation in the hot mantle of the proto-neutron star after core bounce (see, for example, [1]). Owing to the lack of charged-current interactions among the muon and tau neutrinos and antineutrinos, electron neutrino and antineutrinos decouple at lower densities given their larger total interaction cross sections. Decoupling at lower densities and, consequently, lower temperatures results in softer relative spectra for the electron flavor neutrinos [50]. Herein lies the essential relevance of neutrino mixing to the supernova mechanism: If flavor conversion between electron flavor and muon/tau flavor neutrinos were to occur below the supernova shock wave in the neutrino heating epoch after stellar core bounce, the neutrino heating behind the shock, which is mediated predominantly by the charged-current absorption of electron neutrinos and antineutrinos, could be significantly increased [51, 52]. The softer electron neutrino flavor spectra would be replaced by the harder muon/tau neutrino flavor spectra in this region.

The argument to consider neutrino mixing in the core collapse supernova context, particularly with an eye toward the explosion mechanism, has been made even more compelling recently with the discovery that neutrino *and* antineutrino mixing *over the entire neutrino spectrum* may occur deep in the stellar core after bounce at small values of δm^2 [53, 54]. This mixing may arise as a result of the neutrino background in and above the proto-neutron star. Neutral-current neutrino–neutrino forward scattering increases the neutrino effective mass, much as charged-current electron-neutrino scattering increases the electron-neutrino effective mass in the MSW case. Moreover, the net result may be *near-maximal* mixing of neutrino flavors in the environment of the proto-neutron star after bounce [54], with obvious potential ramifications for the supernova mechanism, nucleosynthesis, and neutrino signatures.

While the experimental evidence for neutrino mixing is now clear, and while a number of past exploratory studies have elucidated some of the possible ramifications neutrino mixing may have for core collapse supernova dynamics, the precise impact of such flavor transformation remains to be determined. Neutrino mixing is a coherent, quantum mechanical phenomenon, unlike the incoherent collisional phenomena included in the Boltzmann kinetic equations discussed earlier. A more complete (quantum kinetic) treatment of neutrino transport in stellar cores beyond (classical) Boltzmann transport will be needed if we are to accurately and fully explore the impact of neutrino mixing on core collapse supernova dynamics.

CONCLUSION

The role of magnetic fields and neutrino mixing in core collapse supernovae are still virtually unexplored and may bring great surprises, as ever more sophisticated multidimensional models that include them are developed. As we have shown here, the jump to three dimensions in core collapse supernova models has already yielded surprises. Three-dimensional simulations of the development of the SASI were not only qualitatively different than their two-dimensional counterparts, but the ability to produce a compact object with a spin period ~ 35 ms beginning with spherically symmetric initial conditions is remarkable and forces us to reevaluate our theoretical framework and to rethink the connections we make between theory and observations. And underpinning the entire story are the neutrinos. Regardless of the final mix of ingredients in the supernova mechanism, the neutrinos will remain central to it. They define the dynamics of stellar core collapse, bounce, and shock formation, and they will likely play a significant role in shock revival. There is no longer any question that realistic supernova models will require realistic neutrino transport. Realistic transport (1) is multifrequency (and ultimately multiangle *and* multifrequency), (2) includes *all* terms in the neutrino transport equations in the limit adopted for the simulation (e.g, $O(v/c)$, fully special relativistic, or fully general relativistic), and (3) is conservative for lepton number (for massless neutrinos) and energy (this requires that we identify the conserved quantities and that we carefully construct the discretizations of the neutrino transport equations so as to achieve conservation of lepton number and energy). It is the need to include realistic neutrino transport that by far presents the greatest technical challenge in modeling core collapse supernovae. Without realistic neutrino transport, multidimensional models will remain exploratory.

ACKNOWLEDGMENTS

A.M. and O.E.B.M. are supported at the Oak Ridge National Laboratory, managed by UT-Battelle, LLC, for the U.S. Department of Energy under contract DE-AC05-00OR22725. This work was also supported by a SciDAC grant from the U.S. DOE High Energy, Nuclear Physics, and Advanced Scientific Computing Research Programs. The simulations discussed here were performed on the National Leadership Computing Facility at ORNL. We thank the National Center for Computational Sciences at ORNL and staff members in the Computational Science and Mathematics Division—in particular, Nagi Rao and Steve Carter—for their help in performing these simulations and in managing the large-scale simulation data afterwards.

REFERENCES

1. A. Mezzacappa, *Annual Reviews of Nuclear and Particle Science* **55**, 467–515 (2005).
2. J. R. Wilson, "Supernovae and Post–Collapse Behavior," in *Numerical Astrophysics*, edited by J. M. Centrella, J. M. LeBlanc, and R. L. Bowers, Jones and Bartlett, Boston, 1985, pp. 422–434.
3. H. A. Bethe, and J. R. Wilson, *Astrophysical Journal* **295**, 14–23 (1985).
4. H.-T. Janka, *Astronomy and Astrophysics* **368**, 527–560 (2001).
5. A. Burrows, and J. Goshy, *Astrophysical Journal Letters* **416**, L75–L78 (1993).
6. H.-T. Janka, and E. Müller, *Astronomy and Astrophysics* **306**, 167–198 (1996).
7. O. E. B. Messer, A. Mezzacappa, S. W. Bruenn, and M. W. Guidry, *Astrophysical Journal* **507**, 353–360 (1998).
8. A. Mezzacappa, A. C. Calder, S. W. Bruenn, J. M. Blondin, M. W. Guidry, M. R. Strayer, and A. S. Umar, *Astrophysical Journal* **495**, 911–926 (1998).

9. A. Mezzacappa, M. Liebendörfer, O. E. B. Messer, W. R. Hix, F.-K. Thielemann, and S. W. Bruenn, *Physical Review Letters,* **86**, 1935–1938 (2001).
10. J. R. Wilson, and R. W. Mayle, *Physics Reports* **227**, 97–111 (1993).
11. F. D. Swesty, and J. M. Lattimer, *Astrophysical Journal* **425**, 195–204 (1994).
12. M. Rampp, and H.-T. Janka, *Astrophysical Journal* **539**, L33–L36 (2000).
13. S. W. Bruenn, K. R. DeNisco, and A. Mezzacappa, *Astrophysical Journal* **560**, 326–338 (2001).
14. M. Liebendörfer, A. Mezzacappa, F.-K. Thielemann, O. E. B. Messer, W. R. Hix, and S. W. Bruenn, *Physical Review D* **63**, 103004–1–13 (2001).
15. T. A. Thompson, A. Burrows, and P. A. Pinto, *Astrophysical Journal* **592**, 434–456 (2003).
16. R. Buras, M. Rampp, H.-T. Janka, and K. Kifonidis, *Physical Review Letters* **90**, 241101–1–4 (2003).
17. H.-T. Janka, R. Buras, K. Kifonidis, A. Marek, and M. Rampp, "Core Collapse Supernovae at the Threshold," in *Supernovae*, edited by J. Marcaide, and K. Weiler, Springer Verlag, 2004, p. 253.
18. S. W. Bruenn, and T. Dineva, *Astrophysical Journal Letters* **458**, L71–L74 (1996).
19. S. W. Bruenn, E. A. Raley, and A. Mezzacappa, *Astrophysical Journal* (2006), in press.
20. J. R. Wilson (2004), private communication.
21. M. Liebendoerfer, M. Rampp, H.-T. Janka, and A. Mezzacappa, *Astrophysical Journal* **620**, 840–860 (2005).
22. A. Mezzacappa, and S. W. Bruenn, *Astrophysical Journal* **405**, 669–684 (1993).
23. M. Liebendörfer, O. E. B. Messer, A. Mezzacappa, S. W. Bruenn, C. Y. Cardall, and F.-K. Thielemann, *Astrophysical Journal Supplement* **150**, 263–316 (2004).
24. A. Mezzacappa, M. Liebendörfer, C. Y. Cardall, O. E. B. Messer, and S. W. Bruenn, "Neutrino Transport in Core Collapse Supernovae," in *Stellar Collapse*, edited by C. Fryer, Dordrecht: Kluwer Academic Publishers, 2004, pp. 99–132.
25. C. Y. Cardall, and A. Mezzacappa, *Physical Review D* **68**, 023006–1–26 (2003).
26. C. Y. Cardall, E. J. Lentz, and A. Mezzacappa, *Physical Review D* **72**, 043007–1–8 (2005).
27. S. W. Bruenn, *Astrophysical Journal Supplement* **58**, 771–841 (1985).
28. A. Mezzacappa, and S. W. Bruenn, *Astrophysical Journal* **405**, 637–668 (1993).
29. A. Mezzacappa, and S. W. Bruenn, *Astrophysical Journal* **410**, 740–760 (1993).
30. M. Rampp, and H.-T. Janka, *Astronomy and Astrophysics* **396**, 361–392 (2002).
31. M. Liebendörfer, O. E. B. Messer, A. Mezzacappa, S. W. Bruenn, C. Y. Cardall, and F.-K. Thielemann, *Astrophysical Journal Supplement* **150**, 263–316 (2004).
32. A. Mezzacappa, O. E. B. Messer, and S. W. Bruenn, *Astrophysical Journal* (2005), submitted.
33. J. M. Blondin, A. Mezzacappa, and C. DeMarino, *Astrophysical Journal* **584**, 971–980 (2003).
34. J. M. Blondin, "Capturing Stellar Core Hydrodynamic Instabilities in Core-Collapse Supernovae," in *Open Issues in Core Collapse Supernova Theory*, edited by A. Mezzacappa, and G. M. Fuller, World Scientific, Singapore, 2005, pp. 123–135.
35. J. M. Blondin, "Discovering new dynamics of core collapse supernova shock waves," in *SciDAC 2005, Scientific Discovery through Advanced Computing*, edited by A. Mezzacappa, and et al., Institute of Physics Publishing, Bristol, 2005, pp. 370–379.
36. J. M. Blondin, and A. Mezzacappa, *Nature* (2006), submitted.
37. H.-T. Janka, R. Buras, F. Kitaura Joyanes, A. Marek, M. Rampp, and S. L., *Nucl. Phys. A* **758**, 19–26 (2004).
38. L. Scheck, T. Plewa, H.-T. Janka, K. Kifonidis, and E. Mueller, *Physical Review Letters* **92**, 011103–1–4 (2005).
39. N. Ohnishi, K. Kotake, and S. Yamada (2005), astro-ph/0509765.
40. L. Wang, D. A. Howell, P. Höflich, and J. C. Wheeler, *Astrophysical Journal* **550**, 1030–1035 (2001).
41. J. M. Blondin, and A. Mezzacappa, *Astrophysical Journal* (2006), in press.
42. J. M. LeBlanc, and J. R. Wilson, *Astrophysical Journal* **161**, 541–551 (1970).
43. E. M. D. Symbalisty, *Astrophysical Journal* **285**, 729–746 (1984).
44. J. C. Wheeler, D. L. Meier, and J. R. Wilson, *Astrophysical Journal* **568**, 807–819 (2002).
45. S. A. Balbus, and J. F. Hawley, *Astrophysical Journal* **376**, 214–233 (1991).
46. S. Akiyama, J. C. Wheeler, D. L. Meier, and I. Lichtenstadt, *Astrophysical Journal* **584**, 954–970 (2003).
47. T. Takiwaki, K. Kotake, S. Nagataki, and K. Sato, *Astrophysical Journal* **616**, 1086–1094 (2004).
48. S. Yamada, and H. Sawai, *Astrophysical Journal* **608**, 907–924 (2004).
49. H. Sawai, K. Kotake, and S. Yamada, *Astrophysical Journal* **631**, 446–455 (2005).

50. A. Burrows, and T. A. Thompson, "Neutrino–Matter Interaction Rates in Supernovae," in *Stellar Collapse*, edited by C. Fryer, Dordrecht: Kluwer Academic Publishers, 2004, pp. 133–174.
51. G. M. Fuller, R. W. Mayle, J. R. Wilson, and D. N. Schramm, *Astrophysical Journal* **322**, 795–803 (1987).
52. G. M. Fuller, R. W. Mayle, B. S. Meyer, and J. R. Wilson, *Astrophysical Journal* **389**, 517–526 (1992).
53. Y.-Z. Qian, and G. M. Fuller, *Physical Review* **D52**, 656–660 (1995).
54. G. M. Fuller, and Y.-Z. Qian, *Physical Review* **73**, 023004–1–14 (2006).

Three-dimensional Modeling of Type Ia Supernova Explosions

F. K. Röpke and W. Hillebrandt

Max-Planck-Institut für Astrophysik, Karl-Schwarzschild-Str. 1, D-85741 Garching, Germany

Abstract. Modeling type Ia supernova (SN Ia) explosions in three dimensions allows to eliminate any undetermined parameters and provides predictive power to simulations. This is necessary to improve the understanding of the explosion mechanism and to settle the question of the applicability of SNe Ia in cosmological distance measurements. Since the models contain no tunable parameters, it is also possible to directly assess their validity on the basis of a comparison with observations. Here, we describe the modeling of SNe Ia as thermonuclear explosions in which the flame after ignition near the center of the progenitor white dwarf star propagates outward in the sub-sonic deflagration mode accelerated by the interaction with turbulence. We explore the capabilities of this model by comparison with observations and show in a preliminary approach, how such a model can be applied to study the origin of the diversity of SNe Ia.

Keywords: supernovae – turbulence – computer modeling and simulation – hydrodynamics
PACS: 97.60.Bw; 98.38.Am; 95.75.-z; 95.30.Lz

INTRODUCTION

Type Ia supernovae (SNe Ia) have become one of the major tools in observational cosmology. Yet a sound theoretical understanding of these objects – justifying in particular the calibration techniques applied in distance measurements – is still lacking. First attempts to model SNe Ia were based on one-dimensional numerical simulations. Such models gave valuable insight into the basic mechanism of SN Ia explosions. However, their predictive power is limited due to the fact that underlying physical processes enter the models in a parametrized way. This is overcome by three-dimensional modeling of SNe Ia [1, 2]. As an example, we present a model that is derived from the standard scenario of SN Ia explosions (for a review see [3]). A white dwarf (WD) consisting of carbon and oxygen is assumed to accrete matter from a non-degenerate binary companion until its mass approaches the Chandrasekhar limit. Due to the rapid increase of the central density nuclear reactions ignite giving rise to a stage of convective carbon burning. This stage lasts for several hundred years and terminates once the nuclear energy production cannot be balanced by convective cooling any longer. Subsequently, a thermonuclear runaway of a small temperature fluctuation ignites a thermonuclear flame. The exact mechanism of flame ignition, however, remains controversial. While some studies suggest a flame ignition in multiple sparks distributed around the center of the WD [4, 5, 6, 7], others put forward central single-point ignitions [8].

After ignition the flame propagation is determined by the laws of hydrodynamics. Regarding the flame front as a discontinuity between fuel and ashes, they allow for two distinct modes of flame propagation. In a sub-sonic deflagration burning is mediated by thermal conduction while a super-sonic detonation is driven by a shock wave. On the basis of one-dimensional simulations of a prompt detonation, Arnett [9] ruled out this model for SNe Ia since it drastically underproduces intermediate mass elements observed in spectra of these events.

Starting out as a laminar deflagration, however, the flame propagates too slowly to explain the energy release necessary to explode the WD. Thus, any valid SN Ia model needs to provide means of flame acceleration. Two mechanisms are conceivable here. Firstly, the flame propagation may continue in the deflagration mode being significantly accelerated by the interaction with turbulence. The one-dimensional model *W7* of Nomoto et al. [10] demonstrated that such a model is in principle capable of reproducing the main observational features of SNe Ia. An alternative way to speed up the flame is to assume a deflagration-to-detonation transition (DDT) at later stages of the explosion. The weak point of these delayed detonation models (e.g. [11]) is that a physical mechanism providing a DDT in SNe Ia could not be identified yet [12, 13, 14, 15] and therefore the hypothetical transition of the burning mode enters the model as an undetermined parameter.

A DEFLAGRATION TYPE Ia SUPERNOVAE MODEL

Our goal in the following is to present a SN Ia explosion model that contains no tunable parameters. Therefore we set aside the possibility of a delayed detonation and focus on the turbulent deflagration model. Turbulence is induced here by generic instabilities. The flame propagates from the center of the star outward leaving behind light and hot ashes. Dense and cold fuel in front of the flame gives rise to an inverse density stratification in the gravitational field of the WD. This renders the flame propagation buoyancy unstable and in its nonlinear stage the Rayleigh-Taylor instability leads to the formation of (typically mushroom-shaped) burning bubbles that rise into the cold fuel. At the interfaces of these bubbles strong shear flows emerge. The corresponding Reynolds numbers reach values of the order of 10^{14} and therefore strong turbulence is generated by secondary shear (Kelvin-Helmholtz) instabilities. The turbulent eddies generated on scales of the buoyancy-induced flame features decay to smaller scales forming a turbulent energy cascade and interact with the flame propagation. This stretches and corrugates the flame enlarging its surface area and thus the net burning rate is increased.

The main challenge in numerically implementing this scenario is the vast range of relevant length scales involved. Not only is the width of a thermonuclear flame in the degenerate carbon/oxygen material at the onset of the explosion 9 orders of magnitude below the radius of the WD. The turbulent cascade extends to even smaller scales and interacts with the flame down to the Gibson length at which the laminar flame speed equals the turbulent velocity fluctuations (10^4 cm and decreasing in the explosion process). This problem can be tackled in a Large Eddy Simulation (LES) approach. Here, only the largest scales of the problem are directly resolved (applying the PROMETHEUS implementation [16] for solving the hydrodynamics equations) and turbulence effects on unresolved scales are included via a subgrid-scale model [17, 18]. The thermonuclear flame is modeled as a sharp discontinuity separating the burnt from the unburnt material. Its evolution is followed utilizing the level-set method [19]. Since the structure of the flame is not resolved, the flame propagation velocity must be provided externally. This,

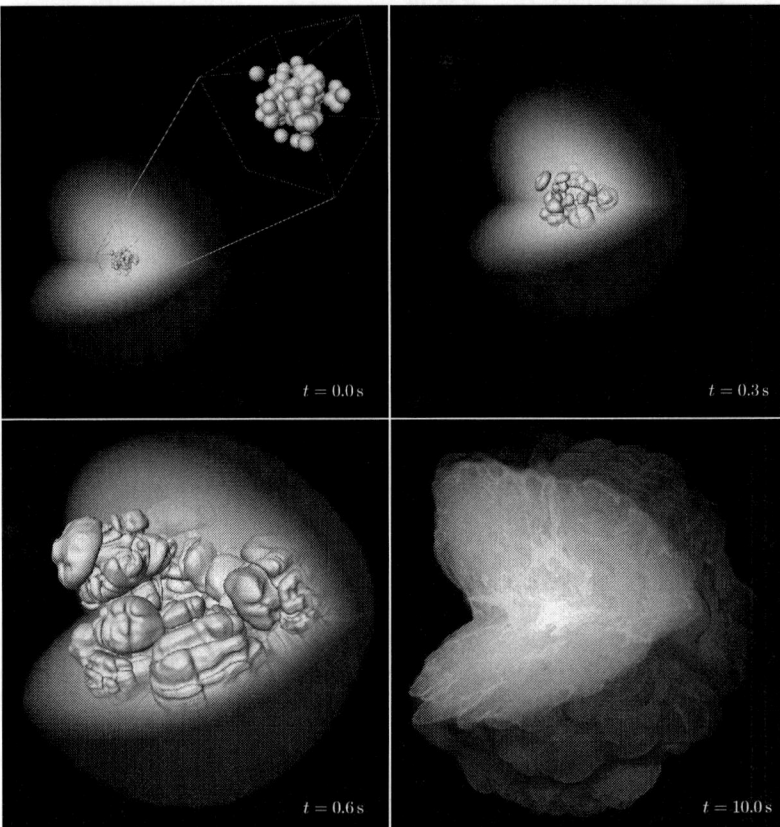

FIGURE 1. Snapshots from a full-star SN Ia simulation starting from a multi-spot ignition scenario. The logarithm of the density is volume rendered indicating the extend of the WD star and the isosurface corresponds to the thermonuclear flame. The last snapshot marks the end of the simulation and is not on scale with the earlier snapshots.

however, does not introduce an undetermined parameter to the model since the theory of turbulent combustion [20] predicts that for most stages of the SN Ia explosion the flame propagation proceeds in the flamelet regime where it completely decouples from the microphysics of the burning and is determined by the turbulent velocity fluctuations that can be derived from the subgrid-scale model. This implementation provides a self-consistent model of SN Ia explosions in the deflagration scenario.

Our description of burning is augmented by a simplified treatment of the nuclear reactions including only five species [21]. It provides the energy release necessary to follow the explosion dynamics. In order to derive observables from the models, however, the chemical composition of the ejecta needs to be known in detail. This is achieved in a nucleosynthesis postprocessing step [22]. Of particular interest is the yield of ^{56}Ni, since its radioactive decay powers the light curve.

TABLE 1. Variation of initial parameters in SN Ia explosion models

Parameter	Range of variation	Effect on ^{56}Ni production	Effect on total energy
$X(^{12}C)$	[0.30, 0.62]	$\leq 2\%$	$\sim 14\%$
ρ_c [10^9 g/cm^3]	[1.0, 2.6]	$\sim 6\%$	$\sim 17\%$
Z [$Z_\odot$]	[0.5, 3.0]	$\sim 20\%$	none

SIMULATION RESULTS

A typical evolution of a SN Ia explosion modeled as described above is shown in Fig. 1. Starting from an ignition in multiple sparks the flame propagates outward. At $t = 0.3$ s, the mushroom-shaped features due to the buoyancy instability are clearly visible. Subsequently, the flame becomes increasingly corrugated and is accelerated by interaction with turbulence. It therefore burns through a large fraction of the WD material. The snapshot at $t = 0.6$ s shows the flame evolution around the peak of energy production due to nuclear burning. Up to this point, the burning terminated in nuclear statistical equilibrium (NSE) and the carbon/oxygen material was primarily converted to iron group elements. The expansion of the WD decreases the fuel density steadily and once it falls below 5×10^7 g cm^{-3} nuclear burning becomes incomplete and produces mainly intermediate mass elements. About 2 s after ignition, expansion quenches the burning and the following evolution is characterized by the relaxation to homologous expansion of the ejecta, which is reached to a reasonable accuracy ~ 10 s after ignition [23]. The density structure of the ejecta at this stage is shown in Fig. 1, where the traces of turbulent flame propagation are clearly visible.

Apart from the initial conditions simulations as described above contain no free parameters. Therefore the question arises whether such models are capable of reproducing observations without any fitting. The explosion energies achievable in the outlined scenario reach up to $\sim 8 \times 10^{50}$ erg and the models produce $\sim 0.4 M_\odot$ of ^{56}Ni. This falls into the range of observational expectations, although on the side of the weaker SN Ia explosions [24, 25]. Nonetheless, the synthetic lightcurves derived from models of the class described here fit the observations in the B and V bands around maximum luminosity rather well [26, 27]. A much harder constraint on the explosion model is posed by spectral observations, since spectra are particularly sensitive to the chemical composition of the ejecta. Kozma et al. [28] pointed out a potential problem of deflagration SN Ia models. In late time "nebular" spectra, unburnt material (transported towards the center in downdrafts due to the large-scale buoyancy-unstable flame pattern) gives rise to a strong oxygen line of low-velocity material which is in conflict with observations. However, the synthetic spectrum of [28] was derived from a simplistic centrally ignited model. Recently, Röpke et al. [29] showed that multi-spot ignition models may succeed to burn out the central parts of the WD, reducing the amount of oxygen at low velocities. A stochastic multi-spot ignition leads to similar results [30]. Detailed spectral observations allow to determine the chemical composition of the ejecta in velocity space [31]. The mixed composition of the ejecta observed there points to a deflagration phase being at least a significant contribution to the SN Ia explosion process. The central parts are found to be clearly dominated by iron group elements, which are mixed out to velocities of about

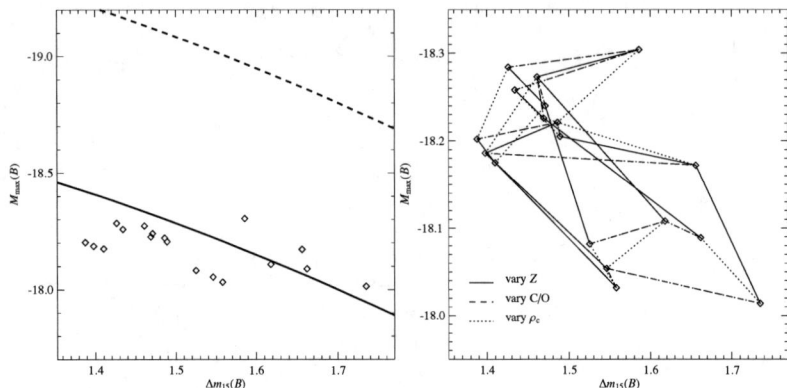

FIGURE 2. Peak luminosity vs. decline rate of the light curve in the B band (diamonds correspond to SN Ia explosion models; the dashed curve in the left panel indicates the original relation by Phillips et al. [34] and a shifted relation is marked as a solid curve)

$12000\,\mathrm{km\,s^{-1}}$. Intermediate mass elements are distributed over a wide range in radii and no unburnt material is found at velocities $\lesssim 5000\,\mathrm{km\,s^{-1}}$. A recent high-resolution full-star deflagration SN Ia simulation [32] demonstrated that such models can get close to these observational constraints.

Although it cannot be ruled out that the pure deflagration model of SNe Ia is incomplete, the results indicate that it may be at least a dominant part of the mechanism. Therefore it is justified to ask how such models are affected by initial parameters of the exploding WD. This may give a hint to the origin of the diversity of SNe Ia. To moderate the computational expenses, simplified setups may be used to study the effects of physical parameters on the explosion models. Such an approach was recently taken by Röpke et al. [33] and resulted in the first systematic study of progenitor parameters in three-dimensional models. The basis of this study was a single-octant setup with moderate (yet numerically converged) resolution. However, the lack in resolution did not allow for a reasonable multi-spot ignition scenario and thus only weak explosions can be expected. It is therefore not possible to set the absolute scale of effects in this approach, but trends can clearly be identified. The parameters chosen for the study were the WD's carbon-to-oxygen ratio, its central density at ignition and its ^{22}Ne mass fraction resulting from the metallicity of the progenitor. All parameters were varied independently to study the individual effects on the explosion process. The results of this survey are given in Tab. 1. To determine the effects of these variations on observables, synthetic light curves were derived from all models [27]. From these, the peak luminosities and decline rates (in magnitudes 15 days after maximum; Δm_{15}) were determined. A comparison with the relation given by Phillips et al. [34] (forming the basis of the calibration of cosmological distance measurements) is provided in the left panel of Fig. 2. Obviously, the absolute magnitude of the *Phillips relation* is not met by our set of models. Moreover, the range of scatter in Δm_{15} is much narrower than that of the set of observations used by Phillips et al. [34]. But there is a trend of our models consistent with the slope of the Phillips relation. The right panel of Fig. 2 shows that this slope is dominated by the variation of the progenitor's metallicity.

ACKNOWLEDGMENTS

F.K.R. gratefully acknowledges the kind invitation to the OMEG05 conference in Tokyo and friutful discussions with K. Nomoto and his group on SN Ia explosions.

REFERENCES

1. M. Reinecke, W. Hillebrandt, and J. C. Niemeyer, *A&A* **391**, 1167–1172 (2002).
2. V. N. Gamezo, A. M. Khokhlov, E. S. Oran, A. Y. Chtchelkanova, and R. O. Rosenberg, *Science* **299**, 77–81 (2003).
3. W. Hillebrandt, and J. C. Niemeyer, *ARA&A* **38**, 191–230 (2000).
4. D. Garcia-Senz, and S. E. Woosley, *ApJ* **454**, 895–900 (1995).
5. S. E. Woosley, S. Wunsch, and M. Kuhlen, *ApJ* **607**, 921–930 (2004).
6. L. Iapichino, M. Brüggen, W. Hillebrandt, and J. C. Niemeyer (2006), *A&A* in press, `astro-ph/0512300`.
7. M. Kuhlen, S. E. Woosley, and G. A. Glatzmaier, *ApJ* **640**. 407–416 (2006).
8. P. Höflich, and J. Stein, *ApJ* **568**, 779–790 (2002).
9. W. D. Arnett, *Ap&SS* **5**, 180–212 (1969).
10. K. Nomoto, F.-K. Thielemann, and K. Yokoi, *ApJ* **286**, 644–658 (1984).
11. V. N. Gamezo, A. M. Khokhlov, and E. S. Oran, *Phys. Rev. Lett.* **92**, 211102 (2004).
12. J. C. Niemeyer, *ApJ* **523**, L57–L60 (1999).
13. A. M. Lisewski, W. Hillebrandt, and S. E. Woosley, *ApJ* **538**, 831–836 (2000).
14. F. K. Röpke, W. Hillebrandt, and J. C. Niemeyer, *A&A* **420**, 411–422 (2004).
15. F. K. Röpke, W. Hillebrandt, and J. C. Niemeyer, *A&A* **421**, 783–795 (2004).
16. B. A. Fryxell, E. Müller, and W. D. Arnett, Hydrodynamics and nuclear burning, MPA Green Report 449, Max-Planck-Institut für Astrophysik, Garching (1989).
17. J. C. Niemeyer, and W. Hillebrandt, *ApJ* **452**, 769–778 (1995).
18. W. Schmidt, J. C. Niemeyer, W. Hillebrandt, and F. K. Röpke (2006), *A&A* in press, `astro-ph/0601500`.
19. S. Osher, and J. A. Sethian, *J. Comp. Phys.* **79**, 12–49 (1988).
20. N. Peters, *Turbulent Combustion*, Cambridge University Press, Cambridge, 2000.
21. M. Reinecke, W. Hillebrandt, and J. C. Niemeyer, *A&A* **386**, 936–943 (2002).
22. C. Travaglio, W. Hillebrandt, M. Reinecke, and F.-K. Thielemann, *A&A* **425**, 1029–1040 (2004).
23. F. K. Röpke, *A&A* **432**, 969–983 (2005).
24. G. Contardo, B. Leibundgut, and W. D. Vacca, *A&A* **359**, 876–886 (2000).
25. M. Stritzinger, B. Leibundgut, S. Walch, and G. Contardo (2005), `arXiv:astro-ph/0506415`.
26. E. Sorokina, and S. Blinnikov, "Light Curves of Type Ia Supernovae as a Probe for an Explosion Model," in *From Twilight to Highlight: The Physics of Supernovae*, edited by W. Hillebrandt, and B. Leibundgut, ESO Astrophysics Symposia, Springer, Berlin Heidelberg, 2003, pp. 268–275.
27. S. I. Blinnikov, F. K. Röpke, E. I. Sorokina, M. Gieseler, M. Reinecke, C. Travaglio, W. Hillebrandt, and M. Stritzinger (2006), *A&A* in press, `astro-ph/0603036`.
28. C. Kozma, C. Fransson, W. Hillebrandt, C. Travaglio, J. Sollerman, M. Reinecke, F. K. Röpke, and J. Spyromilio, *A&A* **437**, 983–995 (2005).
29. F. K. Röpke, W. Hillebrandt, J. C. Niemeyer, and S. E. Woosley (2005), *A&A* **448**, 1–14 (2006).
30. W. Schmidt, and J. C. Niemeyer, *A&A* **446**, 627–633 (2006).
31. M. Stehle, P. A. Mazzali, S. Benetti, and W. Hillebrandt, *MNRAS* **360**, 1231–1243 (2005).
32. F. K. Röpke et al. (2006), in preparation.
33. F. K. Röpke, M. Gieseler, M. Reinecke, C. Travaglio, and W. Hillebrandt (2005), *A&A* in press, `astro-ph/0506107`.
34. M. M. Phillips, P. Lira, N. B. Suntzeff, R. A. Schommer, M. Hamuy, and J. Maza, *AJ* **118**, 1766–1776 (1999).

Gravitational Collapse of Massive Stars

Shoichi Yamada

Department of Physics, Science & Engineering, Waseda University
3-4-1 Okubo, Shinjuku, 169-8555 Tokyo, Japan

Abstract. In this paper, I summarise the recent results of our study on the core-collapse supernova and related phenomena. Among the issues addressed are (1) long-term 1D simulations of core-collapse supernovae, (2) global asymmetry of supernova, and (3) collapse of more massive stars and neutrino signals. In the first topic, I report our latest 1D simulations for more than a second after the bounce and demonstrate that the difference of EOS's manifests itself more clearly in the later phase of the core collapse. In the second part, I discuss hydrodynamic instabilities as a possible cause for the global asymmetry that may be a generic feature of core-collapse supernova. The mode analysis of the non-spherical instability of the standing accretion shock is presented. Inelastic scatterings of neutrino on nuclei are also discussed in this context. Finally, I mention the gravitational collapse of more massive stars which will produce not a neutron star but a black hole. Particular attention is paid to the neutrino signals from these phenomena as a probe of hot dense matter.

Keywords: supernovae, neutrinos, black hole, hydrodynamical instabilities
PACS: 97.60.Bw, 97.60.Jd, 97.60.Lf, 26.50.+x,

INTRODUCTION

The collapse-driven supernova is an explosive phenomenon which is supposed to occur at the end of the evolution of massive stars ($\gtrsim 8 M_\odot$). The mechanism of collapse-driven supernovae has eluded our understanding for more than 40 years.[1] The difficulty in the modelling of collapse-driven supernova arises from the fact that it involves a complex combination of microphysics and macrophysics. In fact, weak interactions and nuclear physics are supposed to be dictating the dynamics in a critical way. It has also become a common sense that the collapse-drive supernova is in general not spherically symmetric. At the moment, we simply do not know what is the crucial element for the supernova explosion.

Although the stellar rotation may be the most naive explanation for the asymmetric explosion, the instability of the standing shock wave against non-spherical perturbations, the so-called SASI (Standing Accretion Shock Instability), [2] is also attracting great interest of researchers not only as an important factor in producing the asymmetric explosion as observed but also as an agent to generate a proper velocity, or a kick, as observed for young pulsars.

The supernova explosion is supposed to occur for massive stars in the mass range of $\sim 10 \cdots \sim 30 M_\odot$.[3] Nowadays more massive stars are attracting much interest of researchers from beyond the supernova community. This is almost entirely due to the recognition that long gamma-ray bursts are somehow caused by the gravitational collapse of massive stars and subsequent black hole formation. It should be also mentioned that the interest in the first generation stars, or population III stars (POP III stars), is also growing rapidly these days thanks to the results of WMAP[4] as well as other observa-

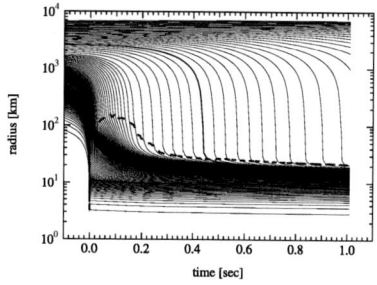 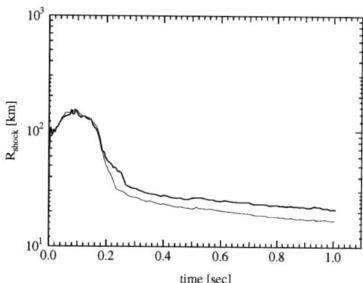

FIGURE 1. The radial trajectories of mass elements for Shen EOS (left panel) and the radial positions of the shock wave (right panel) as a function of time after bounce for 15 $M_\odot$ model. In the left panel, the dahed line indicates the location of the shock wave. In the right panel, the thick and thin lines represent the results for Shen EOS and Lattimer & Swesty EOS, respectively.

tions of old stars.[5] It is generally supposed that not a small fraction of POP III stars had a very high mass of $\sim 100 \cdots \sim 1000 M_\odot$.[6] The future supernova research, thus, should be defined in the greater perspective of the gravitational collapse of massive stars.

1D SIMULATIONS AND NUCLEAR EOS

First I briefly report our latest spherical models. Nowadays the theoretical research of the collapse-driven supernova has been done mainly in the context of the so-called delayed explosion, in which a stalled shock wave is re-energized by neutrinos copiously emitted out of the proto neutron star. It is known that there is a critical neutrino luminosity for the shock revival.[7, 8] Hence the issue is whether this critical value is realized in the evolution or not. The evaluation, however, requires a quantitative numerical modelling with all the relevant neutrino physics taken into account. Such sophisticated simulations have been done by a couple of groups over the years.[9, 10, 11, 12]

For example, Fig. 1 shows our latest results.[12] The evolutions were computed for more than a second after the core bounce for two realistic nuclear EOS's. The left panel shows the mass trajectories for Shen EOS. As is evident, there is no hint of shock revival. In fact, the shock stalls at around 200km from the center and becomes an accretion shock and then starts to recede onto the proto neutron star later on. The evolution is not qualitatively different for the other EOS by Lattimer & Swesty and no explosion was found in this case, either.

The right panl shows the evolutions of the shock wave for the two models. The difference is indeed quite minor for the first 200ms after the bounce, but becomes clearer later. As Lattimer & Swesty EOS we employed here is softer than Shen EOS in the high density regime, the proto neutron star is more compact for the former, and so is the shock radius. The diffefence is also reflected in the neutrino signals. These results suggest that the later evolutionary phase is more suited for the investigation of hot dense matter in the supernova core. This is even more so for more massive stars as will be shown later.

HYDRODYNAMICAL INSTABILITIES OF STANDING ACCRETION SHOCK

Various multi-dimensional aspects of the supernova dynamics have been studied extensively. Here I pay main attention to the global non-sphericity as observed in the recent HST images of SN1987A. Although the rapid rotation of the progenitor appears to be the easiest solution, some stellar evolution models predict otherwise. Furthermore, it is well known that young pulsars are supposed to be born as a slow rotator. Then the natural question is how the asymmetry of explsion as observed was produced. Hydrodynamical instabilities may be the answer.

Figure 2 shows the distributions of entropy (the left half of the panel) and density (the right half) in the meridian section for the models with $L_v = 5.5, 6.0 \cdot 10^{52}$ erg/s after 1% of the $\ell = 1$ single-mode velocity perturbation is added.[13] For both models, we observe the growth of the perturbations. In the case of $L_v = 5.5 \cdot 10^{52}$ erg/s, the shock surface is deformed at first by the increasing amplitude of the non-radial mode and then begins to oscillate with a large amplitude. In the case of $L_v = 6.0 \cdot 10^{52}$ erg/s (right panels), on the other hand, in addition to the oscillations of the shock surface, we observe the substantial increase of the average shock radius as the time passes. In fact, after $t = 400$ ms, the shock radius continues to increase and appears to produce an explosion. Since the model is stable against radial perturbations as mentioned above, the non-radial instability and the neutrino heating therein are responsible for the explosion. We think that this is a reconfirmation of the claim that the instability, whatever the cause, behind the shock is helpful for the shock revival.

For the random multi-mode velocity perturbations, we find that the modes with small ℓ's, especially those with $\ell = 1, 2$, grow rapidly in the linear regime. This is particularly the case for the model without a negative entropy-gradient and the growths of the modes with $\ell > 10$ are negligibly small. With a negative entropy-gradient, the broadening of the spectral ℓ-distribution is observed although the dominance of smaller ℓ modes can be still found. The convective instability may enhance the growth of higher harmonics in the linear phase. The similarity of the two cases suggests again that SASI is dominant over the convection even when the latter is operating.

It is also interesting to note that the modes with $\ell = 1, 2$ are dominant in the nonlinear regime, which begins at ~ 100 ms after bounce. Various modes are amplified by nonlinear couplings with the dominant modes in this phase. The spectra are again broader for the model with a negative entropy-gradient. In both models, however, the dominance of the modes with $\ell = 1, 2$ is remarkable. This should correspond to the large deformations of shock wave found in the numerical simulations. In order to make clear the reason for the dominance of these modes in the nonlinear regime, it will be required to study the nonlinear couplings of various modes in more detail.

We have to wait for realistic simulations before we judge if SASI can give enough boost for the shock revival. The numerical results[14] obtained so far are not very encouraging. If that is really the case, we had better find something more to obtain a successful explosion. Here we consider the inelastic scattering of neutrinos on nuclei as a possible boost.[15] These reactions have not been considered in the supernova simulations simply because nuclei are not abundant in the post-bounce phase, particularly in the

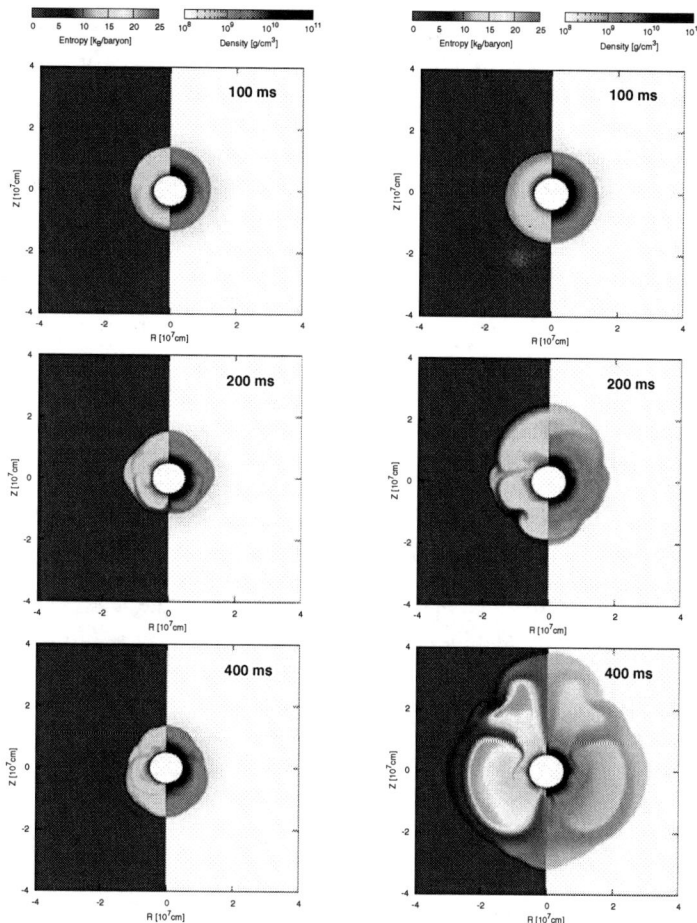

FIGURE 2. Entropy- (the left half of each panel) and density- (the right half) distributions in the meridian section for 1% of the $\ell = 1$ single-mode velocity perturbation. $L_\nu = 5.5 \cdot 10^{52}$ erg/s is assumed for the left panels and $L_\nu = 6.0 \cdot 10^{52}$ erg/s is for the right panels.

spherically symmetric models. In the multi-D models, however, some nuclei re-appear when the shock reaches larger radii compared with the 1D models. In order to see the possible effect of the reactions to the shock revival in SASI, we implement the neutrino-alpha (the most abundant nuclei in the shocked matter except for nucleons) scatterings according to the prescription by Haxton [16] in the above models.

We find that the reactions are rather minor except for the case in which the shock revival barely fails. It is also found that the effect is sensitive to the initial amplitude of perturbations. If we assume 5% of perturbations instead of 1%, the substantial boost of shock heating is observed. Although the reaction rate is sensitive to the energy of incident neutrino, a considerable hardening of the neutrino spectra is needed for the reaction to influence the shock revival. If the reaction rates are larger than the current estimate by a factor of 3 ∼ 10, the interactions cannot be ignored in considering the shock revival.

GRAVITATIONAL COLLAPSES WITH BLACK HOLE FORMATIONS

So far I have discussed theoretical researches on the collapse of $\sim 10 \cdots \sim 30 M_\odot$ stars, which are supposed to produce a supernova explosion eventually. As mentioned already, however, the fate of more massive stars ($\sim 30 \cdots \sim 100 M_\odot$), which will produce a black hole one way or another, is also attracting much interest these days. This is mainly because these massive stars are expected to produce long gamma ray bursts. On the other hand, the interest in POP III stars naturally motivates the study on the fate of very massive stars in the mass range of $\sim 100 \cdots \sim 1000 M_\odot$. In the following, I will show some of our recent results on these subjects.[17, 18]

Stars more massive than $\sim 30 M_\odot$ have large iron cores and will be intrinsically too massive to have stellar explosion. Then, the outcome will be a formation of black hole. The detection of neutrinos is a clear and unique identification of such events. In Fig. 3 we show our simulation of such an event.[17] As shown in the left panel, the second dynamical collapse starts when the enclosed baryon mass reaches $2.66 M_\odot$ (gravitational mass $2.38 M_\odot$) at t_{pb}=1.34 s. The end points in the figure correspond to the formations of apparent horizon, i.e. the births of black hole. The time profile of luminosities is unique. Luminosities are dominated by the contributions from the accreted material. As the proto-neutron star contracts quasi-statically, the luminosities become higher and are dominated by v_e and $\bar{v}_e$ originating from the accreted material. It is remarkable that the luminosities and average energies increase by a factor of two or more toward the formation of black hole. This increase will be used as a signal of black hole formation. The current generation of neutrino detectors will afford to detect the above-mentioned signal of the formation of black hole if it occurs in our own galaxy.

ACKNOWLEDGMENTS

I would like to thank K. Sumiyoshi, K. Kotake, H. Sawai, K. Nakazato, T. Yamasaki and N. Ohnishi for doing numerical computations. The numerical calculations were partially done on the supercomputers in RIKEN and KEK (KEK supercomputer Projects No.02-87 and No.03-92). This work is in part supported by Grants-in-Aid for the Scientific Research from the Ministry of Education, Science and Culture of Japan (No.S14102004, No.14079202, No.17540267), and Grant-in-Aid for the 21st century COE program "Holistic Research and Education Center for Physics of Self-organizing Systems"

REFERENCES

1. K. Kotake, K. Sato, and K. Takahashi, submitted to Rep. Prog. Phys. (2005), astro-ph/0509456.
2. J. M. Blondin, A. Mezzacappa, and C. DeMarino, Astrophys. J. (2003), **584**, 971.

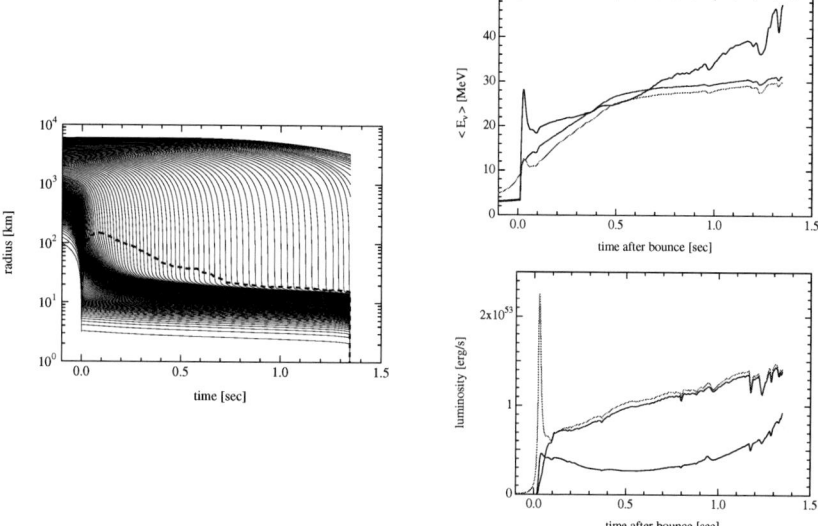

FIGURE 3. The trajectories of mass shells for the collapse of $40 M_\odot$ star (left panel). Shen's EOS is employed. The dashed line represents the location of shock wave. In the right panels, the average energies (upper panels) and luminosities (lowerpanels) of neutrinos are shown. The lines in the upper right panel represent ν_e, $\bar{\nu}_e$ and other neutrino species from bottom to top. The order of lines is the other way around in the lower right panel.

3. K. Nomoto et al. in "The Fate of the Most Massive Stars", ASP Conf. Ser. 332, 374-384 (2005), ed. R. M. Humphreys and K. Z. Stanek, astro-ph/0506597.
4. D. N. Spergel et al., Astrophys. J. Supp. (2003), **148**, 174.
5. A. Frebel et al., Nature (2005), **434**, 871.
6. F. Nakamura and M. Umemura, Astrophys. J. (2001), **548**, 19.
7. H. -Th. Janka and E. Müller, Astron.&Astrophys. (1996), **306**, 167.
8. A. Burrows and J. Goshy, Astrophys. J. Lett. (1993), **416**, L75.
9. M. Liebendöfer et al., Nucl. Phys. (2003), **A719**, 144.
10. M. Rampp and H. -Th. Janka, Astrophys. J. Lett. (2000), **539**, L33.
11. T. A. Thompson, A. Burrows, and P. A. Pinto, Astrophys. J. (2003), **592**, 434.
12. K. Sumiyoshi, S. Yamada, H. Suzuki, H. Shen, S. Chiba, and H. Toki, Astrophys. J. (2005), **629** 922.
13. N. Ohnishi, K. Kotake, and S. Yamada, Astrophys. J. (2005) in print, astro-ph/0509765.
14. H. -Th. Janka et al., Nucl. Phys. (2005), **A758**, 19.
15. N. Ohnishi, K. Kotake, and S. Yamada, in preparation (2005).
16. W. C. Haxton, Phys. Rev. Lett. (1988), **60**, 1999.
17. K. Sumiyoshi, S. Yamada, H. Suzuki, and S. Chiba, submitted to Phys. Rev. Lett. (2005).
18. K. Nakazato, K. Sumiyoshi, and S. Yamada, submitted to Astrophys. J. (2005), astro-ph/0509868.

1.9 NEUTRON-CAPTURE and r-PROCESS NUCLEOSYNTHESIS

Abundance of heavy elements in extremely metal-poor stars

P. François*, E. Depagne †, V. Hill**, M. Spite**, F. Spite**, B. Plez‡, T. C. Beers§, B. Barbuy¶, R. Cayrel‖, J. Andersen††, P. Bonifacio‡‡, P. Molaro§§, B. Nordström¶¶ and F. Primas***

*Paris Observatory,GEPI, France & European Southern Observatory(ESO)
†European Southern Observatory(ESO) & Paris Observatory,GEPI, France
**Paris Observatory ,GEPI, France
‡GRAAL, Université de Montpellier II, France
§Department of Physics & Astronomy, Michigan State University, USA
¶IAG, Universidade de São Paulo, Departamento de Astronomia, Brazil
‖Paris Observatory ,France
††Astronomical Observatory, Copenhagen, Denmark
‡‡Paris Observatory ,GEPI, France & Osservatorio Astronomico di Trieste, Italy
§§Osservatorio Astronomico di Trieste, Italy & Paris Observatory ,GEPI, France
¶¶Lund Observatory, Sweden
***European Southern Observatory (ESO)

Abstract. This paper reports on the abundance determination of neutron-capture elements in 32 extremely metal-poor stars. The study is based on the analysis of high quality spectra obtained with UVES+Kueyen. The results are compared with the most recent analyses of spectra mostly taken with other 10m class telescopes.

Keywords: Abundance, metal poor stars, Chemical evolution of the Galaxy
PACS: 95.75.Fg; 97.10.Tk; 97.20.Tr; 97.20.Wt;98.35.Bd

INTRODUCTION

The solar spectrum contains thousands of useful transitions that can be used to derive accurate abundances. A high spectral resolution is mandatory to resolve the lines and get accurate abundances. The study of solar-type metal poor stars is a powerful tool to obtain information on the detailed chemical composition of the interstellar matter during the early epochs of the galactic history. However, these metal poor stars suffer from the fact that they are rather faint if we want to get high S/N ratios high resolution spectra. The only solution to get reliable abundances is to use large telescopes with efficient spectrographs. Pioneering work has been done by Kodaira [8].

The situation at the end of the last century can be illustrated by the results obtained by Ryan et al. [10] using state of the art 4m class telescopes. The abundance ratios [α/Fe] presented a noticeable dispersion. For example, a peak to peak dispersion by almost a factor of 100 was found for the [Si/Fe] ratio. For the neutron-capture elements, the situation was even more pronounced with a spread of a factor of 1000 found for the ratio [Sr/Fe] in stars with [Fe/H]< −3.

With the help of 10m class telescopes and efficient spectrographs, the reality of this spread could be challenged.

THE ESO LARGE PROGRAMME "FIRST STARS"

38 nights of UVES+Kueyen have been devoted to the large programme "First Stars" between April 2000 and November 2001. The aim of the project was to take advantage of the high resolution spectrograph UVES at the VLT to analyse a sample of extremely old stars in a very homogeneous way. 30 dwarfs and 40 giants were selected in the Beers et al. HK survey [4]. In the sample of giant stars, 32 stars have be used to compute the abundance ratio of neutron-capture elements.

Cayrel et al. [5] have shown that the alpha element showed a very tight relation between [α/Fe] and [Fe/H]. Moreover, they showed that the behaviour of the ratios [light metal/Fe] vs. [Fe/H] vary from one element to the other, indicating that the abundances found in the extremely metal poor stars could not be explained by the mixing of the ejectae from a first generation of stars (Pop III) with interstellar matter with different levels of dilution. These results were very different from the rather large dispersion found by previous studies. These tight relations between light metals and the Fe content of the star are very important to put constraints on the models of the early chemical evolution of the Galaxy.

It becomes then very interesting to see if the spread found for the heavy elements will also be found for these new high quality spectra.

In this talk, I present the abundance determination of 16 neutron-capture elements (Sr, Y, Zr, Ba, La, Ce, Pr, Nd, Sm, Eu, Gd, Dy, Ho, Er, Tm and Yb) in 32 extremely metal-poor stars of the Large programme "First Stars". These observations greatly benefited from the high efficiency of the UVES spectrograph in the blue (8 % at 350 nm). We have excluded from our sample of stars the C-rich stars. This class of objects is particularly well studied by some japanese teams thanks to the SUBARU telescope equipped with HDS.

Analysis

The determination of the abundances has been made assuming LTE using the OS-MARCS models [9]. The surface temperature of the stars has been derived by colour using the Alonso calibration for giants [1]. IR photometry from 2MASS and DENIS has also been used. The accuracy for the temperature determination is of the order of 80 K. Detailed informations can be found in Cayrel et al. [5].

The light heavy elements Sr, Y and Zr

These three elements are mainly built by the s-process. The 3 diagrams of Fig. 1 show the abundance ratios of [Sr/Fe], [Y/Fe] and [Zr/Fe] as a function of [Fe/H] for our sample of stars (in black rectangles) together with results found in the literature. The 3

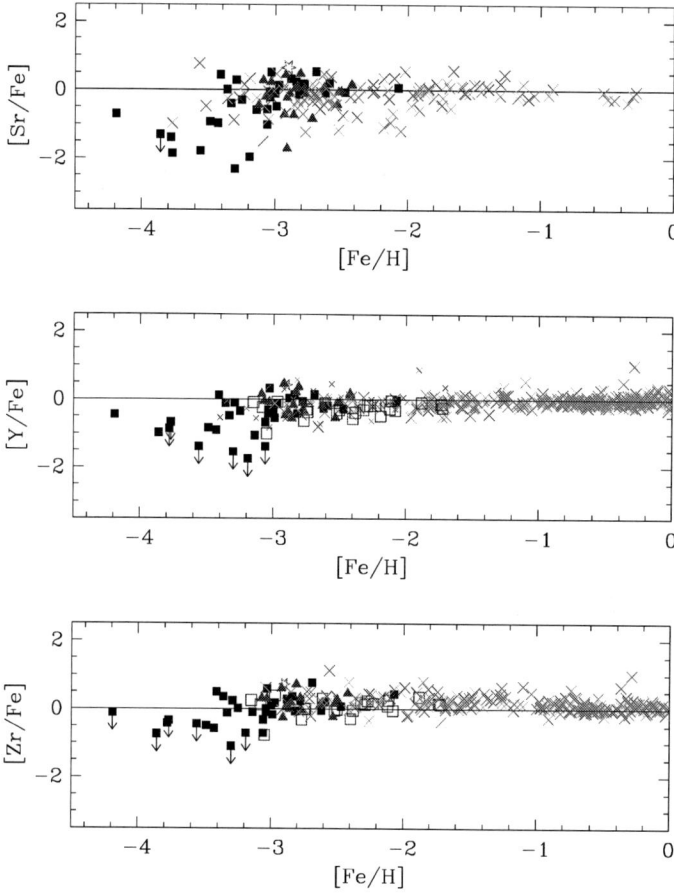

FIGURE 1. [Sr/Fe], [Y/Fe] and [Zr/Fe] as a function of [Fe/H]. Our data are represented by black rectangles. Note the similarities of the three diagrams

plots are rather similar, i.e. a solar ratio for stars with [Fe/H] down to $\simeq -3$. Below this metallicity, the abundance ratios are found highly subsolar.

We confirm the large spread of the ratio [Sr/Fe] in stars with [Fe/H] ≤ -3. It is important to keep in mind that almost no spread is found in the [α/Fe] ratio vs. [Fe/H] for the same stars ([5]).

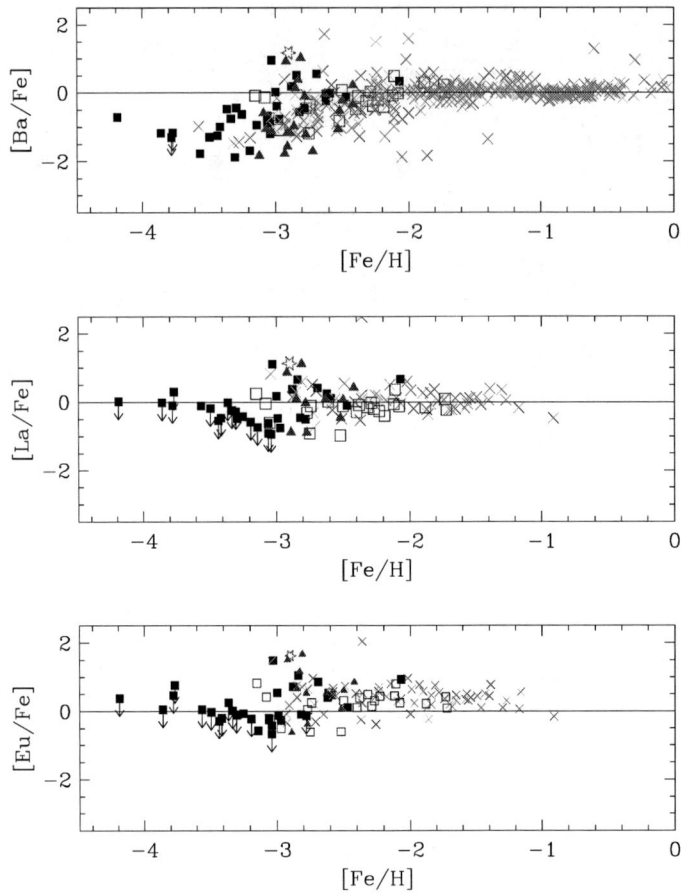

FIGURE 2. [Ba/Fe], [La/Fe] and [Eu/Fe] as a function of [Fe/H]. Our data are represented by black rectangles.

The second peak elements ($56 \leq Z \leq 72$)

In this range of atomic mass, we find the elements Ba, La and Eu which have been well studied. We have been able to measure also the abundance of some more elements belonging to the second peak (Ce, Pr, Nd, Sm, Gd, Dy, Ho, Er and Tm). We have no place here to present all the results and we will only present the results for Ba, La and Eu. All the details concerning the other elements can be found in François et al. [6].

Barium (mostly built by the s-process) is particularly interesting as its abundance has been measured in our sample of stars down to [Fe/H]=-4.2. In the metallicity range -

2.5 to -3.0, the spread found by previous analyses is confirmed. Moreover, the spread is found to increase with decreasing metallicity, a point already raised by Honda et al. [7]. Our study brings new measurements in stars with metallicities down to [Fe/H]=-4.2 for which few precise abundance determinations are available. Our results seem to indicate a flattening of the ratio to a [Ba/Fe] value around -1.

Abundances of La have been measured in stars down to [Fe/H]$\simeq -3.6$. An increasing spread when the metallicity decreases between [Fe/H]=-2 to -3. is also found for this element.

Europium is a most representative element of the r-process (93 % built by r-process as computed by Arlandini et al. [3]). Thanks to the high efficiency of UVES, we have been able to detect and measure the abundance of Europium in stars down to [Fe/H]=-3.25. Our study confirms the results from Honda et al. [7] on a much larger sample and to lower metallicities. It is found an oversolar ratio down to metallicity -3.0, and a large spread of the [Eu/Fe] ratio at metallicities close to -3.0.

The [Eu/Ba] and [Sr/Ba] ratios

Barium is considered as a reference element for the s-process synthesis as 92 % of the barium found in the Sun is formed by s-process [3]. As a classical test (Truran [11]) of the relative contribution of s-process/r-process, we plotted in the upper panel of Fig 3 the ratio [Eu/Ba] as a function of [Fe/H] for our sample of stars together with data from the literature. The [Eu/Ba] found in our sample of stars is similar to the solar system [Eu/Ba] r-process ratio. It is also interesting to note that the dispersion is much smaller than the [Eu/Fe] or the [Ba/Fe] ratio shown in Fig. 2, indicating a correlation of the production of Eu and Ba. In the lower panel of Fig. 3, we have plotted the ratio [Eu/Ba] as a function of [Ba/H] instead of [Fe/H] so that we can compare elements which are built by the same process, i.e. neutron capture. The stars of our sample are populating a large range of [Ba/H] values showing a noticeable correlation between [Eu/Ba] with [Ba/H].

In Fig. 4, we have plotted the [Sr/Ba] ratio vs. [Fe/H] and the [Sr/Ba] ratio vs. [Ba/H] for our sample of stars (filled rectangles) together with results from previous studies. The upper panel shows clearly the existence of an increasing spread as a function of decreasing metallicity in the range -2.0 to -3.4. For the most metal poor stars, we find a decrease of the spread. However, it should be investigated with more data whether this effect is real or whether it reveals a lower probability to find stars with high [Sr/Ba] at extremely low metallicities.

In the lower panel of Fig. 4, we see a clear and tight correlation between [Sr/Ba] and [Ba/H] for stars with [Ba/H] between -4.2 and -2. As the ratio of these two elements varies with [Ba/H] or [Fe/H], this could be the consequence of the onset of another r-process favouring the production of light neutron capture elements relatively to heavier nuclei as it has been already been proposed by Wanajo et al. [12].

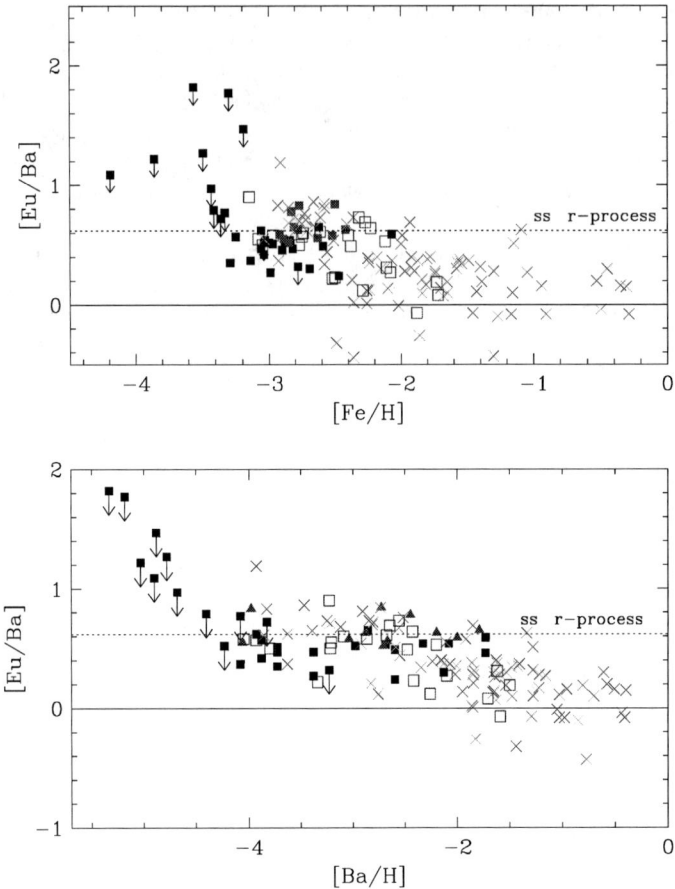

FIGURE 3. upper panel : [Eu/Ba] vs. [Fe/H]; lower panel :[Eu/Ba] vs. [Ba/H].Our data are represented by black rectangles. The dashed line represents the solar system r-process ratio.

SUMMARY

The abundance of 16 neutron-capture elements (only a subset has been presented in this paper) have been measured in 32 extremely metal poor giants. We confirm the large spread of the [n-capture elements /Fe] ratios found in the metal poor stars doubling the sample of stars in the metallicity range -2.7 to -3.2. We found an increasing spread of this ratio as the metallicity decreases. We also have evidences of a lack of stars with high [n-capture elements /Fe] with a [Fe/H] below -3.4. Inhomogeneous models of chemical evolution of the Galaxy which predict the same density of stars at [Fe/H]=-

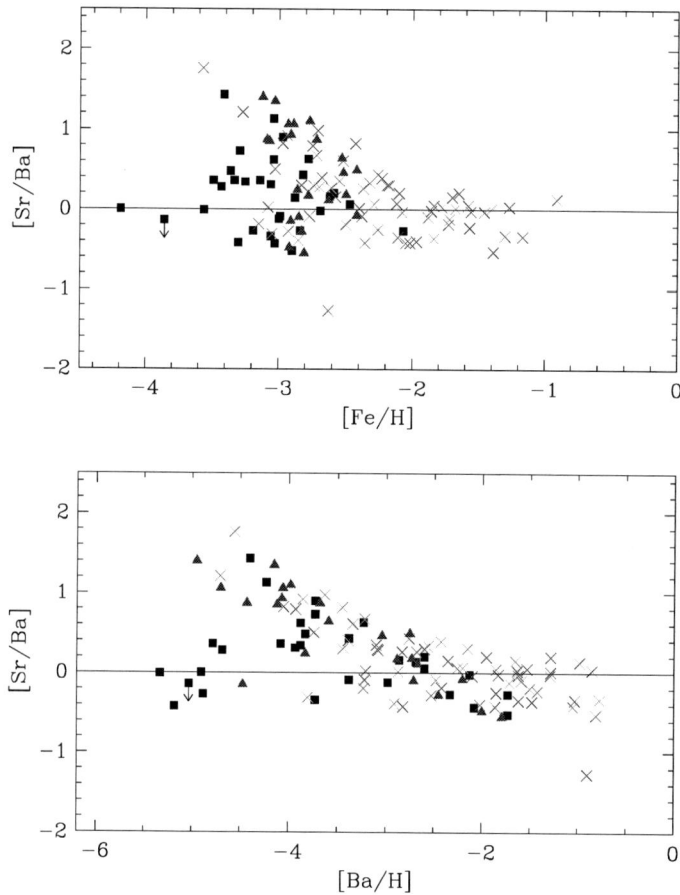

FIGURE 4. [Sr/Ba] vs. [Fe/H]; lower panel :[Sr/Ba] vs. [Ba/H].Our data are represented by black rectangles. Note the tight relation found in the lower panel.

3 and [Fe/H]=-4 ([2]) will require a fine tuning of the mixing process in order to be reconciled with the observations. We also noticed that plotting the abundance of the n-capture elements as a function of Ba instead of Fe significantly reduces the dispersion. Finally, our large set of data permitted us to confirm to a high level of confidence the signature of the r-process in extremely metal-poor stars.

ACKNOWLEDGMENTS

P. François would like to thank the organizers of the OMEG05 international symposium for their kind invitation and their financial support.

REFERENCES

1. Alonso A., Arribas S., Martinez-Roger C. 1999 A&AS 140,261
2. Argast, D.; Samland, M.; Thielemann, F.-K.; Gerhard, O. E. 2002 A&A 388, 842
3. Arlandini, C. Käppeler, F.; Wisshak, K.; Gallino, R.; Lugaro, M.; Busso, M.; Straniero, O. 1999 ApJ 525, 886
4. Beers, T.C.,Preston, G.W., & Schectman, S.A. 1985, AJ, 90, 2089
5. Cayrel, R.; Depagne, E.; Spite, M.; Hill, V.; Spite, F.; François, P.; Plez, B.; Beers, T.; Primas, F.; Andersen, J.; Barbuy, B.; Bonifacio, P.; Molaro, P.; Nordström, B. 2004, A&A 416, 1117
6. François P. et al. (submitted to A&A)
7. Honda, S.; Aoki, W.; Ando, H.; Izumiura, H.; Kajino, T.; Kambe, E.; Kawanomoto, S.; Noguchi, K.; Okita, K.; Sadakane, K.; Sato, B.; Takada-Hidai, M.; Takeda, Y.; Watanabe, E.; Beers, T. C.; Norris, J. E.; Ryan, S. G., 2004, ApJS 152, 113
8. Kodaira, K. 1967, PASJ 1967, 19,550
9. Plez, B., Brett, J.M., Nordlund, Å.1992 A&A 256,551
10. Ryan, S. G.; Norris, J. E.; Beers, T. C 1996, ApJ 471, 254
11. Truran, J. W. 1981 A&A 97, 391
12. Wanajo, S., Kajino, T., Mathews, G. J.; Otsuki, K. 2001 ApJ 554,578

Radioactive Beams and Exploding Stars at ORNL

Michael S. Smith

Physics Division, Oak Ridge National Laboratory, Oak Ridge, TN, 37831-6354, USA

Abstract. Beams of radioactive nuclei from the Holifield Radioactive Ion Beam Facility (HRIBF) at Oak Ridge National Laboratory (ORNL) are being used to make direct and indirect measurements of reactions important in novae, X-ray bursts, supernovae, and our Sun. Experimental results are used in nuclear data evaluations and element synthesis calculations to determine their astrophysical impact. Recent accomplishments include: the first neutron transfer reaction [(d, p)] measurements on nuclei in the r-process path in supernovae; precision measurements with radioactive ^{18}F beams for novae; and a direct ^{7}Be(p,γ)^{8}B measurement relevant for the solar neutrino flux determination.

Keywords: nuclear astrophysics, nucleosynthesis, r-process, rp-process, radioactive beam, stellar explosions, nova, supernova, X-ray burst, thermonuclear reaction rates, nuclear data, visualization, simulations
PACS: 26.30.+k,26.50.+x,26.65.+t,95.30.-k

1. RADIOACTIVE NUCLEI IN ASTROPHYSICS

In the extremely high temperatures and densities characteristic of stellar explosions, unstable nuclei can be formed and undergo subsequent reactions before they radioactively decay. This causes the sequences of reactions in exploding stars to be different than those in other, quiescent stars. To understand these explosions, it is therefore crucial to obtain information on the structure of and reactions involving a wide range of unstable nuclei (Figure 1) [1]. This information is quite difficult to obtain in the laboratory, however. The lifetimes of the species of interest are too short to be used in targets, and beams of these exotic nuclei often suffer from low intensities and purities. For this reason, explosion simulations rely primarily on theoretical estimates of the relevant nuclear physics – estimates which can, in some cases, be incorrect by orders of magnitude. Now, the availability of higher purity and quality beams of some of the nuclei involved in stellar explosions makes it possible to begin to give these models a firm empirical foundation.

2. ORNL HOLIFIELD RADIOACTIVE ION BEAM FACILITY

At the Holifield Radioactive Ion Beam Facility (HRIBF) [3, 4] at Oak Ridge National Laboratory (ORNL), beams of unstable proton- and neutron-rich nuclei are created with the Isotope Separator On-Line (ISOL) technique [1, 5]. Proton-induced fission reactions – via proton bombardment of a UC target – are used to make over 110 species of neutron-rich nuclei with intensities greater than 1000 particles per second (pps). These neutron-rich radioactive beams, reaccelerated up to 4.5 MeV/u, are unique in the world. Several proton-rich beam species are created by transfer reactions. The radioactive beams are

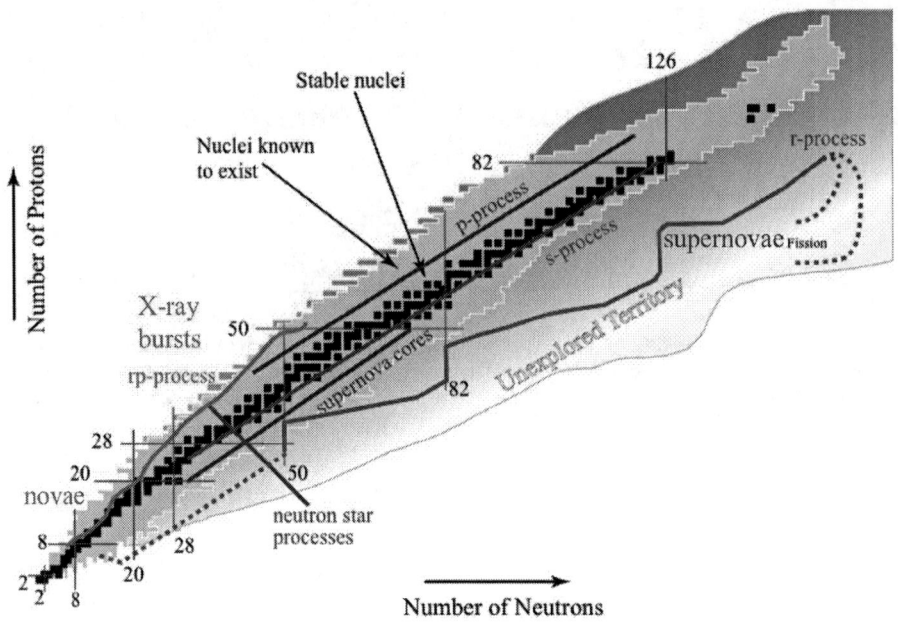

FIGURE 1. Unstable Nuclei in Astrophysical Processes [2]

reaccelerated in a tandem electrostatic accelerator with energies ranging roughly from 0.4 - 10 MeV/u. This enables both direct measurements of nuclear reactions driving explosions (with energies typically less than 1 MeV/u) to indirect techniques where scattering or transfer reactions populate levels in the nuclei of interest, using energies up to 10 MeV/u. The purity of some beams are greatly enhanced by molecular transport techniques [6] in the tandem, while other beams are fully stripped at the tandem exit to 100% purity. A wide range of experimental devices [4], including the Daresbury Recoil Separator [7], the Recoil Mass Separator [8], clover Ge arrays, the Silicon Detector Array (SIDAR) (Figure 2) [9], and others are used to detect the products of reactions with these radioactive beams.

3. EXPERIMENTAL STUDIES FOR THE r-PROCESS

The formation of roughly half of the elements heavier than iron is believed to occur in the r-process in the high-entropy bubble above a newly-born proto-neutron star in a core collapse supernova explosion [10]. This sequence of nuclear reactions involves rapid neutron captures on neutron-rich unstable nuclei (Figure 1). Simulations of the r-process require nuclear structure information (masses, lifetimes, level structure, decay properties) on thousands of nuclei out to the neutron drip line. Additionally, nuclear reaction information is needed, especially near the $N = 50$ and 82 closed neutron shells [11] where the abundances peak.

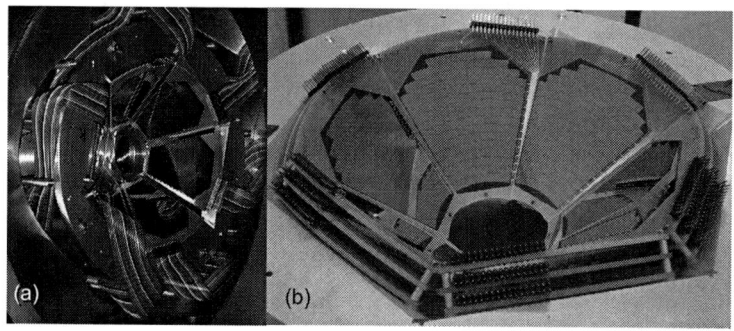

FIGURE 2. Silicon Detector Array (SIDAR) [9] setup for (a) ^{82}Ge(d,p)^{83}Ge and (b) ^{7}Be(d,t)^{6}Be.

The first studies of neutron-transfer reactions onto neutron-rich radioactive nuclei in the r-process path were recently performed at HRIBF. The (d,p) reaction in inverse kinematics was used to "simulate" the neutron capture on the N=50 closed shell nuclei ^{82}Ge and ^{84}Se [12]. For ^{82}Ge(d,p)^{83}Ge, a 430 μg/cm^2 (CD$_2$)$_n$ target was bombarded with a 330 MeV, 15% pure beam of 10^4 radioactive ^{82}Ge nuclei per second. Protons from the (d,p) reaction were detected in the annular array of silicon detectors, SIDAR [9], while the ^{83}Ge heavy products were detected in coincidence (within 80 ns) in a gas ionization counter. This kinematically complete measurement enables the impure ^{82}Ge beam (with major contaminant ^{82}Se) to be used. Our experimental energy resolution was approximately 300 keV, primarily limited by the thickness of the target needed to obtain a reasonable yield. Peak centroids (and reaction Q-values) were determined with relative energy resolution of 20 keV and an absolute precision of 70 keV when systematic uncertainties (e.g., the ^{82}Ge mass uncertainty) are included. The mass and neutron separation energy of ^{83}Ge were determined from the Q-value. Angular distributions of protons in coincidence with ^{83}Ge recoils were measured with SIDAR (Figure 3). This enabled us to determine the spins, parities, and spectroscopic strengths of the ^{83}Ge ground state and 280-keV first excited state by comparison to DWBA calculations, as shown in Figure 3 [12].

A very similar technique was used to measure the ^{84}Se(d,p)^{85}Se reaction. A 4.5 MeV/u, 8% pure beam of 8000 radioactive ^{84}Se nuclei per second bombarded a 200 μg/cm^2 (CD$_2$)$_n$ target, and 4 levels were populated in ^{85}Se. Spin-parities and spectroscopic factors were determined for the ground and first excited states [13]. Combining the results of the two experiments with previous results enables trends in properties of N=51 isotones (e.g., excitation energy, spectroscopic strength) to be examined.

In the future, the ^{130}Sn(d,p)^{131}Sn and ^{132}Sn(d,p)^{133}Sn reactions will be measured at HRIBF to learn about neutron captures on N = 82 neutron-rich radioactive nuclei in the r-process. To prepare for these measurements, the ^{124}Sn(d,p)^{125}Sn reaction was successfully measured with a stable ^{124}Sn beam. Spectroscopic factors for three levels in ^{125}Sn [14] were extracted, and these agreed with previous measurements using a ^{2}H beam and a ^{124}Sn target. Efforts are currently underway to overcome some of the limitations in using (d,p) reactions to determine (n,γ) reaction rates. For example, higher

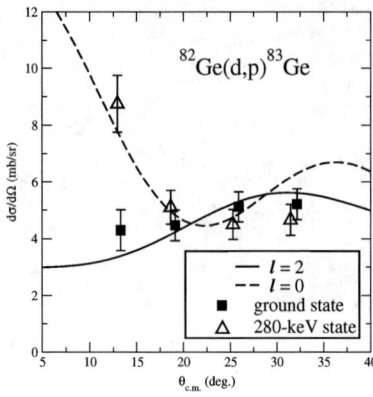

FIGURE 3. Angular distributions of protons from the ^{82}Ge(d,p)^{83}Ge reaction [12].

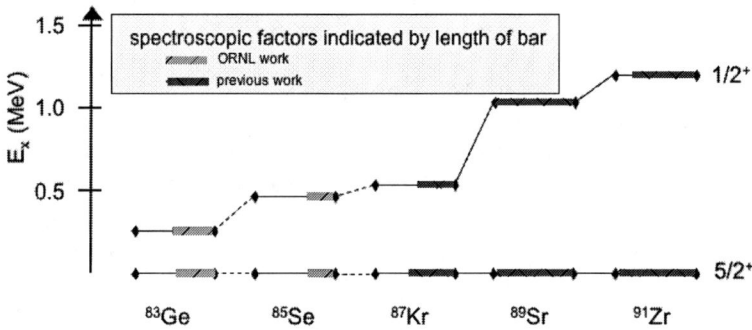

FIGURE 4. Excitation energies and spectroscopic strengths of N=51 isotones [12].

gain preamplifiers are being investigated; a low gain currently limits the lowest-energy detectable protons to those from states below the neutron threshold in the final nucleus. Also, a detector system with a very large solid angle is being designed to enable the use of thinner targets – thereby improving the energy resolution. The Oak Ridge - Rutgers University Barrel Array (ORRUBA [15]) will employ two annular rings of 12 dE - E, position sensitive, 4-strip silicon detector telescopes arranged in a barrel surrounding the $(CD_2)_n$ target. The array is currently under construction, and the first detector segments are being tested with stable beams. Finally, the (d,pγ) coincidence reaction has been utilized to significantly improve our energy resolution. Using a 10^7 pps stable ^{80}Se beam, a Q-value resolution of 200 keV (similar to previous work) was obtained with charged-particle detection, while a γ-ray resolution of 25 keV [16] was measured. This resolution is sufficient to resolve levels in neutron-rich unstable nuclei of interest for the r-process. The challenge for this coincidence technique will be to devise detector configurations to obtain reasonable event rates with currents of 10^5 pps characteristic of radioactive beams.

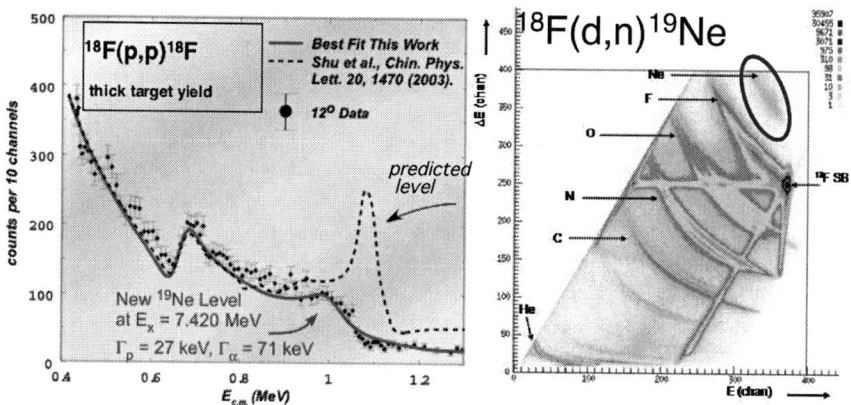

FIGURE 5. Data from measurements of (a) ^{18}F(p,p)^{18}F [22] and (b) ^{18}F(d,n)^{19}Ne [24].

4. EXPERIMENTAL STUDIES FOR THE rp-PROCESS

Proton capture reactions on proton-rich unstable nuclei generate the energy that drives nova explosions and X-ray bursts. To understand the relevant reaction sequence, the rp-process [1] (Figure 1), it is necessary to measure the structure properties and reactions of radioactive nuclei with an excess of protons [17]. The synthesis in novae of some long-lived radioactive nuclei such as ^{18}F, ^{22}Na, and ^{26}Al may provide a diagnostic tool to understand these explosions – via the comparison of predicted synthesized abundances with observations of their characteristic decay radiation [18]. At HRIBF, an extensive series of measurements to understand reactions that produce and destroy the ^{18}F nova observable have been made. These include direct measurements of the ^{18}F(p,α)^{15}O reaction at the 330-keV [19] and 665-keV [20] resonances, measurements of ^{18}F(p,p) with thin [21] and thick [22] targets, measurement of the ^{17}F(p,p) reaction [9], a measurement of the ^{18}F(d,p)^{19}F reaction [23], and a very recent measurement of the ^{18}F(d,n)^{19}Ne reaction [24]. Representative plots of the ^{18}F(p,p)^{19}F and ^{18}F(d,n)^{19}Ne reactions are shown in Figure 5. These studies have dramatically improved the knowledge of the level structure in ^{19}Ne and ^{19}F that are important for reactions that destroy ^{18}F in novae. Investigations of the impact of this work is discussed briefly below.

5. EXPERIMENTAL STUDIES FOR SOLAR BURNING

The ^{7}Be(p,γ)^{8}B reaction determines the flux of high-energy neutrinos from the core of our sun. Improved precision of the rate of this reaction is needed to better determine the properties of neutrinos derived from comparisons of the standard solar model flux predictions with neutrino observations in underground terrestrial detectors [25]. A direct measurement of the ^{7}Be(p,γ)^{8}B reaction has been made with a radioactive ^{7}Be beam [26] (Figure 6) which has a different set of systematic errors than the measurements using a ^{7}Be target or using the Coulomb dissociation of a ^{8}B beam. Using a technique first

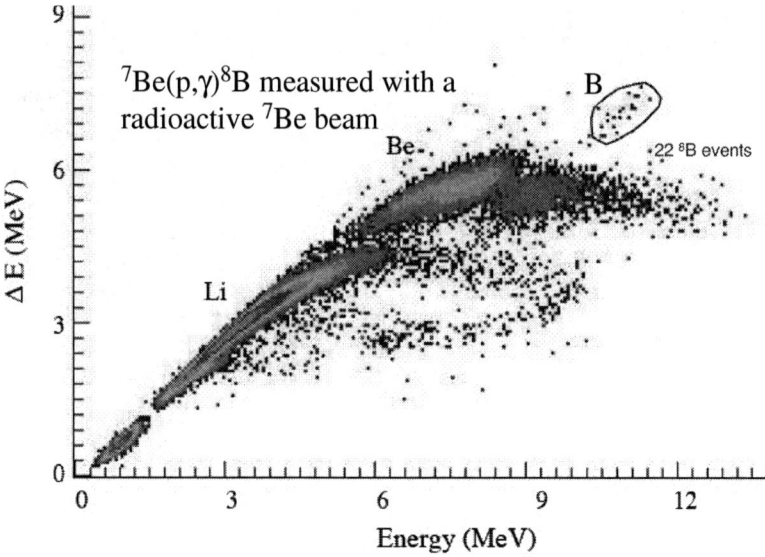

FIGURE 6. ^{8}B events in the focal plane of the Daresbury Recoil Separator [7] from a direct measurement of the ^{7}Be(p,γ)^{8}B reaction with a radioactive ^{7}Be beam [26].

developed in 1991 [27], the recoiling ^{8}B nuclei were detected in the focal plane of the Daresbury Recoil Separator [7]. For this first experiment, the systematic uncertainties were less than 9%, but the statistical uncertainties were 23%. This preliminary result of 26.8±6.5 eV b is consistent with recent capture measurements using a ^{7}Be target. Future measurements are planned for 2006. In related work, the ^{7}Be(p,p)^{7}Be elastic and inelastic scattering reactions have been studied with a thin (CH$_2$)$_n$ target to aid in the extrapolation of the ^{7}Be proton capture from laboratory energies down to stellar energies. Preliminary results of this work suggests a very large inelastic scattering yield, which may significantly modify the interpretation of previous thick target scattering measurements (e.g., [28]). Finally, the ^{7}Be(d,t)^{6}Be reaction was recently used to search for a subthreshold resonance that could be responsible for a rise in the ^{3}He(^{3}He,2p)^{4}He astrophysical S-factor at very low energies. The data is still being analyzed, but preliminary results show no evidence for such a resonance [29].

6. SYNERGISTIC EFFORTS IN NUCLEAR DATA EVALUATIONS AND ELEMENT SYNTHESIS CALCULATIONS

To examine the astrophysical impact of HRIBF measurements, evaluations are performed to determine the "best" value of a property (e.g., cross section, spectroscopic factor) by combining ORNL results with all relevant published results. This information is then converted into thermonuclear reaction rates, the input required by codes that calculate the synthesis of elements in astrophysical environments. For example, measurements at HRIBF to understand the ^{18}F(p,α)^{15}O reaction – listed above in Section 4 – have been used to refine the properties of roughly 20 levels in ^{19}Ne and ^{19}F [30, 31] that are important in the ^{18}F + p thermonuclear fusion reaction rates.

The results of our data evaluations are being used in post-processing element synthesis calculations to determine the astrophysical consequences of both completed and planned measurements. This work is now greatly streamlined by the **Computational Infrastructure for Nuclear Astrophysics** [32]. This is a unique suite of computer codes, freely available online at **nucastrodata.org**, that enables anyone to quickly – with a few mouse clicks – incorporate new nuclear physics results in astrophysical simulations, run the simulations, and visualize the results. Furthermore, the suite enables users to share large datasets (rate libraries, atrophysical simulations) with each other in an online community. The nucleosynthesis calculations utilized the reaction network code of Hix and Thielemann [33]. More features are continually being added to this suite, many on the basis of user recommendations. The suite was recently used, for example, to determine that the latest HRIBF ^{18}F(p,α)^{15}O reaction rate changed the production of ^{18}F in a particular model of a nova outburst [34] by a factor of 6 in the innermost (hottest) ejected zone of the outburst, and by a factor of 1.6 when all 28 zones in the ejected white dwarf envelope are considered.

7. SUMMARY AND FUTURE OUTLOOK: THE RARE ISOTOPE ACCELERATOR

Only a small fraction of the relevant nuclei and reactions important in stellar explosions have been, and will be, accessed at HRIBF and at other existing and planned radioactive beam facilities in the world. While this work is significantly improving our understanding of exploding stars, it is hoped that the Rare Isotope Accelerator (RIA) [35, 36] will be constructed in the next decade in the U.S. to provide access to the vast majority of nuclei in *all* astrophysical processes. By utilizing three complementary techniques, RIA promises to produce the world's most intense beams of all nuclear species. The Astrophysics at RIA (ARIA) Working Group [36, 37] was established to develop and promote future nuclear astrophysics research at RIA. At HRIBF, techniques that will be invaluable at RIA and other radioactive beam facilities around the world are now being developed and used for studies of stellar explosions. Examples include designing experiments for very low beam intensities and purities, developing (d,p) and (d,pγ) measurement techniques, and developing techniques for direct measurement of capture reactions with recoil separators. This work will continue in parallel with efforts to build up an astrophysics program at RIA.

ACKNOWLEDGMENTS

This paper describes the work of the RIBENS Collaboration (Radioactive Ion Beams for Explosive Nucleosynthesis Studies), specifically work by: J.S. Thomas [^{82}Ge(d,p)^{83}Ge and ^{84}Se(d,p)^{85}Se]; K.L. Jones and R.L. Kozub [^{124}Sn(d,p)^{125}Sn]; M. Johnson [^{80}Se(d,p)^{81}Se]; S.D. Pain and J.C. Blackmon (ORRUBA); D.W. Bardayan

[^{18}F(p,α)^{15}O]; C.R. Brune [^{18}F(d,n)^{19}Ne]; R. Fitzgerald [^{7}Be(p,γ)^{8}B]; R.J. Livesay [^{7}Be(p,p')^{7}Be]; K. Chae [^{7}Be(d,t)^{6}Be]; E.J. Lingerfelt and J.P. Scott (Computational Infrastructure for Nuclear Astrophysics); W.R. Hix (element synthesis calculations); and C.D. Nesaraja and N. Shu [^{18}F(p,α)^{15}O evaluations]. ORNL is managed by UT-Battelle, LLC, for the U.S. Department of Energy under contract DE-AC05-00OR22725.

REFERENCES

1. M.S. Smith and K.E. Rehm, *Ann. Rev. Nucl. Part. Sci.*, **51**, 91–130 (2001).
2. F.X. Timmes, H. Schatz, M. Smith, M. Wiescher, U. Greife, in *"Rare Isotope Accelerator Brochure"* (2005), http://usnuclearscience.org/ria/ria.html.
3. B.A. Tatum, J.R. Beene, in Proc. Particle Accelerator Conference (PAC'05), Knoxville, TN, 2005, http://accelconf.web.cern.ch/accelconf/ 2005
4. http://www.phy.ornl.gov/hribf/
5. H. Ravn, *Phys. Rep.* **54**, 201 (1979).
6. D. Stracener *et al.*, Proceedings XXVI Symposium on Nuclear Physics, Taxco, Mexico, Jan. 6-9, 2003, *Rev. Mex. Fis.* (Suppl.) **49** 92-96 (2003).
7. R. Fitzgerald *et al.*, *Nucl. Phys. A* **748**, 351 (2005).
8. C.J. Gross *et al.*, *Nucl. Instrum. Methods Phys. Res. A* **450**, 12-29 (2000).
9. D.W. Bardayan *et al.*, *Phys. Rev. C* **62**, 0155804 (2000).
10. S.E. Woosley *et al.*, *Astrophys. J.* **433**, 229 (1994).
11. R. Surman, J. Engel, *Phys. Rev. C*, **64**, 035801 (2003).
12. J.S. Thomas *et al.*, *Phys. Rev. C* **71**, 021302(R) (2005).
13. J.S. Thomas *et al.*, in preparation (2006); J.S. Thomas, Ph.D. Thesis, Rutgers Univ. (2005).
14. K.L. Jones *et al.*, *Phys. Rev. C* **710**, 067602 (2004).
15. S.D. Pain, private communication (2005); http://www.orau.org/stewardship/
16. M. Johnson *et al.*, in preparation (2006); M. Johnson *et al.*, BAPS **50**, no. 6, 162 (2005).
17. S. Starrfield, W.M. Sparks, J.W. Truran, M.C. Wiescher, *Astrophys. J. Suppl.* **127**, 485 (2000).
18. M. Hernanz, J. Jose, *New Astronomy Rev.* **48**, 35 (2003).
19. D.W. Bardayan *et al.*, *Phys. Rev. Lett.* **89**, 262501 (2002).
20. D.W. Bardayan *et al.*, *Phys. Rev. C* **63**, 065802 (2001).
21. D.W. Bardayan *et al.*, *Phys. Rev. C* **62**, 042802(R) (2000).
22. D.W. Bardayan *et al.*, *Phys. Rev. C* **70**, 015804 (2004).
23. R.L. Kozub *et al.*, *Phys. Rev. C* **71**, 032801(R) (2005).
24. C.R. Brune *et al.*, in preparation (2006).
25. S.N. Ahmed *et al.*, *Phys. Rev. Lett.* **92**, 181301 (2004).
26. R. Fitzgerald, Ph.D. Thesis, Univ. North Carolina (2005); R. Fitzgerald *et al.*, in preparation (2006).
27. M.S. Smith, C. Rolfs, C.A. Barnes, *Nucl. Inst. Meth.* **306**, 233 (1991).
28. G.V. Rogachev *et al.*, *Phys. Rev. C* **64**, 061601 (2001).
29. K. Chae, private communication (2005).
30. N. Shu, D.W. Bardayan, J.C. Blackmon, Y.S. Chen, R.L. Kozub, P.D. Parker, M.S. Smith, *Chin. Phys. Lett.* **20**, 1470 (2003).
31. C.D. Nesaraja *et al.*, *Phys. Rev. C.* (in preparation) (2006).
32. M.S. Smith *et al.*, *"Computational Infrastructure for Nuclear Astrophysics"*, this proceedings.
33. W.R. Hix, F.-K. Thielemann, *J. Comp. Appl. Math*, **109**, 321 (1999).
34. M. Politano, S. Starrfield, J.W. Truran, W.M. Sparks, *Astrophys. J.* **448**, 807 (1995).
35. http://www.orau.org/ria
36. M.S. Smith, H. Schatz, F.X. Timmes, M. Wiescher, U. Greife, *"Astrophysics at RIA (ARIA) Working Group"*, this proceedings.
37. http://ariaweb.org

Subaru/HDS studies of r-process elements in metal-poor stars from near UV-spectra

Satoshi Honda[*], Wako Aoki[*], Yuhri Ishimaru[†], Shinya Wanajo[**] and Sean G Ryan[‡,§]

[*]*National Astronomical Observatory, Mitaka, Tokyo, 181-8588, Japan*
[†]*Academic Support Center, Kogakuin University*
[**]*Research Center for the Early Universe, Graduate School of Scie nce, University of Tokyo*
[‡]*Department of Physics and Astronomy, The Open University*
[§]*Centre for Astrophysics Research, STRI, University of Hertfordshire*

Abstract. We present detailed abundance measurements of neutron-capture elements for very metal-poor stars. We obtained very high quality, near-UV spectra of four bright metal-poor stars using Subaru/HDS, and determined the abundances of many neutron-capture elements. We investigated the detailed abundance patterns of moderately neutron-capture-enhanced-stars and neutron-capture-poor-stars, and clarified the difference in the abundance patterns. We focus on the well-known metal-poor giant HD122563, in which 19 neutron-capture elements including Nb, Mo, Ru, Pd, Ag, Pr, and Sm have been detected by the present work. Our new results clearly demonstrate that the elemental abundances more steeply decrease with increasing atomic number than those of the r-process component in solar-system material. Especially, the tendency that the abundance continuously decreases in the region of $38 \leq Z \leq 47$ does not agree with any neutron-capture processes known (i.e., main s-process, weak s-process, or main r-process). Since the abundance pattern of such very metal-poor stars, in contrast to that of solar system material, is expected to be formed by a small number of nucleosynthesis events, another process yielding elements with $38 \leq Z \leq 47$ would be required. This result will give a new, strong constraint on the modeling of the process that has provided light neutron-capture elements in the very early Galaxy.

Keywords: r-process, elemental abundances, neutron-capture elements
PACS: 97.10.Tk, 97.20.Tr

INTRODUCTION

The rapid neutron capture process (r-process) is known to be responsible for about half of the abundances of the elements heavier than the iron-group in solar system material. However, the astrophysical site of the r-process is still unclear. An effective approach to understand this process from astronomical observations is to measure the abundances of heavy metals in the recently discovered very metal-poor stars enriched with r-process elements.

Sneden et al. (1996) have studied the abundances of the extremely r-process rich metal-poor star CS 22892-052. They found that the abundance pattern of heavy ($56 \leq Z \leq 72$) neutron-capture elements in this object is in agreement with that of the solar system r-process component. This phenomenon is seen in other extremely r-process rich stars (Hill et al. 2002, Christlieb et al. 2004). Therefore, this phenomenon is sometimes called "universality" of the r-process. Our recent study (Honda et al. 2004) using the Subaru Telescope has determined the detailed abundance patterns of neutron-capture

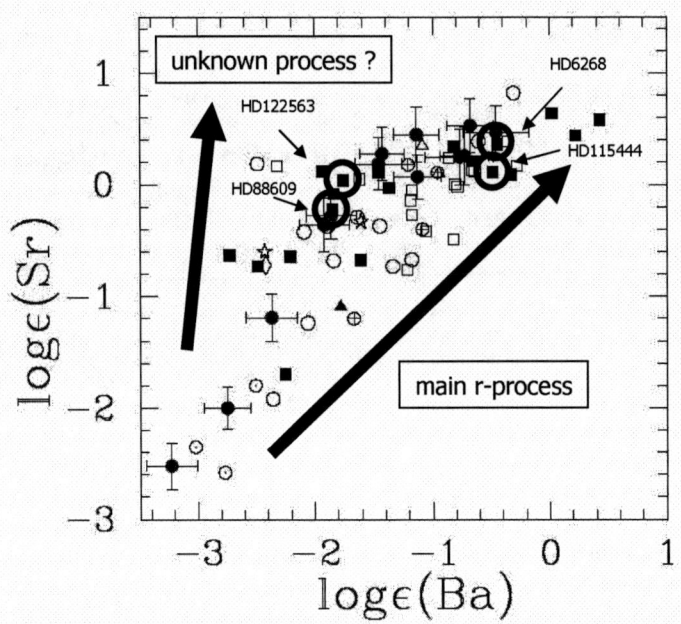

FIGURE 1. Sr abundances as a function of the Ba abundance for very metal-poor stars with [Fe/H] ≤ −2.5 from McWilliam 1998; Burris et al 2000; Carretta et al. 2002; Depagne et al. 2002; Johnson & Bolte 2002; Ishimaru et al. 2004; Cohen et al. 2004; Honda et al. 2004, and Aoki et al. 2005. Stars having significant excesses of carbon and s-process elements are excluded. The targets discussed in the present work are labelled.

elements of seven r-process rich stars, and confirmed the universality in heavy r-process elements. The discovery of the universality of the r-process has had a large impact on the studies of the nature of r-process, and its astrophysical site.

However, from these observations, the abundance patterns of light neutron-capture elements exhibit clear deviations from that of the solar system r-process component. Honda et al. (2004) and Aoki et al. (2005) investigated correlation of Sr ($Z = 38$) and Ba ($Z = 56$) for many sample, in order to investigate this phenomenon (see Figure 1). As a result, a correlation between the Sr abundance and the Ba abundance was found. There is no object which is Ba-rich and Sr-poor, and also the scatter of Sr abundances decreases with increasing Ba abundance.

This trend can be explained by assuming the existence of two processes (Sneden et al. 2000, Truran et al. 2002). One synthesize both Sr and Ba in similar proportions, while the other synthesize Sr with little Ba. The former is known as "main r-process", while the later is an unknown process.

In order to understand these two processes, measurements of the detailed abundances of neutron-capture elements for objects which were affected by these processes. We selected two objects that have high Sr and Ba abundances, and two other stars having Sr abundance with low Ba one (see Figure). We here focus on the stars with low Ba

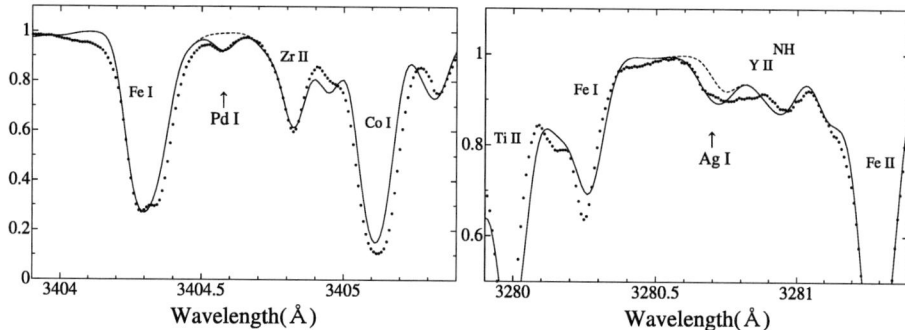

FIGURE 2. The sample of the spectra for HD 122563. Dot indicates the observations, and solid line indicates the synthetic spectrum. The dashed line indicate the synthetic spectra with no contribution from the line of interest.

abundances, in particular the well-known bright metal-poor star HD 122563.

OBSERVATIONS AND ANALYSIS

Observations were carried out with the Subaru Telescope High Dispersion Spectrograph (HDS ; Noguchi et al. 2002). Wavelength coverage is from 3000 Å to 4600 Å with resolving power of $R = 90,000$. The bright very metal-poor stars HD 88609 and HD 122563 are selected as the target. The metallicity of HD 88609 and HD 122563 is −3.07 and −2.77, respectively. It is known that these stars have high Sr abundance and low Ba abundance by previous studies. The signal to noise ratios of the obtained spectra are 120 at 3500 Å for HD 88609 and 480 at 3500 Å for HD 122563.

Observations in near-UV region, where many lines of neutron-capture elements exist, is effective for measurements of light neutron-capture elements. However, the observation with a ground-based telescope is very difficult because of atmospheric absorptions. To obtain high quality data, we need long exposure time and high resolving power. Examples of the obtained spectra are shown in Fig 2. We have detected several light neutron-capture elements which were not so much detected by previous studies.

ATLAS9 of Kurucz model atmosphere is used for the abundance analysis, and we adopted the model atmosphere parameters derived by Honda et al. (2004). Recent progress in measurements of the transition probabilities, including hyperfine splitting effect (e.g., Lawler et al. 2001, Den Hartog et al. 2003), enables us to determine accurate abundances. The effects of hyperfine and isotopic splitting are taken into account in the analysis of Ba, La, Eu, and Yb.

RESULTS AND DISCUSSION

Figure 3 shows the abundance pattern of HD 122563, comparing with that of the solar system r-process component. This figure clearly shows that the abundances of light neutron-capture elements ($38 \leq Z \leq 47$) are much higher than those of heavy ones in

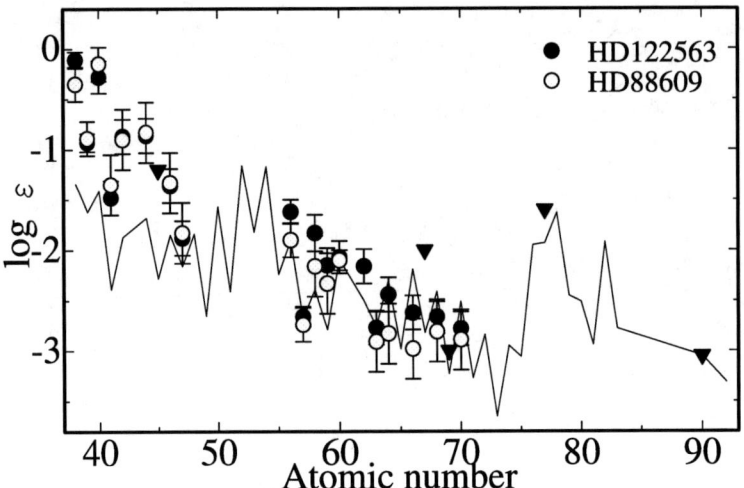

FIGURE 3. Abundance patterns of HD 88609 and HD 122563. The line is the solar system r-process abundance pattern (Burris et al. 2000), scaled to fit the measured Eu abundance of HD 122563. The triangles represent upper limits for several elements.

HD 122563, compared to the solar-system r-process pattern. In addition to this object, HD 88609 also shows the same trend. This behavior is very different from that found in r-process enhanced stars (e.g. CS 22892-052, CS 31082-001; Sneden et al. 2003, Hill et al. 2002). These two stars have abundance patterns of heavy neutron-capture elements that agree with the solar system r-process pattern, while their abundances of light neutron-capture elements are lower than the solar ones if the abundance pattern is normalized for heavy neutron-capture elements. Our new measurements for HD 88609 and HD 122563 clearly demonstrate that the elemental abundances more steeply decreases with increasing atomic number. The abundance patterns of heavy neutron-capture elements ($56 \leq Z \leq 70$) also show a small deviation from the solar system r-process curve. Our study revealed, for the first time, such abundance pattern by obtaining very high S/N spectra at the near UV range.

This now raises a question what is the source of neutron-capture elements in these objects. The main r-process , at least, is not the source of the light neutron-capture elements in these two stars. The main s-process is also excluded from the source of neutron-capture elements in those objects, because if the effect of main s-process is large, the much higher values of Ba, La, and Ce abundances are expected. We note that these objects are not C-rich stars, and very metal-poor ([Fe/H] ≤ -2.5) stars, in which the effect of main s-process is small (if any). There is also little possibility of weak s-process, because the weak s-process makes a rapidly decreasing pattern from Zr to heavier elements. Thus, the nucleosynthesis processes which are known well by previous studies can not explain the abundance patterns found in HD 122563 and HD 88609, and we need to assume a process that provides light neutron-capture elements at the early Universe. The existence of such process was shown by the recent observations of metal-poor stars (c.f., Aoki et al. 2005). This process was called the "Light Element

Primary Process" (LEPP) by Travaglio et al. (2004) who evaluated its contribution to the solar abundance. Wanajo et al. (2001) have shown that such light r-process nuclei (up to $A \sim 130$) are produced in neutrino winds as a result of "weak" (or failed) r-processing (see also Wanajo & Ishimaru 2005). However, it is difficult to explain the whole abundance pattern from light to heavy element observed for our objects, and further studies are clearly required. Further observations of the very metal-poor stars with no excess of heavy neutron-capture elements are also desired.

CONCLUSIONS

We have obtained the abundance patterns for wide atomic number range for metal-poor stars HD 88609 and HD 122563 by near-UV spectroscopy. These objects show a gradually decreasing trend of abundances as a function of atomic number, compared to the abundances pattern measured for r-process enhanced stars. This trend does not agree with any neutron-capture process known to date. This result gives a new constraint on the modeling of the process that provided light neutron-capture elements at the vary early Universe.

ACKNOWLEDGMENTS

This study is based on data collected at the Subaru Telescope, which is operated by the National Astronomical Observatory of Japan.

This work was supported in part by a Grant-in-Aid for the Japan-France Integrated Action Program (SAKURA), awarded by the Japan Society for the Promotion of Science, and Scientific Research (17740108) from the Ministry of Education, Culture, Sports, Science, and Technology of Japan.

REFERENCES

1. W. Aoki, et al. *ApJ*, **632**, 611–637 (2005)
2. D. L. Burris et al. *ApJ*, **544**, 302–319 (2000)
3. E. Carretta et al. *AJ*, **124**, 481–506 (2002)
4. N. Christlieb, et al. *A&A*, **428**, 1027–1037 (2004)
5. J. G. Cohen et al. *ApJ*, **612**, 1107–1135 2004
6. E. A. Den Hartog, et al. *ApJS*, **148**, 543–566 (2003)
7. E. Depagne *A&A*, **390**, 187–198 (2002)
8. V. Hill, et al. *A&A*, **387**, 560–579 (2002).
9. S. honda, et al. *ApJ*, **607**, 474–498 (2004)
10. S. honda, et al. *ApJ*, Submitted
11. Y. Ishimaru et al. *ApJL*, **600**, 47–50 2004
12. J. A. Johnson, & M. Bolte *ApJ*, **579**, 616–625 2002
13. R. L. Kurucz, Kurucz CD-ROM, No.13 (Harvard-Smithsonian Center for Astrophysics) (1993)
14. J. E. Lawler, et al. *ApJ*, **556**, 452–460 (2001)
15. A. McWilliam *AJ*, **115**, 1640–1647 (1998)
16. K. Noguchi et al. *PASJ*, **54**, 855–864 (2002)
17. C. Sneden, et al. *ApJ*, **467**, 819–840 (1996)
18. C. Sneden, et al. *ApJL*, **533**, 139–142 (2000)

19. C. Sneden, et al. *ApJ*, **591**, 936–953 (2003)
20. C. Travaglio, et al. *ApJ*, **601**, 864–884 (2004)
21. J. W. Truran, et al. *PASP*, **114**, 1293–1308 (2002)
22. S. Wanajo et al. *ApJ*, **554**, 578–586 (2001)
23. S. Wanajo & Y. Ishimaru Nucl. Phy. A, in press (2006)

Origin of the main r-process elements

K. Otsuki[*], J. Truran[*], M. Wiescher[†], J. Gorres[†], G. Mathews[†], D. Frekers[**], A. Mengoni[‡], A. Bartlett[§] and J. Tostevin[§]

[*]*Department of Astronomy and Astrophysics, University of Chicago*
[†]*University of Notre Dame*
[**]*University of Münster*
[‡]*CERN*
[§]*University of Surrey*

Abstract. The r-process is supposed to be a primary process which assembles heavy nuclei from a photo-dissociated nucleon gas. Hence, the reaction flow through light elements can be important as a constraint on the conditions for the r-process. We have studied the impact of di-neutron capture and the neutron-capture of light ($Z<10$) elements on r-process nucleosynthesis in three different environments: neutrino-driven winds in Type II supernovae; the prompt explosion of low mass supernovae; and neutron star mergers. Although the effect of di-neutron capture is not significant for the neutrino-driven wind model or low-mass supernovae, it becomes significant in the neutron-star merger model. The neutron-capture of light elements, which has been studied extensively for neutrino-driven wind models, also impacts the other two models. We show that it may be possible to identify the astrophysical site for the main r-process if the nuclear physics uncertainties in current r-process calculations could be reduced.

Keywords: r-process nucleosynthesis, reaction rates, supernovae, neutron star mergers
PACS: 26.30.+k

INTRODUCTION

In spite of decades of study, the astrophysical site for the main r-process component is still unknown. There are three dominant candidates: neutrino-driven winds in Type II supernovae; the prompt explosion of low-mass supernovae; and neutron star mergers. The conditions for the r-process nucleosynthesis, such as the temperature and density profile, are significantly different in each of these candidate environments. although there have been many theoretical nucleosynthesis studies aimed at testing the viability of those models none of them has reached a definitive conclusion. This uncertainty is due in part to the difficulty of modeling of such explosive events.

On the other hand, many new determinations of r-process elemental abundances have been reported, in part due to the availability of a new generation observational facilities. Specifically, studies of r-process abundances in metal-poor halo stars have been made because they provide clues to the origin of r-process elements and the evolution of the early Galaxy. One of the most interesting results is the robustness of r-process abundance distributions(e.g., [1]). The r-process abundance distribution for elements heavier than Ba in a dozen studied metal-poor halo stars show the same pattern. Furthermore, this abundance pattern is in good agreement with the solar r-process abundance pattern. This fact suggests that the r-process for elements heavier than Ba are always generated in same pattern from the time of Galaxy formation to the present epoch. Hence, the r-process is metallicity independent. This means that the r-process is a primary nucleosynthesis process (i.e. seed elements for the r-process are generated when the r-process occurs). In a primary r-process, light element reactions for seed production are as important as the r-process itself, especially for testing the reliability of theoretical models. This is because light-element reactions often consume neutrons and affect the neutron-to-seed ratio at the beginning of the r-process.

The importance of the neutron-capture reactions of light elements in neutrino-driven wind models was pointed out by Terasawa et al. [2]. Although the neutron capture reaction rates of such light elements have large uncertainties, those reactions can have a large impact on the final calculated abundances. We have studied the importance of those reactions for the other two r-process models using same reaction rates.

The reaction flow of ^{4}He(2n, γ)^{6}He(α,n)^{9}Be has been neglected in previous r-process calculations because the ^{4}He(2n,γ)^{6}He reaction rate is very small and this reaction flow has been thought to be insignificant. The study in Bartlett et al. [3], however, shows that di-neutron capture can increase the ^{4}He(2n,γ)^{6}He reaction rate by orders of magnitude. We performed r-process nucleosynthesis calculations to see if this reaction flow has any impact on the final r-process abundances.

MODELS

Our nucleosynthesis code is based on the dynamical network described in Meyer [4], which has been extended by Terasawa and Orito [2][6]. This code calculates dynamically the r-process and its seed production simultaneously. We terminated our calculation when the neutron abundance Y_n became less than 10^{-15}, by which point the abundance distribution is no longer affected by neutron capture. We adopted a primitive fission recycling model, which assumes that all elements with A=260 immediately break into two symmetric nuclei. For each of the models considered, we have calculated the nucleosynthesis yields with three networks: 1) without the neutron capture of light elements nor the di-neutron capture of ^{4}He; 2) with the neutron capture of light elements but without di-neutron capture; and 3) with both the neutron capture of light elements and di-neutron capture.

We have assumed simple schematic parameterizations for each of the environments as described below. We chose parameter set for each environment such that the ratio of the actinide-to-third peak elements in the final abundances was consistent with observed abundances in metal-poor stars.

1. Neutrino-driven wind in core-collapse supernovae

The proto-neutron stars born in Type II supernovae, release their energy via neutrinos during their Kelvin-Helmholtz cooling phase. Those neutrinos heat up material on the surface and eject them into a high entropy bubble above the neutron star. This is the so-called neutrino-driven wind (e.g., [7][8]). Since the energy of anti-electron neutrinos is higher than electron neutrinos, the wind material becomes slightly neutron-rich. If the entropy is high enough, it becomes a suitable environment for the main r-process. Although this is a popular model, there are several problems in this model. For example, it is still controversial at to whether such high entropy is actually realized in the wind (e.g., [8] and reference there in). In addition, Meyer et al. [14] pointed out that neutrino interactions with nuclei during the r-process increases the electron fraction. Higher entropy and/or shorter timescale may thus be necessary to realize suitable conditions for the r-process in this model when neutrino effects are considered.

We assume an exponential adiabatic expansion model for the neutrino-driven wind. This model is often invoked in parametric studies of high-entropy environments such as neutrino-driven winds. In this model, the density and temperature profiles are given as;

$$T_9 = 9.0\exp(-t/t_{\exp}) + 0.6 \tag{1}$$

$$\rho = 3.3 \times 10^5 T_9^3/S, \tag{2}$$

where T_9 is the temperature in units of 10^9, $t_{\exp}$ is the expansion timescale, ρ is the baryon matter density in g/cm^{-3}, and S is the entropy per baryon. We chose a parameter set of S=300, $t_{\exp}$=0.05 sec, and Y_e=0.45 for this study as it provides a reasonable reproduction of the solar r-process abundance curve with a single profile.

2. Prompt explosions of low-mass supernovae

Low mass core-collapse SNe ($8 - 12M_\odot$) are another candidate. If such low-mass supernovae explode, they would occur via a prompt explosion. During such prompt explosions, relatively low entropy ($\sim$15k), and a low electron fraction ($Y_e \sim$0.2) can be realized (e.g.,[9][10]). These are also ideal conditions for the r-process. The main objection to this model is the explosion mechanism itself. It is still unclear as to whether such light supernovae actually explode. This is exacerbated by the fact that, there is no convincing observational evidence of the remnants of such explosions.

For these calculations, we have assumed that the material expands exponentially, $\rho \propto \exp(-t/t_{exp})$. Here, t_{exp} is the expansion timescale. Corresponding temperatures are obtained from the equation of state of Timmes et al. [11]. As used in previous studies, we adopted the entropy to be S=15, and Y_e=0.2. We calculated the case with $t_{\exp}$=0.05 sec for $T_9 > 1.0$ and $t_{\exp}$=0.3 sec when the temperature drops below $T_9 < 1.0$. We also did calculations with different timescales. These, however, did not change the conclusions discussed here.

3. Neutron star mergers Another neutron-rich environment could be realized in neutron star mergers (e.g., [12]). In this model the entropy is very low, but the electron fraction is also very low. It is the most neutron-rich environment among the three models considered here. Although theoretical calculations show reasonable yields, this model encounters some difficulties in explaining the chemical enrichment history of the Galaxy [13].

For the other two models, the heating from nucleosynthesis can be shown to be negligible. However, in the neutron star merger model, heating from beta-decays and fission can significantly affect the temperature profiles. Since the current version of our network code does not take heating from nucleosynthesis into account, we have assumed a constant temperature T_9=1.0. In other r-process calculations in the neutron star merger model, the temperature varies between 0.2 to 1.2 $\times 10^9$ K during the r-process [12]. Hence, a constant temperature is a reasonable assumption for our purposes of testing the impact of new nuclear reaction flows. Initial conditions for this model are from Tables 1 and 2 in [5]. Here, we assume that the material expands on a free fall timescale ($\alpha=1.0$) and the electron fraction is initially set to $Y_e=0.15$.

DISCUSSION

Neutron capture on light elements is effective for all of the models considered here. Since neutrons are consumed by the formation of lightr elements, the final actinide abundances are reduced in the neutrino-driven wind and low-mass supernova models. Especially in the case of low mass supernovae, these reactions make the ratio of actinide elements to the third peak elements much smaller. Even if we increase the entropy to 50, we cannot reproduce the higher actinide abundances which have been observed in metal-poor stars. If we believe all of the reaction rates which were involved in the present calculations, low mass supernovae model should be discarded. Although their reaction rates have large uncertainties, neutron-capture of light elements seems to have a large impact on all three models. It is indeed a high priority to obtain more reliable neutron-capture reaction rates for the relevant light elements.

On the other hand, the effect of the new reaction flow ^{4}He(n^2,γ)^{6}He(α, n)^{9}Be is not significant in the neutrino driven wind models or low-mass supernovae. For the neutron star merger model, the effects of both reactions appear at $A < 130$, but are not significant for the production of heavy element abundances. This is because the environment is so neutron rich that fission recycling obscures the effects of all processes that occurred at an earlier stage of nucleosynthesis.

Fission recycling plays a role in all models, but it is most effective for the neutron-star merger models. Note, that there is no significant difference for elements heavier than the second r-process peak in the neutron-star merger model. If we compare snapshots of each calculations, there are large differences at the early stage of nucleosynthesis. But those differences are obscureded by fission recycling. This effect of fission recycling could be a hint to explain the apparent robustness of the r-process. Since our fission model is very primitive, more realistic, systematic studies of fission are needed for further discussion.

The final computed abundances for the three different environments considered here are compared in Fig. 2. They are scaled to agree at the third r-process peak. The most obvious difference appears between the first and second peak. Unfortunately, our results are based upon a single trajectory in density and temperature. In a realistic calculation one must sum over different trajectories ejected at different times. These trajectories can also affect on this region. In addition, contributions from the weak r-process can also affect the same region. Other differences appear at the third peak. Although yields from the neutron-star merger model is distinguishable from the others, the yield from different types of supernovae are similar in this region. However, neutrino-nucleus interactions (e.g., [15]) have not been included for this calculation. If neutrino-nucleus interactions have a large impact on the final abundances, This may be detectable by observations. If so, we could use this detection to identify the origin of the main r-process. There are still large uncertainties from nuclear physics. We need to reduce those uncertainties if we are ever to identify the origin of the main r-process via a comparison between observed

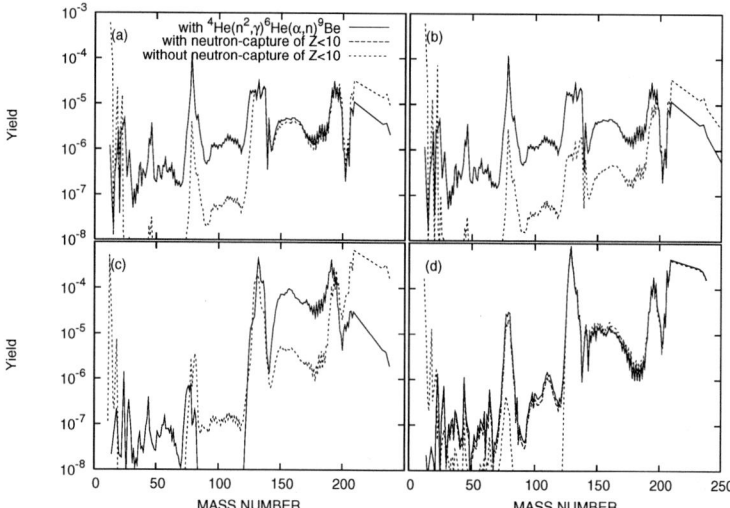

FIGURE 1. Calculated final abundances in the models of (a) neutrino-driven winds, (b) neutrino-driven winds without fission, (c) low-mass supernovae, and (d) neutron star mergers. For each models, the case with di-neutron capture (solid lines), without it (dashed lines), and without any ^{6}He reactions (dotted lines).

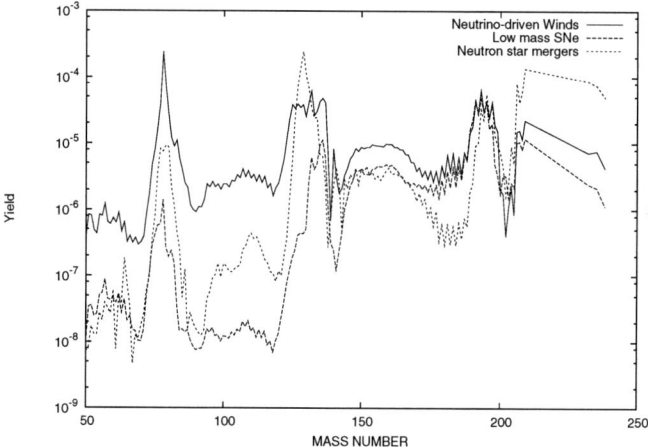

FIGURE 2. Comparison of final abundances of r-process calculations in three different environments, neutrino-driven winds (solid line), low mass supernovae (dashed lines), neutron star mergers (dotted lines).

abundance distributions and theoretical calculations.

Our results suggest that the light element reactions play important roles for nucleosynthesis in the main r-process. We need to pay attention to those light elements when we discuss the viability of environments based upon theoretical calculations.

ACKNOWLEDGMENTS

This work is supported at the University of Chicago in part by the National Science foundation under Grant PHY 02-16783 for the Physics Frontier Center "Joint Institute for Nuclear Astrophysics(JINA)".

REFERENCES

1. S. Honda et al., *ApJS*, **152**, 113–128 (2004).
2. Terasawa, M., Sumiyoshi, K., Kajino, T., Mathews, G.M., and Tanihata, I., *ApJ*, **562**, 470–479 (2001).
3. Bartlett, A., Görres, J., Mathews, G.J., Otsuki, K., Wiescher, M., Frekers, D., Mengoni, A., and Tostevin, J., *submitted to Phys. Rev.C*
4. Meyer, B.S., *ARA&A*, **32**, 153–190 (1994).
5. Meyer B.S., *ApJ*, **343**, 254–276 (1989).
6. Orito, M., Kajino, T., Boyd, R.N., and Mathews, G.J., *ApJ*, **488**, 515–523 (1997).
7. Qian. Y.-Z., & Woosley, S.E., *ApJ*, **471**, 331–351 (1996).
8. Otsuki, K., Tagoshi, H., Kajino, T., and Wanajo, S., *ApJ*, **533**, 424–439 (2000).
9. Wheeler, J.C., Cowan, J.J., & Hillebrandt, W., *ApJ*, **493**, L101–L104 (1998).
10. Wanajo, S., Tamamura, M., Itoh, N., Nomoto, K., Ishimaru, Y., Beers, T.C., & Nozawa, S., *ApJ*, **593**, 968–979 (2003).
11. Timmes, F.X.,& Swesty, F.D., *ApJS*, **126**, 501–516 (2000).
12. Freiburghaus, C., Rosswog, S., & Thielemann, F.-K., *ApJ*, **525**, L121–L124 (1999).
13. Argast, D., Samland, M., Thielemann, F.-K., & Qian, Y.-Z., *A&A*, **416**, 997–1011 (2004).
14. Meyer B.S., McLaughlin, G. C. & Fuller, G. M., *Phys. Rev. C*, **58**, 3696–3710 (1998)
15. Hektor, A., Kolbe, E., Langanke, K., & Toivanen, J., *Phys. Rev. C*, **61**, 55803 (2000).

1.10 NEUTRON CAPTURE and s-PROCESS NUCLEOSYNTHESIS

Neutron capture in massive stars - the challenge of the weak s process

M. Heil and F. Käppeler

Forschungszentrum Karlsruhe, Institut für Kernphysik, Postfach 3640, 76021 Karlsruhe, Germany

Abstract. Neutron capture nucleosynthesis in massive stars has regained considerable interest for the analysis of abundance patterns in very early, metal-poor halo stars as well as for a quantitative picture of galactic chemical evolution. This so-called weak component, which is responsible for the s abundances between Fe and Y, turned out to be very sensitive to the stellar neutron capture cross sections of the isotopes in this mass region. However, these data are incomplete and exhibit large discrepancies between different experiments. New facilities and experimental techniques have to be invoked to meet these challenges.

Keywords: neutron capture cross sections, s process nucleosynthesis, massive stars, activation technique
PACS: 25.40.Lw, 97.10.Cv

NUCLEOSYNTHESIS OF THE HEAVY ELEMENTS

Already in the work of B²FH [1] the concept of neutron capture nucleosynthesis was formulated with the s and r processes as the dominant mechanisms responsible for the abundance distribution between Fe and the actinides. The s process is characterized by relatively low neutron densities, resulting in neutron capture times much longer than typical β-decay half-lives. Accordingly, the s-process reaction path is restricted to the stability valley, and the s abundances are determined by the respective (n,γ) cross sections averaged over the stellar neutron spectrum. Cross sections and s abundances are anti-correlated such that the small cross sections of neutron magic nuclei give rise to the sharp s-process peaks in the abundance distribution at $A = 88, 140,$ and 208.

Since the r process occurs in regions of extremely high neutron density (presumably during stellar explosions in supernovae) neutron captures are much faster than β-decays. Therefore, the r-process path is driven off the stability valley until nuclei with neutron separation energies of ≈ 2 MeV are reached. At these points, (n,γ) and (γ,n) reactions are in equilibrium, and the reaction flow has to wait for β-decay to the next higher element. Hence, the primary r-process yields are proportional to the half-lives of these waiting point nuclei. It should be noted that the odd-even effects in the primary yields are flattened by the effect of β-delayed neutron emission and by neutron capture during the freeze-out phase, when the neutron density in the rapidly expanding explosion vanishes rapidly.

While the observed abundances are dominated by the s and r contributions, which both account for approximately 50% of the abundances in the mass region A>60, the rare proton-rich nuclei cannot be produced by neutron capture reactions. This minor part of the abundance distribution is ascribed to the p process that is assumed to occur in the explosively burning Ne/O layers of supernovae [2, 3].

Among these processes, the s process is the one that can be best studied in laboratory experiments as well as by stellar models and astronomical observations [4]. In contrast, attempts to describe the explosive r and p processes are hampered by the large uncertainties in the nuclear physics data far from stability [5]. If the nuclear reaction and decay rates could be more reliably constrained [6, 7] the respective abundance distributions can be used for a critical discussion of information related to the stellar explosion itself, i.e. whether and to which extent the nucleosynthesis zones are affected by 3D asymmetries, rotation, and magnetic fields, effects which are not yet considered in present r- and p-process models.

THE CASE OF THE s PROCESS

Main and weak component

It was recognized early on that the s-process abundance distribution is composed of two parts, a main component, which is responsible for the mass region from Zr to Bi, and a weak component, which contributes to the region from Fe to Sr. It is important to note that the main component can be assigned to long-lived low mass stars with $1 \leq M/M_\odot \leq 3$, whereas the weak component occurs in massive stars with $M \leq 8 M_\odot$ ($M_\odot$ stands for the mass of the sun). Because massive stars evolve much quicker, the galactic enrichment with s-process material starts with the lighter s elements.

Stellar models for the main s-process component in the mass range $A \geq 90$ refer to helium shell burning in thermally pulsing low mass AGB stars [8]. This scenario is characterized by the subsequent operation of two neutron sources during a series of helium shell flashes. First, the $^{13}C(\alpha, n)^{16}O$ reaction occurs under radiative condition during the intervals between convective He-shell burning episodes. While the ^{13}C reaction provides most of the neutron exposure at low temperatures ($kT \sim 8$ keV) and neutron densities ($n_n \leq 10^7$ cm^{-3}), the resulting abundances are modified by a second burst of neutrons from the $^{22}Ne(\alpha, n)^{25}Mg$ reaction, which is marginally activated during the next convective instability, when high peak neutron densities of $n_n \geq 10^{10}$ cm^{-3} are reached at $kT \sim 23$ keV. Although this second neutron burst accounts only for a few percent of the total exposure, it is essential for adjusting the final abundance patterns of the s-process branchings.

The s process in massive stars with $M \geq 8 M_\odot$ operates in two evolutionary stages, first during core He burning and subsequently during carbon shell burning. Neutrons are produced by the $^{22}Ne(\alpha,n)^{25}Mg$ reactions in both cases, but at rather different temperatures and neutron densities. Because of the lower temperatures of $T_8 = 2-3$, neutron densities during core He burning are limited to $\leq 10^6$ cm^{-3}, whereas peak values of 10^{12} cm^{-3} are reached at the higher temperatures during carbon shell burning [9, 10].

In both s-process scenarios, the stellar (n, γ) cross sections of the involved isotopes constitute the essential nuclear physics input, but with an important difference: The high neutron exposure during the main component is sufficient for establishing equilibrium in the reaction flow, resulting in the so-called local approximation

$$\langle\sigma\rangle N_s = constant.$$

This implies that the emerging s abundances are inversely proportional to the stellar cross sections and that cross section uncertainties are affecting only the respective isotope abundance. In contrast, the neutron exposure in massive stars is too small to achieve flow equilibrium, and this means that cross section uncertainties are not only influencing the abundance of a particular isotope but have a potentially strong propagating effect on the subsequent isotopes.

This wave effect has been discovered because large discrepancies in the existing TOF data for the ^{62}Ni$(n,\gamma)^{63}$Ni reaction [11] were found to result in a bottle neck behavior for the reaction flow. The fact that even the neutron capture rate of a single nucleus can have a significant impact on the abundances of many subsequent isotopes in the reaction chain bears enormous consequences for the reliable decomposition of the r and s contributions to the observed abundance distribution, particularly in view of the limited quality of the stellar cross sections below mass number 120.

COMPARISON WITH OBSERVATIONS

Part of these consequences are related to the present discussion of the surface composition of ultra metal poor (UMP) halo stars [12, 13], which were shown to scale exactly with the solar r-process abundances. The latter are obtained by subtraction of the s-process distribution from the solar composition,

$$N_r = N_\odot - N_s.$$

The perfect agreement between the scaled solar r residuals with the abundance patterns of the very metal-poor halo stars was interpreted as evidence for a robust, primary r process, independent of the metallicity of the precursor star.

However, it turned out that the agreement was limited to the mass region above barium, whereas the observed abundances below barium were found to exhibit a systematic deficiency of about 20% compared to the solar values as shown in Fig. 1. In this figure, the abundance contributions from the r, s, and p process are presented in terms of a sum rule in order to verify how well this sum matches 100% of the respective solar system values. For this comparison the r-process part has been adopted from the elemental abundances in the UMP star CS-22892-052 [13] (normalized to solar Eu), and the s-process part was taken from the s-process calculations normalized to the ensemble of s-only isotopes [14, 15]. In general, the p contribution could be neglected, except for the sizable contribtions to the Mo and Ru abundances.

The uncertainties that are contributed from the r-pocess part due to the UMP observations and from the the s-process part due to the stellar (n,γ) cross sections are plotted in the left and right panel of Fig. 1, respectively. Note that below mass number 120 the error bars in the two panels are similar in size, right where the sum rule exhibits the 20% deficiency.

This 20% discrepancy was assumed to indicate the existence of at least two different r process mechanisms, which dominate the mass regions below and above barium [16].

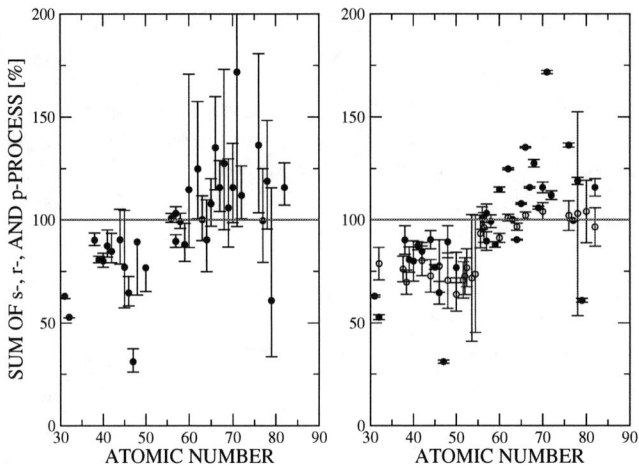

FIGURE 1. Sum of the abundance contributions from the r, s, and p process, where the r-process part corresponds to the element pattern in the UMP star CS-22892-052 normalized at Eu, the s-process part has been adopted from Refs.[14, 15], and the p contribution is considered only for Mo and Ru. Uncertainties refer to the UMP observations (left panel) and to the stellar (n, γ) cross sections reflected in the s-process subtraction (right panel). Below mass number 120, where the sum rule exhibits a 20% deficiency, the error bars in the two panels are similar in size. Note, that the abundances of the s-only isotopes (right panel, open symbols) correlate with the overall distribution.

There are two important aspects to be considered in this discussion, (i) the role of cross section uncertainties and (ii) the fact that the s-only isotopes exhibit exactly the same deficiency as the r residuals below barium. If the missing abundances below barium are produced in a separate process, the second point is important since it strongly suggests that this must be related to an s-process type scenario. Whether an additional process is required at all refers to the first point as discussed below.

The influence of cross section uncertainties could either be directly due to the correlation with the s abundances of the main component or due to the propagation effect in the weak s-process component. In fact there is evidence from new experimental resonance data on Zr [17] that a number of stellar (n, γ) rates in the mass region $A \leq 120$ may be systematically too high because of underestimated corrections for scattered neutrons. If confirmed by improved measurements, this would imply larger s abundances, and hence lower r residuals. In addition, the propagation effect of cross section changes in combination with a reduced effect from neutron poisons may well cause enhanced abundance contributions from massive stars up to the mass 120 region. The latter effect can be illustrated at the example of the minor neutron poison ^{23}Na. The reduction of only this single cross section by a factor of 2 results already in a 5-10% enhancement of the abundances in the weak s component. First activation measurements on F, Na, and Al confirm, indeed, that these cross sections had been significantly overestimated in the past [18]. These results underline the importance of renewed experimental efforts to determine the very small cross sections of the light neutron poisons with improved accuracy.

NEUTRON CROSS SECTIONS - STATUS AND FURTHER QUESTS

The status of (n,γ) data for the s process [19] indicates that experimental techniques have reached a stage, where the 1% accuracy level required for meaningful abundance analyses can meanwhile be met. However, this has been achieved so far only for a minority of the relevant isotopes. In addition to the remaining key nuclei, also a large number of cross sections with uncertainties in excess of 10% await improvement. In contrast to the fairly stable situation of the s process, explosive nucleosynthesis scenarios of the r and p process imply complex reaction paths far from stability, which are described by huge networks including several thousand reactions. By far most of these reaction rates have to be obtained by statistical model calculations [20, 21]. Nevertheless, experimental data for stable and as many unstable isotopes as possible are required for testing the necessary extrapolation to the region of unstable nuclei.

The present status of the cross sections for s-process nucleosynthesis calculations is illustrated in Fig. 2, showing the respective uncertainties as a function of mass number. Though the necessary accuracy has been locally achieved, further improvements are clearly required in the mass region below $A = 120$. The remaining requests concentrate on (n,γ) measurements in the following areas:

- The cross sections of the key isotopes for s-process investigations should be determined with uncertainties of $\approx 1\%$. This goal has been reached only for half of the 33 s-only nuclei between ^{70}Ge and ^{204}Pb.
- Meaningful analyses of the characteristic signatures preserved in presolar grains require also accurate cross sections with uncertainties of $\approx 1\%$. The present status is far from being adequate, particularly for the lighter elements, where about 70 isotopes are concerned.
- Nuclei at or near magic neutron numbers N=50, 82, and 126 act as bottlenecks for the reaction flow in the main s-process region between Fe and Bi. For the majority of these data the necessary uncertainties of $\leq 3\%$ have not been reached.
- The cross sections of abundant light isotopes below Fe, which may constitute crucial neutron poisons for the s-process, need to be improved. Of particular importance are ^{16}O, ^{18}O, and ^{22}Ne.
- Cases where Direct Capture contributes significantly to the astrophysical reaction rate are of particular interest, because this effect plays an important role in neutron-rich nuclei. Interesting examples are ^{208}Pb, ^{14}C, ^{16}O, ^{88}Sr, and ^{138}Ba.
- Nuclei, which still constitute white spots in the s-process chain or which exhibit very uncertain cross sections, are found in the mass region below Fe, around A=100, and near the end of the s-process region. These gaps in the experimental data need to be closed by measurements at the 5% level.
- Last, but not least, enhanced efforts should be directed to measurements on unstable nuclei. In addition to the activation technique, the very high neutron fluxes available at spallation neutron sources appear to be promising options for such studies. Priority should be given to the important branch points ^{79}Se, ^{147}Pm, ^{163}Ho, ^{170}Tm, ^{171}Tm, ^{179}Ta, ^{204}Tl, and ^{205}Pb.

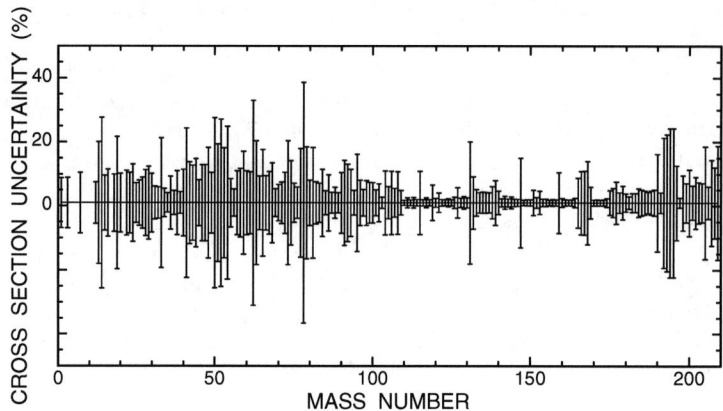

FIGURE 2. Quoted uncertainties for the stellar (n,γ) cross sections required for s-process nucleosynthesis [19].

From the items listed above, the cross sections of the isotopes in the mass region between Cr and Kr as well as of the potential neutron poisons between C and Ca are clearly the most important with respect to the requests for defining the contribution of massive stars to the surface abundances of UMP stars.

ACTIVATION MEASUREMENTS

Activation in a quasi-stellar neutron spectrum provides a completely independent approach for the determination of stellar (n,γ) rates, but is restricted to those cases, where neutron capture produces an unstable nucleus. This method has the twofold advantage that measurements can be performed on sub-μg samples and that isotopically enriched samples are not required. The spectrum produced via the ^{7}Li$(p,n)^7$Be reaction [22, 23] yields an energy distribution very similar to a Maxwell-Boltzmann spectrum for $kT = 25$ keV, typical for He shell flashes in thermally pulsing low-mass AGB stars [24] and for He core burning in massive stars [10]. The complementary reactions ^{18}O$(p,n)^{18}$F [25] and ^{3}H$(p,n)^3$He [26] allow one to simulate conditions close to the temperatures typical for the dominant ^{13}C$(\alpha,n)^{16}$O neutron source in thermally pulsing low-mass AGB stars [24], and during carbon shell burning in massive stars [10].

The possibility to use minute samples makes the activation technique an attractive tool for investigating unstable nuclei of relevance for s-process branchings. The outstanding sensitivity of this method has been illustrated by a measurement of the ^{147}Pm cross section ($t_{1/2}$=2.62 yr), which was performed with a sample of only 28 ng corresponding to $\approx 10^{14}$ atoms. This reduction in sample mass is essential for minimizing the sample activity and for keeping the radiation hazard manageable [27].

The activation method represents often the only way to obtain reliable stellar cross sections, e.g. for determining the partial cross sections to isomeric states, or the very small cross sections of neutron poisons, where the direct capture mechanism (DC) contributes significantly to the stellar average. Moreover, activation data are important for resolving discrepancies among previous TOF data, in particular if one is dealing with resonance dominated cross sections, which can be affected by uncertain scattering corrections.

In view of the cross section uncertainties related to the weak s-process component, a number of activation measurements has been carried out using the ^{7}Li$(p,n)^7$Be reaction for simulating the stellar spectrum for a thermal energy of $kT = 25$ keV. Among the investigated neutron capture cross sections, i.e. for ^{45}Sc, ^{58}Fe, ^{59}Co, ^{63}Cu, ^{65}Cu, ^{79}Br, ^{81}Br, and ^{87}Rb, severe discrepancies of up to a factor of two were obtained. In this series of activations the example of the neutron poison ^{23}Na was investigated as well, and also in this case the cross section was found to be 30% smaller than deduced from previous experiments [19]. Given the sensitive dependence of the abundance yields of the weak s process on the involved (n, γ) cross sections, these results indicate the urgent need for more and in particular, more accurate, measurements.

In this respect, the activation technique represents a useful tool. Common limitations of activation measurements result from the fact that the reaction products need to be unstable with suitable half-lives and detectable decay radiations. However, in many cases, the limits set by standard γ and β spectroscopy can be considerably extended by means of accelerator mass spectroscopy (AMS).

These prospects have recently been demonstrated at the example of the stellar (n, γ) cross section of ^{62}Ni. The decay of the product nucleus ^{63}Ni cannot be detected by spectroscopy of the emitted radiation because the induced activities are very low due to the long half-life of 100 yr, and because the β^- decay leads directly to the ground state of ^{63}Cu with an end point energy of only 66 keV. Any attempt of a quantitative analysis is, therefore, subject to large absorption effects with unacceptable systematic uncertainties. The way out of this dilemma was to detect the produced ^{63}Ni in the irradiated nickel samples by AMS as has been demonstrated successfully in Ref. [28]. It is remarkable that already this first AMS application to an s-process problem yielded an accuracy of 10%, a result that can possibly be further improved to reach the 5% level.

SUMMARY

The origin of the chemical elements, in particular between Fe and the actinides, represents a fascinating astrophysical challenge. Quantitative studies require accurate reaction rates, which have to be collected in laboratory experiments. Comparison of the abundance patterns obtained in such studies with observational data can provide detailed insight in the mechanisms of the corresponding processes. In this respect the s process has been subject to rather quantitative analyses.

Of the two s-process sites, thermally pulsing low mass AGB stars (responsible for the main s component) and massive stars, the latter, which produces the weak s component, has been found to be very sensitive to persisting cross section uncertainties of the involved isotopes in the mass region between Cr and Kr as well as to those of the neutron poisons between C and Ca.

Accurate TOF measurements in the required mass range are difficult because these light isotopes exhibit pronounced resonance structures in the relevant energy range of the s process. Therefore, a first important step would be to use the activation method, which has been shown to be of comparable accuracy, for establishing a reliable grid of improved data. However, the activation method is limited to reactions with an unstable product nucleus as well as by the fact that stellar neutron spectra can be simulated for a few temperatures only. Hence, both methods have to be combined for achieving a satisfactory solution of the nuclear physics input to the s-process models of massive stars.

REFERENCES

1. E. Burbidge, G. Burbidge, W. Fowler, and F. Hoyle, *Rev. Mod. Phys.* **29**, 547 (1957).
2. D. Lambert, *Astron. Astrophys. Rev.* **3**, 201 (1992).
3. M. Arnould and S. Goriely, *Phys. Rep.* **384**, 1 (2003).
4. F. Käppeler, *Prog. Part. Nucl. Phys.* **43**, 419 (1999).
5. F. Käppeler, F.-K. Thielemann, and M. Wiescher, *Ann. Rev. Nucl. Part. Sci.* **48**, 175 (1998).
6. K.-L. Kratz et al., *Astrophys. J.* **403**, 216 (1993).
7. W. Rapp et al., *Astrophys. J.*, submitted.
8. O. Straniero et al., *Astrophys. J.* **440**, L85 (1995).
9. C. Raiteri, M. Busso, R. Gallino, and G. Picchio, *Astrophys. J.* **371**, 665 (1991).
10. C. Raiteri et al., *Astrophys. J.* **419**, 207 (1993).
11. T. Rauscher, A. Heger, R. Hoffman, and S. Woosley, *Astrophys. J.* **576**, 323 (2002).
12. C. Sneden, *Nature* **409**, 673 (2001).
13. C. Sneden et al., *Astrophys. J.* **591**, 936 (2003).
14. C. Arlandini et al., *Astrophys. J.* **525**, 886 (1999).
15. C. Travaglio et al., *Astrophys. J.* **521**, 691 (1999).
16. C. Sneden et al., *Astrophys. J.* **467**, 819 (1996).
17. C. Moreau and et al., in *Nuclear Data for Science and Technolgy*, edited by R. Haight, M. Chadwick, T. Kawano, and P. Talou, AIP Conference Proceedings 769, American Institute of Physics, New York, 2005, pp. 880 – 883.
18. E. Uberseder et al., in preparation.
19. Z. Bao et al., *Atomic Data Nucl. Data Tables* **76**, 70 (2000).
20. T. Rauscher and F.-K. Thielemann, *Atomic Data Nucl. Data Tables* **75**, 1 (2000).
21. S. Goriely, in *Long term needs for nuclear data development*, edited by M. Herman, International Atomic Energy Agency, report INDC(NDS)-428, Vienna, 2001, pp. 83 – 94.
22. H. Beer and F. Käppeler, *Phys. Rev. C* **21**, 534 (1980).
23. W. Ratynski and F. Käppeler, *Phys. Rev. C* **37**, 595 (1988).
24. R. Gallino et al., *Astrophys. J.* **497**, 388 (1998).
25. M. Heil et al., *Phys. Rev. C* **71**, 025803 (2005).
26. F. Käppeler, A. Naqvi, and M. Al-Ohali, *Phys. Rev. C* **35**, 936 (1987).
27. R. Reifarth et al., *Astrophys. J.* **582**, 1251 (2003).
28. H. Nassar et al., *Phys. Rev. Lett.* **94**, 092504 (2005).

1.11 NUCLEAR ASTROPHYSICS and NUCLEOSYNTHESIS (I)

Electron screening in metallic environments: a plasma of the poor man

C. Rolfs

Experimentalphysik III, Ruhr-Universität Bochum, Germany

For the astrophysically important class of charged-particle-induced fusion reactions, there is a repulsive Coulomb barrier in the entrance channel of height $E_c = Z_1 Z_2 e^2/r$, where Z_1 and Z_2 are the integral nuclear charges of the interacting particles, e is the unit of electric charge, and r is the radius. Due to the tunneling effect through the Coulomb barrier, the cross section $\sigma(E)$ of the fusion reaction drops nearly exponentially with decreasing energy E:

$$\sigma(E) = S(E)\, E^{-1} \exp(-2\pi\eta), \qquad (1)$$

where $\eta = 2\pi Z_1 Z_2 e / hv$ is the Sommerfeld parameter (h = Planck constant, v = relative velocity). The function S(E) defined by this equation contains all nuclear effects and is referred to as the nuclear or astrophysical S(E) factor. It is commonly used to extrapolate available data to the relevant thermal energies in stars and other astrophysical objects [1], i.e. $E \approx 0.01\, E_c$. In this extrapolation of the cross section using equation 1, it is assumed that the Coulomb potential of the target nucleus and projectile is that resulting from bare nuclei. However, for nuclear reactions studied in the laboratory, the target nuclei and the projectiles are usually in the form of neutral atoms or molecules and ions, respectively. The electron clouds surrounding the interacting nuclides act as a screening potential: the projectile effectively sees a reduced Coulomb barrier, both in height and radial extension. This, in turn, leads to a higher cross section for the screened nuclei, $\sigma_s(E)$, than would be the case for bare nuclei, $\sigma_b(E)$. There is, in fact, an enhancement factor [1,2],

$$f_{lab}(E) = \sigma_s(E)/\sigma_b(E) \approx \exp(\pi\eta U_e/E) \geq 1, \qquad (2)$$

where U_e is an electron-screening potential energy. This energy can be calculated, for example, from the difference in atomic binding energies between the compound atom and the projectile plus target atoms of the entrance channel, or from the acceleration of the projectiles by the atomic electron cloud: e.g. for the d(d,p)t reaction one finds an acceleration of $U_e = 2 \times 13.6$ eV = 27.2 eV due to the atomic electrons at the Bohr radius. For energy ratios $E/U_e > 1000$, shielding effects are negligible, and laboratory experiments can be regarded as essentially measuring the bare cross section: $\sigma(E) \equiv \sigma_b(E)$. However, for $E/U_e < 100$, shielding effects begin to become important for understanding and extrapolating low-energy data. Relatively small enhancements arising from electron screening at $E/U_e \approx 100$ can cause significant errors in the extrapolation of cross sections to lower energies, if the curve of the cross section is forced to follow the trend of the enhanced cross sections, without correction for the screening. Note that for a stellar plasma, the value of the bare cross section $\sigma_b(E)$ must be known because the screening in the plasma could be quite different from that in the laboratory nuclear-reaction studies, i.e. $\sigma_p(E) = f_p(E)\, \sigma_b(E)$, where the plasma enhancement factor $f_p(E)$ must be explicitly included for each situation. A good understanding of electron-screening effects in the laboratory is needed to arrive at reliable $\sigma_b(E)$ data at low energies. An improved understanding of laboratory electron screening may also help eventually to improve the corresponding understanding of electron screening in stellar plasmas, such as in our sun.

Experimental studies of reactions involving light nuclides ([3] and references therein) have shown the expected exponential enhancement of the cross section at low energies (equation 2). However, the observed enhancements were in some cases larger (up to about a factor 2) than could be accounted for from available atomic-physics models, i.e. the adiabatic limit U_{ad}. Recently, the electron screening in d(d,p)t has been studied for deuterated metals and insulators, i.e. 58 samples in total [4-6]. As compared to measurements performed with a gaseous D_2 target ($U_e = 25\pm5$ eV [7], $U_{ad} = 27.2$ eV), a large screening was observed in all metals (of order $U_e = 300$ eV, i.e. higher by one order of magnitude than U_{ad}), while a small (gaseous) screening was found for the insulators. An explanation of the surprisingly large screening in metals was suggested by the Debye plasma model applied to the quasi-free metallic electrons. The electron Debye radius around the deuterons in the lattice is given by

$$R_D = (\varepsilon_0\, kT\, /\, e^2\, n_{eff}\, \rho_a)^{1/2} = 69\, (T\, /n_{eff}\, \rho_a)^{1/2}\ [m] \qquad (3)$$

with the temperature T of the quasi-free electrons in units of K, n_{eff} the number of these electrons per metallic atom, and the atomic density ρ_a in units of atoms/m^3. With the Coulomb energy of the Debye electron cloud and a deuteron projectile at R_D set equal to $U_e \equiv U_D$, one obtains

$$U_D = 2.09 \times 10^{-11} (n_{eff}\, \rho_a\, /\, T)^{1/2}\ [eV]. \qquad (4)$$

For $T = 293$ K, $\rho_a = 6\times10^{28}$ m^{-3}, and $n_{eff} = 1$ one obtains a radius R_D, which is about a factor 10 smaller than the Bohr radius of a hydrogen atom; as a consequence, one obtains $U_D = 300$ eV, the order of magnitude of the observed U_e values. A comparison of the calculated and observed U_e values led to n_{eff} values, which were for most metals of the order of one. The acceleration mechanism of the incident ions leading to the high observed U_e values is thus the Debye electron cloud at the rather small radius R_D. The n_{eff} values were compared also with those derived from the Hall coefficient: they agreed within experimental uncertainties for all metals with known Hall coefficient. Another critical test of the Debye model was the predicted temperature dependence, $U_D \propto T^{-1/2}$, i.e. a decrease of U_D with increasing temperature, which was experimentally verified for $T = 260$ to 670 K. Furthermore, the Debye energy U_D should scale with the nuclear charge Z_t of the target atoms, $U_D \propto Z_t$: the prediction was verified [8-10] in ^{7}Li(p,α)α and ^{6}Li(p,α)^{3}He ($Z_t = 3$), ^{9}Be(p,α)^{6}Li and ^{9}Be(p,d)^{8}Be ($Z_t = 4$), ^{50}V(p,n)^{50}Cr ($Z_t = 23$), and ^{176}Lu(p,n)^{176}Hf ($Z_t = 71$), always for pure metals and alloys. The data demonstrated that the enhanced electron screening occurs across the periodic table and is not restricted to reactions among light nuclides studied so far. The two reactions with neutrons in the exit channel demonstrated furhermore that the electron screening is an effect in the entrance channel of the reaction and not influenced by the ejectiles of the exit channel, i.e. by the charged particles of the exit channel studied so far. The results for ^{7}Li(p,α)α and ^{6}Li(p,α)^{3}He demonstrated an isotopic independence of the effects of electron screening, as expected. Finally, the Debye model predicts a dependence on the nuclear charge of the ion, $U_D \propto Z_i$; the prediction was verified in the d(^{3}He,p)^{4}He studies in metals ($Z_i = 2$): taking a typical value of $U_e = 300$ eV for the d+d fusion reaction in metals at T = 293 K, one expects for d(^{3}He,p)^{4}He the Debye value to be $U_D = Z_i U_e(d+d) = 600$ eV, consistent with observation ($U_e = 680 \pm 60$ eV). It should be noted that the Debye model is used to calculate the effects of electron screening on fusion reactions in a stellar plasma, $f_p(E)$. Using a metallic plasma the Debye model was tested (in the reports just discussed) successfully with respect to all parameters entering the model. One may thus call metals "a plasma of the poor man". An improved theory is highly desirable to explain why the simple Debye model appears to work so well. Without such a theory, one may consider the Debye model as a parametrisation of the data, with an excellent predictive power.

There is another important prediction of the Debye model concerning radioactive decay of nuclides in a metallic environment. In general, for the α-decay and β$^+$-decay one expects a shorter halflife due to the acceleration mechanism of the Debye electrons for these positively charged particles similar as for the protons, deuterons or ^{3}He in the fusion reactions, while for the β$^-$-decay and e-capture process one predicts a longer halflife (here: deceleration of the negatively charged particles). For example, if the α-decay ^{210}Po→α+^{206}Pb with $E_α = 5.30$ MeV and $T_{1/2} = 138$ days occurs in a metal cooled to T = 4 K, one arrives at $U_D = Z_α Z_t U_e(d+d)(293/4)^{1/2}$ = 2x82x300eVx8.5 = 420 keV, where we used again a typical value of $U_e = 300$ eV for the d+d fusion reaction in metals at T = 293 K and assumed the relation $U_D \propto T^{-1/2}$ to be valid also below T = 260 K. The enhancement factor then gives $f_{lab} = 265$, and thus the halflife is shortened to 0.5 days. For the biologically dangerous transuranic waste ^{226}Ra→α+^{222}Rn ($E_α = 4.78$ MeV, $T_{1/2} = 1600$ years) an analogous calculation leads to $T_{1/2} = 1.3$ years. Experiments are in progress to test these predictions. If they should also be verified, one may have a solution to remove the transuranic waste (involving all an α-decay) of used-up rods of fission reactors in a time period of a few years. Finally, a reduced halflife of α-emitters such as ^{238}U and ^{232}Th in a metallic environment may have important corrections in their use as cosmo-chronometers [2] (i.e. the age of the elements) as well as in understanding the flux of geo-neutrinos using the Kamland detector [11] (i.e. the energy source of the earth).

REFERENCES

1. H.J. Assenbaum et al.: Z. Phys. A327, 461, (1987).

2. C. Rolfs, W.S. Rodney: Cauldrons in the Cosmos (University of Chicago Press, 1988)
3. F. Strieder et al.: Naturwissenschaften 88, 461, (2001).
4. F. Raiola et al.: Eur.Phys.J. A13, 377, (2002).
5. F. Raiola et al.: Eur.Phys.J. A19, 283, (2004).
6. F. Raiola et al.: Eur.Phys.J. A31, 1141, (2005).
7. U. Greife et al.: Z.Phys. A351, 107, (1995).
8. J. Cruz et al.: Phys.Lett. B624, 181, (2005).
9. D.Zahnow et al.: Z.Phys. A359, 211, (1997).
10. K. U. Kettner et al.: Eur.Phys.J. (in press)
11. T. Araki et al.: Nature 436, 499, (2005).

Study of astrophysical (α, n) reactions using light-neutron rich radioactive nuclear beams

Hironobu Ishiyama, Yutaka Watanabe, Nobuaki Imai, Yoshikazu Hirayama, Hiroari Miyatake, Masa-Hiko Tanaka, Nobuharu Yoshikawa, Sunchan Jeong, Yoshihide Fuchi, Ichiro Katayama, Toru Nomura, Tomoko Ishikawa[*], Suranjan K. Das[#], Yutaka Mizoi[#], Tomokazu Fukuda[#], Takashi Hashimoto[**], Katsuhisa Nishio[**], Shinichi Mitsuoka[**], Hiroshi Ikezoe[**], Makoto Matsuda[**], Shinichi Ichikawa[**], Tadashi Shimoda[!],

Institute of Particle and Nuclear Studies, High Energy Accelerator Research organization (KEK), Tsukuba, Ibaraki, 305-0801 Japan
[*]*Department of Physics, Tokyo University of Science, Noda, Chiba, 278-8510 Japan*
[#]*Osaka Electro-Communication University, Neyagawa, Osaka, 572-8530 Japan*
[**]*Japan Atomic Energy Agency (JAEA), Tokai, Ibaraki, 319-1195 Japan*
[!]*Department of Physics, Osaka University, Toyonaka, Osaka, 560-0043 Japan*

Abstract. A systematic study of astrophysical reaction rates of (α, n) reactions on light neutron-rich nuclei using low-energy radioactive nuclear beams is in progress at the tandem facility of Japan Atomic Energy Agency. Exclusive measurements of ^{8}Li(α, n)^{11}B and ^{12}B(α, n)^{15}N reaction cross sections have been performed successfully. Their excitation functions together with the experimental method are presented.

Keywords: Nuclear astrophysics. Radioactive nuclear beam. Nucleosynthesis.
PACS: 25.10.+s, 25.60.-t, 26.30.+k.

INTRODUCTION

Where the rapid neutron capture (r-) process takes place has been a long-standing puzzle in the history of theoretical studies of element synthesis in the universe. One of the most probable sites for the r-process often discussed today is the so-called "hot bubble" in a supernova explosion. The r-process is considered to occur in the region between the surface of a pre-neutron star and the outward-moving shock wave during the explosion. The nuclear statistical equilibrium favors abundant free neutrons and alpha particles in this region as long as the relevant temperature is high. When we follow the paper of Ref. [1], even seed nuclei for the r-process can be produced in this region in the α-capture process at around $T_9 = 3$. When the temperature and density become lower and charged-particle induced reactions almost cease, the usual r-process starts from seed nuclei and a large number of free neutrons. Therefore, nuclear reactions such as (α, n) on light neutron-rich nuclei play important roles as the r-process starting point [1].

However, there is little experimental data on reaction cross-sections on light neutron-rich nuclei. We are therefore proceeding an experimental project to measure cross sections for (α, n) reactions on ^{6}He, ^{8}Li, ^{12}B, ^{15}C and ^{16}N [2, 3] at the tandem facility of Japan Atomic Energy Agency (JAEA). Direct measurements of ^{8}Li$(\alpha, n)^{11}$B, ^{12}B$(\alpha, n)^{15}$N and ^{16}N$(\alpha, n)^{19}$F reaction rates have been already carried out.

The present method of production of light-neutron rich radioactive nuclear beams (RNBs), some characteristics of the detection system as well as the resultant excitation functions of ^{8}Li$(\alpha, n)^{11}$B and ^{12}B$(\alpha, n)^{15}$N reactions are presented in this paper.

RNB PRODUCTION

One of the most powerful methods used to produce low-energy RNBs is nuclear transfer reactions on light targets, since the emitted secondary particles are focused within a forward scattering cone defined by inversion kinematics conditions. On this method, it is important to avoid impurities originating from the primary beam particles [4]. In order to suppress them, a recoil mass separator (RMS) existing at the tandem facility [5, 6] was utilized. It consists of two electric dipoles and a magnetic dipole as shown in Fig. 1. The RNB can be separated from the primary beam using the difference of the magnetic and electric rigidities.

The ^{8}Li beam was produced via ^{9}Be(^{7}Li, ^{8}Li) reaction. A 42μm thick ^{9}Be foil was set at the target position. In order to make energy resolution of the ^{8}Li beam better, the target foil was tilted at 130 degree with respect to the beam axis. The initial energy of the primary ^{7}Li beam was 24 MeV. The intensity of ^{8}Li was 1.4x10^4 pps for 10 pnA ^{7}Li beam at the focal plane of the RMS. Its energy and resolution were 14.6 MeV and 5%, respectively. Although a little amount of ^{6}He particles were mixed in the secondary ^{8}Li–beam, no ^{7}Li impurities were observed. The purity of ^{8}Li RNB was 99%.

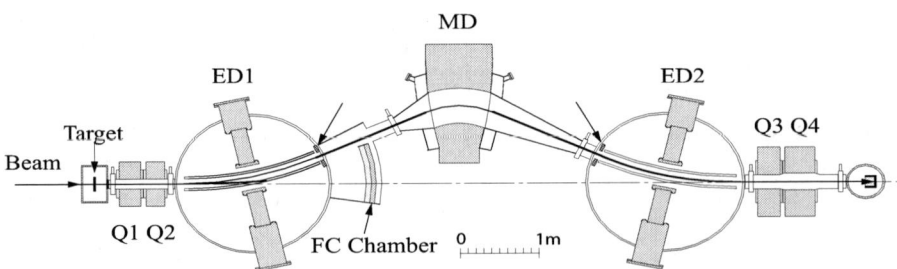

FIGURE 1. Ion optical configuration of the RMS. Q, ED and MD stand for magnetic quadrupole, electric dipole and magnetic dipole, respectively.

The ^{12}B beam was produced via d(^{11}B, ^{12}B)p reaction. A gas target filled with D_2 gas was set at the target position. The D_2 gas pressure was set to be 50 kPa with 5cm thickness.

TABLE 1. The measured yield and purity of RNBs.

RNB	production reaction	E [MeV]	Typical yield	Contaminant (fraction)
^{6}He	d(^{7}Li, ^{6}He)^{3}He	10	5.9x10^3	t (42%)
^{8}Li	^{9}Be(^{7}Li, ^{8}Li)^{8}Be	14.6	1.4x10^4	^{6}He (1%)
^{12}B	d(^{11}B, ^{12}B)p	24	7.8x10^3	^{12}C, ^{11}B (<2 %)
^{16}N	d(^{18}O, ^{16}N)α	32	4.7x10^3	^{18}O (1.5%)

The intensity of the ^{12}B beam was 7.8x10^3 pps for 10 pnA ^{11}B beam. The measured energy and its spread were 24 MeV and 7.2%, respectively. The purity of the ^{12}B beam was 98%.

Table 1 summarizes RNBs so far produced. On the detailed report about our RNB-production methods, please see the Ref. [7].

EXPERIMENT

The present detector system is schematically shown in Fig. 2. It consists of a beam pick-up detector system, a "multi-sampling and tracking proportional chamber" (MSTPC) [8] placed at the focal-plane of the RMS, and a neutron-detector array, in which the first one is composed of a multi-channel plate (MCP) and a parallel-plate avalanche counter (PPAC). The absolute energy of RNB is determined by the time-of-flight (TOF) information between MCP and PPAC. The RNB is injected into the MSTPC filled with gas of He+CO_2 , which works as counter gas and as gas target.

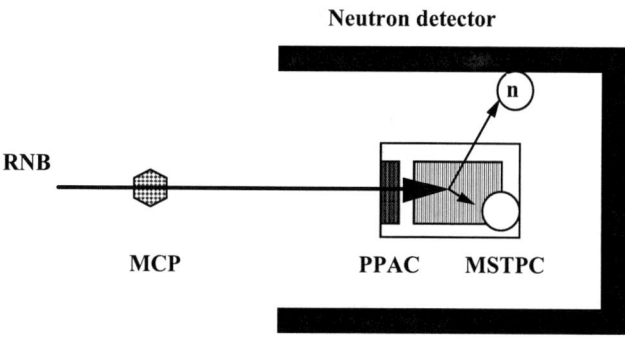

FIGURE 2. Schematic view of the experimental setup. The RNB provided from the RMS is directly injected into the MSTPC filled with He +CO_2 gas, which acts as counter gas and as gas target. A neutron-detector array, consisting of 28 plastic scintillators, surrounds the MSTPC to detect emitted neutrons.

When a nuclear reaction takes place inside the MSTPC, the energy loss (dE/dx) changes largely due to the change of the relevant atomic numbers. Therefore, we can determine a reaction point by finding these changes and the reaction energy by evaluating the energy losses through the gases.

A neutron-detector array to detect neutrons emitted from nuclear reactions surrounds the MSTPC as much as possible. It consists of 28 pieces of BC408 plastic scintillator bars, covering 31.4% of 4 π. The typical efficiency for a 5 MeV neutron measured by using a ^{252}Cf-fission source is about 40%.

About our experimental setup in detail, please see the reference [9].

As for the measurement of the ^{8}Li(α, n)^{11}B reaction, we performed two experiments in order to obtain reaction cross sections in different energy regions. The recent experimental result, whose energy region covers from 0.45 MeV to 1.75 MeV, will be presented in these proceedings [10]. Reaction cross sections of ^{12}B(α, n)^{15}N has also been measured at the energy region of E_{cm} = 1.1 – 3.7 MeV.

TABLE 2. Some typical parameters for the measurement of the ^{12}B(α, n) reaction.

parameter	Value
electrode of MCP	1.5 µm Mylar evaporated gold
entrance window of PPAC	2.0 µm Mylar, 40 mm in diameter
electrode of PPAC	1.5 µm Mylar evaporated aluminum
exit window of PPAC	7.5 µm Mylar, 40 mm in diameter
gas in PPAC	iso-butane, 6 Torr
length between MCP and PPAC	900 mm
gas in MSTPC	He + CO_2 (10 %), 140 Torr
roof plate voltage of MSTPC	-1005 V
gating grid voltage (V_0) of MSTPC	-172 V
gating grid voltage ($\pm\Delta V$) of MSTPC	60 V
anode wire voltage of MSTPC	661.1 V
length between MSTPC and plastic scintillators	900 mm
size of a typical plastic scintillator	50 x 150 x 1500 mm

RESULTS

The measured excitation function of the ^{8}Li(α, n) ^{11}B reaction is shown in Fig. 3. Black and white circles indicate our experimental cross sections and the other symbols show previous experimental data. Present data covers almost of T_9 = 1-3 (T_9 = 10^9 k). Triangles and inversed triangles indicate the data from Boyd et al. [11] and Gu et al. [12] based on the inclusive measurements without neutron detection. Lozenges indicate from Cherubini et al. [13] on the inclusive measurement without ^{11}B detection. White squares indicate the data from Mizoi et al. [14] based on the exclusive measurement using a detector system similar to that in the present work.

Present results have at least ten times better statistics compared with the previous exclusive measurement [14] and it is smaller than results of previous inclusive measurements. The difference is by a factor of more than 2 in the E_{cm} region lower than 1.5 MeV. This result may suggest that the previous data include background events, like elastic scatterings. On our data, we could clearly see a resonance-like structure located around 0.85 MeV, which may correspond to a resonance state located at 10.9 MeV in ^{12}B reported in Ref. [15].

The measured excitation function of the ^{12}B(α, n)^{15}N reaction is shown in Fig. 4. Present data covers almost of T_9 = 2-5. This result is still preliminary since analysis

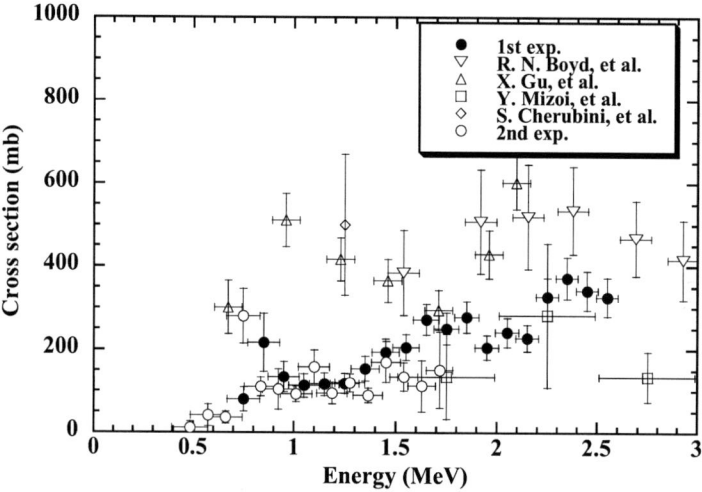

FIGURE 3. The measured excitation function of ^{8}Li(α, n)^{11}B reaction (black and white circles). The horizontal axis is the center-of-mass energy and the vertical one is the cross section in unit of mb.

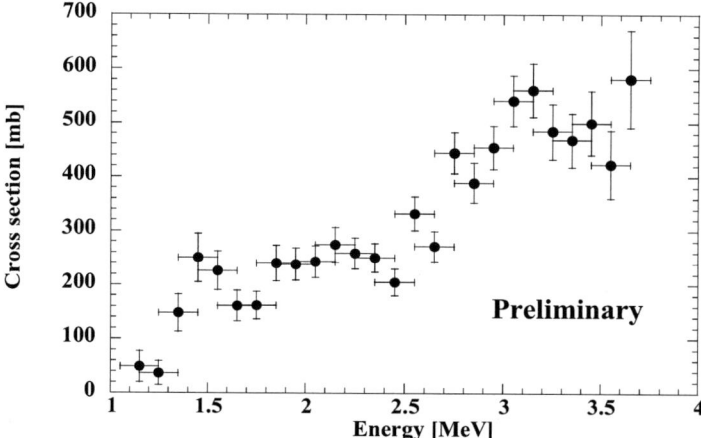

FIGURE 4. The measured excitation function of ^{12}B(α, n)^{15}N reaction (black circles). The horizontal axis is the center-of-mass energy and the vertical one is the cross section in unit of mb.

is in progress. But, we could clearly see some resonance-like structures, especially, at E_{cm} = 1.5-1.6 MeV, which may correspond to a resonance state located at 11.61 MeV in ^{16}N reported in Ref. [16].

SUMMARY

Exclusive measurements of the (α, n) reaction cross sections on light neutron-rich nuclei are in progress using low-energy radioactive nuclear beams. The excitation function of the ^{12}B(α, n)^{15}N reaction has been measured in the energy region of E_{cm} = 1.1 – 3.7 MeV, corresponding to Gamow peak at T_9 = 2 – 5. A resonance-like structure was clearly observed at E_{cm} = 1.5-1.6 MeV, which may correspond to the state at 11.61 MeV in ^{16}N. The excitation function of the ^{8}Li(α, n)^{11}B reaction in the energy region of E_{cm} = 0.4 – 2.6 MeV are also reported. A resonance-like structure was clearly observed at E_{cm} = 0.85 MeV, which may correspond to the state at 10.9 MeV in ^{12}B.

ACKNOWLEDGMENTS

The authors wish to thank Prof. Y. Tagishi and Dr. T. Komatsubara at University of Tsukuba for their helpful support in the performance test of the MSTPC. We also thank the staff members of the JAEA tandem facility for their kind operation of the tandem accelerator.

REFERENCES

1. Terasawa, M. et. al., *Nucl. Phys.* **A688**, 581c(2001).
2. Ishiyama, H. et. al., *Nucl. Phys.* **A718**, 481c (2003).
3. Ishikawa, T. et. al., *Nucl. Phys.* **A718**, 484c(2003).
4. Becchetti, F. D. et. al., *Nucl. Instrum. Meth. Phys. Res.* **B56/57**, 554 (1991).
5. Ikezoe, H. et al., *Nucl. Instrum. Meth. Phys.* **A376**, 470 (1996).
6. Kuzumaki, T. et al., *Nucl. Instrum. Meth.* **A437**, 107 (1999).
7. Ishiyama, H. et al., to be published in *Nucl. Instrum. Meth.* **A**
8. Mizoi, Y. et al., *Nucl. Instrum. Meth.* **A431**, 112 (1999).
9. Hashimoto, T. et al., *Nucl. Instrum. Meth.* **A556**, 339 (2006).
10. Das. S.K. et al., *in this proceedings*.
11. Boyd, R.N. et al., *Phys. Rev. Lett.* **68**, 1283 (1992).
12. Gu, X. et al., *Phys. Lett.* **B343**, 31 (1995).
13. S. Cherubini, et al., *Eur. Phys. J.* **A20**, 355(2004).
14. Mizoi, Y. et al., *Phys. Rev.* **C62**, 065801 (2000).
15. N. Soic, et al., *Europhys. Lett.* **63**, 524(2003).
16. D.R. Tilley, et al., *Nucl. Phys.* **A564**, 1(1993).

Indirect techniques in nuclear astrophysics

A. M. Mukhamedzhanov*, G. Rogachev[†], L. D. Blokhintsev**, S. Brown[†],
V. Burjan[‡], S. Cherubini[§], V. Z. Gol'dberg[¶], B. F. Irgaziev[∥,††], E. Johnson[†],
K. Kemper[†], V. Kroha[‡], A. Momotyuk[†], R. G. Pizzone[§], B. Roeder[†], S.
Romano[§], C. Spitaleri[§], R. E. Tribble* and A. Tumino[§]

*Cyclotron Institute, Texas A&M University, College Station, TX, 77843, USA
[†]State University of Florida, Tallahassee, FL, USA
**Institute of Nuclear Physics, Moscow State University, Moscow, Russia
[‡]Nuclear Physics Institute of Czech Academy of Sciences, Prague-Řež, Czech Republic
[§]DMFCI, Università di Catania, Catania, Italy and INFN - Laboratori Nazionali del Sud, Catania, Italy
[¶]Cyclotron Institute, Texas A&M Univeristy, College Station, TX, 77843, USA
[∥]Faculty of Engineering Sciences, GIK Institute of Engineering Sciences and Technology, Topi-23640, N.W.F.P., Pakistan
[††]Physics Department, National University, Tashkent, Uzbekistan

Abstract. We address two important indirect techniques, the asymptotic normalization coefficient (ANC) and the Trojan Horse (TH) methods. We discuss the application of the ANC technique to determine the astrophysical factor for the $^{13}C(\alpha,n)^{16}O$ reaction which is one of the neutron generators for the s processes in the AGB stars. The TH method is unique indirect technique allowing one to measure astrophysical rearrangement reactions down to astrophysically relevant energies. Using a simple model, we demonstrate that off-energy-shell and Coulomb effects in the entry channel and the final state nuclear interaction do not change the energy dependence of the astrophysical factor extracted from the TH reaction.

Keywords: Indirect methods, Asymptotic normalization coefficient, $^{13}C(\alpha,n)^{16}O$, Trojan Horse
PACS: 26.20.+f, 21.10.Jx, 25.55.Hp, 27.20.+n

INTRODUCTION

For better understanding stellar evolution, cross sections of astrophysically relevant nuclear reactions should be known at the Gamow energy. The presence of the Coulomb barrier for colliding charged nuclei makes nuclear reaction cross sections at astrophysical energies so small that their direct measurements in laboratories is very difficult, or even impossible. That is why direct measurements are being done at higher energies and then extrapolated down to the Gamow energy. Such an extrapolation procedure can cause an additional uncertainty. Also for nuclear reactions studied in laboratory, the electron clouds surrounding the interacting nuclei lead to a screened cross section which is larger than the "bare" nucleus one (see [1, 2] and references therein). We address here two important indirect techniques: the asymptotic normalization coefficient (ANC) method [3, 4] and Trojan Horse (TH) [5, 2].

ANC AND ASTROPHYSICAL FACTOR FOR THE $^{13}C(\alpha,N)^{16}O$ REACTION

The ANC method has been suggested in [3, 4] and can be used to determine the astrophysical factors for radiative capture processes populating loosely bound final states. Such direct radiative capture proceses are peripheral and the overall normalization of their cross sections is entirely determined by the ANC of the projection of the final bound state wave function into the two-body channel corresponding to the colliding particles [6].

The ANC technique turns out to be very productive for analysis of the astrophysical processes in the presence of the subthreshold state [7]. Here we address the first application of the ANC method to determine the astrophysical factor for the $^{13}C(\alpha,n)^{16}O$ reaction. At astrophysically relevant energies this reaction proceeds through the s-wave suthreshold resonance $(1/2^+, 6.356 \text{ MeV})$ in ^{17}O. The overall normalization of the astrophysical factor for this reaction is determined by the product of two widths- the width corresponding to the formation of the subthreshold resonance and its decay width. The first one is determined by the ANC [7]. The ANC $C^F_{xAl_Fj_F}$ for the virtual decay (synthesis) $F \leftrightarrow A + x$ defining the amplitude of the tail of the radial overlap function $I^F_{xAl_Fj_F}(r)$ is given by

$$I^F_{xAl_Fj_F}(r) \stackrel{r>R_N}{\longrightarrow} C^F_{xAl_Fj_F} \frac{W_{-\eta_F,l_F+1/2}(2\kappa_{xA}r)}{r}, \tag{1}$$

where R_N is the nuclear interaction radius between x and A, $W_{-\eta_F,l_F+1/2}(2\kappa_{xA}r)$ is the Whittaker function describing the asymptotic behavior of the bound state wave function of two charged particles x and A with the orbital angular momentum of the relative motion l_F, $\kappa_{xA} = \sqrt{2\mu_{xA}\varepsilon_F}$ is the wave number of the bound state $F = (xA)$, μ_{xA} is the reduced mass of particles x and A, ε_F is the binding energy of the bound state (xA) and $\eta_F = Z_x Z_A e^2 \mu_{xA}/\kappa_{xA}$ is the Coulomb parameter of the bound state $F = (xA)$, $Z_i e$ is the charge of particle i. We use the system of units such that $\hbar = c = 1$.

The amplitude of the reaction $x + A \to b + B$ proceeding through the subthreshold resonance F is given in the R-matrix approach by [7]

$$M \sim \sqrt{\frac{P_{l_F}(k_{xA}, r_{0(i)})}{\mu_{xA} r_{0(i)}}} \tilde{W}_{-\eta_F, l_F+1/2}(2\kappa_{xA} r_{0(i)}) \frac{\tilde{C}^F_{xAl_Fj_F} \Gamma_f^{1/2}(E_{bB}, r_{0(f)})}{E_{xA} + \varepsilon_F + i\Gamma_f(E_{bB}, r_{0(f)})/2}. \tag{2}$$

Here, $P_{l_F}(k_{xA}, r_{0(i)})$ ($P_{l_f}(k_{bB}, r_{0(f)})$) is the Coulomb-centrifugal barrier penetration factor in the entry (exit) channel, $\tilde{W}_{-\eta_F, l_F+1/2}(2\kappa_{xA} r_{0(i)}) = W_{-\eta_F, l_F+1/2}(2\kappa_{xA} r_{0(i)})\Gamma(l_F + 1 + \eta_F)$ the Coulomb modified Whittaker function, $r_{0(i)}$ ($r_{0(f)}$) the channel radius for the initial (final) channel, $\tilde{C}^F_{xAl_Fj_F} = C^F_{xAl_Fj_F}/\Gamma(l_F+1+\eta_F)$ stands for the Coulomb modified ANC for the virtual decay (synthesis) $F \leftrightarrow x + A$ and $\Gamma_f(E_{bB}, r_{0(f)})$ is the resonance width for the decay to the final channel $b + B$; we assume also that the total width of the resonance F is equal to Γ_f. In the R-matrix method the resonance width is given by $\Gamma_f(E_{bB}, r_{0(f)}) = (P_{l_f}(k_{bB}, r_{0(f)})/P_{l_f}(k_{bB(R)}, r_{0(f)}))\Gamma_f(E_{bB(R)})$. Also $E_{iJ} = k_{ij}^2/(2\mu_{ij})$ is the relative kinetic energy of particles i and j; hence, $E_{bB} = k_{bB}^2/(2\mu_{bB}) = E_{xA} + Q$,

$Q = m_x + m_A - m_b - m_B$, m_i is the mass of particle i. The resonance energy in the initial (final) channel is $E_{xA(R)} = -\varepsilon_F$ ($E_{bB(R)} = k_{bB(R)}^2/(2\mu_{bB}) = -\varepsilon_F + Q$).

Until now the ANC method has been applied to determine the astrophysical factors for radiative capture processes. Here we present the first case of application of the ANC method to determine the astrophysical factor for the $^{13}C(\alpha,n)^{16}O$ reaction. In order for the s-process to operate in the AGB stars there must be a source of neutrons. The $^{13}C(\alpha,n)^{16}O$ reaction, which occurs as a result of the injection of ^{13}C into the He-burning shell, is one of the reactions producing neutrons for the s processes. At $E_{xA} \equiv E < 300$ keV the reaction proceeds through the subthreshold state $^{17}O(1/2^+, 6.356$ MeV$)$ with the 3 keV binding energy for the virtual decay into the channel $^{13}C + \alpha$. The $^{17}O(1/2^+, 6.356$ MeV$)$ state is the s-wave resonant state in the channel $^{16}O + n$. $\Gamma_n = 124 \pm 12$ keV is known very well, hence, according to Eq. (2), to find the astrophysical factor for the $^{13}C(\alpha,n)^{16}O$ reaction it is enough to determine the ANC. The advantage of the ANC technique is that it allows one to determine the astrophysical factor at zero energy and its energy dependence at low energies. To determine the ANC the *alpha* transfer reaction $^{13}C(^6Li,d)^{17}O$ was measured at sub-Coulomb relative kinetic energy. The measurements have been performed at the Florida State University Tandem-LINAC facility. Inverse kinematics has been used for this measurements. The energy of ^{13}C beam was 8.5 MeV. Litium targets were prepared and transported to the scattering chamber in vacuum. Targets were 50 $\mu g/cm^2$ thick with ^6Li enrichment of 98%. Deuterons from α transfer reaction $^{13}C(^6Li,d)$ were identified in array of silicon telescopes. Thickness of ΔE detectors used in this experiment was 25 μm. Spectrum of deuterons, obtained in telescope at 35° (120° in the c.m. system) is shown in figure 1. Transition to the $^{17}O(1/2^+, 6.356$ MeV$)$ state is clearly seen and the cross section was measured to be 63.0 ± 6.0 $\mu b/sr$ at this angle. The sub-Coulomb α transfer reaction is peripheral at backward scattering angles. That is why dominant contribution to the reaction comes from the tail of the overlap function of the bound state wave functions of $^{17}O(1/2^+, 6.356$ MeV$)$, ^{13}C and α-particle. Variation of the number of nodes in the $(^{13}C\alpha)_{1/2^+}$ s-wave bound state wave function practically dos not change the calculated DWBA cross section. Variation of the optical potentials changes the DWBA cross section by 20%. The ANC for $^6Li \rightarrow \alpha + d$ is very well established [8]. Hence by normalization of the DWBA cross section (FRESCO code) to the experimental one at backward angles we determined the Coulomb modified ANC squared $(\tilde{C}_{\alpha^{13}C 11}^{17O(1/2^+)})^2 = 0.87 \pm 0.19$ fm^{-1} [1] The uncertainty of our ANC is determined by 6% statistical uncertainty, 7% systematic uncertainty and 20% optical potentials dependence. Using the obtained ANC we calculated the astrophysical factor for the $^{13}C(\alpha,n)^{16}O$ reaction proceeding through the subthreshold $^{17}O(1/2^+, 6.356$ MeV$)$. Our astrophysical factor $S(0) = (2.36 \pm 0.52) \times 10^6$ MeVb, Fig. 2, almost order of magnitude smaller then the adopted $S(0)$ given in the NACRE compilation [9] and approximately 2.6 times larger then the astrophysical factor $S(0)$ obtained in [10].

---

[1] Note that the standard ANC squared due to the extremely small binding energy and large $\eta_F = 60.38$ is $(5.00 \pm 1.10) \times 10^{168}$ fm^{-1}. Hence for the weakly bound states it is more convenient to operate with the Coulomb modifed ANC and Whittaker function.

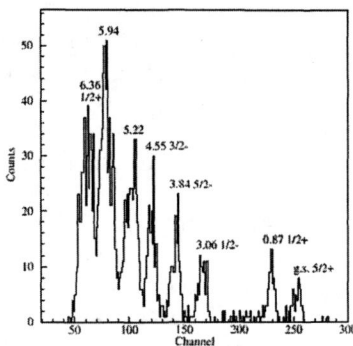

FIGURE 1. Spectrum of deuterons obtained at 120° in the c. m. system.

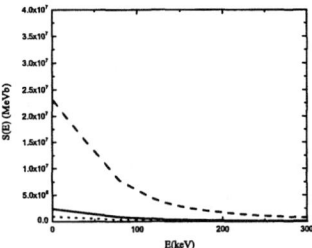

FIGURE 2. The astrophysical factor for the $^{13}C(\alpha,n)^{16}O$ reaction at energies $E \leq 300$ keV. The solid line represents the calculated $S(E)$ factor using our measured ANC, the dashed line is the NACRE compilation [9] and the dotted line is the astrophysical factor obtained in [10].

TROJAN HORSE

The Trojan Horse (TH) method is a powerful indirect technique which allows one to determine the astrophysical factor for rearrangement reactions. The TH method, first suggested by Baur [5], has been advanced and practically applied by a group from the Universitá di Catania working at the INFN-Laboratori Nazionali del Sud in Catania in collaboration with other Institutions (see [2] and references therein). This method involves obtaining the cross section of the binary $x+A \rightarrow b+B$ process at astrophysical energies by measuring the two-body to three-body ($2 \rightarrow 3$) process, $a+A \rightarrow y+b+B$, in the quasifree (QF) kinematics regime, where the "Trojan Horse" particle, $a = (xy)$, is accelerated at energies above the Coulomb barrier. After penetrating through the Coulomb barrier, nucleus a undergoes breakup leaving particle x to interact with target A while projectile y flies away. From the measured $a+A \rightarrow y+b+B$ cross section, the energy dependence of the binary subprocess, $x+A \rightarrow b+B$, is determined.

The main advantage of the TH method is that the extracted cross section of the binary

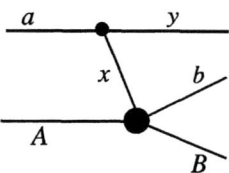

FIGURE 3. The pole diagram describing the quasi-free mechanism of TH reaction $a+A \to y+b+B$.

subprocess does not contain the Coulomb barrier factor. Consequently the TH cross section can be used to determine the energy dependence of the astrophysical factor, $S(E)$, of the binary process, $x+A \to b+B$, down to zero relative kinetic energy of the particles x and A without distortion due to electron screening [1, 11]. The absolute value of $S(E)$ must be found by normalization to direct measurements at higher energies. At low energies where electron screening becomes important, comparison of the astrophysical factor determined from the TH method to the direct result provides a determination of the screening potential which is difficult to determine from direct measurements.

The THM has already been applied many times to reactions connected with fundamental astrophysical problems such as $^7\text{Li}(p,\alpha)^4\text{He}$, $^6\text{Li}(d,\alpha)^4\text{He}$, $^6\text{Li}(p,\alpha)^3\text{He}$, and many others, (see [2] and references therein). Nevertheless there are still reservations about the reliability of the method due to two potential modifications of the yield from off-shell effects and initial and final state interactions in the TH $2 \to 3$ reaction. In the TH reaction, shown schematically in Fig. 3, particle x in the binary subprocess $x+A \to b+B$ is virtual (off-energy-shell). In the standard analysis, the virtual nature of x is neglected and the plane wave approximation is used [11, 12]. Here we address the reliability of these assumptions. We also consider scattering between particles a and A in the initial channel, particles b and B in the final channel of the TH reaction and the dominance of the QF mechanism. Although the TH reaction is a many-body process and its strict analysis requires many-body techniques some important features of the TH method can be addressed in a simple model. Let $\mathbf{p}_{xj}$ to be the relative momentum of the virtual particle x and particle j. The internal particle x of the diagram shown in Fig. 3 is virtual, i.e. $E_x \neq p_x^2/2m_x$. From energy-momentum conservation laws in the three-ray vertex $a \to x+y$ and the four-ray vertex $x+A \to b+B$, we get $\sigma_x = p_{xy}^2/(2\mu_{xy}) + \varepsilon_a = p_{xA}^2/(2\mu_{xA}) - E_{xA}$, where $\varepsilon_a = m_x + m_y - m_a$ is the binding energy for the virtual decay $a \to x+y$. Thus in a TH reaction $p_{xA}^2/(2\mu_{xA}) > E_{xA} = k_{xA}^2/(2\mu_{xA})$ holds. The reaction amplitude corresponding to the diagram is given by (all particles are assumed, for simplicity, to be spinless) $M_p = M_{2\to 2}^{(HOF)} \varphi_{xy}(p_{xy})$. Here $M_{2\to 2}^{(HOF)}(\sigma_x, E_{xA}, z)$ is the half-off-energy-shell (HOF) reaction amplitude for the binary subprocess $x+A \to b+B$ which depends on the additional variable σ_x due to the virtual nature of particle x. Also $z = \hat{\mathbf{p}}_{xA} \cdot \hat{\mathbf{k}}_{bB}$, with $\hat{\mathbf{k}} = \mathbf{k}/k$. The Fourier transform of the s-wave bound-state wave function for $a = (xy)$ can be written as $\varphi_{xy}(p_{xy}) = -W_{xy}(p_{xy})/(\sigma_x)^{1-\eta_a}$, where η_a is the Coulomb parameter of the bound state $a = (xy)$ and W_{xy} is the amplitude for the virtual decay $a \to x+y$, which is regular at the singular point $\sigma_x = 0$. This singularity is a branch point for the charged particles x and y. For $a = d = (pn)$, $\varphi_{xy}(p_{xy})$ has a pole at $\sigma_x = 0$, corresponding to the real (on-energy-shell (ON)) particle x. Due to this pole the name of the diagram in Fig. 3 is the pole diagram. The modulus of the amplitude $|M_p|$ of the pole diagram has a maximum at $p_{xy} = 0$ called the QF peak and the condition corresponding to $p_{xy} = 0$ is called QF kinematics. The QF peak is a trace of the pole located at $p_{xy}^2 < 0$. The QF kinematics are ideal for the TH method since small p_{xy} corresponds to large separation distance between particles x and y, thereby allowing particle y to be treated as a spectator.

By dropping all the terms containing particle y and its interaction with other nuclei, the half-off-shell post-form amplitude of the direct binary reaction, $A(x,b)B$, can be extracted from the exact amplitude for the TH reaction, $A(a,yb)B$. The result is given by

$$M_{2\to 2}^{(HOF)}(\sigma_x, E_{xA}, z) = \lambda <\psi_{bB}^{(-)}\varphi_b\varphi_B|\hat{O}|\varphi_x\varphi_A p_{xA}>. \quad (3)$$

Here λ is the kinematical factor, $\psi_{bB}^{(-)}$ is the distorted wave in the final state of the binary process, φ_i is the bound state wave function of nucleus i, $\hat{O} = \Delta V_{bB}[1 + G_{xA}^{(+)}\Delta V_{xA}]$ is the transition operator, $\Delta V_{ij} = V_{ij} - U_{ij}$, V_{ij} (U_{ij}) is the interaction (optical) potential between nuclei i and j and $G_{xA}^{(+)}$ is the Green's function of the system $x + A$ (or $b + B$). The half-off-shell amplitude contains the off-shell plane wave $|p_{xA}>$ which describes the relative motion of the virtual particle x and A in the initial channel of the binary reaction rather than the distorted wave describing the initial state in the on-shell amplitude. Hence the half-off-shell amplitude does not contain a Coulomb barrier factor. First we estimate the contribution from the transition operator $[\Delta V_{bB} G_{xA}^{(+)} \Delta V_{xA}]$ to $M_{2\to 2}^{(HOF)}$. Consider two reactions previously analyzed using the TH method, ^{6}Li$(d,\alpha)^4$He ($Q_{2\to 2} = 21.64$ MeV) [11] and ^{7}Li$(p,\alpha)^4$He ($Q_{2\to 2} = 16.56$ MeV) [13]. The first reaction was treated as a deuteron transfer and the second as a triton transfer. Nuclei A and b are bound states $A = (Bt)$ and $b = (xt)$ with constituent particles x, t, B, where t is the transferred particle. In this simple model the half-off-shell amplitude takes the form (all particles are assumed to be spinless)

$$M_{2\to 2}^{(HOF)}(\sigma_x, E_{xA}; z) = \lambda <\psi_{bB}^{(-)}\varphi_{xt}|\hat{O}|\varphi_{Bt} p_{xA}>. \quad (4)$$

The bound state wave functions have been approximated by their tails, $\Delta V_{bB} \approx V_{Bt}$ and the zero-range approximation has been used for $V_{Bt}\varphi_{Bt}$. At $E_{xA} < 1$ MeV, only the Coulomb interaction needs to be included in the transition operator ΔV_{xA} and in the Green's function. Also at low E_{xA} the dominant contribution comes only from the s-wave in the channel $x + A$ due to the large $Q_{2\to 2}$ value for both reactions. Consequently the z dependence of $M_{2\to 2}^{(HOF)}$ can be neglected. The contribution from the transition operator $[\Delta V_{bB} G_{xA}^{(+)}\Delta V_{xA}]$ is negligible for both reactions as is the case for the on-shell processes. It constitutes a background for the DWBA with the transition operator given by ΔV_{bB}.

The on-shell amplitude to be compared with the half-off-shell one is

$$M_{2\to 2}^{(ON)}(E_{xA}) = \lambda <\psi_{bB}^{(-)}\varphi_{xt}|\Delta V_{bB}|\varphi_{Bt}\psi_{xA}^{(+)}>. \quad (5)$$

Here, $\psi_{xA}^{(+)}$, the distorted wave describing the scattering of real particles x and A in the initial channel, can be approximated by the pure Coulomb scattering wave function at low energies. Consequently, $\psi_{xA}^{(+)} = N_{xA} exp(i\mathbf{k}_{xA}\cdot\mathbf{r}_{xA})\,_1F_1$, where N_{xA} is the Gamow normalization factor and $_1F_1$ is the hypergeometric function, which has a very weak

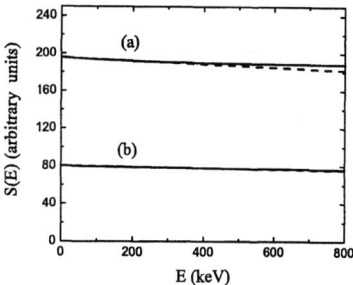

FIGURE 4. Energy dependence ($E \equiv E_{xA}$) of the half-off-shell (dashed line) and on-shell (solid line) astrophysical factors for (a) the $^7\text{Li}(p,\alpha)^4\text{He}$ reaction ($a=d, x=p, y=n$); (b) the $^6\text{Li}(d,\alpha)^4\text{He}$ reaction ($a=\,^6\text{Li}, x=d, y=\alpha$).

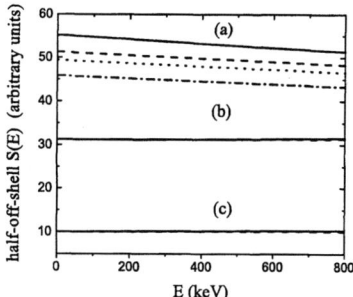

FIGURE 5. (a) Energy dependence ($E \equiv E_{d^6\text{Li}}$) of the half-off-shell astrophysical factor for the $^6\text{Li}(d,\alpha)^4\text{He}$ reaction as a function of σ_d: $\sigma_d = 0$- solid line (on-shell kinematics); $\sigma_d = \varepsilon_{d\alpha}^{^6\text{Li}}$- dashed line (QF kinematics); $\sigma_d = 1.5\,\varepsilon_{d\alpha}^{^6\text{Li}}$ - dotted line; $\sigma_d = 2.5\,\varepsilon_{d\alpha}^{^6\text{Li}}$ - dashed-dotted line. (b) Comparison of the energy dependence of the half-off-shell astrophysical factor for the $^6\text{Li}(d,\alpha)^4\text{He}$ reaction determined from the $^6\text{Li}(^6\text{Li},\alpha\alpha)^4\text{He}$ TH reaction with (solid line) and without (dashed line) inclusion of the initial $^6\text{Li}-\,^6\text{Li}$ Coulomb interaction. (c) The same as in (a) but with the final-state interaction described by the Ali-Bodmer potential: $\sigma_d = 0$- solid line (on-shell kinematics); $\sigma_d = \varepsilon_{d\alpha}^{^6\text{Li}}$- dashed line (QF kinematics). In both cases (b) and (c) the dashed line is normalized to the solid line at $E = 1$ keV.

energy dependence at small k_{xA}. The only difference in the energy dependence between the half-off-shell and on-shell astrophysical factors comes from the the use of the half-off-shell plane wave in the initial state in Eq. (4) and the on-shell distorted wave in Eq. (5). From Fig. 4 it is clear that the energy dependence of the half-off-shell (calculated in the QF kinematics) and on-shell astrophysical factors at low energies are practically identical. It justifies the TH method. Since only the energy dependence is of interest, the half-off-shell result in Fig. 4 has been normalized to the on-shell one at an energy $E_{xA} = 1$ keV for ease of comparison. In Fig. 5 (a) we compare the energy dependence of the half-off-shell astrophysical factor for the $^6\text{Li}(d,\alpha)^4\text{He}$ reaction ($x=d; A=\,^6\text{Li}; b=B=\alpha$) for different values of σ_d. The deuteron becomes farther from the energy-shell as σ_d

grows. It is clear from Fig. 5 that the size of the S factor changes but the energy dependence does not change. This justifies the procedure of disregarding the virtual nature of the entry particle in the analysis of the binary process $A(x,b)B$ [11, 12].

The plane wave approximation used in the TH method [12] ignores the interaction between particles a and A in the entry channel. To test the importance of the Coulomb interaction in the initial state, the energy dependence of the half-off-shell astrophysical factor for the binary $^6\text{Li}(d,\alpha)^4\text{He}$ reaction, determined from the TH reaction $^6\text{Li}(^6\text{Li},\alpha\alpha)^4\text{He}$ at $E_{^6\text{Li}^6\text{Li}} = 3.14$ MeV, was calculated with and without the initial Coulomb interaction in the TH reaction amplitude. To extract the half-off-shell amplitude for the binary process from the TH amplitude in QF kinematics with the initial Coulomb interaction we have to divide the later by the factor $R = <\varphi_{xy}(\mathbf{k}_{yF}-(m_y/m_a)\mathbf{p}_{aA})|\psi^{(C)(+)}_{\mathbf{k}_{aA}}(\mathbf{p}_{aA})>$. Integration is performed over $\mathbf{p}_{aA}$ and $\psi^{(C)(+)}_{\mathbf{k}_{aA}}(\mathbf{p}_{aA})$ is the Fourier component of the $a-A$ Coulomb scattering wave function. The results, as seen in Fig. 5 (b), show that the energy dependence of both astrophysical factors is nearly identical. As also seen from Fig. 5 (c), the final state interaction (here described by the Ali-Bodmer $\alpha-\alpha$ potential) does not affect the energy behavior of the on-shell and half-off-shell astrophysical factors for the binary process $^6\text{Li}(d,\alpha)^4\text{He}$. We have demonstrated that the energy dependence of the half-off-shell and on-shell astrophysical factors are nearly identical when analyzed in the same model. Validating this makes it clear why the TH method is such a powerful indirect technique for nuclear astrophysics.

ACKNOWLEDGMENTS

This work was supported in part by the U.S. DOE under Grant No. DE-FG02-93ER40773 and the U.S. NSF under Grant No. PHY-0140343.

REFERENCES

1. H. J. Assenbaum, K. Langanke, and C. Rolfs, *Z. Phys.*, **A 327**, 461-468, (1987).
2. C. Spitaleri *et al.*, *Phys. Rev.*, **C 69**, 055806(1-7) (2004).
3. A. M. Mukhamedzhanov and N. K. Timofeyuk, *JETP. Lett.*, **51**, 282-284 (1990).
4. H. M. Xu *et al.*, *Phys. Rev. Lett.*, **73**, 2027-2030 (1994).
5. G. Baur, *Phys. Lett.*, **B 178**, 135-138 (1986).
6. C. A. Gagliardi *et al.*, *Phys. Rev.*, **C 59**, 1149-1153 (1999).
7. A. M. Mukhamedzhanov, R. E. Tribble, *Phys. Rev.*, **C 59**, 3418-3424 (1999).
8. L. D. Blokhintsev *et al.*, Phys. Rev. **C 48**, (1993) 2390-2394.
9. C. Angulo *et al.*, *Nucl. Phys.*, **A656**, 3-183 (1999).
10. S. Kubono *et al.*, *Phys. Rev. Lett.*, **90**, 062501(1-4) (2003).
11. C. Spitaleri *et al.*, *Phys. Rev.*, **C 63**, 055801(1-7) (2001).
12. S. Typel and G. Baur, *Ann. Phys.*, **305**, 228-265 (2003).
13. M. Lattuada *et al.*, *Astrophys. J.*, **562**, 1076-1080 (2001).

Trojan Horse Method: Recent Experiments

S. Cherubini [a], C. Spitaleri [a], V. Crucillà [a], M. Gulino [a], M. La Cognata [a], L. Lamia [a], R.G. Pizzone [a], S. Romano [a], S. Tudisco [a], A. Tumino [a], A. Mukhamedzhanov [b], L. Trache [b], R. Tribble [b], C. Rolfs [c], S. Typel [d]

[a] *DMFCI, Università di Catania and INFN-Laboratori Nazionali del Sud, Catania, Italy*
[b] *Cyclotron Institute, Texas A&M University, College Station, TX, 77843, USA*
[c] *Ruhr-Universität, Bochum, Germany*
[d] *Gesellschaft für Schwerionenforschung mbH, Darmstadt, Germany*

Abstract. The Trojan Horse Method allows for the measurements of cross sections in nuclear reactions between charged particles at astrophysical energies. The basic features of the method are discussed and recent applications are presented.

Keywords: Nuclear Astrophysics, Indirect Methods, Trojan Horse Method.
PACS: 26.20+f, 24.50+g

INTRODUCTION

Nuclear reactions in stellar environments take generally place at energies much lower than the Coulomb barrier, E_{CB}, existing between the colliding nuclei. Actually, the relevant center of mass energy range called "Gamow window", E_G, is typically of the order of a few tens of keV (non-explosive burning) or hundreds of keV (explosive stellar burning) at most, while E_{CB} spans the MeV region.

Nuclear fusion reactions proceed therefore via tunnel effect with an exponential decrease of the cross section: $\sigma(E) \approx \exp(-2\pi\eta)$, where η is the Sommerfeld parameter [1]. This implies that the cross sections values are generally very small, of the order of micro- and even nanobarns. This results in serious difficulties when making a measurement in the Gamow window region. Measurements were therefore performed at higher energies and extrapolated to the Gamow ones.

In performing this extrapolation, owing to the strong exponential suppression of the cross section mentioned above, the astrophysical factor $S(E)=E\,\sigma(E)\exp(2\pi\eta)$ is introduced, the inverse of the Gamow factor removing the dominant energy dependence in $\sigma(E)$

The extrapolation is therefore easier but it is not completely safe and often errors occurred in the evaluation of $S(E)$ at E_G energies, e.g. because of the presence of unknown resonances in the energy range of interest.

Even recent experiments, performed directly in the Gamow energy regions, couldn't avoid an extrapolation procedure as the values of the *bare* nucleus cross section, $\sigma^{bare}(E)$, needed by astrophysicists are obtained from measured data after removing the electron screening effects using an extrapolation of higher energy data. These effects are not as negligible as they were thought to be before these measurements were carried out. A discussion of this important issue [2] goes beyond the scopes of this paper.

In order to overcome some of the difficulties outlined above, a number of indirect methods were introduced. Among them, the Trojan Horse Metod (THM) allows for the measurement of thermonuclear reaction cross sections. Recent results are presented here.

BASIC THEORY OF THE TROJAN HORSE METHOD

The study of quasi-free processes performed by our group in the last decades has shown that a two-body process cross-section, say $a(x,c)d$ could be indirectly extracted from a three-body process, say $a(b,cd)s$, mainly proceeding via a quasi-free reaction mechanism where b virtually decays into $x+s$, the virtual particle x interacts with the nucleus a while s remains a spectator to the process.

The nucleus b is chosen because of a dominant $x+s$ cluster structure possibly associated to a low binding energy for this configuration.

A polar diagram as shown in Figure 1 describes this three-body process.

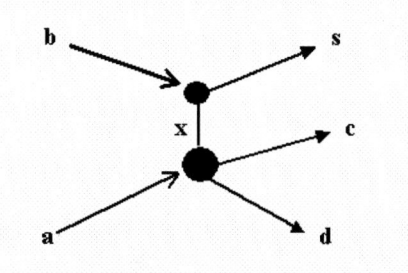

FIGURE 1. This is the pole diagram representing the quasi-free process $a(b,cd)s$. The upper pole represents the virtual decay $b\rightarrow x+s$, while the lower one represents the two-body virtual reaction $a(x,c)d$.

The cross section for this process is written in Plane Wave Impulse Approximation (PWIA) [3] as

$$\frac{d^3\sigma}{dE_c d\Omega_c d\Omega_d} = KF|\Phi(p_s)|^2 \left(\frac{d^2\sigma(E_{2b})}{d\Omega}\right)_{cm}.$$

This factorization reflects the pole diagram of Figure 1: $d^2\sigma/d\Omega$ is the two-body virtual cross section (lower pole), the momentum distribution $\Phi(p_s)$ is related to the virtual decay (upper pole) of b particle and it is well known in the case of the nuclei used in the experiments while KF is simply a kinematical factor. The subscript cm in the two-body cross section reminds that this cross section is related to the real two-body process as measured in its center of mass (cm) system. The energy in the two-body cm system, E_{2b}, is obtained from the post-collision prescription: $E_{2b}=(E_{cd}-Q_{2b})$, where E_{cd} is the relative energy of particles c and d as measured in the three-body reaction and Q_{2b} is the q-value of the real two-body process.

Note that the energy of the three-body reaction is by no means supposed to be lower than the Coulomb barrier so that the three-body process can proceeds with the cross section values that are typical for direct break-up reactions (order of the millibarn - tens of millibarn).

Once demonstrated the dominance of the quasi-free channel, it is quite simple in principle to extract a two-body cross section is hence in principle quite simple: one has to measure the three-body cross-section $d^3\sigma/dE_c d\Omega_c d\Omega_d$, calculate KF and $\Phi(p_s)$ and divide the former by these latter quantities. In practice, the selection of the events that correspond to the quasi-free channel is a quite delicate task while the division mentioned above often requires Monte Carlo calculations.

Examples of two-body cross sections obtained from a three-body one are reported in [4, 5]

When this procedure is applied to study cross sections at energies below the Coulomb barrier [6,7] it is essential to take into account that 1) the extracted cross section will not show the exponential decrease affecting direct measurements (see A. Mukhamedzanov et al. contribution to this conference for a detailed discussion) and that 2) the results will not be affected by the electron screening problem.

From 1) it is evident that the THM derived cross sections for process of astrophysical interest cannot be compared with those coming from direct measurement unless one inserts the suppression factor due to the barrier penetration "by hand":

$$\sigma^{bare} = \sigma^{THM} G^C_l$$

where G^C_l is the penetration coefficient. In this sense one can say that the THM allows for measuring the *bare nucleus* cross section. This allows for contributing to the study of the electron screening that affects directly measured cross sections.

RECENT RESULTS

In the following, some recent results obtained by applying the THM will be presented. In particular results for the astrophysical processes $^{11}B+p\rightarrow {}^{8}Be+\alpha$, $^{3}He+d\rightarrow p+\alpha$ and $^{15}N+p \rightarrow {}^{12}C+\alpha$ will be analyzed.

$^{11}B+p\rightarrow {}^{8}Be+\alpha_0$

This reaction, which is connected with the problem of determining the light elements abundances, has been studied by applying the THM to the $^{11}B+d\rightarrow {}^{8}Be+\alpha+n$ reaction. The experiment was performed at the INFN – Laboratori Nazionali del Sud in Catania (Italy). A beam of 2-5 nA of ^{11}B at 27 MeV produced by the HV tandem accelerator was used to bombard a CD_2 isotopically enriched target of 200 µg/cm^2 thickness.

The detection set-up consisted in a pair of split silicon PSD detectors that recorded alpha particles coming from the decay of the emitted ^{8}Be while another standard silicon PSD was used on the opposite side with respect to the beam direction to detect the third alpha particle in coincidence with the other two. Data corresponding to the quasi-free mechanism were selected imposing various cuts in the phase space. The result obtained by applying a simple PWIA [8] is shown in Figure 2. The discrepancy with respect to direct measurement [9] is quite evident and a new calculation using a modified PWIA [10] is in progress.

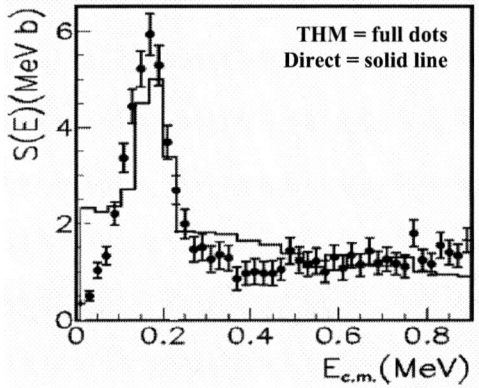

FIGURE 2. The $^{11}B+p\rightarrow {}^{8}Be+\alpha_0$ cross section obtained by applying the THM to the three-body $^{11}B+d\rightarrow {}^{8}Be+\alpha_0+n$ process. Full dots are extracted data points while the line represents the directly measured excitation function rebinned to take into account the THM energy resolution.

$^{3}He+d \rightarrow p+\alpha$

The $^{3}He+d\rightarrow P+\alpha$ reaction was studied as a part of the primordial nucleosynthesis chain and also because its measurement via THM could be helpful in studying the electron screening potential that seems to be much larger than the theoretical limit for this reaction [11].

The experiment was performed at the Dynamitron Tandem Laboratorium in Bochum (Germany). A ^{3}He beam of 5 and 6 MeV was used to bombard a lithium fluoride target ^{6}Li (^{6}Li ≈ 95%). Silicon telescopes with position sensing capabilities were placed at symmetric angles on the two opposite sides with respect to the beam direction to detect the outgoing protons and alpha particles. Details of the experiment are described in [11].

Data collected in this experiment allowed for the extraction of the astrophysical factor. Figure 3 shows a comparison between the result of this experiment and the directly measured one [12].

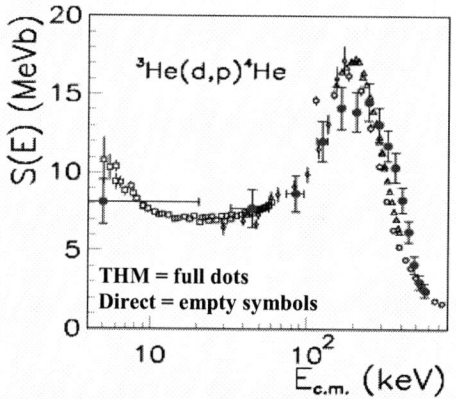

FIGURE 3. The ^{3}He+d→ p +α astrophysical factor obtained by applying the THM to the three-body ^{3}He+^{6}Li→p+α+α process. Full dots are extracted data points while the line represents the directly measured excitation function rebinned to take into account the THM energy resolution

^{15}N+p→ ^{12}C+α

The ^{15}N+p→^{12}C+α process is very important in the AGB scenario. This process was studied at the K500 superconducting Cyclotron of the Texas A&M University in College Station (USA) using the THM applied to the ^{15}N+d→^{12}C+α+n reaction. A 60 MeV ^{15}N beam was used to bombard a CD_2 target. A pair of ΔE-E telescopes consisting in gas ionization gas chambers and silicon position-sensing detectors recorded the ^{12}C and α particles produced in the reaction in coincidence. The astrophysical factor obtained as a preliminary result of this study [13] turns out to be much smaller than other previous results [14] as shown in Figure 4.

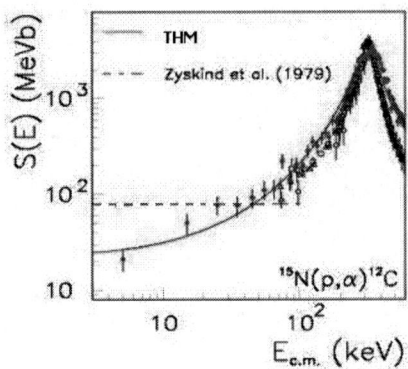

FIGURE 4. The astrophysical factor for the $^{15}N+p \rightarrow {}^{12}C +\alpha$ reaction obtained via the THM (solid line) compared to results from Zyskind et al. (dashed line).

REFERENCES

1. C. Rolfs and W.S. Rodney, *Cauldrons in the cosmos*, University of Chicago Press, Chicago (1988)}
2. F. Streider, C.Rolfs, C.Spitaleri, P.Corvisiero, Naturwissenchaften 88, 461, (2001) and references therein
3. G.F. Chew, Phys. Rev. 80, 196 (1950)
4. M. Zadro et al. Phys. Rev. C 40, 181 (1989)
5. G. Calvi et al., Phys. Rev. C 49, 1848 (1990)
6. C. Spitaleri, Problems of Fundamental Modern Physics II, World Scientific,p. 21, 1990
7. S. Cherubini et al., Ap. J. 457, 855 (1996) and references therein
8. C. Spitaleri et al., Phys. Rev. C 69, 055806 (2004)9. H. W. Becker, C. Rolfs, and H. P. Trautvetter, Z. Phys. A 327, 341 (1987)
10. S. Typel and G. Baur, *Ann. Phys.*, **305**, 228 (2003)
11. La Cognata et al., Phys. Rev. C, in press
12. F. C. Barker, Nucl. Phys. 707, 277 (2002) and references therein
13. La Cognata et al., Nucl. Phys. A 758, 98 (2005)
14. J. L. Zyskind and P. D. Parker, Nucl. Phys. A 320, 4040 (1979).

Determination of S_{17} from ^{8}B breakup by means of the method of continuum-discretized coupled-channels

K. Ogata*, S. Hashimoto*, Y. Iseri†, M. Kamimura* and M. Yahiro*

Department of Physics, Kyushu University, Fukuoka 812-8581, Japan
†*Department of Physics, Chiba-Keizai College, Todoroki-cho 4-3-30, Inage, Chiba 263-0021, Japan*

Abstract. The astrophysical factor for ^{7}Be$(p,\gamma)^8$B at zero energy, $S_{17}(0)$, is determined from an analysis of ^{208}Pb$(^8$B, $p+^7$Be$)^{208}$Pb at 52 MeV/nucleon by means of the method of continuum-discretized coupled-channels (CDCC) taking account of all nuclear and Coulomb breakup processes. The asymptotic normalization coefficient (ANC) method is used to extract $S_{17}(0)$ from the calculated breakup cross section. The main result of the present paper is $S_{17}(0) = 20.9^{+2.0}_{-1.9}$ eV b. This value of $S_{17}(0)$ differs from the one extracted with the first-order perturbation theory including Coulomb breakup by dipole transitions: 18.9 ± 1.8 eV b. It turns out that the difference is due to the inclusion of the nuclear and Coulomb-quadrupole transitions and multi-step processes of all-order in the present work. Our main result of $S_{17}(0)$ is consistent with the value obtained from a precise measurement of the p-capture reaction cross section extrapolated to zero incident energy, $S_{17}(0) = 22.1 \pm 0.85$ eV b. Thus, the agreement between the values of $S_{17}(0)$ obtained from direct ^{7}Be$(p,\gamma)^8$B and indirect ^{8}B-breakup measurements is significantly improved.

Keywords: Solar neutrino problem, Nuclear and Coulomb breakup, CDCC
PACS: 24.10.Eq, 25.60.Gc, 25.70.De, 26.65.+t

1. INTRODUCTION

The solar neutrino problem is one of the central issues in the neutrino physics [1]. The cross section $\sigma_{p\gamma}(E)$ of the p-capture reaction ^{7}Be$(p,\gamma)^8$B at about zero incident energy, i.e. $E \sim 0$, plays an essential role in the solar-neutrino phenomenology, since the observed flux of ^{8}B neutrino is proportional to it; $\sigma_{p\gamma}(E)$ is customarily expressed by the astrophysical factor $S_{17}(E) \equiv \sigma_{p\gamma}(E) E \exp[2\pi\eta]$, where η is the Sommerfeld parameter. The required accuracy of $S_{17}(0)$, to determine the neutrino oscillation parameters with sufficient accuracy, is the error within 5% [1].

Recently, precise measurement of $\sigma_{p\gamma}(E)$ was carried out by Junghans *et al.* [2] at energies of 116–2460 keV, which are low but still higher than stellar energies (~ 20 keV). Extrapolating the measured $S_{17}(E)$ to $E = 0$ using a three-cluster model [3] for ^{8}B structure, they derived $S_{17}(0) = 22.1 \pm 0.6$ (expt) ± 0.6 (theo) eV b. The three-body model, however, did not simultaneously reproduce the magnitude and the energy-dependence of $S_{17}(E)$ sufficiently well. Moreover, as pointed out in Ref. [4], the uncertainty of the s-wave p-^{7}Be scattering length (with about 50% error) prevents one from determining $S_{17}(0)$ with very high accuracy.

Because of the difficulty of the direct measurement of $\sigma_{p\gamma}(E)$ at stellar energies, alternative indirect measurements were proposed. Coulomb dissociation [5, 6, 7, 8] of ^{8}B is one of such indirect measurements. So far, extraction of $S_{17}(0)$ from these experiments has been based on the virtual photon theory with the assumption of ^{8}B dissociation by virtual electric dipole (E1) photon absorption. Nuclear interaction and absorption of quadrupole (E2)- and multi-photons were not taken into account. The value thus extracted from the RIKEN experiment at 52 MeV/nucleon was $S_{17}(0) = 18.9 \pm 1.8$ eV b [5]. In the analysis of the MSU experiment [8] measured at 44 and 83 MeV/nucleon, the E2 contribution to one-step process was estimated from the parallel-momentum-distribution of ^{7}Be fragment and subtracted from the breakup spectrum of ^{8}B. As a result, the extracted $S_{17}(0)$ was $17.8^{+1.4}_{-1.2}$ eV b, which is smaller than the value obtained at RIKEN mentioned above. However, the analysis of the angular distribution of the ^{8}B breakup cross section of the RIKEN experiment showed no contribution of E2 transitions. Moreover, the recent experiment at rather high energy, 250 MeV/nucleon, done at GSI [7] showed that the E2 contribution was negligibly small; the resulting $S_{17}(0)$ was $18.6 \pm 1.2 \pm 1.0$ eV b. Even though the significance of E2 transitions can be energy-dependent, the conclusions from the MSU and RIKEN measurements at similar energies look inconsistent and roles of the E2 component are still not clear. More seriously, there exists a non-negligible discrepancy of about 15% between the values of $S_{17}(0)$ mentioned above that are derived from direct p-capture and indirect ^{8}B-breakup measurements.

Very recently, it was shown by a semiclassical calculation of ^{208}Pb(^{8}B, $p+^7$Be)^{208}Pb at 52 MeV/nucleon that the discrepancy mentioned above was significantly reduced by taking account of nuclear-breakup components, E2 and higher-order Coulomb breakup processes [9]. Motivated by this result, we attempt in the present paper to extract a reliable value of $S_{17}(0)$ analyzing the ^{8}B dissociation experiment measured at RIKEN [5], with the method of continuum-discretized coupled-channels [10] (CDCC), assuming a $p+^7$Be+^{208}Pb three-body model of the system. The result of the analysis is then used to extract $S_{17}(0)$ by means of the asymptotic normalization coefficient (ANC) method [11]. CDCC has been successful in describing various processes in which effects of projectile-breakup are essential [12, 13]. It has been successfully applied also to ^{8}B nuclear and Coulomb breakup processes [14, 15, 16]. In the CDCC calculation of the present work, we include all nuclear and Coulomb breakup processes, and take account of the intrinsic spins of both p and ^{7}Be using the channel spin representation and the consistency of the p-^{7}Be interaction potential used in the CDCC calculation with the s-wave p-^{7}Be scattering length. We then use the calculated breakup cross section and the p-^{7}Be separation energy in the ^{8}B nucleus to obtain $S_{17}(0)$ by the ANC method. An important advantage of the ANC method is that there is no restriction of the reaction mechanisms. In addition, the uncertainty of $S_{17}(0)$ due to the use of the ANC method can quantitatively be evaluated [13, 16].

In Sec. 2 we analyze the ^{8}B breakup cross section measured at RIKEN with CDCC and extract $S_{17}(0)$ by the ANC method. In Sec. 3, the value of the extracted $S_{17}(0)$ is compared with the result of the virtual photon theory and the roles of nuclear breakup, E2 transitions, and higher-order processes are discussed. Finally summary and conclusions are given in Sec. 4.

2. ANALYSIS OF ^{8}B BREAKUP EXPERIMENT

We calculate the cross section of the ^{8}B breakup reaction ^{208}Pb(^{8}B, $p+^7$Be)^{208}Pb at 52 MeV/nucleon measured at RIKEN [5] with CDCC. Details of the CDCC calculation are shown in Ref. [17]. Figure 1 shows the results of the χ^2-fit of the calculated

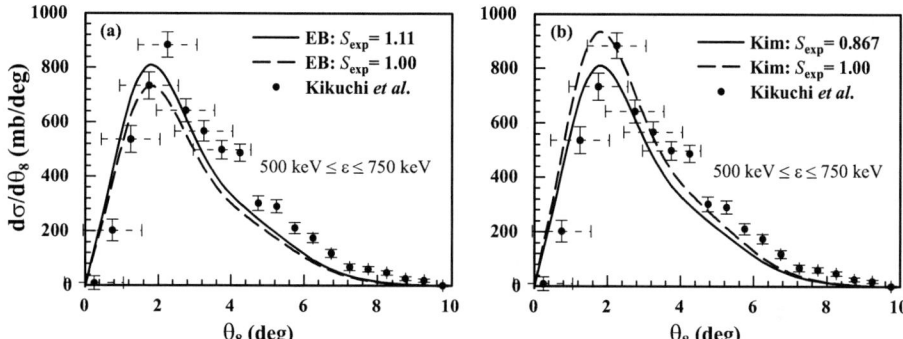

FIGURE 1. (a) The result of the χ^2-fit of the breakup cross section calculated with the ^{8}B single-particle potential of Esbensen and Bertsch (EB) [18] to the experimental data [5]. The dashed line is the result of CDCC with the normalized ^{8}B single-particle wave function, i.e. with $S_{exp} = 1.00$, and the solid line is the one multiplied by the spectroscopic factor $S_{exp} = 1.11$. (b) Same as in (a) except that the single-particle wave function of Kim et al. [19] is used; the resulting value of S_{exp} is 0.867.

breakup cross sections of ^{8}B by ^{208}Pb at 52 MeV/nucleon to the experimental data. The calculated and measured breakup cross sections, integrated over the excitation energy ε of ^{8}B measured from the $p+^7$Be threshold energy from 500 keV to 750 keV, are shown as a function of the ^{8}B scattering angle θ_8. The left (right) panel corresponds to the CDCC calculation with the ^{8}B single-particle potential of Esbensen and Bertsch [18] (Kim et al. [19]). In each panel the dashed line is the result of CDCC with a normalized ^{8}B single-particle wave function and the solid line is the result obtained by the χ^2-fit to the data. We use the eight data points below 4° in the fitting procedure, since the quantitative comparison between the present CDCC calculation and the experimental data is possible only in this angular region [17]. It should be noted that each horizontal bar put on the data points below 4° does not represent a statistical error but it shows the range of θ_8 in which the breakup cross sections contribute to each data point [5].

The values of S_{exp} corresponding to the left and right panels are 1.11 and 0.867, respectively, which show the strong dependence of S_{exp} on the choice of the model of ^{8}B. On the other hand, the values of ANC, which is the ratio of the ^{8}B wave function to the Whittaker function in the tail region, is found to be almost independent of it; the resulting value of ANC is 0.740 fm$^{-1/2}$ (0.741 fm$^{-1/2}$) when the ^{8}B single-particle wave function of Esbensen and Bertsch (Kim et al.) is used. This result shows that in the present case only the tail of the ^{8}B wave function contributes to the breakup process. We thus conclude that the ANC method works with very high accuracy in the present analysis. Once the value of the ANC is obtained, one can determine $S_{17}(0)$ by using the ANC method. After careful estimation of theoretical ambiguities [17] and inclusion of 8.4% systematic experimental error, we obtain $S_{17}(0) = 20.9^{+1.0}_{-0.6}$ (theo) $\pm$ 1.8 (expt) eVb. It should be noted that the reaction concerned is quite peripheral and fragmentation of p or ^{7}Be from the target is negligible. Core excitation of ^{7}Be in the ^{8}B projectile is found to have no effect on the value of $S_{17}(0)$ above [17].

3. DISCUSSION OF THE EXTRACTED $S_{17}(0)$

The main result of the present paper shown above is significantly larger than $S_{17}(0) = 18.9 \pm 1.8$ eV b [5] obtained in the previous analysis of the same experiment with the first-order perturbation theory. In order to clarify the reason for the difference, we discuss roles of nuclear interaction, E2 transitions, and higher-order processes. In

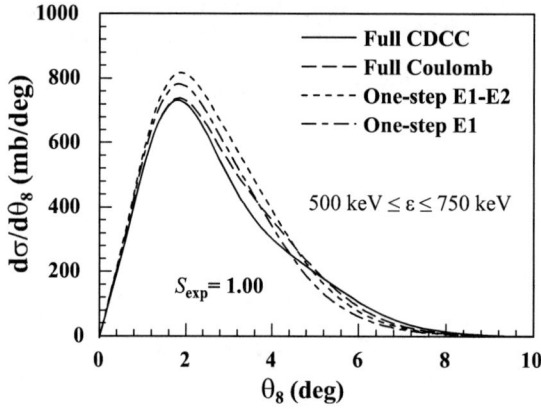

FIGURE 2. Same as in Fig. 1 (a) with different assumptions for CDCC. See the text for details.

Fig. 2 the solid line is the same as the dashed line in Fig. 1 (a) that shows the result of full CDCC. The dashed line in Fig. 2 corresponds to the calculation without breakup components of nuclear coupling potentials. One sees that the dashed line agrees with the solid line below 2° but deviates from it for $\theta_8 > 2°$. As a result, the value of $S_{17}(0)$ obtained, $S_{17}(0) = 20.0^{+1.9}_{-1.8}$ eV b, is smaller than $S_{17}(0) = 20.9^{+1.0}_{-0.6}$ (theo) ± 1.8 (expt) eV b obtained with the full CDCC calculation mentioned above. The dash-dotted line in Fig. 2 represents the result of first-order iterative solutions of CDCC, designated as one-step CDCC, without nuclear breakup, including only E1 component of the Coulomb interaction; this calculation is essentially the same as the first-order perturbation theory (virtual photon theory) used in the previous analysis of the experimental data [5]. It overestimates the solid line above 1.5° and the resulting value of $S_{17}(0)$ is $19.0^{+1.9}_{-1.7}$ eV b, which agrees well with the value 18.9 ± 1.8 eV b obtained in Ref. [5]. If one includes the E2 component in the one-step CDCC calculation of Coulomb breakup, the dotted line in Fig. 2 is obtained; the E2 coupling potentials are artificially multiplied by 0.7, following the analysis of the MSU data [8]. One sees from Fig. 2 that inclusion of the reduced E2 component somewhat increases the breakup cross section, which results in further decrease of $S_{17}(0)$ to $S_{17}(0) = 17.9^{+1.7}_{-1.6}$ eV b. This result is consistent with the conclusion of Ref. [8], in which $S_{17}(0) = 17.8^{+1.4}_{-1.2}$ eV b was derived with first-order perturbation theory including both E1 and the reduced E2 components. If the E2 component is not scaled, the resulting value of $S_{17}(0)$ is found to be 16.7 eV b. This value is about 20% less than the result of full CDCC, 20.9 eV b. The difference is due to the nuclear and higher-order Coulomb breakup processes.

Thus, description of ^{8}B breakup process with nuclear and Coulomb breakup processes of both the E1 and E2 transitions and higher-order processes is a key to solve a puzzle recognized so far of the discrepancy between the values of $S_{17}(0)$ extracted from direct p-capture reactions, $S_{17}(0) = 22.1 \pm 0.6$ (expt) ± 0.6 (theo) eV b, and indirect ^{8}B dissociation experiments. In order to clarify the role of these components in ^{8}B breakup reaction in general, it will be necessary to carry out analyses to wider range of data, such as those measured at MSU [8] and GSI [6, 7], including the data on other quantities than that dealt with in the present work. Analysis of parallel momentum distribution of ^{7}Be-fragment after breakup of ^{8}B is particularly important, since the role of E2 component was determined from it [8]. Unfortunately, detailed information on the resolution and efficiency of the MSU experiment, which is necessary to carry out quantitative analysis of the data, is not available to the present authors. At this stage, therefore, we can quantitatively extract $S_{17}(0)$ with CDCC and the ANC method only from the ^{8}B breakup experiment done at RIKEN.

4. SUMMARY

The principal result of the present paper is the value of $S_{17}(0)$ of $20.9^{+1.0}_{-0.6}$ (theo) ± 1.8 (expt) eV b obtained by an analysis of the cross section of the ^{8}B breakup reaction ^{208}Pb(^{8}B, $p+^7$Be)^{208}Pb at 52 MeV/nucleon measured at RIKEN [5] by means of the method of continuum-discretized coupled-channels [10] (CDCC) combined with the

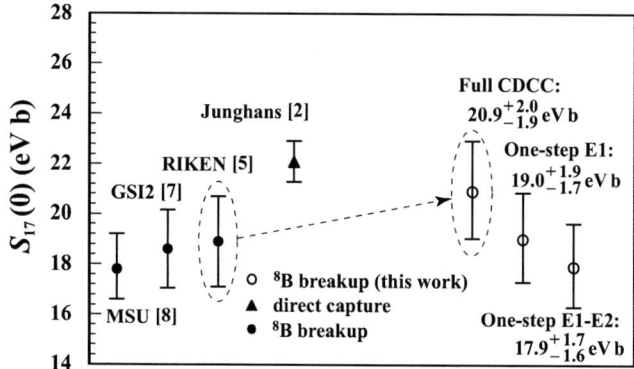

FIGURE 3. The values of $S_{17}(0)$ obtained from three-types of CDCC analysis of ^{8}B breakup combined with the ANC method are shown by the open circles. The error of each result is obtained by adding the theoretical and experimental errors in quadrature. The results are compared with $S_{17}(0)$ extracted from the precise measurement of direct capture (p,γ) cross section [2] (closed triangle) and those obtained from ^{8}B dissociation with first-order perturbation theory (closed circles).

asymptotic normalization coefficient (ANC) method [11]. The value is consistent with the one extracted from the precise measurement of the cross section of direct capture $^7\text{Be}(p,\gamma)^8\text{B}$, $S_{17}(0) = 22.1 \pm 0.6$ (expt) ± 0.6 (theo) eV b [2]. Calculations of $S_{17}(0)$ with some simplified assumptions are summarized in Fig. 3. The results of first-order iterative solutions of CDCC, designated as one-step CDCC, correspond to and agree with those of first-order perturbation theory in the previous works [5, 8]. The inclusion of the Coulomb quadrupole (E2) transitions, scaled by 0.7, in the one-step calculation decreases $S_{17}(0)$ by about 6%, and multistep processes increase it by about 20%. This shows the crucial importance of accurate description of the ^8B breakup process by CDCC including nuclear and Coulomb E1 and E2 transitions and all higher-order processes. It will be interesting to apply the method of the present paper to analyses of wider range of experiments such as those in Refs. [6, 7, 8]. It may, however, be that the three-body CDCC used in the present paper is not valid in general and the use of CDCC with four-body model such as the one in Ref. [20] becomes necessary.

ACKNOWLEDGMENTS

The authors would like to thank T. Motobayashi for helpful discussions and providing detailed information on the experiment. The authors also thank M. Kawai for fruitful discussions. This work has been supported in part by the Grants-in-Aid for Scientific Research of Monbukagakusyou of Japan.

REFERENCES

1. J. N. Bahcall *et al.*, Astrophys. J. **555**, 990 (2001) and references therein; J. N. Bahcall, Nucl. Phys. B Proc. Suppl. **118**, 77 (2003).
2. A. R. Junghans *et al.*, Phys. Rev. C **68**, 065803 (2003).
3. P. Descouvemont and D. Baye, Nucl. Phys. **A567**, 341 (1994).
4. P. Descouvemont, Phys. Rev. C **70**, 065802 (2004).
5. T. Motobayashi *et al.*, Phys. Rev. Lett. **73**, 2680 (1994); T. Kikuchi *et al.*, Phys. Lett. B **391**, 261 (1997); T. Kikuchi *et al.*, Eur. Phys. J. A **3**, 209 (1998).
6. N. Iwasa *et al.*, Phys. Rev. Lett. **83**, 2910 (1999).
7. F. Schümann *et al.*, Phys. Rev. Lett. **90**, 232501 (2003).
8. B. Davids *et al.*, Phys. Rev. Lett. **86**, 2750 (2001); Phys. Rev. C **63**, 065806 (2001).
9. H. Esbensen *et al.*, Phys. Rev. Lett. **94**, 042502 (2005).
10. M. Kamimura *et al.*, Prog. Theor. Phys. Suppl. **89**, 1 (1986); N. Austern *et al.*, Phys. Reports. **154**, 125 (1987).
11. A. M. Mukhamedzhanov and N. K. Timofeyuk, Yad. Fiz. **51**, 679 (1990) [Sov. J. Nucl. Phys. **51**, 431 (1990)]; H. M. Xu *et al.*, Phys. Rev. Lett. **73**, 2027 (1994).
12. M. Yahiro *et al.*, Phys. Lett. **141B**, 19 (1984); Y. Sakuragi *et al.*, Nucl. Phys. **A480**, 361 (1988); Y. Iseri *et al.*, Nucl. Phys. **A490**, 383 (1988); J. A. Tostevin *et al.*, Phys. Rev. C **66**, 024607 (2002).
13. K. Ogata *et al.*, Phys. Rev. C **67**, 011602(R) (2003).
14. J. A. Tostevin *et al.*, Phys. Rev. C **63**, 024617 (2001).
15. J. Mortimer *et al.*, Phys. Rev. C **65**, 064619 (2002).
16. K. Ogata *et al.*, Nucl. Phys. A **738c**, 421 (2004).
17. K. Ogata *et al.*, Phys. Rev. C **73**, 024605 (2006).
18. H. Esbensen and G. F. Bertsch, Nucl. Phys. **A600**, 66 (1996).
19. K. H. Kim *et al.*, Phys. Rev. C **35**, 363 (1987).
20. T. Matsumoto *et al.*, Phys. Rev. C **70**, 061601(R) (2004).

Proton resonance scattering of ^{7}Be

H. Yamaguchi*, A. Saito*, J.J. He*, Y. Wakabayashi*, G. Amadio*,
H. Fujikawa*, S. Kubono*, L.H. Khiem*, Y.K. Kwon*,†, M. Niikura*,
T. Teranishi**, S. Nishimura‡, Y. Togano§, N. Iwasa¶ and K. Inafuku¶

*Center for Nuclear Study (CNS), University of Tokyo, RIKEN campus, 2-1 Hirosawa, Wako, Saitama 351-0198, Japan
†Department of Physics, Chung-Ang University, Seoul 156-756, South Korea
**Department of Physics, Kyushu University, 6-10-1 Hakozaki, Fukuoka 812-8581, Japan
‡The Institute of Physical and Chemical Research (RIKEN), 2-1 Hirosawa, Wako, Saitama 351-0198, Japan
§Department of Physics, Rikkyo University, Tokyo 171-8501, Japan
¶Department of Physics, Tohoku University, Aoba, Sendai, Miyagi 980-8578, Japan

Abstract.
We have studied the proton resonance scattering of ^{7}Be by using a pure ^{7}Be beam produced at CRIB (CNS Radioactive Ion Beam separator; CNS stands for Center of Nuclear Study, University of Tokyo). The excitation function of ^{8}B was measured up to the excitation energy of 6.8 MeV, with the thick-target method. The excited states of ^{8}B higher than 3.5 MeV were not known by the past experiments. This proton elastic scattering is also of importance in relation with the ^{7}Be$(p,\gamma)^8$B reaction, which is a key reaction in the standard solar model.

Keywords: proton resonance scattering, astrophysical S-factor, solar neutrino problem
PACS: 25.40.Ny, 27.20.+n, 21.10.Hw, 26.65.+t

INTRODUCTION

The astrophysical S-factor $S_{17}(E)$ is one of the most important parameters in the standard solar model, and is defined as

$$S_{17}(E) = E\sigma_{17}(E)\exp(2\pi\eta), \tag{1}$$

where $\sigma_{17}(E)$ is the cross section of the ^{7}Be$(p,\gamma)^8$B reaction, and η is the Sommerfeld parameter. This S_{17} value at the solar energy is directly related to the flux of the ^{8}B neutrino. ^{8}B neutrinos are only less than 0.01 % of the total neutrinos emitted from the sun, but they are the majority of the detected neutrinos in most of the neutrino detectors such as Super-Kamiokande and Sudbury Neutrino Observatory (SNO). Due to this reason, $S_{17}(E)$ is regarded as an important factor for the solar neutrino problem in the standard solar model. Although great efforts were spent by many experimental groups [1], the experimental precision remains still around 10 %, because of the small reaction cross section. It is claimed that the determination of the S_{17} below 300 keV with a precision better than 5% may make a major contribution to our knowledge for the solar model. [2].

The existence of excited levels of ^{8}B may affect the determination of S_{17}. However, we do not have sufficient knowledge of the nuclear structure of ^{8}B. Only the lowest two states at 0.77 MeV and 2.32 MeV were clearly observed in many past experiments.

Another excited state around 3 MeV was observed as an unexpectedly wide resonance, and this was explained as a low-lying 2s state [3]. The reason why a 2s state appeared at such a low energy is also an interesting topic, and there are theoretical investigations [4, 5, 6]. This kind of wide states may affect the measurement of ^{7}Be$(p,\gamma)^8$B cross section even at very low energies (much less than 1 MeV). In the same measurement, an indication of 1^+ state at 2.8 MeV was also reported. On the other hand, in another recent experiment [7] they could not observe the 1^+ state at 2.8 MeV nor yet another 1^+ state at 1.5 MeV, the latter of which was theoretically proposed in [8]. The wide state was not directly observed, but they concluded that their spectrum is consistent if it is located at 3.5 MeV with a width of 4 MeV or more. Thus we intended to measure the resonances of ^{8}B, to evidently observe the 3.5 MeV resonance reported in the past measurement, and to explore the totally unknown region $E > 3.5$ MeV, where we may find new resonances. The "thick target method", by which we can measure proton elastic resonance scatterings, was suitable for our purpose.

The ^{7}Be$(p,\gamma)^8$B reaction and ^{8}B structure are important topics also in the nuclear synthesis. In a standard nuclear synthesis theory, the triple-α process is considered as the dominant process to pass over the stability gap at A=8. However, in a special environment such as a very metal-poor, high temperature star, the proton or α capture process of ^{7}Be might play a significant role, and thus how they compete each other is an interesting problem. We are able to obtain information about these processes by measuring the ^{7}Be$(p,p)^7$Be elastic scatterings.

METHOD

The measurement was performed at CRIB, which is an RI beam facility of CNS, University of Tokyo [9, 10]. CRIB can produce RI beams with the in-flight method, using primary heavy-ion beams from the AVF cyclotron of RIKEN (K=70). Fig. 1 shows the whole structure of CRIB. The primary beam used in this measurement was ^{7}Li^{3+} of 8.76 MeV/u, with the beam current of about 100 pnA. The RI-beam production target was pure hydrogen gas, which was in an 8 cm-long cell, 760 Torr, and at room temperature ($\sim$ 300 K).

At CRIB, the produced secondary beam (^{7}Be^{4+} in this measurement) can be separated from other isotopes in two stages. The first stage is by two analysing dipole magnets, and the second is by a Wien filter. After the production, secondary beams go through the two dipoles, both of which have a maximum magnetic rigidity of 1.1 T m. There is a momentum dispersive focal plane, called F1, in between the two dipoles. At this focal plane, the momentum of the particle can be selected (resolution $\Delta p/p$ is 1/850) by a movable slit. After the second dipole magnet, the beam is achromatically focused onto the focal plane, called F2. In the Wien filter section, secondary beams can be separated by a horizontal electric field and a vertical magnetic field, according to their velocities. The electric field is produced by applying high voltages of opposite polarities at two parallel plates, which are 1.5m-long and 8 cm distant from each other. The maximum applicable voltages are ± 150–200 kV, corresponding to the electric field of 38–50 kV/cm. The beam energy used in this measurement was 53.8 MeV, which enabled us to measure events with the center-of-mass energy up to 6.7 MeV.

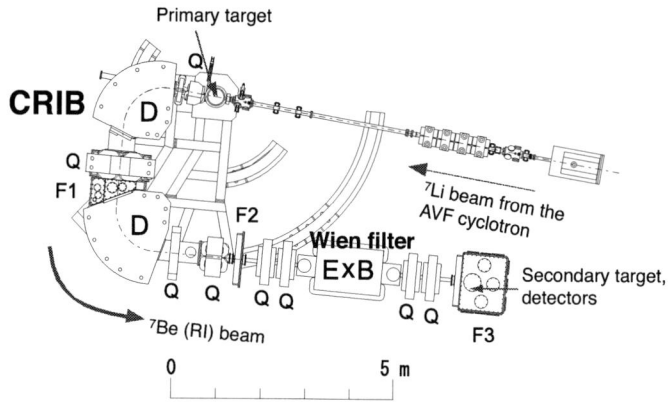

FIGURE 1. CRIB overview. D, Q, and F represent a dipole magnet, a quadrupole magnet, and a focal plane, respectively.

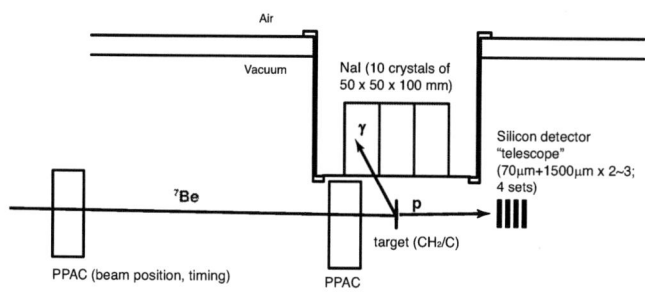

FIGURE 2. Setup at the experimental chamber (schematic).

We have used a standard experimental method for the proton elastic resonance scattering, well-established at CRIB [11]. A main feature of this method is the thick target, which makes it possible to measure the cross section of various excitation energies at the same time, and so is suitable for the elastic resonance scattering.

The targets and detectors for the scattering experiment were in a chamber located at the downstream of the Wien filter. Fig. 2 shows a schematic setup in the experimental chamber. Two PPACs (Parallel-Plate Avalanche Counters [12]) measured the timing and position of the incoming ^{7}Be beam. The timing information was used for making event triggers, and also for the particle identification by the time-of-flight (TOF) method. The beam position and its incident angle at the target were determined by extrapolating the positions measured by the PPACs. The targets were foils of 39 mg/cm^2-thick polyethylene (CH$_2$), and 54 mg/cm^2-thick carbon. Carbon foils were used for evaluating backgrounds by carbon in the polyethylene target. These target foils were thick enough to stop all the ^{7}Be beam. Multi-layered silicon detector sets (called ΔE-E telescopes) were used for measuring the energy and angular distribution of the recoil protons. They were placed ~23 cm distant from the target, and they covered scattering angle of up to 45 degrees in the laboratory frame. Each ΔE-E telescope consisted of a ΔE counter and two or three E counters, all of which had an area of 50 mm x 50 mm. The ΔE counters were 60–75 μm thick and were divided into 16 strips for both sides. The E counters were 1.5 mm thick and were put in rear of the ΔE counters. The recoil proton energy was 23 MeV at maximum at zero degrees, and a proton having this energy stopped in the E counters. With these ΔE-E telescopes, we identified recoil protons from other particles. NaI detectors were used for measuring 429 keV gamma rays from inelastic scatterings, $p(^7Be, ^7Be^*)p$. Each NaI crystal has a geometry of 50 mm x 50 mm x 100 mm, and ten crystals were used to cover ~20% of the total solid angle.

Compared to the past measurements with similar methods [3, 7], this measurement has three major advantages. The first one is the energy range. We used a high-energy ^7Be beam, and we could measure up to 6.8 MeV in the excitation energy of ^8B. The second is the covered scattering angle. We measured data with almost full coverage between 0 and 45 degrees in laboratory frame, and thus we obtained complete information about the angular distribution. The final advantage is that we used the NaI detectors to evaluate the inelastic scattering events. The contribution of the inelastic scattering events should be taken into account for the excitation function.

RESULT

Beam production

We used full-stripped lithium ($^7Li^{3+}$) beam from AVF cyclotron to obtain a high energy secondary beam. Although this was the first trial of $^7Li^{3+}$ beam in RIKEN, it was successfully operated and used in this measurement. Fig. 3 shows the time of flight and energy of the secondary beams detected at near the F2 focal plane. The time of flight was determined by the timing of PPAC signal against the RF signal of the AVF cyclotron. $^7Be^{4+}$ had the highest energy among the other beams, and the beam purity (the number ratio of $^7Be^{4+}$ to total) was 56%. The largest background was from the primary beam, $^7Li^{3+}$. The other particles, such as proton, deuteron, alpha, and ^6Li, are also seen in this figure, of which positions agree well with our energy-loss calculations. The intensity of the produced $^7Be^{4+}$ beam was 5×10^5 particles per second at F2. For low-energy and relatively light ions as in this experiment, the Wien filter worked efficiently. The Wien filter was operated at high voltages of ±39 kV with the corresponding magnetic field of 0.025 T. With these conditions, we had a beam purity of almost 100% at the experimental target.

Measured recoil protons

Fig. 4 shows the relation of the energy deposit in the first layer and total energy, for the particles measured with one of the silicon ΔE-E telescopes. Comparing these with energy-loss calculations, we performed a clear particle identification, as indicated

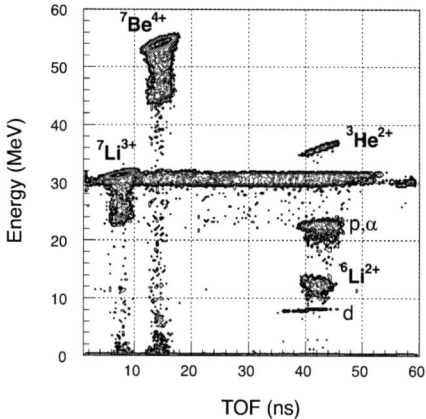

FIGURE 3. Secondary beams detected at the F2 focal plane.

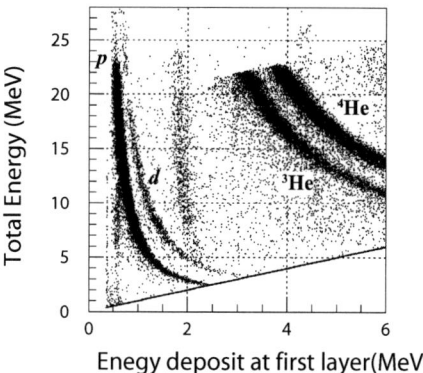

FIGURE 4. ΔE-E relationship of the detected particles with a silicon telescope.

in the figure. Proton was the most frequently detected particle, but we also detected considerable amount of ^{3}He and ^{4}He, which should be originated from ^{7}Be. Using these data, we were able to select proton events from other particles clearly.

The proton energy at the reaction point can be calculated from the measured proton energy, by considering the energy loss in the target. Then the proton energy E_p can be converted into the center-of-mass energy E_{cm}, with the following formula,

$$E_{cm} = E_p \frac{m_1 + m_7}{4 m_7 \cos^2 \theta}, \quad (2)$$

where m_1 and m_7 are the masses of proton and ^{7}Be, θ is the scattering angle. The excitation energy is the sum of center-of-mass energy of the elastic resonance scattering and the proton threshold energy,

$$E_{ex} = E_{cm} + E_{th}. \qquad (3)$$

By calculating the cross section from the counted numbers of the proton events, an excitation function will be obtained. The analysis is still going on, and the excitation function will be presented in a future publication.

ACKNOWLEDGMENTS

We are grateful to RIKEN accelerator staff for their help. This work was supported by the Grant-in-Aid for Young Scientists (B) (Grant No. 17740135) of JSPS.

REFERENCES

1. C. Angulo, *et al.*, *Nucl. Phys. A* **656**, 3–187 (1999).
2. E. Adelberger, *et al.*, *Rev. of Mod. Phys.* **70**, 1265–1291 (1998).
3. V. Gol'dberg, G. Rogachev, M. Golovkov, V. Dukhanov, I. Serikov, and V. Timofeev, *JETP Lett.* **67**, 1013–1017 (1998).
4. A. van Hees, and P. Glaudemans, *Z. Phys. A* **314**, 323 (1983).
5. A. van Hees, and P. Glaudemans, *Z. Phys. A* **315**, 223 (1984).
6. F. Barker, and A. Mukhamedzhanov, *Nucl. Phys. A* **673**, 526–532 (2000).
7. G. Rogachev, J. Kolata, F. Becchetti, P. DeYoung, M. Hencheck, K. Helland, J. Hinnefeld, B. Hughey, P. Jolivette, L. Kiessel, . M. L. H.-Y. Lee, T. O'Donnell, G. Peaslee, D. Peterson, D. Roberts, P. Santi, , and S. Shaheen, *Phys. Rev. C* **64**, 061601(R) (2001).
8. A. Csótó, *Phys. Rev. C* **61**, 024311 (2000).
9. S. Kubono, Y. Yanagisawa, T. Teranishi, S. Kato, T. Kishida, S. Michimasa, Y. Ohshiro, S. Shimoura, K. Ue, S. Watanabe, and N. Yamazaki, *Eur. Phys. J.* **A13**, 217 (2002).
10. Y. Yanagisawa, S. Kubono, T. Teranishi, K. Ue, S. Michimasa, M. Notani, J. He, Y. Ohshiro, S. Shimoura, S. Watanabe, N. Yamazaki, H. Iwasaki, S. Kato, T. Kishida, T. Morikawa, and Y. Mizoi, *Nucl. Instrum. Methods Phys. Res., Sect. A* **539**, 74–83 (2005).
11. T. Teranishi, S. Kubono, S. Shimoura, M. Notani, Y. Yanagisawa, S. Michimasa, K.Ue, H.Iwasaki, M. Kurokawa, Y. Satou, T. Morikawa, A. Saito, H. Baba, J. Lee, C. Lee, Zs.Fulop, and S. Kato, *Phys. Lett. B* **556**, 27–32 (2003).
12. H. Kumagai, A. Ozawa, N. Fukuda, K. Sümmerer, and I. Tanihata, *Nucl. Instrum. Methods Phys. Res., Sect. A* **470**, 562 (2001).

Coulomb Dissociation of ^{27}P for Study of ^{26}Si(p,γ)^{27}P Reaction

Y. Togano*, T. Gomi*, T. Motobayashi[†], Y. Ando*, N. Aoi[†], H. Baba*,
K. Demichi*, Z. Elekes**, N. Fukuda[†], Zs. Fülöp**, U. Futakami*,
H. Hasegawa*, Y. Higurashi[†], K. Ieki*, N. Imai[†], M. Ishihara[†],
K. Ishikawa[‡], N. Iwasa[§], H. Iwasaki[¶], S. Kanno*, Y. Kondo[‡], T. Kubo[†],
S. Kubono[∥], M. Kunibu*, K. Kurita*, Y. U. Matsuyama*, S. Michimasa[∥],
T. Minemura[†], M. Miura[‡], H. Murakami*, T. Nakamura[‡], M. Notani[∥],
S. Ota[††], A. Saito*, M. Serata*, S. Shimoura[∥], T. Sugimoto[‡], E. Takeshita*,
S. Takeuchi[†], K. Ue[¶], K. Yamada*, Y. Yanagisawa[†], K. Yoneda[†] and
A. Yoshida[†]

Department of Physics, Rikkyo University, Tokyo 171-8501, Japan
[†]*RIKEN, Saitama 351-0198, Japan*
**ATOMKI, 4001 Debrecen, Hungary*
[‡]*Department of Physics, Tokyo Institute of Technology, Tokyo 152-8551, Japan*
[§]*Department of Physics, Tohoku University, Miyagi, 980-8578, Japan*
[¶]*Department of Physics, University of Tokyo, Tokyo 113-0033, Japan*
[∥]*Center for Nuclear Study (CNS), University of Tokyo, Saitama 351-0198, Japan*
[††]*Department of Physics, Kyoto University, Kyoto 606-8502, Japan*

Abstract. The Coulomb dissociation of the proton-rich nuclei ^{27}P was studied experimentally using a ^{27}P beam at 57 MeV/nucleon with a lead target. The γ decay widths of the low-lying excited states in ^{27}P were deduced. The resonant capture reaction rate of stellar ^{26}Si(p,γ)^{27}P through these states was estimated using the measured γ decay width. The astrophysical implications derived from the estimated reaction rate is discussed.

Keywords: Reaction induced by unstable nuclei, Coulomb excitation, Nucleosynthesis in novae, supernovae and other explosive environments, $20 \leq A \leq 38$
PACS: 25.60.-t, 25.70.De, 26.30.+k, 27.30.+t

INTRODUCTION

The nucleosynthesis process of ^{26}Al has attracted much attention in connection with the cosmic evolution such as the X-ray bursts and novae. This was triggered by the cosmic γ ray observation, where the characteristic γ rays of ^{26}Al were found in the interstellar medium [1]. The information obtained from this observation is quite useful for investigating ongoing nucleosynthesis processes, since the lifetime of ^{26}Al (10^6 yr) is much shorter than the timescale of the cosmic evolution (10^{10} yr). For this purpose a lot of theoretical studies have attempted to reproduce the ^{26}Al γ ray yield, even assuming a variety of stellar sites as the origin [2]. Nevertheless, they were not successful, and the primary source of the uncertainty is the insufficient of nuclear database, especially for unstable nuclei including ^{26}Al and around.

In most of the stellar sites, the isotope ^{26}Al is produced via the reaction sequence ^{24}Mg(p,γ)^{25}Al(β^+ ν)^{25}Mg(p,γ)^{26}Al. In the sites which have high stellar density and temperature, the proton capture on ^{25}Al becomes faster than its β decay and then produces the unstable nuclei ^{26}Si which β decays to ^{26}Al. The ground state of ^{26}Al β decays to the first excited state in ^{26}Mg, giving rise to a 1.8 MeV γ ray from the deexcitation. The isomeric level ($T_{1/2}$ = 6.3 s) at E_x = 228 keV β decays predominantly to the ^{26}Mg ground state and therefore, is of no relevance to the astrophysical observations. However, in the condition whose temperature is higher than 0.4 GK, the thermal equilibrium is achieved among low lying states in ^{26}Al [3]. The β decay of ^{26}Si feeds the only isomeric state and thus affects the synthesis of ^{26}Al through the equilibration. Therefore, ^{26}Si destruction by proton capture is important to determine the amount of the ground state in ^{26}Al produced on the basis of the equilibrium.

Under such high temperature conditions the capture reaction rate, $N_A \langle \sigma v \rangle$, is expected to be dominated by the resonant capture via the unbound excited states in ^{27}P. Resonant capture reaction rates for isolated, narrow resonances are given by

$$N_A \langle \sigma v \rangle = 1.54 \times 10^{11} (\mu T_9)^{-3/2} \omega\gamma \exp\left[-11.605 \frac{E_R}{T_9}\right]. \quad (1)$$

with N_A Avogadro's number, μ the reduced mass in the atomic mass unit, T_9 the temperature in unit of GK, $\langle \sigma v \rangle$ the thermally averaged nuclear cross section, and E_R the resonance energy in MeV [4]. The resonance strength $\omega\gamma$ (in units [MeV]) is given by

$$\omega\gamma = \frac{2J+1}{(2J_p+1)(2J_T+1)} \frac{\Gamma_p \Gamma_\gamma}{\Gamma_{tot}}, \quad (2)$$

where J is the spin of the compound nucleus, and J_p and J_T are the spins of the proton and target, respectively. Γ_p and Γ_γ are the proton widths and the γ decay widths of the resonance. The total width Γ_{tot} is the sum of the partial widths.

The Coulomb dissociation at intermediate energies is an alternative method to study the radioactive capture reactions of astrophysical interests at low energies [5, 6]. The reaction process can be regarded as a photodisintegration by virtual photons, which is essentially the inverse of the radiative capture reaction [7]. The extracted electromagnetic transition probabilities directly determine the Γ_γ values.

The present paper reports on an experimental study of the dissociation of ^{27}P in the Coulomb field of a Pb nucleus. The Γ_γ values of the low-lying excited states in ^{27}P, which have the largest contributions to the resonant capture rate of ^{26}Si(p,γ)^{27}P, has been extracted.

EXPERIMENT

The experiment was performed at the RIKEN Accelerator Research Facility. A ^{27}P beam was produced via the projectile fragmentation of a 115-A-MeV ^{36}Ar beam incident on a 460 mg/cm^2 thick Be target, and was separated by the RIKEN Projectile-fragment Separator (RIPS) [8]. The typical ^{27}P intensity was about 2800 counts per second and the energy was 57 A MeV with an energy spread of about 1.5%. It bombarded a 125 mg/cm^2

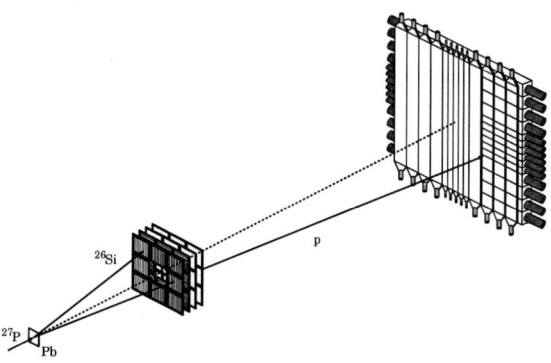

FIGURE 1. A schematic view of the experimental setup. The entire system is in vacuum.

thick lead target. The isotopic purity of ^{27}P in the secondary beam is about 1%. The major contributions are ^{26}Si, ^{25}Al, and ^{24}Mg. The particle identification for secondary beams was performed event-by-event by means of the time-of-flight (TOF) -ΔE method using a 0.5 mm-thick plastic scintillator (F2PL) located at the second focal plane of the RIPS. Two sets of parallel plate avalanche counters (PPACs) were also placed at the final focal plane of the RIPS to extrapolate the position and angle of the beam at the target.

Nuclei ^{27}P were excited on the lead target and then disintegrated to ^{26}Si and proton. Figure 1 shows the detector system of the breakup products.

The emission angles of these products were measured at a position sensitive silicon telescope located at 48 cm downstream of the target. The silicon telescope consists of four layers of detectors with 0.5 mm thickness. Each layer was of eight silicon detectors with 50×50 mm^2 effective areas on 56×56 mm^2 frames. The eight detectors in a layer formed a 3×3 matrix with a hole in the center. The count rate in the telescope was suppressed to 3000 counts per second by introducing the hole whereas the total beam rate was about 3×10^5 counts per second. The detectors in the first and second layers have 5 mm wide strip electrodes, which enable to measure the hit positions of the products. The energy of the ^{26}Si was also measured by the silicon telescope.

The energy of the proton, which penetrated the silicon telescope, was determined with a plastic scintillator hodoscope placed at the 2.8 m downstream of the target by measuring the TOF. The hodoscope with an active area of 1×1 m^2, consists of thirteen 5 mm thick ΔE- and sixteen 60 mm thick E-plastic scintillators. The outgoing proton was stopped in the E counters after passing through the ΔE counters.

The relative energy E_{rel} between ^{26}Si and proton was extracted by combining the positions and energies of the products. The energy E_{rel} corresponds to the center-of-mass energy, and thus the excitation energy of ^{27}P E_x can be expressed as,

$$E_x = E_{rel} + E_s, \qquad (3)$$

where E_s represents the separation energy for the decay channel (0.859 MeV for ^{27}P). Therefore, by measuring the E_{rel} the excitation energy E_x can be determined.

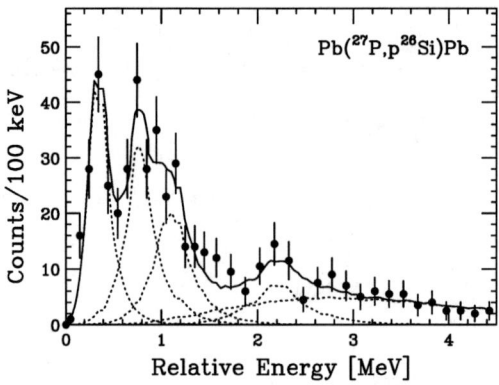

FIGURE 2. Relative energy spectrum of the ^{27}P breakup on Pb (filled circle). The data was fitted by the detector responses simulated using GEANT4 code [9]. The dashed curves and solid curve represents the each components and sum of the elements.

RESULTS AND DISCUSSIONS

The relative energy spectrum is shown in Fig. 2. The filled circles represent the experimental data. The solid curve represents the best fit with five contributions shown by the dashed curves. The detector responses were simulated by the GEANT4 code [9]. The peak at 0.31 MeV corresponds to the known first excited state at 1.2 MeV in ^{27}P. The bump at around 1 MeV may be respectively due to the known second excited state at 1.6 MeV and an unknown one at 2.0 MeV. The peak at 2.0 MeV is also an unknown one at 3.2 MeV. The direct breakup component, which distributes from 0.8 MeV to 4.5 MeV, corresponds to the non-resonant proton capture process. We assumed that the non-resonant capture is dominated by the E1 transition and the astrophysical S-factor is independent of the energies.

The Coulomb dissociation cross section for the first excited state in ^{27}P was determined to be 14.8 ± 2.7 mb from the yield of the peak. The error includes the statistical one and ambiguity of the detection efficiency. Supposing the spin and parity of the state is $3/2^+$ from the level scheme of the mirror nucleus ^{27}Mg [10], the transition between the first excited state and ground state ($1/2^+$) is induced by the M1 and E2 multipolarities. This means that the Γ_γ is a sum of a E2 component ($\Gamma_\gamma(E2)$) and a M1 component ($\Gamma_\gamma(M1)$). Since the Coulomb dissociation is highly sensitive to the E2 component compared with the M1 transition [11], the experimental cross section is expected to be exhausted by the E2 excitation. $\Gamma_\gamma(E2)$ is determined to be $(2.8 \pm 0.5) \times 10^{-5}$ eV form the measured cross section. To deduce the $\Gamma_\gamma(M1)$, the E2/M1 mixing ratio δ of the first excited state in ^{27}P was estimated by the combination of the known mixing ratio [12] of the mirror nucleus ^{27}Mg and the double ratio R defined as $R = \delta_{T_z+}/\delta_{T_z-}$, where δ_{T_z+} and δ_{T_z-} represent the mixing ratios of the mirror pair. To estimate the ratio R, we performed shell model calculations using the USD effective interaction [13] with conventional values of the effective charges in sd-shell region, $e_p = 1.3e$ and $e_n = 0.5e$ [14]. The reliability of the R calculated by the shell model was evaluated to be 60% (2σ) by

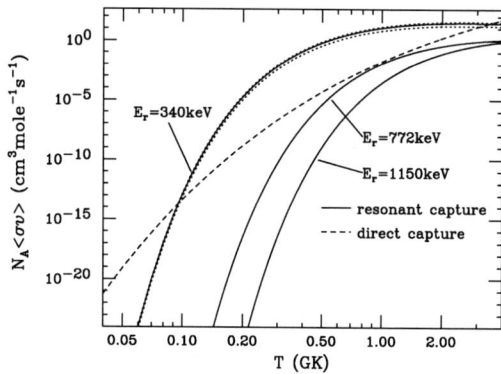

FIGURE 3. The reaction rate of the ^{26}Si(p,γ)^{27}P as a function of temperature of stars obtained in present study. The solid line represent the resonant capture component of each excited state and dotted line shows the margin of its error. Dashed curve denotes the direct capture component estimated by J. A. Caggiano et al [15].

comparing with the experimentally known values. From the calculated R and the known δ of ^{27}Mg, the mixing ratio for ^{27}P was estimated to be 0.020 ± 0.012. The $\Gamma_\gamma(M1)$ was derived from the mixing ratio and thus the total Γ_γ estimated to be $(3.7 \pm 2.2) \times 10^{-3}$ eV. The error includes the experimental error and also ambiguity from the estimation of M1 contribution. This value is consistent with the previous estimation [15].

The cross sections for the second and third excited states were obtained to be 18.6 mb and 13.4 mb, respectively. Supposing the spin and parity of the states is 5/2$^+$ from the mirror analogy [10], their Γ_γ values were determined directly from the experimental cross sections alone because the transitions to the ground state are induced by almost pure E2 transitions. They were determined to be $(1.82 \pm 0.28) \times 10^4$ eV and $(3.77 \pm 0.74) \times 10^4$ eV, respectively.

The reaction rate of the ^{26}Si(p,γ)^{27}P was calculated using the extracted Γ_γ values. Fig. 3 shows the temperature dependence of the reaction rate. The resonance parameters other than Γ_γ were adopted from Caggiano et al. [15]. The solid and dotted curves represent the present result of the reaction rate and range of its error. The dashed line denotes the direct capture component of reaction rate calculated by J. A. Caggiano et al. [15] based on a shell model. This figure indicates that the resonant capture via the first excited state is the dominant process above 0.1 GK, the temperature region of novae and X-ray bursts temperature [16], and through the second and third excites states give negligible contributions.

The competition between ^{26}Si(p,γ)^{27}P reaction and ^{26}Si β decay can be discussed by using the extracted reaction rate. The competition depends on the density, temperature, and mass fraction of proton in stars. The solid line in fig. 4 represents the temperature and density condition for which (p,γ) reaction and the competing β decay are of equal strength. The dot-dashed lines show the error of the present estimate. The solid curve were calculated using our result and assuming a hydrogen mass fraction of $X_H = 0.5$. In the region above the solid curve the proton capture reaction dominates, while below

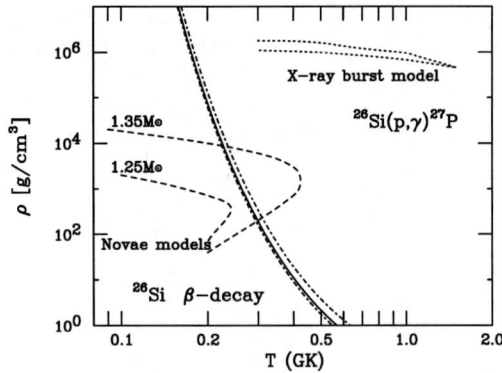

FIGURE 4. Temperature-density boundary (solid line) at which the ^{26}Si(p,γ)^{27}P reaction and the competing β decay are of equal strength assuming the hydrogen mass fraction of $X_H = 0.5$. Dot-dashed curves represents the range of error. The dashed curves and dotted line indicate T-ρ profiles of two novae sequences and an X-ray burst, respectively.

the solid line the nuclei ^{26}Si are exhausted by β decay. The dashed lines denotes temperature-density profiles for the two novae sequences, whose masses are 1.25 M$_\odot$ and 1.35 M$_\odot$ [17, 18], and the dotted line shows one for an X-ray burst [19] of a 1.4 M$_\odot$ neutron star. The novae sequence evolves in time from larger to smaller densities, while the one for X-ray burst evolves from smaller to larger densities. It can be seen that the (p,γ) reaction dominates in the heavy novae when they are at around the peak temperature, whereas the reaction hardly occurs in the light novae. On the other hand, the X-ray burst is dominated by the (p,γ) reaction in their all phases.

REFERENCES

1. R. Diehl et al., *Astron. & Astrophys. Suppl. Ser.* **97**, 181 (1993).
2. G. Meynet et al., *Astrophys. J. Suppl.* **92**, 441 (1994).
3. A. Coc et al., *Phys Rev. C* **61**, 015801 (1999).
4. W. A. Fowler et al., *Ann. Rev. Astron. Astrophys.* **5**, 525 (1967).
5. T. Motobayashi et al., *Phys. Rev. Lett.* **73**, 2680 (1994).
6. N. Iwasa et al., *Phys. Rev. Lett.* **83**, 2910 (1999).
7. G. Baur et al., *Nucl. Phys. A* **458**, 188 (1986).
8. T. Kubo et al., *Nucl. Instr. and Meth. B* **70**, 309 (1992).
9. S. Agostinelli et al., *Nucl. Inst. and Meth. A* **506**, 250 (2003).
10. P. M. Endt, *Nucl. Phys. A* **521**, 1 (1990).
11. K. Kanganke et al., *Phys. Rev. C* **49**, 1771 (1994).
12. M. J. A. de Voigt et al., *Nucl. Phys. A* **186**, 365 (1972).
13. B. H. Windenthal et al., *Prog. Part. Nucl. Phys.* **11**, 5 (1984).
14. B. A. Brown et al., *Annu. Rev. Nucl. Sci.* **38**, 29 (1988).
15. J. A. Caggiano et al., *Phys. Rev. C* **64**, 025802 (2001).
16. C. Iliadis et al., *Astrophys. J.* **524**, 434 (1999).
17. J. Jose et al., *Astrophys. J.* **520**, 347 (1999).
18. C. Iliadis et al., *Astrophys. J. Supp.* **142**, 105 (2002).
19. O. Koike et al., *Astron. & Astrophys.* **342**, 464 (1999).

1.12 X-RAYS, GAMMA RAYS and COSMIC RAYS

Studies of Isotopic Abundances through Gamma-Ray Lines

Roland Diehl

Max Planck Institut für extraterrestrische Physik, D-85748 Garching, Germany

Abstract. Cosmic gamma-ray lines convey isotopic information from sites of nucleosynthesis and from their surrounding interstellar medium. With recent space-borne gamma-ray spectrometers of high resolution (INTEGRAL, RHESSI), new results have been obtained for ^{44}Ti from the Cas A core-collapse supernova, from long-lived radioactive ^{26}Al and ^{60}Fe, and from positron annihilation in our Galaxy: ^{44}Ti ejection from Cas A may be on the low side of previously-reported values, and/or at velocities >7000 km s^{-1}. ^{26}Al sources apparently share the Galactic rotation in the inner Galaxy, and thus allow to estimate a total mass of ^{26}Al in the Galaxy of 2.8 M$_\odot$ from the measured flux. The ^{60}Fe production in massive stars appears lower than predicted by standard models, as constrained by the recent, though marginal, ^{60}Fe detections. Positron annihilation in the Galaxy shows a remarkable bulge component, which is difficult to understand in terms of nucleosynthetic production of the positrons.

Keywords: stars: late stages of evolution; nucleosynthesis; nuclear processes (astrophysics); Interstellar matter: Milky Way
PACS: 97.60.-s; 97.10.Cv ; 26.20.+f; 26.30.+k; 95.30.Cq;98.38.-j

COSMIC GAMMA-RAY LINE OBSERVATIONS

Nucleosynthesis in cosmic sources can be observed through a variety of, mostly indirect, measurements; examples are stellar photospheric absorption lines, or mass spectrometry of meteoritic inclusions. Gamma-rays from radioactive by-products of nucleosynthesis ejecta provide a rather direct measurement, in comparison, as their decay gamma-ray measurements with satellite-borne telescopes provide direct isotopic constraints to the physics of nuclear burning regions inside these sources. Yet, the technique of gamma-ray telescopes is complex, and less precise than the alternatives for cosmic abundance measurements, mainly from two reasons: Spatial resolutions of ~degrees and signal-to-background ratios of ~% restrict contributions from gamma-ray astronomy to nearby sources in the Galaxy (up to 10 Mpc for SNIa ^{56}Ni radioactivity). There are, however, advantages to gamma-ray astronomical data: They provide *isotopic* information, are unaffected by physical conditions in/around the source such as temperature or density, and gamma-rays are nearly un-attenuated along the line-of-sight due to their penetrating nature (attenuation length ~few g cm^{-2}).

In recent years, the launches of high-resolution solid state detectors (RHESSI [1], INTEGRAL/SPI [2,3,4]) into space have added a new quality to this field (Fig.1): Spectroscopy of these nuclear lines allows to better identify lines above background,

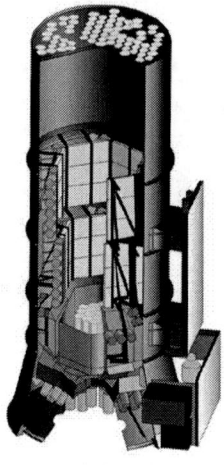 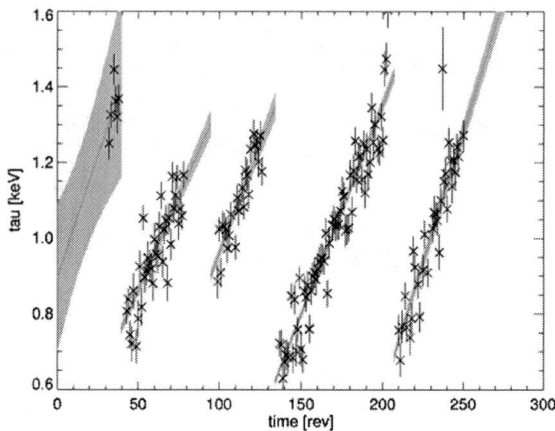

FIGURE 1. The SPI instrument[3,4] is built around a 19-element camera of Ge detectors; incident gamma-rays will cast a characteristic shadow onto this camera due to the coded mask, which allows to discriminate sources against instrumental background (left). Cosmic-ray bombardement in space destroys the charge-collection properties of the Ge detectors. Periodic annealing cures these defects, such that the high spectral resolution can be maintained over years. Shown is the degradation parameter τ versus time in units of INTEGRAL's 3-day orbits[2] (right)

and, astrophysically most importantly, is capable to constrain the kinematics of the isotopes in the gamma-ray emission region. This has been successfully demonstrated with solar-flare data[5,6] showing the signatures of beaming in the accelerated-particle flow, and will be shown below to constrain supernova nucleosynthesis ejecta velocities (rather than envelope ejecta velocities), as well as velocities in the turbulent interstellar medium around massive stars. The special processes involved in positron annihilation gamma-ray production[7] shape the annihilation gamma-ray line, which can be exploited to constrain ionization state and temperature of the interstellar gas in the annihilation region.

Candidate sources of characteristic gamma-ray lines are supernovae and novae, but also the winds from massive stars[8,9]. Freshly-produced nuclei from explosive layers near the surface of compact stars and from stellar interiors are ejected into interstellar space, where gamma-rays from their decay can be observed directly. Radioactivity which is still embedded within the source (i.e. below an envelope more massive than several g cm^{-2}) will lead to X-ray/low-energy gamma-ray continuum emission due to Comptonization, and therefore, e.g., early radioactive energy in novae and supernovae is completely thermalized and available for indirect bolometric measurements only. Once ejected, long-lived radio-isotopes will decay in interstellar space; their characteristics will therefore be influenced by the state of circumstellar and interstellar gas around the nucleosynthesis source. Young supernova remnants will lead to modified decay histories for ^{44}Ti (decay time 89 years, decay by electron capture only) through its ionization state; longer-lived isotopes such as ^{26}Al ($\tau \sim 1.04$ Myrs) will reflect the kinematic properties in an interstellar medium around massive star sources, which is otherwise hard to study due to its hot and diluted nature. Positrons from decays of isotopes on the proton-rich side of the valley of isotopic stability will typically be produced at MeV energies[7]; their propagation before their annihilation is more complex, being controlled by density and magnetic field morphologies near the nucleosynthesis sources[7,10].

In this paper, we will discuss what has been learned from the first mission years of RHESSI and INTEGRAL with respect to nucleosynthesis sources in our Galaxy.

LESSONS FROM RECENT RESULTS

Supernovae and ^{44}Ti

The light curve of supernovae is powered by ^{56}Ni radioactivity; gamma-ray lines from this decay are therefore the prominent target of all gamma-ray astronomy experiments. Unfortunately, the bright supernovae of type Ia (thermonuclear) are rarely close enough to provide a chance of detecting their ^{56}Ni decay chain gamma-rays [11], and the only two opportunities (SN1991T and SN1998bu) led to conflicting results, with a marginal detection of SN1991T [12] and an upper limit for SN1998bu[13], the former above, the latter below expectations from models. Core collapse supernovae produce much less radioactive ^{56}Ni, on average, than thermonuclear supernovae. Moreover this radioactivity is burried inside a massive stellar envelope. Therefore such supernovae would have to be much closer than SNIa to be detectable as Ni decay gamma-ray line sources. SN1987A was such an exceptional event. But nuclear burning in core collapse supernovae, through α-rich freeze-out from nuclear statistical equilibrium, is expected to produce long-lived ^{44}Ti ($\tau\sim86$ y) in its inner region close to the compact remnant star [14]. The ^{44}Ti which manages to escape fallback onto the remnant star will decay in the dilute young supernova remnant, thus should be observable from core-collapse events throughout

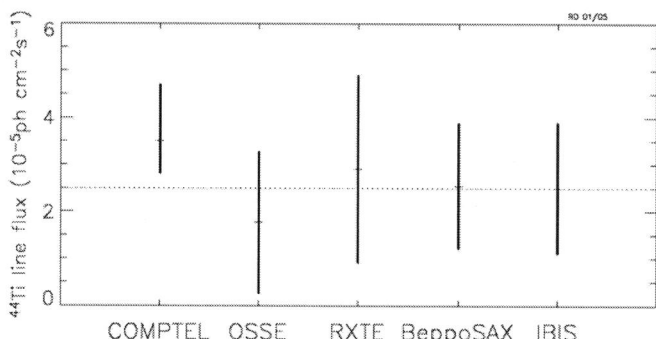

FIGURE 2. ^{44}Ti decay gamma-ray measurements from different experiments for the Cas A supernova remnant yield a flux of ~2.5 10^{-5} ph cm^{-2} s^{-1}, corresponding to a ^{44}Ti mass of 1.6 10^{-4} M$_\odot$. (Figure data updated from [17])

the Galaxy. At a rate of ~2 such events per century, a few detected ^{44}Ti sources thus would be expected in the surveys of the Compton Observatory and of INTEGRAL. Presently just one such soure is firmly established: The ~340-year old Cas A supernova remnant has now been measured with three instruments [15,16,17] in different lines originating from ^{44}Ti decay (Fig.2). Although each measurement carries its own systematic uncertainties from large instrumental backgrounds, the ^{44}Ti amount derived from these data for the Cas A supernova is 1.6 ±0.3±0.3 10^{-4} $M_\odot$. This is within the range of expectations, especially if a possible reduction of the decay rate due to ionization in the young supernova remnant is also considered.

The fact that Cas A has been seen as a source of ^{44}Ti, while in the inner Galaxy, where most of the core collapse events should appear, none of such sources could be found[17,18], suggests that Cas A is not representative for the 'average' core collapse event in our Galaxy, and that spherical asymmetries may be needed to lead to substantial ejected amounts of ^{44}Ti [19].

Diffuse Gamma-Ray Line Emission in the Galaxy

^{26}Al Radioactivity

Radioactive decay from the long-lived ^{26}Al isotope (τ~1 Myr) is driven by the weak interaction, the excited state of the ^{26}Mg daughter is very short, and therefore environmental parameters do not play a role: ^{26}Al gamma-rays directly reflect the kinematic properties of the decaying nuclei. From previous studies with the imaging Compton telescope instrument aboard the Compton Observatory (1991-2000 [20,21,22]), ^{26}Al emission has been mapped all along the plane of the Galaxy. The conclusions from these measurements were that massive stars dominate ^{26}Al

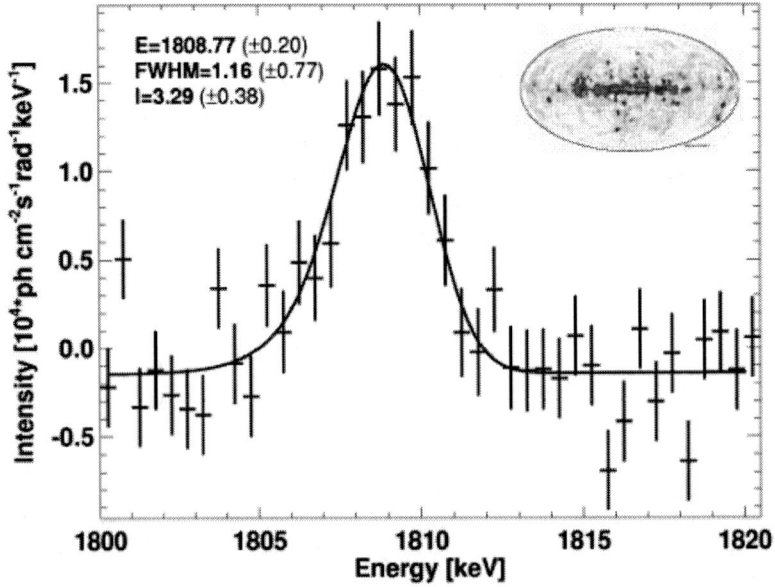

FIGURE 3. ^{26}Al line measurement of the SPI instrument, from 2 years of data [29].

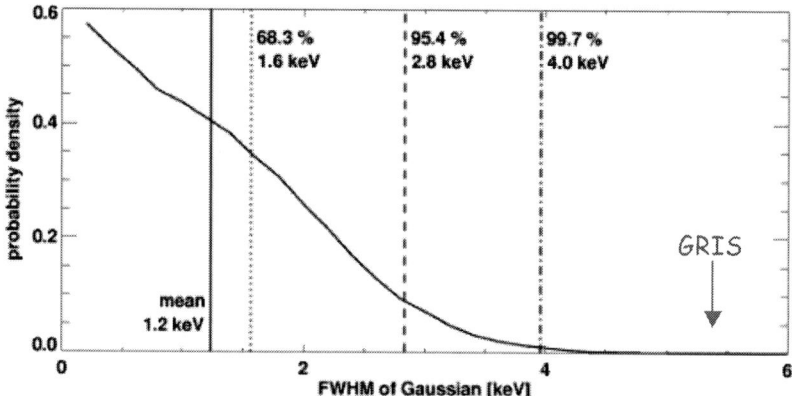

Figure 4. Constraints on celestial broadening of the ^{26}Al line[29]: The probability distribution is asymmetric, and values smaller than the formal mean of 1.2 keV are most probable. The GRIS value of 5.4 keV lies above 99.997% of all values.

production, and bright regions such as Cygnus suggest that even before the core collapse substantial ^{26}Al is ejected in the Wolf-Rayet phase of massive stars [23,24]. Then the GRIS balloon experiment had reported [25] a line width of 6.4 keV, the celestial broadening of which would have corresponded to interstellar gas velocities of ~500 km s^{-1}. This was difficult to understand 26], further spectroscopic measurements were important. Early results from RHESSI[27] and SPI[28] on INTEGRAL then showed that such a spectacularly-large line width was not real.

With 2 years of INTEGRAL/SPI observations, the measurement of ^{26}Al from the inner Galaxy is sufficient (~16σ, Fig.3) for an analysis of the line shape details [29]. Modelling the shape with a convolution of the expected instrumental response and a Gaussian for a possibly broadened celestial line, it is found that such additional celestial broadening must be small. Progress was made from recent improvements of the spectral response treatment of SPI, together with the excellent spectral resolution

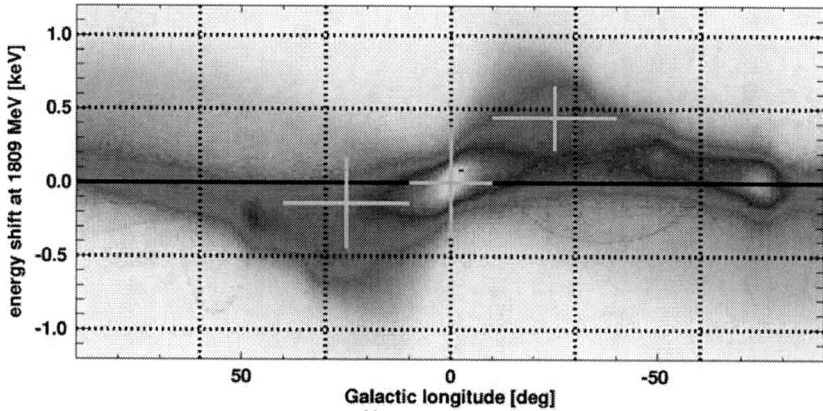

Figure 5. The space-resolved spectra of the ^{26}Al line along the plane of the Galaxy show the expected signature from Doppler shifts of the line centroid due to Galactic rotation [30].

of SPI, maintained through periodic annealings against degradation of detectors (Fig. 1). This now yields an upper limit (2σ) on celestial line broadening of 2.8 keV (Fig. 4), constraining the line width to more moderate and plausible values of interstellar-medium velocities (1 keV corresponds to 122 km s^{-1}) [29].

Moreover, recent results indicate ^{26}Al line shifts with Galactic longitude, which are consistent with expectations from differential Galactic rotation [30]. This is a remarkable finding, because it allows us to conclude that the ^{26}Al sources which we see towards the inner Galaxy indeed are populating the inner Galaxy as expected from candidate source distribution models. That means that the integrated flux from this region can be taken as a representative measurement of a source population throughout the Galaxy, and hence can be converted into a total Galactic ^{26}Al mass produced by massive stars assuming a steady state (i.e. the star formation and supernova rates have remained constant over the past few mission years). We derive a Galactic amount of 2.8 ±0.8 M$_\odot$ of ^{26}Al. This translates into a star formation rate of 3 ±1.4 M$_\odot$ yr^{-1} or a core-collapse supernova rate of 1.9 ±1.1 events per century [30]. The global interstellar isotopic ratio ^{26}Al/^{27}Al is 8.4 10^{-6}, comparing to a value for the early solar system of 4.5 10^{-5}. Those values are in agreement with a range of alternative methods, yet, unlike those, they are derived solely from measurements of our own Galaxy, and are free from major corrections for observational biases. The price paid in our case is a rather uncertain flux measurement due to the large instrumental background, and some dependency on the assumed Galactic source distribution model; both effects add to the substantial uncertainties we have to attach to our measurement. Our recent imaging studies with SPI, from three years of data from a survey of the inner Galaxy and Cygnus region, indicate however a consistency of the emission mapping along the plane of the Galaxy, and hence support the conclusions obtained from COMPTEL results on large-scale source distributions [31]. We hope that the INTEGRAL mission will be extended into the next decade, to allow for substantial improvements through improved background determination.

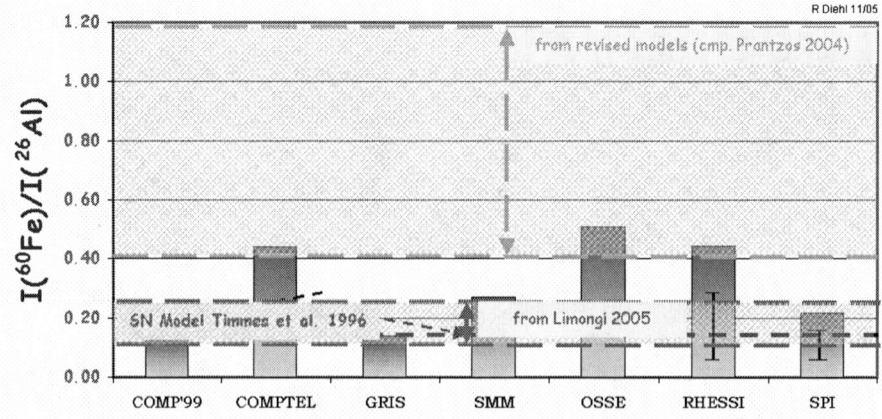

Figure 6. Constraints on the ^{60}Fe/^{26}Al gamma-ray intensity ratio from different gamma-ray measurements, and different evaluations of nucleosynthesis models, for comparison.

^{60}Fe Production in Massive Stars

Both the RHESSI and SPI instrument have reported long-awaited detections of ^{60}Fe decay gamma-rays from the inner Galaxy [32,33,34]. Models of massive star nucleosynthesis have long predicted [35] that the massive stars producing ^{26}Al should also be sources of ^{60}Fe, and that the gamma-ray line intensities should be in the same order of magnitude. Predictions ten years ago showed a ^{60}Fe/^{26}Al gamma-ray intensity ratio of 0.14 [35], while later studies showed that values above 1.0 seemed plausible [36]. Recent re-evaluation of models for the supernova yields into the interstellar medium from a population of massive stars appear in agreement with the recently-revised gamma-ray results [37]. But uncertainties are large both in theoretical models and in gamma-ray data, more homework is needed (Fig. 5).

Positron Annihilation

Positron annihilation emission in the Galaxy has been imaged with SPI on INTEGRAL for the first time [38,39]. The morphology of this image had been indicated by earlier measurements [40] with OSSE on CGRO, and is now established[39] to be dominated by a rather symmetric, bulge-shaped, extended source (FWHM ~8°) centered in the inner Galaxy. It is unclear at present, which source would be responsible for producing the positrons, which annihilate in the inner Galaxy at a rate of 1.5 10^{43} s^{-1}. Radioactive β^+ decay of ^{26}Al explains the observed (but comparatively low) level of annihilation emission from the disk of the Galaxy. But as the sources of ^{26}Al, also most candidate sources of positrons (supernovae of type Ia producing ^{56}Co positrons, pulsars, microquasars and accreting binaries producing pair (e$^-$e$^+$) plasma) should not be confined to the bulge itself, but also show a significant

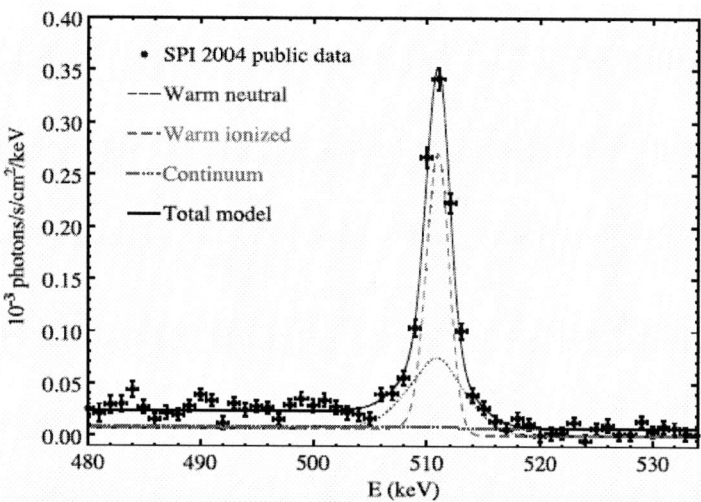

Figure 7: The 511 keV line shape as measured by SPI, and the spectral components attributed to different components (from []). Annihilation predominantly occurs in warm, partially-ionized interstellar gas (see also []).

disk component. Therefore, annihilation of dark matter in the gravitational field of the Galaxy has been suggested [41] as an alternative source.

Whatever the source of positrons, the annihilation occurs through intermediate stages of Positronium formation, which leads to either 2- or 3-photon annihilation due to its singlet and triplet states [42]. Therefore, from the comparison of the triplet continuum gamma-ray emission to the line intensity, and from the line width itself, constraints of the temperature and ionization state of the interstellar gas in the annihilation region have been derived [43,44]. Early studies explained the rather narrow line with a relatively hot, while significantly ionized interstellar medium, but had to add a substantial contribution of annihilation on dust grains to produce a sufficiently narrow line [45]. With recent assessment of all annihilation channels involved [], this is no longer required, and a partially-ionized medium in the temperature range 4000-8000K (warm neutral and warm ionized interstellar medium) is assessed [,] as the most plausible interstellar-medium environment of annihilation (Fig. 6). It remains unclear if this is the environment also of the positron source regions, or if positrons have propagated away from their sources: Thermalization times of positrons with their typical energies of ~MeV or above are ~10^5 y. It depends on the magnetic-field configurations of their source environments [46] if they can leave those and escape into the Galactic halo, where possibly magnetic fields might be suitable to direct them towards the inner Galaxy [47].

ACKNOWLEDGMENTS

This contribution results from the collaborative work with the members of the INTEGRAL and SPI Teams, and fruitful discussions with many other colleagues; I am grateful for their collaboration, in particular to Hubert Halloin, Karsten Kretschmar and Andrew Strong at MPE, Pierre Jean, Jürgen Knödlseder, and Jean-Pierre Roques at CESR, Trixi Wunderer at SSL Berkeley, Bonnard Teegarden at GSFC Greenbelt, Jacco Vink at SRON Utrecht, Nikos Prantzos at IAP, Marco Limongi and Alessandro Chieffi at CNR Frascati, Dieter Hartmann at Clemson University, and Stan Woosley at UC Santa Cruz. I am grateful to the OMEG conference organizers for their great hospitality and support. INTEGRAL is an ESA project with instruments and science data centre funded by ESA member states (especially the PI countries: Denmark, France, Germany, Italy, Switzerland, Spain), Czech Republic and Poland, and with the participation of Russia and the USA. The SPI spectrometer has been completed under the responsibility and leadership of CNES/France, its anticoincidence system is supported by the German government through DLR grant 50.0G.9503.0. We acknowledge the support of INTEGRAL from ASI, CEA, CNES, DLR, ESA, INTA, NASA and OSTC.

REFERENCES

1. Lin R.P., et al., *Sol.Phys.* **210**, 3 (2003)
2. Winkler, C., Courvoisier, T.J.-L., DiCocco, G., et al., *A&A*, **411**, L1-L6 (2003)
3. Vedrenne, G., Roques, J.-P., Schönfelder, V. et al., *A&A*, **411**, L63-L70 (2003)
4. Roques, J.-P., Schanne, S., von Kienlin, A.. et al., . *A&A*, **411**, L91-L100 (2003)
5. Murphy, R.J., Share, G.H., *Adv.Sp.Res.*, **35**, 1825 (2005)
6. Gros, M. et al., *ESA-SP* **552**, 669 (2005)
7. Guessoum, N., Jean, P., Gillard, W., *A&A*, **436**, 171 (2005)
8. Diehl, R., and Timmes, F.X., *PASP*, **110, 748**, 637 (1998)
9. Diehl, R., Prantzos, N., von Ballmoos, P., *accepted for publication in Nucl.Phys.A* (arXiv:astro-ph/0502324) (2005)
10. Murphy, R.J., and Share, G.H., *ApJS*, 161, 495 (2005)
11. Isern, J., Bravo, E., Hirschmann, A., *Adv.Sp.Res. (35th COSPAR Mtg, Paris 2004), in press* (2005)
12. Morris, D., et al., *AIP Conf.Proc.*, **410**, 1087 (1997)
13. Georgii, R., et al., *A&A*, **394**, 517 (2002)
14. Timmes, F.X., Woosley, S.E., Hartmann, D.H., et al., *ApJ*, **464**, 332 (1996)
15. Iyudin, A.F., et al., *A&A*, **284**, L1 (1994)
16. Vink, J., et al., *ApJ* **560**, L79 (2001)
17. The, L.-S., et al., *AIP Conf. Proc.*, **510**, 64 (2000)
18. Renaud, M., et al., *EdP Conf Series*, **SF2A-2004**, 393 (2004)
19. The, L.S., Clayton, D.D., Diehl, R., et al., *ApJ, in press* (2006)
20. Diehl, R., Dupraz, C., Bennett, K., et al., *A&A*, **298**, 445 (1995)
21. Knödlseder, J., Dixon, D.D., Bloemen, H., et al., *A&A*, **345**, 813 (1998)
22. Plüschke, S., Diehl, R., Schönfelder, V., et al., *ESA-SP*, **459**, 55 (2001)
23. Prantzos, N., and Diehl, R., *Phys.Rep,* **267, 1**, 1 (1996)
24. Knödlseder, J., *ApJ*, **510**, 915
25. Naya, J.E., et al., *Nature*, **384**, 44 (1996)
26. Chen, W., et al., *ESA-SP* **382**, 105 (1997)
27. Smith, D., *ApJ* **589**, L55 (2003)
28. Diehl, R., et al., *A&A* **411**, L451 (2003)
29. Diehl, R., Halloin, H., Kretschmer, K., et al., *A&A*, **449**, 1025 (2006)
30. Diehl, R., Halloin, H., Kretschmer, K., et al., *Nature*, 439, 45 (2006)
31. Halloin, H., et al., *in preparation for A&A* (2006)
32. Smith, D., *ApJ* **589**, L55 (2003)
33. Smith, D., *ESA-SP* **552**, 45 (2005)
34. Harris M. J., Knödlseder J., Jean P., et al., *A&A,* **433**, L49 (2005)
35. Timmes, F.X., Woosley, S.E., Hartmann, D.H., et al., *ApJ*, **449**, 204 (1995)
36. Prantzos, N., *A&A*, **420**, 1033 (2004)
37. Limongi, M., this volume (2006)
38. Knödlseder, J., et al., A&A **411**, L457 (2003)
39. Knödlseder, J., et al., *A&A,* **441**, 513 (2005)
40. Purcell, W., et al., *ApJ* **491**, 725 (1997)
41. Boehm, C., et al., *Phys.Rev.Lett* **92**, 1301 (2004)
42. Guessoum, N., et al., A&A **436**, 171 (2005)
43. Churazov, E., et al., *Mon.Not.Roy.Astron.Soc.*, **357**, 1377 (2005)
44. Jean, P., et al., *A&A,* **445**, 579 (2006)
45. Guessoum, N., *ESA-SP* **552**, 57 (2004)
46. Milne, P., et al., *ApJ* **559**, 1019, (2001)
47. Prantzos, N., *A&A*, 449, 869 (2006)

Probing Galactic ^{26}Al with Exotic Ion Beams

Alan A. Chen*, for the DRAGON collaboration and CRIB collaboration

Department of Physics and Astronomy, McMaster University, Hamilton ON L8S 4M1, Canada

Abstract. The goal of understanding the production of galactic ^{26}Al brings together progress in nuclear astrophysics from observations, theory, meteoritics, and laboratory experiments. In the case of experimental work, nuclear reactions involving unstable isotopes are being studied to elucidate the production of ^{26}Al in stellar explosive nucleosynthesis. We discuss a direct measurement of the ^{26}Al(p,γ)^{27}Si reaction with the DRAGON collaboration at TRIUMF, and a measurement of ^{25}Al+p elastic scattering with the CRIB (CNS-U.Tokyo) collaboration, toward constraining the ^{25}Al(p,γ)^{26}Si reaction.

Keywords: stars, nucleosynthesis, nuclear reactions, radioactive ion beams
PACS: 26.30.+k

INTRODUCTION

The origin of galactic ^{26}Al is a long-standing question in nuclear astrophysics. The gamma ray line at 1.809 MeV, associated with the β$^+$ decay of the ground state ^{26}Al to the first excited state in ^{26}Mg, has been decidedly observed by orbiting telescopes, such as COMPTEL [1] and, more recently, RHESSI [2]. These observations give clear indication of recent nucleosynthesis in our galaxy and provide clues to conditions within the galactic ISM. Furthermore, the RHESSI observatory has also detected gamma rays from ^{60}Fe decay in amounts that, when combined with the observed gamma ray flux from ^{26}Al, has led to ongoing discussions on the possible stellar origins of ^{26}Al [3]. Since then, further theoretical work has sought to address this issue.

In meteoritic studies, through measurements of excess ^{26}Mg in presolar inclusions, the isotope ^{26}Al has also served as a probe for understanding the conditions under which the solar system formed [4]. These measurements give the ^{26}Al/^{27}Al ratio in the local ISM just before the collapse of the solar nebula, roughly 4.5 billion years ago.

In determining the stellar source(s) of ^{26}Al, the global analysis of the gamma-emission maps should be complemented with a deeper understanding of specific nucleosynthesis sites with regard to their ^{26}Al production. The former suggests that Type II supernovae and Wolf-Rayet stars could be the dominant contributors of ^{26}Al to the measured gamma ray map [5], while present models of novae and AGB stars also result in significant ^{26}Al production.

The work to be described focuses in particular on elucidating the means of ^{26}Al nucleosynthesis in the context of explosive stellar environments. In novae, for example, the production of ^{26}Al happens in the MgAl hydrogen burning cycle [6].

We will focus in particular on two proton-capture reactions in this cycle that strongly affect the ^{26}Al yield, namely, the ^{26}Al(p,γ)^{27}Si and ^{25}Al(p,γ)^{26}Si reactions. The first has the obviously important role of depleting the yield of ground state ^{26}Al. The latter reaction's impact is more subtle, but no less important: the ^{26}Si produced in the reaction decays to ^{26}Al, but through its isomeric state (^{26m}Al) and not its ground state. This has the effect of bypassing the production of the ground state ^{26}Al, whose decay results in the 1.8 MeV β-delayed gamma ray. (The decay of ^{26m}Al, in turn, does not produce this gamma ray.) The impact of these two reactions in nova nucleosynthesis was recently corroborated in a study that systematically investigated the impact of nuclear reaction rate uncertainties on the nucleosynthesis, using density-temperature profiles from hydrodynamic nova models [7].

Furthermore, one should mention that the importance of the ^{25}Al(p,γ)^{26}Si reaction to ^{26}Al production at higher temperatures up to a few GK is qualitatively different from that found in the relatively lower temperatures characteristic of novae. While gamma decay from the isomer to the ground state is suppressed due to the large difference in spins, the two states however can "communicate" with each other through thermal excitations. In this scenario, the ^{25}Al(p,γ)^{26}Si reaction rate can be sufficiently high to produce effectively most of the ground state ^{26}Al through transitions from the isomer [8].

Both reaction rates are dominated by narrow isolated resonances in the compound nuclei ^{27}Si and ^{26}Si, whose energies and resonance strengths need to be known. For experimental studies, the ideal scenario is one in which a direct measurement of the reaction is performed. Since ^{25}Al and ^{26}Al are both unstable, radioactive ion beams of high intensity are required for direct measurements, while lower intensities are useful in indirect approaches, such as elastic scattering or transfer reactions, to determine level parameters. In the following, we discuss early results from two experiments: (1) a direct measurement of the ^{26}Al(p,γ)^{27}Si reaction with a ^{26}Al beam; and (2) a measurement of the ^{25}Al+p elastic scattering using a beam of ^{25}Al.

EXPERIMENTS

Measurement of the ^{26}Al(p,γ)^{27}Si reaction with DRAGON

The dominant contribution to the ^{26}Al(p,γ)^{27}Si reaction rate at T ~ 0.1-0.4 GK comes from a resonance located at excitation energy $E_x(^{27}Si)$ = 7652 keV and with resonance energy E_r = 188 keV (see Fig.1). The uncertainty in the rate at these temperatures is determined largely by the adopted range for the resonance strength of this state: ωγ = 64 μeV, with upper and lower limits of 290 μeV and 0.0099 μeV, respectively [9]. This strength was previously measured to be 55 ± 9 μeV in an experiment with a radioactive ^{26}Al target, but the results have not been published [10]. The adopted strength is consequently an estimate based on results from indirect approaches.

In view of the wide range of uncertainty in the strength of the E_r = 188 keV resonance, the goal of this experiment was to perform a measurement of this strength,

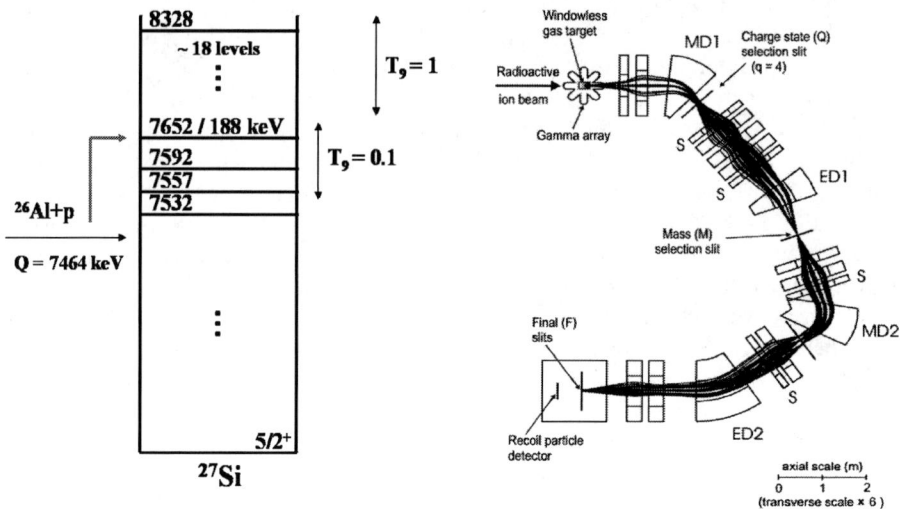

FIGURE 1. Left panel: Energy level scheme of ^{27}Si, showing low-lying resonances above the proton threshold and Gamow windows for T = 0.1 GK and T = 1 GK (energy spacing not to scale). Right panel: Schematic of the DRAGON recoil separator (see text for details).

taking advantage of high-intensity ^{26}Al beams from the TRIUMF-ISAC facility and the DRAGON recoil separator (Fig.1), which has been built specifically for measurements of radiative capture reactions of importance to nuclear astrophysics, such as the one under discussion.

The ^{26}Al beam was produced with the Isotope Separation On-Line (ISOL) method, by impinging 65 μA of 500 MeV protons from the TRIUMF cyclotron onto a high-power SiC target. The long-lived ^{26}Al diffuses out of the target, and a new laser-based ionization system (TRILIS) was utilized to improve the ionization selectivity and the beam intensity. The beam was accelerated to the desired energies through the ISAC RFQ-LINAC accelerator. For measurements on resonance, the beam energy was 201 keV/u with intensities of about 5×10^9 particles per second.

The beam ions impinged on a windowless H$_2$ gas target, which is the first component of the DRAGON recoil separator (Detector of Recoils And Gammas Of Nuclear reactions) [11]. The beam energy and the target pressure are selected to ensure that the resonant reactions occur at the center of the gas target. The beam intensity is monitored by a detector, inside the gas of the target, that detects protons from ^{26}Al + p elastic scattering. The target is surrounded by an array of BGO detectors, which detect gamma rays from the (p,γ) reactions. The ^{27}Si recoils, along with beam ions that did not react in the gas, emerge from the target with an equilibrium distribution of charge states and enter the electromagnetic separator, whose main purpose is the separation between recoils and beam ions at the level of about one part in $10^{9\text{-}10}$. The separator comprises two stages, each consisting of a mag-

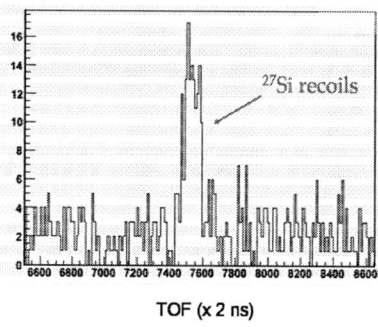 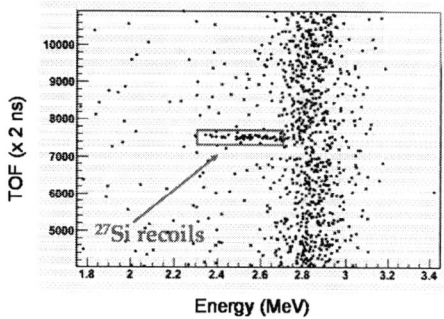

FIGURE 2. Spectra from the ^{26}Al(p,γ)^{27}Si measurement: the left panel is a histogram of the time of flight (TOF) of the recoils through the DRAGON separator. The events in the peak correspond to coincidence ^{27}Si events. The right panel shows a two-dimensional histogram of energy vs. time of flight. The box represents the region corresponding to coincidence ^{27}Si-gamma events.

netic dipole and an electrostatic dipole. One charge state (4$^+$) of the Si recoils and beam ions is selected after the first magnetic dipole. Most of this remaining beam is stopped at a set of slits immediately after the first electrostatic dipole. The second stage serves to remove additional beam ions that passed through the first stage. At the final focus of the separator, a double-sided silicon strip detector was used to measure the energies of the ^{27}Si recoils and of any remaining "leaky" beam ions. The latter are further suppressed by a coincidence requirement between gamma-ray and recoil detections. In addition, the time-of-flight through the separator (21m in length) for the recoils is also measured.

The left panel in Fig.2 shows a time-of-flight spectrum for coincident events, in which a peak corresponding to the ^{27}Si recoils appears prominently. The right panel shows the locus of these events in a two-dimensional histogram of time of flight versus energy. More than 100 recoil-gamma coincidence events for the E_r = 188 keV resonance were observed in the experiment. The data analysis is currently in progress, with the aim of extracting the resonance strength.

Measurement of ^{25}Al+p elastic scattering with CRIB

This experiment used elastic scattering as a tool to probe resonances in ^{26}Si of potential importance to the ^{25}Al(p,γ)^{26}Si reaction. The reaction rate at nova temperatures is also dominated by narrow isolated resonances (Fig. 3), whose energies have been studied in recent transfer reaction experiments [12-14]. One would ultimately like to measure the strengths of these key resonances directly, but ^{25}Al beams of sufficient intensity ($\geq 10^8$ particles per second) are currently unavailable. In the interim, elastic scattering, which can be measured with much lower beam intensities, can be profitably used to search for new levels and to determine the properties of known states. The technique is particularly useful as a probe for states at higher excitation energies, which can be important for this reaction due to the onset of

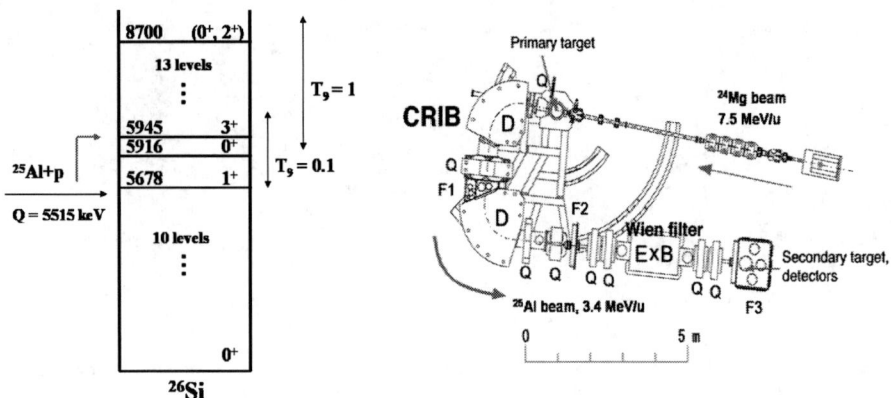

FIGURE 3. Left panel: Energy level diagram for ^{26}Si, showing low-lying resonances and Gamow windows for T = 0.1 GK and T = 1 GK (energy spacing not to scale). Right panel: Schematic of the CRIB separator (see text for details).

thermal equilibration between ^{26}Al isomeric and ground states at high temperatures, as discussed earlier.

The experiment was performed with the CRIB (CNS low-energy Radioactive Ion Beam separator) [15], which is operated and managed by the Center for Nuclear Study (CNS) of the University of Tokyo and is located at RIKEN. CRIB produces beams of light proton-rich nuclei with energies of up to 5 MeV/u with the in-flight technique. The elastic scattering was therefore measured in inverse kinematics with a CH_2 target using the thick-target method [16], with protons detected at forward angles. The target thickness was chosen such that the beam ions were stopped in the target.

The ^{25}Al secondary beam was produced using the ^{24}Mg(d,n)^{25}Al reaction. The ^{24}Mg primary beam was delivered from the AVF cyclotron with an energy of 7.5 MeV/u, a charge state of 8^+, and an intensity of 25 pnA. The secondary beam production target was a D_2 target at 1 atm pressure. Following momentum selection and mass separation in CRIB, ^{25}Al beam intensities of about 3×10^5 pps were attained at the CH_2 target, with a purity of about 50% and energy of 3.4 MeV/u on target. Two Parallel Plate Avalanche Counters (PPACs) provided event-by-event beam ion identification, timing, and position on the CH_2 target. The CH_2 target had a thickness of 6.5 mg/cm^2, sufficiently thick to stop the beam and scan a center-of-mass energy region of about 3.3 MeV above the proton threshold in ^{26}Si. An array of NaI detectors was placed directly above the CH_2 target to measure gamma rays from ^{25}Al + p inelastic scattering. The protons emerging from the target were detected in three ΔE-E silicon telescopes placed at forward angles, which also provided particle discrimination among protons, alphas, and other background events. The ΔE detectors were position sensitive and had a thickness of 75 μm, while the E detectors were 1500 μm thick. Data were also taken with a carbon target to quantify the background from the carbon in the CH_2 target. The calibration of the silicon detectors was performed using protons of known energies from CRIB and alpha particles from radioactive sources.

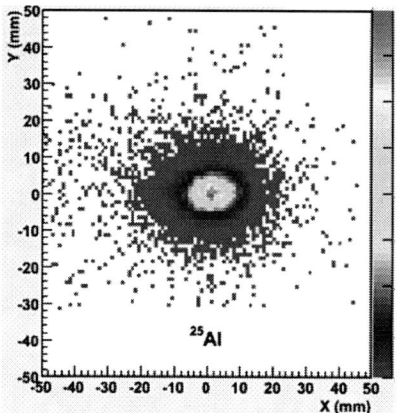
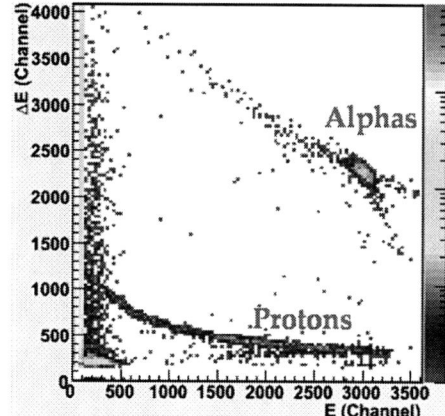

FIGURE 4. Spectra from the p(^{25}Al,p)^{25}Al measurement. Left panel: two-dimensional ^{25}Al beam profile near the target location, derived from gating on the time-of-flight. Right panel: two-dimensional histogram of E versus ΔE for the silicon telescope at zero degrees.

Figure 4 displays the ^{25}Al beam profile near the target location, and a representative ΔE-E spectrum for the silicon telescope at zero degrees, showing the loci of the proton and alpha groups. The data analysis is currently in progress, with remaining work including background subtraction, center-of-mass energy reconstruction, cross section section calculation and R-matrix fits.

ACKNOWLEDGMENTS

AAC acknowledges support from the National Science and Engineering Research Council of Canada.

REFERENCES

1. R. Diehl et al., *Astron. and Astrophys.* **97** 181 (1993).
2. D.M. Smith, *New Ast. Rev.* **48** 87 (2004).
3. N. Prantzos, *Astron. and Astrophys.* **420** 1033 (2004).
4. G. Wasserburg, in *Protostars and Planets II*, edited by D.C. Black and M.S. Matthews, Tucson: University of Arizona Press, 1985, p.703.
5. J. Knödlseder, *Ap. J.* **510** 915 (1999).
6. J. José, A. Coc, and M. Hernanz, *Ap. J.* **520** 347 (1999).
7. C. Iliadis et al., *Ap. J. Suppl.* **142** 105 (2002).
8. R. Runkle, A.E. Champagne, and J. Engel, Ap. J. **556** 970 (2001).
9. C. Angulo et al., *Nucl. Phys.* A **656** 3 (1999).
10. R.B. Vogelaar, Ph.D. Thesis, California Institute of Technology (1989).
11. D.A. Hutcheon et al., *Nucl. Inst. Meth.* A **498** 190 (2003).
12. D.W. Bardayan et al., *Phys. Rev.* C **65** 032801(R) (2002).
13. J.A. Caggiano et al., *Phys. Rev.* C **65** 055801 (2002).
14. Y. Parpottas et al., *Phys. Rev.* C **70** 065805 (2004).
15. S. Kubono et al., *Eur. Phys. J.* A **13** 217 (2002).
16. T. Teranishi et al., *Phys. Lett.* B **556** 27 (2003).

The Abundance Of Live ^{60}Fe In The Early Solar System

Shogo Tachibana[*], Gary R. Huss[†], Noriko T. Kita[¶,‡], Gen Shimoda[¶] and Yuichi Morishita[¶]

Department of Earth and Planetary Science, University of Tokyo, 7-3-1 Hongo, Tokyo 113-0033, Japan.
†*Hawaii Institute of Geophysics and Planetology, University of Hawaii at Manoa, 1680 East-West Road, Honolulu, HI 96822, USA.*
¶*Geological Survey of Japan, National Institute of Advanced Industrial Science and Technology, Tsukuba 305-8567, Japan.*
‡*Department of Geology and Geophysics, University of Wisconsin-Madison, 1215 W. Dayton Street, Madison, WI 53706, USA.*

Abstract. Iron-60 decays to ^{60}Ni with a half-life of 1.49 million years (Myr). Because ^{60}Fe is produced only in stars, its initial abundance in the solar system provides a constraint on the stellar contribution of radionuclides to the solar system and on the nature of the stellar source. The initial abundance of ^{60}Fe in the solar system has been only loosely constrained from sulfides, of which ^{60}Fe-^{60}Ni systems are easily disturbed, in primitive chondrites. In this study, we show that ^{60}Fe was present in chondrules, silicate spherules commonly found in chondrites, with initial ^{60}Fe/^{56}Fe ratios ((^{60}Fe/^{56}Fe)$_0$) of (2.2-3.7) x 10^{-7} at the time of their formation. By applying the time difference of 1.5-2.0 Myr between formation of the oldest known solar-system solids (Ca-Al-rich inclusions) and chondrules, a (^{60}Fe/^{56}Fe)$_0$ ratio for the initial solar system of (5-10) x 10^{-7} is estimated. This new solidly based (^{60}Fe/^{56}Fe)$_0$ ratio is consistent with predictions for nucleosynthesis in a supernova or in an intermediate-mass AGB star just before the solar-system formation, but too high for the source to have been a low-mass AGB star. Because encounters between a molecular cloud and an AGB star are rare, our results appear to be the best evidence found to date for a contribution of material from a nearby supernova to the solar system.

Keywords: short-lived radionuclides; iron-60; meteorites; chondrules; solar system formation; stellar necleosynthesis
PACS: 26.30.+k, 91.67.Qr, 96.30.Za

INTRODUCTION

Isotopic analyses of meteorites have revealed that several radionuclides with short half lives ($t_{1/2} < \sim100$ Myr) once existed in the early solar system; ^{41}Ca, ^{36}Cl, ^{26}Al, ^{60}Fe, ^{10}Be, ^{53}Mn, ^{107}Pd, ^{182}Hf, ^{129}I, ^{92}Nb, ^{244}Pu, and ^{146}Sm [e.g., 1]. They are produced either by stellar nucleosynthesis or by energetic-particle irradiation. Beryllium-10 cannot be produced in stars, so its presence in the early solar system indicates a contribution from energetic-particle irradiation. On the other hand, ^{60}Fe that decays into ^{60}Ni by the β⁻ emission is produced only efficiently in stars, and its presence suggests that stellar nucleosynthesis also contributed to the solar-system short-lived radionuclides. Although the presence of ^{60}Fe in the early solar system was found in differentiated meteorites [2], clear evidence of ^{60}Fe in primitive chondrites have been recently found from sulfides and magnetites in unequilibrated ordinary chondrites [3, 4]. The inferred initial ratios, $(^{60}\text{Fe}/^{56}\text{Fe})_0$, from chondritic materials range from ~1 x 10^{-7} to >1 x 10^{-6} and are significantly higher than the steady-state value in interstellar medium (~2.6 x 10^{-8} [5]). This indicates that stellar production of ^{60}Fe shortly before the solar system formation is required. However, the $(^{60}\text{Fe}/^{56}\text{Fe})_0$ ratio for the initial solar system has been only loosely constrained from chondritic sulfides, and both low- and high-mass stars could be the source of ^{60}Fe. To make matters worse, the ^{60}Fe-^{60}Ni systems in sulfides are easily disturbed even by mild thermal metamorphism or aqueous alteration on the parent body [6].

In this work, we estimate the initial abundance of ^{60}Fe in the solar system from isotopic study of chondrules, major silicate constituents in chondrites formed 1-2 Myr after the oldest solar-system solid objects (Ca-Al-rich inclusions). Because silicates are less susceptible to thermal and aqueous processes and are apparently unaltered in the most primitive chondrites, chondrule silicates should preserve $(^{60}\text{Fe}/^{56}\text{Fe})_0$ at the time of chondrule formation.

ISOTOPIC ANALYSES

Thin sections of Semarkona and Bishunpur, both of which are the least metamorphosed ordinary chondrites, were examined using a scanning electron microscope equipped with energy-dispersive spectrometer to identify chondrules with FeO-rich silicates. The Fe/Ni ratios of chondrule silicates were checked using the Cameca ims-6f ion microprobe at Arizona State University. We selected four ferromagnesian chondrules containing abundant pyroxene ((Mg,Fe)SiO$_3$) with Fe/Ni elemental ratios up to ~3x10^4 for isotopic analysis (Fig. 1).

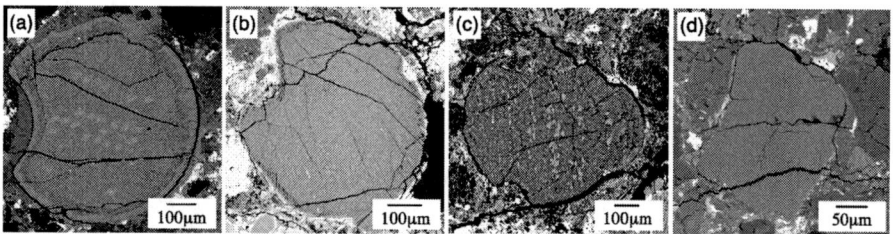

FIGURE 1. Back-scattered electron images of pyroxene-rich chondrules with high Fe/Ni ratios in Semarkona (SMK1-4, SMK2-1, and SMK2-4) and Bishunpur (BIS21). (a) SMK1-4. A fine-grained radiating pyroxene chondrule. (b) SMK2-1. A fine-grained radiating pyroxene chondrule. (c) SMK2-4. A barred pyroxene chondrule with Fe-rich olivine grains between the bars. (d) BIS-21. An irregular-shaped fine-grained pyroxene chondrule, which may be a fragmental of a larger chondrule.

The ^{60}Fe-^{60}Ni systems in chondrules were analyzed using the Cameca imfs-1270 ion microprobe at the Geological Survey of Japan. A ~15-µm, 1 nA, primary O$_2^-$ beam was used to sputter the samples. The secondary mass spectrometer was operated at 10 kV with a 50 eV energy window and a mass resolving power (MRP) of ~4500. Secondary ions ($^{57}Fe^+$, $^{60}Ni^+$, $^{61}Ni^+$ and $^{62}Ni^+$) were counted on an electron multiplier. Although a MRP of 4500 is insufficient to resolve interferences from hydrides and molecular ions of oxides ($^{44}Ca^{16}O$, $^{45}Sc^{16}O$, and $^{46}Ti^{16}O$), the contributions of such interferences were confirmed to be <0.1 %. The Fe/Ni sensitivity factor obtained for KL2-G basalt glass was used for instrumental sensitivity correction. Instrumental mass fractionation for the measured $^{60}Ni/^{61}Ni$ was corrected internally using $^{62}Ni/^{61}Ni$.

RESULTS

Chondrule SMK1-4 shows excesses of ^{60}Ni that are correlated with $^{56}Fe/^{61}Ni$, and the $(^{60}Fe/^{56}Fe)_0$ inferred for this chondrule is $(2.7 \pm 0.8) \times 10^{-7}$ (Fig. 2). Chondrules SMK2-1, SMK2-4, and BIS-21 also show ^{60}Ni excesses correlated with $^{56}Fe/^{61}Ni$, and their $(^{60}Fe/^{56}Fe)_0$ ratios are $(2.2 \pm 1.0) \times 10^{-7}$, $(2.8 \pm 2.1) \times 10^{-7}$, and $(3.7 \pm 1.9) \times 10^{-7}$, respectively (Fig. 2)

The $(^{60}Fe/^{56}Fe)_0$ ratios inferred for the four chondrules are larger than those obtained for sulfides in Bishunpur and Krymka [3] and for magnetite in Semarkona [4], but are smaller than that inferred for Semarkona sulfides [4].

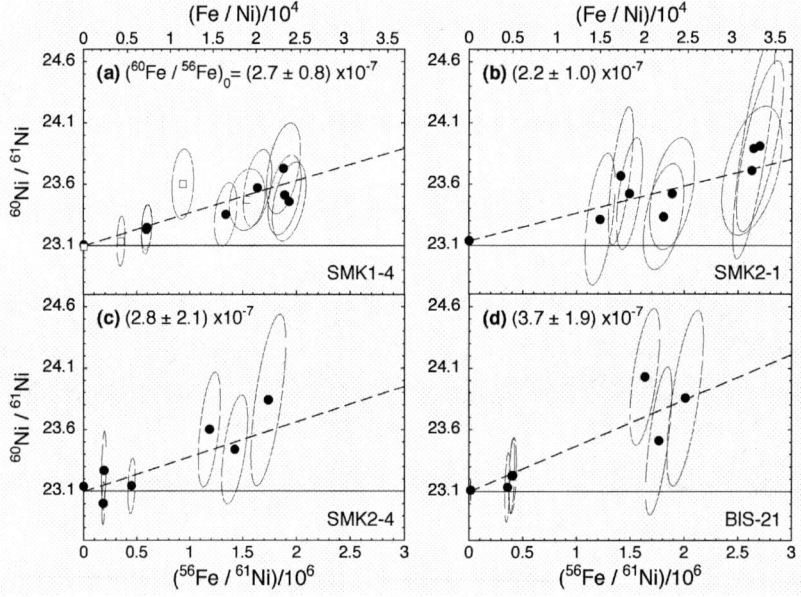

FIGURE 2. Isochron diagrams of ^{60}Fe-^{60}Ni systems in chondrules. The presence of now-extinct ^{60}Fe in an early-solar-system object is demonstrated by a correlation between the abundance of radiogenic ^{60}Ni and the Fe/Ni elemental ratio. In an undisturbed object, the $^{60}Fe/^{56}Fe$ ratio at the time the object formed, $(^{60}Fe/^{56}Fe)_0$, can be obtained from the slope of a linear correlation between $^{60}Ni/^{61}Ni$ and $^{56}Fe/^{61}Ni$ because $(^{60}Ni/^{61}Ni) = (^{60}Ni/^{61}Ni)_0 + (^{60}Fe/^{56}Fe)_0 \times (^{56}Fe/^{61}Ni)$. The 2σ uncertainty is shown as an error ellipse. (a) SMK1-4. Open squares are data obtained as weighted means of several measurements [7]. (b) SMK2-1. (c) SMK2-4. (d) BIS-21.

DISCUSSION

Semarkona and Bishunpur experienced only mild metamorphism at temperatures no higher than ~530 K and ~570 K, respectively [8-10]. Because no cation diffusion is expected in pyroxene at such low temperatures [11], the initial ratios for chondrule pyroxenes must represent those for last melting events they experienced during chondrule formation. If this is the case, we can estimate the initial $(^{60}Fe/^{56}Fe)_0$ for the solar system by correcting our measured $(^{60}Fe/^{56}Fe)_0$ for the chondrules for the ~1.5-2.0 Myr time delay from the formation of Ca-Al-rich inclusions. From this correction, we estimate $(^{60}Fe/^{56}Fe)_0$ for the initial solar system to have been $(5-10) \times 10^{-7}$, which can be a significantly tighter constraint than provided by previous data from sulfides [3, 4].

The estimated $(^{60}Fe/^{56}Fe)_0$ ratio and the inferred $(^{26}Al/^{27}Al)_0$ ratio (5×10^{-5}) and abundances for other short-lived radionuclides estimated for the initial solar system (e.g., ^{41}Ca) can be compared with the abundance ratios for nucleosynthesis for different types of stars, allowing time for the nuclides to enter the solar system. Non-exploding Wolf-Rayet stars cannot produce the estimated initial $(^{60}Fe/^{56}Fe)_0$ ratio at all $((^{60}Fe/^{56}Fe)_0 \ll (1-2) \times 10^{-8})$ [12]. Models for nucleosynthesis in low-mass AGB stars predict $(^{60}Fe/^{56}Fe)_0$ ratio up to a few $\times 10^{-7}$ [e.g., 13, 14], but this is insufficient to explain our estimated $(^{60}Fe/^{56}Fe)_0$ ratio. Models for intermediate-mass AGB stars (5 $M_\odot$ with the solar metallicity or 3 $M_\odot$ with 1/3 × the solar metallicity) predict $(^{60}Fe/^{56}Fe)_0$ of $(1-2) \times 10^{-6}$ for the initial solar system [14]. This seems to be consistent with our estimates, but the rarity of encounters between molecular clouds and AGB stars (probability of $<3 \times 10^{-6}$; [15]) makes it implausible for an AGB star to be a source of short-lived radionuclides in the solar system.

The $(^{60}Fe/^{56}Fe)_0$ ratio for the solar system, predicted by nucleosynthesis models for type II supernovae, ranges from 3×10^{-7} to $>1 \times 10^{-5}$, depending on the mass and metallicity of the star, the nuclear reaction rates, and the kinetic energy of explosion [16-19]. Current model calculations imply that a >25 $M_\odot$ and/or slightly low-metallicity (~0.3-0.5 × the solar metallicity) star that exploded ~1 Myr before the solar-system formation could explain abundances of ^{60}Fe as well as ^{26}Al and ^{41}Ca. However, the $(^{53}Mn/^{55}Mn)_0$ ratio for the solar system inferred for type II supernovae is 10-100 times larger than that estimated from meteorites. Provided that a massive star exploded with a less kinetic energy, this discrepancy could be explained by fallback of most of the ^{53}Mn onto a collapsing stellar core [e.g., 20]. Although additional theoretical work is clearly required, a supernova is currently the best option for the source that contributed to the inventory of short-lived radionuclides in the early solar system.

If a supernova explosion near the newly forming solar system provided short-lived radionuclides in the solar system, it may constrain the environment where the sun formed. Most stars form in massive molecular clouds where hundreds of stars with a wide range of masses form at approximately the same time [21]. Massive stars have short lifetimes and can explode as supernovae in only a few million years, strongly affecting the star-forming region around them. A long-standing idea for the birth of the solar system is that the explosion of a supernova may have triggered the gravitational collapse of the solar system and may have injected supernova materials into the collapsing system [22]. Alternatively, intense stellar winds of the pre-supernova massive star may have triggered the collapse of the solar system, after which the massive star exploded and injected newly synthesized material into the Sun's accretion disk [e.g., 23]. Although further work for short-lived radionuclides, stellar nucleaosynthesis, and star formation is surely required, the high $(^{60}Fe/^{56}Fe)_0$ ratio inferred from chondrules may provide a clear indication for the contribution of a massive star to the solar-system formation. The solar system may have formed in a region where hundreds of stars were forming at once, much like the Orion nebula is today.

ACKNOWLEDGMENTS

We thank R. Gallino for providing a preprint of reference [14] prior to publication. We also thank K. Nomoto, H. Umeda, N. Iwamoto, and N. Tominaga for discussion on stellar nucleosynthesis. This work is supported by NASA grants NAG5-11543 and NNG05GG48G (GRH).

REFERENCES

1. N. T. Kita, G. R. Huss, S. Tachibana, Y. Amelin, L. E. Nyquist and I. D. Hutcheon, "Constraints on the Origin of Chondrules and CAIs from Short-lived and Long-lived Radionuclides" in *Chondrites and the Protoplanetary Disk* edited by A. N. Krot et al., Astronomical Society of the Pacific, San Francisco, 2005, pp. 558-587.
2. A. Shukolyukov and G. W. Lugmair, *Science* **259**, 1138-1142 (1993).
3. S. Tachibana and G. R. Huss, *Astrophys. J.* **588**, L41-L44 (2003).
4. S. Mostefaoui, G. W. Lugmair, and P. Hoppe, *Astrophys. J.* **625**, 271-277 (2005).
5. G. J. Wasserburg, M. Busso, and R. Gallino, *Astrophys. J.* **466**, L109-L113 (1996).
6. Y. Guan, G. R. Huss and L. A. Leshin, *Appl. Surf. Sci.* **231-232**, 899-902 (2004).
7. G. R. Huss and S. Tachibana, *Lunar Planet. Sci.* **35**, 1811 (2004).
8. E. R. Rambaldi and J. T. Wasson, *Geochim. Cosmochim. Acta* **45**, 1001-1015 (1981).
9. C. M. O'D. Alexander, D. J. Barber and R. Hutchison, *Geochim. Cosmochim. Acta* **53**, 3045-3057 (1989).
10. G. R. Huss and R. S. Lewis, *Meteoritics* **29**, 811-829 (1994).
11. J. Ganguly and V. Tazzoli, *Am. Mineral.* **79**, 930-937 (1994).
12. M. Arnould, G. Paulus and G. Maynet, *Astron. Astrophys.* **321**, 452-464 (1997).
13. G. J. Wasserburg, M. Busso, R. Gallino and C. M. Raiteri, *Astrophys. J.* **424**, 412-428 (1994).
14. G. J. Wasserburg, M. Busso, R. Gallino and K. M. Nollett, *Nuclear Physics A* (in press).
15. J. H. Kastner and P. C. Meyers, *Astrophys. J.* **421**, 605-614 (1994).
16. G. J. Wasserburg, R. Gallino and M. Busso, *Astrophys. J.* **500**, L189-L192 (1998).
17. S. E. Woosley and T. A. Weaver, *Astrophys. J. Suppl. Ser.* **101**, 181-235 (1995).
18. T. Rauscher, A. Heger, R. D. Hoffman and S. E. Woosley, *Astrophys. J.* **576**, 323-348 (2002).
19. A. Chieffi and M. Limongi, *Astrophys. J.* **608**, 405-410 (2004).
20. B. S. Meyer, "Synthesis of Short-lived Radioactivities in a Massive Star" in *Chondrites and the Protoplanetary Disk* edited by A. N. Krot et al., Astronomical Society of the Pacific, San Francisco, 2005, pp. 515-526.
21. C. J. Lada and E. A. Lada, *Annu. Rev. Astron. Astrophys.* **41**, 57-115 (2003).
22. A. G. W. Cameron and J. W. Truran, *Icarus* **30**, 447-461 (1977).
23. J. J. Hester, S. J. Desch, K. R. Healy and L. A. Leshin, *Science* **304**, 1116-1117 (2004).

1.13 METEORITIC ABUNDANCES

Presolar Graphite from the Murchison Meteorite: Imprint of Nucleosynthesis and Grain Formation

Sachiko Amari[1], Roberto Gallino[2], Marco Limongi[3,4] and Alessandro Chieffi[5]

1 *Laboratory for Space Sciences and the Physics Department, Washington University, St. Louis, MO 63130-4899, USA (sa@wuphys.wustl.edu)*
2 *Dipartimento di Fisica Generale, Universita' di Torino, I-10125 Torino, Italy (gallino@ph.unito.it)*
3 *INAF Osservatorio Astronomico di Roma, I-00040 Rome, Italy (marco@oa-roma.inaf.it)*
4 *Centre for Stellar and Planetary Astrophysics, Monash University, Australia*
5 *Instituto di Astrofisica Spaziale e Fisica Cosmica (CNR), I-00133, Roma, Italy (achieffi@rm.iasf.cnr.it)*

Abstract. Presolar graphite is the carrier of Ne-E(L) and most ^{22}Ne in Ne-E(L) had long been attributed to radiogenic decay of ^{22}Na from novae. Of presolar graphite grains with a range of density (1.6-2.2g/cm^3), low-density graphite grains extracted from the Murchison meteorite are characterized by ^{18}O excesses and Si isotopic anomalies and are believed to have formed in supernova ejecta. From noble gas analyses of low-density graphite grains, we conclude that ^{22}Ne in the grains is from the *in situ* decay of ^{22}Na ($T_{1/2}$=2.6a) produced in the C-burning zone in presupernova stars. The grains also contain Kr that was produced by neutron capture, either in the He-burning zone or the C-burning zone during hydrostatic burning. The ^{22}Ne of a ^{22}Na origin indicates that the grains formed shortly after the explosion. The presence of ^{22}Ne of a ^{22}Na origin and Kr, and the absence of ^{22}Ne of a non-radiogenic origin might give us a further clue for graphite formation in supernova ejecta.

Keywords: supernovae; nucleosynthesis; slow neutron capture process; dust; meteorite; Ne-E(L)
PACS: 26.30.+k; 95.30.Wi; 96.30.Za; 97.10.Tk; 97.60.Bw; 98.58.Mj

INTRODUCTION

Noble gases in meteorites have played an important role in cosmochemistry. They are literally rare, thus a small addition of isotopically anomalous noble gas components could be easily detected. The anomalous components include Ne-E (enriched in ^{22}Ne), Kr-S (*s*-process Kr), Xe-S (*s*-process Xe) and Xe-HL (enriched in both light and heavy Xe isotopes) [see 1]. Ne-E was first discovered when Black and Pepin [2] analyzed the Ne isotopic composition in a fragment of the Orgueil meteorite by stepwise heating. Stepwise heating is a method commonly applied for noble gas studies. Temperature is incrementally increased and the released noble gases are analyzed in each step. The concept behind this technique is that different noble gas components are trapped in different sites in the minerals (e.g., surface or interior) or different minerals and thermal properties of the minerals/trapping sites are different, thus different noble gas components are released at different temperatures. The Ne components commonly observed in meteorites are called Ne-A, Ne-B and Ne-S [e.g., see Fig. 13.1.1 of reference 1]. If the Ne in the Orgueil meteorite was a mixture of these components, Ne isotopic ratios of all the temperature steps should have fallen in the triangle bounded by the components. However, in high (900 – 1000°C) temperature steps, the ratios fell below the triangle, indicating the presence of a ^{22}Ne-rich component. Black [3] named the component Ne-E. Subsequently, Jungck [4] separated the Orgueil meteorite into several fractions and found that there are two kinds of Ne-E. One was released at *low* temperatures (500 – 700°C) and was concentrated in *low*-density fractions (2.2 – 2.5 g/cm^3), whereas the other was released at *high* temperatures (1200 – 1400°C) and was enriched in a *high*-density fraction (2.5 – 3.1 g/cm^3). They were dubbed Ne-E(L) and Ne-E(H), respectively.

The identification of carriers of the anomalous noble gas components came almost two decades after the discovery of the components in meteorites. Diamond was identified as the carrier of Xe-HL in 1987 [5]. Subsequently, silicon carbide (SiC), the carrier of Ne-E(H), Kr-S and Xe-S, was isolated and found [6, 7]. The ^{22}Ne of Ne-E(H) is from the He-shell in asymptotic giant branch (AGB) stars [8-10]. Finally, graphite was identified as the carrier of Ne-E(L) in 1990 [11]. Its ^{22}Ne is predominantly from ^{22}Na with an addition of ^{22}Ne from He-shell in AGB stars [12].

The carrier grains exhibit huge isotopic anomalies not only in noble gases but also other elements, indicating that they formed in the ejecta of supernovae or in the circumstellar envelopes of asymptotic giant branch (AGB) stars. They are called presolar grains because they formed before the solar system. Extensive studies of presolar grains have provided new information on nucleosynthesis in stars, mixing in supernova ejecta, Galactic chemical evolution and grain formation in the outflow of stars [e., g., 13, 14, 15]. In this paper, we will reexamine noble gas data on presolar graphite grains from the Murchison meteorite and discuss the implications of the data on nucleosynthesis and grain formation.

DATA SOURCES

Most studies on presolar graphite have been performed on four graphite-rich separates extracted from the Murchison meteorite [16]. They are KE1 (1.6–2.05 g/cm^3), KFA1 (2.05–2.10 g/cm^3), KFB1 (2.10–2.15 g/cm^3) and KFC1 (2.15–2.20 g/cm^3). The separate KE1 was further purified, yielding KE3 (1.65–1.72 g/cm^3) [17]. Concerning isotopic signatures, KE1 and KE3 can be regarded as the same separate.

There exist two sets of noble gas analyses of the separates. Amari et al. [12] analyzed Ne, Ar, Kr and Xe in bulk (=aggregates of millions of grains) samples by stepwise heating. Nichols et al. [18] measured ^{4}He and ^{20,22}Ne in single grains of known C and Si isotopic ratios from KE3, KFB1 and KFC1.

In this paper, we will focus on the lowest-density separates KE1 and KE3.

DISCUSSION

Neon

Low-density Graphite

Low-density graphite grains from the separate KE3 have been most extensively studied with ion probe: their large grain size (median grain size: 4.9 μm) and high trace element concentrations made it possible to analyze isotopic ratios of multiple elements [17, 19, 20]. Many grains (~50 %) have ^{18}O excesses and Si isotopic anomalies, mainly in the form of ^{28}Si excess. Highest ^{26}Al/^{27}Al ratios inferred from ^{26}Mg excesses are ~0.1. ^{28}Si excesses and high ^{26}Al/^{27}Al ratios are also observed in SiC grains of type X (X grains), which are believed to have formed in Type II supernovae [21, 22]. Proof for their supernova origin came from the presence of ^{44}Ti ($T_{1/2}$ = 60a) in a few low-density graphite and SiC X grains [23, 24] because ^{44}Ti is produced only by explosive nucleosynthesis.

In order to quantitatively examine whether the graphite grain data could be explained with supernova models, Travaglio et al. [20] used yields of Type II supernova models by Woosley and Weaver [25] and mixed different zones. They found that general isotopic features of the low-density graphite grains are explained if a small amount of material from the inner Si-rich zone is ejected and mixed at a microscopic scale before grain formation into the outer He-rich zones.

^{22}Ne-rich low-density grains

Of 21 KE3 grains analyzed for He and Ne, nine grains contained measurable amount of ^{22}Ne. ^{4}He and ^{20}Ne were under the detection limit in all grains. One ^{22}Ne-rich grain was not analyzed for its C, O, and Si isotopic ratios. Seven ^{22}Ne-rich grains show elevated ^{18}O/^{16}O ratios (6.83×10^{-3} to 0.119) above the solar ratio (2×10^{-3}). Grain KE3a-573 has the normal ^{18}O/^{16}O ratio, but ^{28}Si excess {δ^{29}Si/^{28}Si = −235±97 ‰, δ^{30}Si/^{28}Si = −327±128 ‰, where δ^{i}Si/^{28}Si ≡ [(^{i}Si/^{28}Si)$_{grain}$/(^{i}Si/^{28}Si)$_{solar}$ − 1] × 1000.} This grain was so enriched in ^{22}Ne that an upper limit of its ^{20}Ne/^{22}Ne (<0.01) could be determined. Two grains with ^{18}O excesses show the presence of ^{44}Ti [^{44}Ti/^{48}Ti = (1.06±0.22)×10^{-3} and (2.40±1.10)×10^{-3}]. Thus all ^{22}Ne-rich grains bear supernova signatures such as ^{18}O and ^{28}Si excesses, and the presence of ^{44}Ti, indicating that they formed in supernovae.

In Type II supernovae, the lowest ^{20}Ne/^{22}Ne ratio is found in the partial He burning zone (or He/C zone, according to the terminology by Meyer et al. [26], whose name indicates most abundant elements) where ^{14}N is converted into ^{22}Ne. The predicted ^{20}Ne/^{22}Ne ratio in the zone in a 25M$_O$ star of solar metallicity is 0.088 by Chieffi and Limongi [27] and 0.096 by Heger et al. [28]. They are much higher than the upper limit observed in grain KE3a-573. This implies that the ^{22}Ne in the grain was not implanted because if it was the case ^{20}Ne should have been also implanted, resulting in a much higher upper limit than 0.01. This leaves ^{22}Na as a sole source of the ^{22}Ne in the grain.

Although upper limits of ^{20}Ne/^{22}Ne ratios of the other grains are not available, the portion of ^{22}Ne from ^{22}Na can be determined from the Ne isotopic ratios of the separate KE1 (^{20}Ne/^{22}Ne = 0.0301±0.0018, ^{21}Ne/^{22}Ne = 0.000118±0.000017). Assuming that the Ne in KE1 is a mixture of ^{22}Ne from ^{22}Na, Ne from the He/C zone, and solar Ne, more than 99% of the ^{22}Ne in KE1 originated from ^{22}Na. Thus, the ^{22}Ne-rich grains most likely contain ^{22}Ne solely from the decay of ^{22}Na, as first suggested by Nichols et al. [18]. ^{22}Na is produced in the C convective shell (O/Ne zone) during hydrostatic C burning by ^{21}Ne(p,γ)^{22}Na, where ^{21}Ne is produced by ^{20}Ne(n,γ)^{21}Ne and protons are produced by ^{12}C(^{12}C,p)^{23}Na [27].

Krypton

^{86}Kr/^{82}Kr and ^{80}Kr/^{82}Kr ratios are very sensitive indicators for nucleosynthetic conditions in stars. Neutron capture on ^{84}Kr feeds both the ground state and the isomeric state of ^{85}Kr [29]. The unstable isotope ^{85}Kr decays to ^{85}Rb with the half-life of 11 years when it is at the ground state. At the isomeric state, its decay rate is much faster ($T_{1/2}$ = 4.48h). During convective core He burning, the ground and the isomeric states are not thermalised and need to be treated independently. During shell C burning, there is full thermalization between the ground state and the isomeric state. In any case, ^{86}Kr yields depend on neutron density. Selenium-79 is at a branching point of the s-process and its behavior is critical to ^{80}Kr yields. The half-life of ^{79}Se strongly depends on temperature: it is much shorter at stellar conditions (one month at ~1×10^{9}K) than in terrestrial conditions (650,000 years) [30]. As a consequence, ^{80}Kr yields depend on neutron density and temperature.

TABLE 1. Krypton isotopic ratios.

Separate/Zone	80/82	83/82	84/82	86/82
KE1+KFA1	0.070±0.045	≡0.334	1.26±0.65	0.02±0.26
KE1+KFA1	0.127±0.023	≡0.623	2.86±0.33	0.67±0.14
KFC1	0.030±0.047	≡0.375	2.58±0.41	4.43±0.46
O/C*	0.405	0.334	1.41	0.0938
O/Ne*	0.0656	0.623	1.97	0.726

* A 25M$_\odot$ model by Chieffi and Limongi [27]

Amari et al. [12] have found that the four Murchison separates are enriched s-process Kr (Kr-S) and that in a ^{86}Kr/^{82}Kr–^{83}Kr/^{82}Kr plot (Fig. 1), KE1+KFA1 and KFC1 form two distinct lines, indicating that the Kr in the separates is a mixture of close-to-normal Kr (on the right in Fig. 1) and Kr-S (on the left) and that there are two Kr-S components: Kr-SH in KFC1 and Kr-SL in KE1+KFA1. To infer the isotopic composition of Kr-SH, we assumed (^{83}Kr/^{82}Kr)s ≡ 0.375 instead of 0.30 used by Amari et al. [12], reflecting the improvement of precisions of analyses in neutron capture cross sections (Table 1). Kr-SH, with a high ^{86}Kr/^{82}Kr ratio, most likely originated from low-metallicity AGB stars as concluded by Amari et al. [12]. Kr-SL was originally associated with high-metallicity AGB stars or a mixture of AGB and massive stars, thus its isotopic ratios were inferred by assuming $^{83}Kr/^{82}Kr = 0.30$. The $^{83}Kr/^{82}Kr$ ratio was used to infer other Kr isotopic ratios because it is predicted to be constant and defined by the inverse ratio of their neutron capture cross sections at relatively low temperature (kT ~ 30 keV). However, at the higher temperature [~10^9K (~90 keV)] realized during convective shell C-burning in the O/Ne zone, deviations from the classical 1/v rule (v: thermal velocity) for the cross section of ^{83}Kr become significant. Thus, it is necessary to reevaluate the $^{83}Kr/^{82}Kr$ ratio to apply for low-density graphite grains, which formed in supernovae. Since the lowest $^{86}Kr/^{82}Kr$ and $^{83}Kr/^{82}Kr$ ratios observed in the grains are 1 and 0.7, respectively (Fig. 1), Kr-SL must have smaller ratios than those.

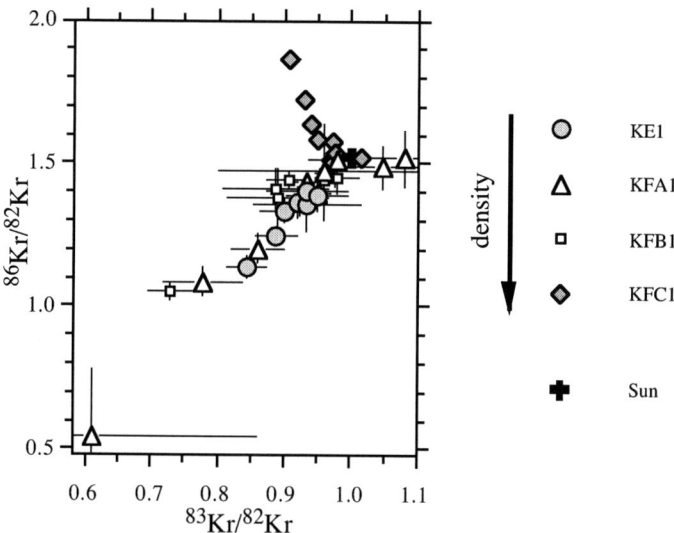

FIGURE 1. Krypton isotopic ratios of the graphite separates from the Murchison meteorite. KFC1 and KE1+KFA1 form two distinct lines. Errors are 1 σ. Data from [12].

Chieffi and Limongi [27] constructed a set of explosive yields of massive stars of solar metallicity in the mass range of 11–120M_o, using the latest version of the FRANEC code (5.05218) that includes a nuclear network extending to ^{98}Mo. Comparing the grain data and their predicted ratios, the 15M_o model can be excluded from a source of the Kr in the low-density grains (KE1+KFA1) because the $^{86}Kr/^{82}Kr$ ratio in the O/Ne zone (C burning zone) is 5.05 and the $^{83}Kr/^{82}Kr$ ratio of the bottom of the O/C zone (the relic of the He convective core) is 2.504, which are much higher than 1 and 0.7, respectively. In their 25M_o and 35M_o models, the $^{83}Kr/^{82}Kr$ ratio is ~0.3 (0.334 in 25M_o and 0.310 for 35M_o) in the O/C zone (6.2–6.6M_o in Fig. 2), and ~0.6 (0.623 in 25M_o and 0.597 in 35M_o) in the O/Ne zone (2.6–6.2M_o in Fig. 2). The average neutron density in the O/C zone is ≤ 10^6 n/cm^3 [31], whereas that in the O/Ne zone reaches 10^{12} n/cm^3 [32], which by far exceeds the range of a classical notion of the s-process. The KE1+KFA1 data were extrapolated to $^{83}Kr/^{82}Kr = 0.334$ and 0.623

to examine the zone where the Kr in the grains originated (Table 1). When extrapolated to 0.334 (hence assuming the Kr was produced in the O/C zone), both ^{80}Kr/^{82}Kr and ^{86}Kr/^{82}Kr ratios are close to zero, whereas the model predicts a much higher ^{80}Kr/^{82}Kr ratio. When extrapolated to 0.623, the ^{80}Kr/^{82}Kr ratio from the grains is still higher than the ratio from the model.

It is difficult to further narrow down which zone is responsible for the Kr in the grains because of huge uncertainties in ^{86}Kr and ^{80}Kr yields. The neutron capture cross section of ^{85}Kr is theoretically estimated with ~80 % uncertainty [29]. The cross section of ^{79}Se is theoretically determined with a huge uncertainty up to 50 %. Moreover, as the half-life of ^{79}Se strongly depends on temperature, a slight difference in temperature can result in a big difference in ^{80}Kr yields.

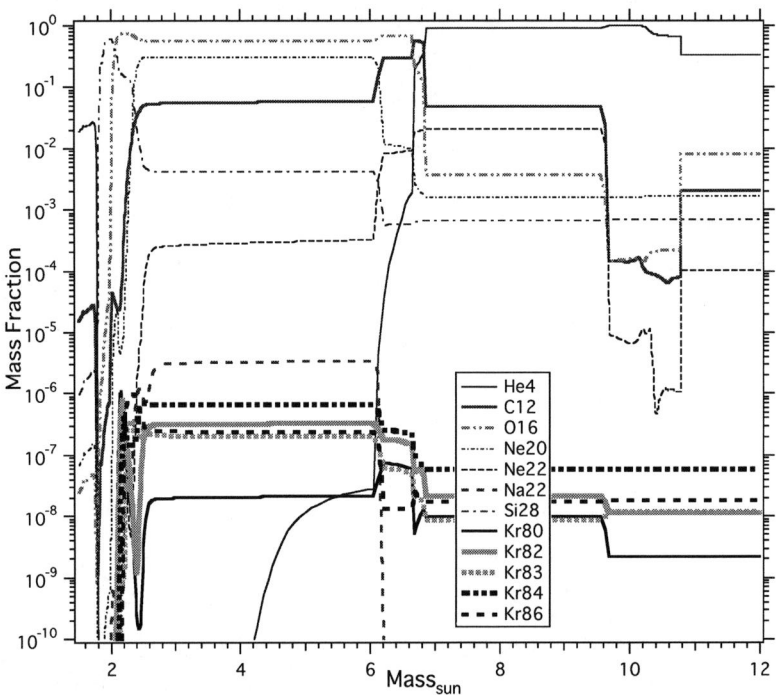

FIGURE 2. Elemental yields for a 25M$_\odot$ model with the solar metallicity 6.8×10^6 seconds after the explosion [27].

Implications for Grain Formation

The noble gas data of low-density grains reflect conditions of grain formation. The presence of the ^{22}Ne of a ^{22}Na origin indicates that the grains formed within a few years after the explosion before ^{22}Na completely decayed. This does not contradict to observations of supernovae: in SN 1987A formation of dust grains has been shown to occur in the ejecta about 400 days after the explosion, as implied by the sudden decrease in the visible light accompanied by a huge infrared counterpart [33].

Furthermore, the presence of the s-process Kr and absence of ^{22}Ne of a non-radiogenic origin in the grains could provide a further clue. Neon and Kr are inert and do not form compounds with other elements, thus implantation was the only way that the grains acquired these gases. In the O/C and O/Ne zones where the Kr in the grains was synthesized, ^{20}Ne yields are more than four orders of magnitude higher than ^{84}Kr yields [27].

When supernova ejecta expand, it is generally assumed that elements in the same zones have the same velocity. At the same velocity, Kr is implanted into a much deeper region than Na and Ne. For example, if their velocity relative to the grains is 3100 km/s, the penetration depth of Ne into carbonaceous grains is 0.65 μm and that of Kr is 3.4 μm, respectively. If Ne was lost by diffusion loss, Ne of a ^{22}Na origin should have been equally lost: penetration depths of Na and Ne are almost identical.

Grain formation in supernova ejecta is a complex process and we need to take various things into account to explain the formation conditions. When supernova ejecta hit the interstellar medium or a circumstellar shell that had been expelled from a progenitor star, the ejecta were heated by the reverse shock and elements in the same zones obtained the same energies thus different velocities. It is observed that there are zones with a different degree of ionization [34]. A work to disentangle the clues on grain formation is in progress.

CONCLUSIONS

Low-density graphite grains, characterized by ^{18}O excesses and Si isotopic anomalies (mainly ^{28}Si excesses), formed in supernova ejecta. They contain ^{22}Ne from the decay of ^{22}Na and s-process Kr. The former was produced in the O/Ne zone (C convective shell) during C burning, while the latter was produced either in the O/Ne zone (C convective shell) or the O/C zone during core He burning. The presence of the initial ^{22}Na and Kr, and the absence of non-radiogenic ^{22}Ne may provide a further clue on graphite grain formation in supernova ejecta.

ACKNOWLEDGMENTS

This work is supported by NASA grants NNG04GG13G and NNG05GF81G (SA), and MIUR-FIRB project "The astrophysical origin of heavy elements beyond Fe" (RG). Penetration depths of Na, Ne and Kr were calculated using the SRIM (Stopping and Range of Ions in Matter) program available on http://www.srim.org/. We thank Kevin Croat for calculating the penetration depths with his computer.

REFERENCES

1. E. Anders "Circumstellar material in meteorites: noble gases, carbon and nitrogen", In *Meteorites and the Early Solar System,* edited by J. F. Kerridge and M. S. Matthews, Tucson: University of Arizona Press, 1988, pp. 927.
2. D. C. Black and R. O. Pepin, *Earth Planet. Sci. Lett.,* 6, 395 (1969).
3. D. C. Black, *Geochim. Cosmochim. Acta,* 36, 377 (1972).
4. M. H. A. Jungck, "Pure ^{22}Ne in the Meteorite Orgueil", München: E. Reinhardt, 1982, pp. 80.

5. R. S. Lewis, M. Tang, J. F. Wacker, E. Anders and E. Steel, *Nature,* 326, 160 (1987).
6. T. Bernatowicz, G. Fraundorf, M. Tang, E. Anders, B. Wopenka, E. Zinner and P. Fraundorf, *Nature,* 330, 728 (1987).
7. M. Tang and E. Anders, *Geochim. Cosmochim. Acta,* 52, 1235 (1988).
8. R. S. Lewis, S. Amari and E. Anders, *Nature,* 348, 293 (1990).
9. R. S. Lewis, S. Amari and E. Anders, *Geochim. Cosmochim. Acta,* 58, 471 (1994).
10. R. Gallino, M. Busso, G. Picchio and C. M. Raiteri, *Nature,* 348, 298 (1990).
11. S. Amari, E. Anders, A. Virag and E. Zinner, *Nature,* 345, 238 (1990).
12. S. Amari, R. S. Lewis and E. Anders, *Geochim. Cosmochim. Acta,* 59, 1411 (1995).
13. T. J. Bernatowicz and E. Zinner, "Astrophysical Implications of the Laboratory Study of Presolar Materials", AIP Conf. Proc. 402, AIP, New York, 1997, pp. 750.
14. E. Zinner, *Ann. Rev. Earth Planet. Sci.,* 26, 147 (1998).
15. K. Lodders and S. Amari, *Chem. Erde,* 65, 93 (2005).
16. S. Amari, R. S. Lewis and E. Anders, *Geochim. Cosmochim. Acta,* 58, 459 (1994).
17. S. Amari, E. Zinner and R. S. Lewis, *Astrophys. J.,* 447, L147 (1995).
18. R. H. Nichols, Jr., K. Kehm, C. M. Hohenberg, S. Amari and R. S. Lewis, *Geochim. Cosmochim. Acta,* submitted (2006).
19. S. Amari, E. Zinner and R. S. Lewis, *Astrophys. J.,* 470, L101 (1996).
20. C. Travaglio, R. Gallino, S. Amari, E. Zinner, S. Woosley and R. S. Lewis, *Astrophys. J.,* 510, 325 (1999).
21. L. R. Nittler et al., *Astrophys. J.,* 453, L25 (1995).
22. P. Hoppe, R. Strebel, P. Eberhardt, S. Amari and R. S. Lewis, *Meteorit. Planet. Sci.,* 35, 1157 (2000).
23. L. R. Nittler, S. Amari, E. Zinner, S. E. Woosley and R. S. Lewis, *Astrophys. J.,* 462, L31 (1996).
24. P. Hoppe, R. Strebel, P. Eberhardt, S. Amari and R. S. Lewis, *Science,* 272, 1314 (1996).
25. S. E. Woosley and T. A. Weaver, *Astrophys. J. Suppl.,* 101, 181 (1995).
26. B. S. Meyer, T. A. Weaver and S. E. Woosley, *Meteoritics,* 30, 325 (1995).
27. A. Chieffi and M. Limongi, in preparation (2006).
28. A. Heger, S. E. Woosley, T. Rauscher and R. D. Hoffman, in preparation (2006).
29. Z. Y. Bao, H. Beer, F. Käppeler, F. Voss, K. Wisshak and T. Rauscher, *At. Data Nucl. Data Tables,* 76, 70 (2000).
30. N. Klay and F. Käppeler, *Phys. Rev. C,* 38, 295 (1988).
31. C. M. Raiteri, R. Gallino, M. Busso, D. Neyberger and F. Käppeler, *Astrophys. J.,* 419, 207 (1993).
32. S. Bisterzo, R. Gallino, M. Pignatari, L. Pompeia, K. Cunha and V. Smith, *Mem. Soc. Astron. It.,* 75, 741 (2004).
33. R. McCray, *Ann. Rev. Astron. Astrophys.,* 31, 175 (1993).
34. R. A. Chevalier and C. Fransson, *Astrophys. J.,* 420, 268 (1994).

Stardusts in Meteorites –Precursors of Planets–

Hisayoshi Yurimoto

Department of Natural History Sciences, Hokkaido University (HokuDai), Sapporo 060-0810, Japan

Abstract. Presolar circumstellar grains have been surveyed in 18 primitive meteorites. More than hundreds presolar grains have been identified. Presolar silicates are the most abundant species among presolar grains. The typical size is ~300 nm and the abundance is ~50 ppm in the most primitive chondrites. Main source of silicate presolar grains is from AGB and red giant stars. The average O-isotopic composition of presolar silicates is enriched in ^{17}O relative to the solar composition. The counterpart to form solar isotope ratios having ^{17}O-depleted compositions are missing in the chondrites. The missing matter would be supernovae ejecta but it is difficult to identify because the grain size is expected to be ~10 nm.

Keywords: presolar grain, silicate, carbon, meteorites, SIMS, isotopes
PACS: 96.30.Za, 32.10.Bi, 68.49.Sf, 82.80.Ms, 91.65.Dt, 97.10.-q, 81.05.Uw

INTRODUCTION

Meteorites are basically classified into primitive and differentiated ones. The primitive meteorites are named chondrites. Among the chondrites, petrographic type 3 chondrites, which are characterized as the matrix is less altered and less metamorphosed, are often named primitive chondrites.

The primitive chondrites are mainly composed by mechanical aggregates of dusts existed in the protoplanetary disk around the proto-sun. If the most parts of the disk have maintained in low temperature below evaporation temperatures of rocky dusts, most dusts in the matrix corresponds to circumstellar dusts that mean presolar grains. Isotope ratios of circumstellar dusts would be distinctive each other because isotope ratios of star are diverse among the stellar types. However, chondrite matrices have homogeneous isotope ratios within % order variations down to micrometer scale [1].

Recently it is discovered that circumstellar silicates have survived in matrices of primitive carbonaceous chondrites [2, 3]. The circumstellar silicates are characterized by clearly distinct isotope ratios of oxygen having more than % order variations from the solar composition. Because of the very small grain size of hundreds nanometers and small abundances of tens ppm, existence of the circum stellar silicates is consistent with the isotopically homogeneous feature of the matrix in micrometer scale.

Here we report that circumstellar silicates have been ubiquitous in various chemical groups of chondrites.

EXPERIMENTAL

A HokuDai isotope microscope system (Cameca ims-1270 + SCAPS; originally installed in Tokyo Institute of Technology and now in Hokkaido Univ. (HokuDai)) [4] has been used for in-situ survey of presolar grains. We obtained secondary ion images of $^{12}C^-$, $^{13}C^-$, $^{12}C^-$, $^{27}Al^-$, $^{28}Si^-$, $^{16}O^-$, $^{18}O^-$, $^{16}O^-$, $^{17}O^-$, and $^{16}O^-$ for one analytical sequence. We used a 50 µm contrast aperture (CA) except for C isotopes. A 150 µm CA was used for C isotopes. Beam irradiation time for the sequence was ~1 hour. The primary beam intensity was adjusted to ~0.3 nA. The sputtering depth was less than 100 nm for the sequence. An image processing method of moving average with 3 x 3 pixels was applied to reduce the statistical error of an isotope image (isotopograph). The other analytical methods for the isotopography were same as those in [3]. The selection criterion for distinguishing presolar grain is that one of their isotopic ratio is >2σ away from the 3σ ellipse of the distribution of isotopically normal matrix.

Thin sections of primitive chondrites were prepared for the in situ survey. The mineralogical and petrographical characterization of matrices was conducted using a scanning electron microscope (JEOL JSM-5310LV or JEOL JSM-7000F) equipped with energy dispersive X-ray spectrometer (Oxford LINK ISIS or Oxford INCA). This characterization was operated before and after in situ survey of circumstellar materials.

RESULTS AND DISCUSSION

Survey of presolar circumstellar grains has been performed for 13 carbonaceous chondrites (Murchison CM2, Targish Lake C2, LEW 85332 C3ung, MAC 87300 C3ung, Acfer 094 C3ung, Adelaide C3ung, ALHA 77307 CO3.0, Y-81025 CO3.0, Vigarano CV3, NWA 530 CR2, SAH 00182 CR3, Acfer 214 CH3 and HaH 237 CB), and 5 ordinary chondrites (Semarcona LL3.0, Krymka LL3.1, Bishunpur LL3.1, JaH 026 and Dhofer 008 H/L3.2/3.3). The total survey area is ~1 mm^2, corresponding to ~7000 sets of isotope ratio images), of matrices in these primitive meteorites. We found 60 carbonaceous-, 77 silicate- and 2 oxide- presolar circumstellar grains from the survey areas. Electron microscope images of presolar circumstellar grains are shown in Fig. 2. Relatively larger grains are selected for the figures in order to show clear shape images. Grain size distributions of presolar silicates are shown in Fig. 3. Presolar silicates are smaller than 1 µm and typically ~300 nanometers. Size limit of smaller side of the distribution is ended at 100 nm. The small size limit may be due to spatial resolution limit of the measurements. The maximum abundance for presolar silicates is observed in Adelaide and in Y-81025 for ~50 ppm, and for carbonaceous grains is in Murchison for ~10 ppm.

Most presolar silicate grains from the primitive chondrites are enriched in ^{17}O relative to the solar composition, categorizing into group 1 of presolar oxide grain defined by [10] (Fig. 3). Some grains are depleted in ^{17}O and ^{18}O categorizing into group 3. Minor grains are enriched in ^{18}O and are categorized into group 4. The characteristics for the most presolar silicate grains enriched in ^{17}O would be an origin of O-rich red giant or AGB stars. Grains highly enriched in ^{18}O and depleted in ^{17}O and ^{18}O, suggesting an origin of metal-rich stars or super novae and of metal-poor red giant or AGB stars, respectively [10]. Therefore, O-rich red giant and AGB stars are the main source for presolar silicates found in meteorites. This conclusion from the isotope ratios is supported by the size distribution of the circumstellar silicates which size is estimated theoretically to be ~200 nm [11].

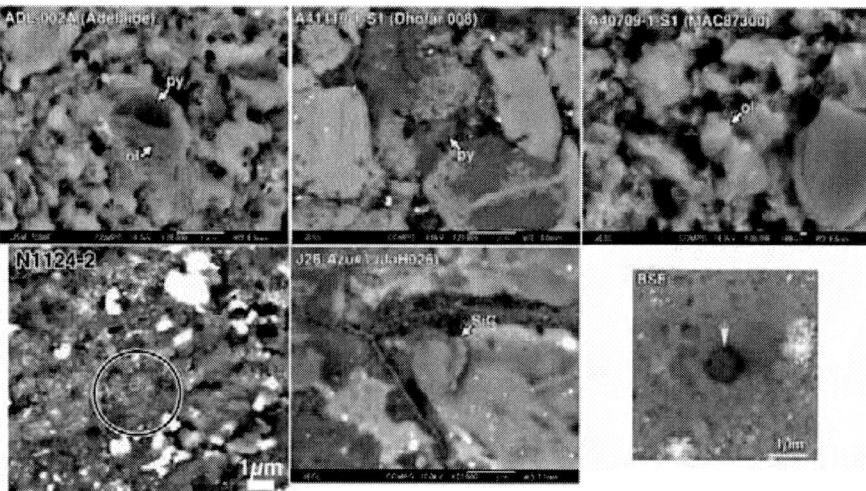

FIGURE 1. Back-scatter electron images of presolar circum stellar grains. (Top left) MgSiO$_3$ pyroxene from Adelaide. Scale bar: 1 µm. (Top center) MgSiO$_3$ pyroxene from Dhofer 008. Scale bar: 1 µm. (Top right) Mg$_2$SiO$_4$ olivine from MAC 87300. Scale bar: 100 nm. (Bottom left) Amorphous silicate from NWA 530. Scale bar: 1 µm. (Bottom center) SiC from JaH 026. Scale bar: 1 µm. (Bottom right) Graphite from NWA 530. Scale bar: 1 µm.

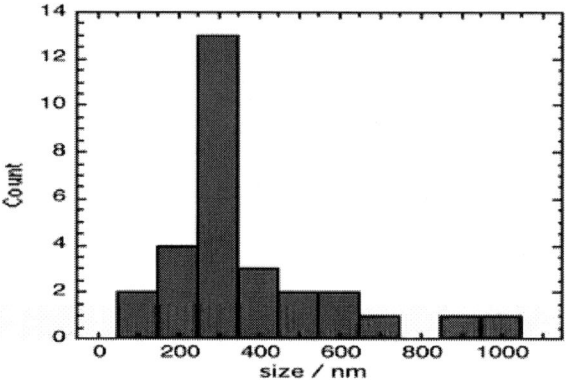

FIGURE 2. Size distribution of presolar circumstellar silicates. Data from [2, 3, 5, 6, 7, 8, 9] and this study.

Figure 3 shows that average O-isotopic composition of presolar circumstellar grains is clearly apart from the solar composition. This is a serious puzzle because all oxygen atoms composed of solar system were synthesized in stars. The circumstellar grains in meteorites are not an appropriate representative for interstellar oxygen. Supernova ejecta are a possible candidate for the missing counterpart because average O isotopic composition of supernovae is expected to be depleted in ^{18}O. Therefore, the contribution of supernovae for solar system oxygen should be comparable to those of red giant and AGB stars. However, silicate grain size formed in supernova ejecta is theoretically estimated to be ~7 nm [12]. Therefore, an isotope nanoscope with real nano-scale spatial resolution power should be necessary to detect presolar grains formed in supernova ejecta and discuss the contribution to the solar system formation quantitatively.

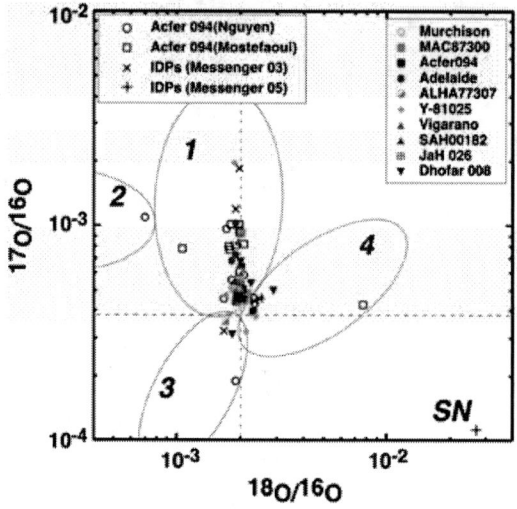

FIGURE 3. Oxygen isotopic compositions of presolar silicates in primitive chondrites. Dashed lines correspond to the solar isotopic compositions. Data from [2, 7, 9] and this study.

ACKNOWLEDGMENTS

I thank K. Nagashima, S. Kobayashi, N. Sakamoto, A. Tonotani for their collaboration. I also thank A. N. Krot and S. S. Russell for providing meteorite samples. This work supported by Monka-sho grants.

REFERENCES

1. T. Kunihiro, K. Nagashima and H. Yurimoto, *Geochim. Cosmochim. Acta* **69**, 763-773 (2005).
2. A. N. Nguyen and E. Zinner, *Science* **303**, 1496-1499 (2004).
3. K. Nagashima, A. N. Krot and H. Yurimoto, *Nature* **428**, 921-924 (2004).
4. H. Yurimoto, K. Nagashima and T. Kunihiro *Appl. Surf. Sci.* **203-204**, 793-797 (2003).
5. S. Mostefaoui, P. Hoppe, K. K. Marhas and E. Groner, *Meteorit. Planet. Sci.* **38**(Supplement), A99 (2003).
6. S. Mostefaoui, K. K. Marhas and P. Hoppe, *Lunar and Planetary Science* **XXXV**, #1593 (2004).
7. S. Messenger, L. P. Keller, F. J. Stadermann, R. M. Walker and E. Zinner, *Science* **300**, 105-108 (2003).
8. S. Messenger, L. P. Keller, *Meteorit. Planet. Sci.* **39**(Supplement), A68 (2004).
9. C. Floss and F. J. Stadermann, *Lunar and Planetary Science* **XXXV**, #1281 (2004).
10. L. R. Nittler, *Earth Planet. Sci. Lett.* **209**, 259-273 (2003).

11. H.-P. Gail and E. Sedlmayr, *Astron. Astrophys.* **347**, 594–616 (1999).
12. T. Kozasa, H. Hasegawa and K. Nomoto, *Astron. Astrophys.* **249**, 474-482.

Eu isotopic analyses of SiC grains from the Murchison Meteorite

K. Terada[a], T. Yoshida[b], N. Iwamoto[c], W. Aoki[d] and I.S. Williams[e]

[a] *Department of Earth and Planetary Systems Science, Hiroshima University, JAPAN.*

[b] *Astronomical Institute, Tohoku University, JAPAN.*

[c] *Japan Atomic Energy Agency, JAPAN.*

[d] *National Astronomical Observatory, JAPAN*

[e] *Research School of Earth Sciences, The Australian National University, AUSTRALIA*

Abstract. We report Eu isotopic analyses of single SiC grains from primitive meteorites using the Sensitive High Resolution Ion Microprobe. The results are compared with Eu isotopic ratios predicted from the recently determined ^{151}Sm(n, γ) cross sections and the thermally pulsed s-process model of AGB stars. The observed Eu isotopic compositions of SiC grains place constraints on s-process conditions such as the temperature and neutron densities in AGB stars.

Keywords: Presolar grains; Mass spectrometry; AGB star; *s*-process; neutron capture process
PACS: 96.50.Mt; 26.20.+f; 07.75.+h; 97.10.Cv

INTRODUCTION

Since the monumental study by B^{2}FH in 1957 [1], an enormous amount of work on nucleosynthesis during stellar evolution has been reported (for a recent review, see 2]). It is considered that the formation of most of the heavy elements occurs by the slow (s-process) and/or rapid (r-process) neutron capture process, that are characterized by their neutron-capture timescales in comparison with lifetimes of β-decays from unstable nuclei encountered along the neutron capture path. It is also generally agreed that Asymptotic Giant Branch stars (AGB stars) are the main contributors of s-process elements. Recent versions of the s-process model are based on AGB stellar structure computed by evolutionary codes with the artificial introduction of a parameterized ^{13}C pocket [3-6]. According to these stellar evolution models of thermally pulsing AGB stars, the best candidate reactions for the neutron source are ^{13}C(α, n)^{16}O and/or ^{22}Ne(α, n)^{25}Mg. The former reaction operates at temperatures from about 8 keV up to 10keV in the ^{13}C-pocket at the top of the He inter-shell during the inter-pulse phase, where the neutron density is low (up to 10^7 m^{-3} in solar-metallicity stars). This phase is followed by He-shell flash episodes which are characterized by higher temperatures of 25–30 keV. In this flash phase, the

^{22}Ne(α, n)^{25}Mg reaction is activated and gives rise to a higher neutron density (10^8–10^{11} cm^{-3} in the case of solar-metallicity stars).

Several approaches based on the chemical/isotopic composition of the bulk of Solar system materials have been used to try and understand the "origin of the s-process isotopes in the Solar system" [e.g. 7-10]. For example, based on the σ N curve of isotopes produced only by the s-process, Howard et al. [7] calculated average neutron densities of N_n=1.1 (+0.6, -0.3)x10^8 cm^{-1} at an optimum temperature of T = (2.7±0.3) x 10^8 K and a mean neutron exposure of τ_0 = 0.26 ± 0.01 mbarn^{-1}. Wisshak et al. [8] and Toukan et al. [9] suggested that neutron densities were possibly N_n = (4.1±0.6)x10^8 cm^{-3} and N_n =(3.8±0.6)x10^8 cm^{-3}, respectively, based on the ^{148}Sm/^{150}Sm isotopic ratio of bulk solar materials. Moreover, Wisshak et al. [10] suggested a temperature of 28–33 keV based on the ^{152}Gd/^{154}Gd ratio of solar materials. They suggested that this higher temperature and neutron density might result from the ^{22}Ne(α, n)^{25}Mg reaction being the neutron source.

In contrast, a series of studies of relict presolar grains in primitive meteorites has provided a different point of view of nucleosynthesis in AGB stars (for a general review, see [11]). For example, the enhancements of ^{25}Mg expected as a result of the ^{22}Ne(α, n)^{25}Mg reaction are uncommon, even in the presolar grains rich in s-process elements [12], indicating that the ^{13}C(α, n)^{16}O reaction is a more likely neutron source. Nicolussi et al. [13] suggested, based on Mo and Zr measurements, that most of the SiC grains condensed around low-mass AGB stars where the ^{13}C(α, n)^{16}O reaction is the principal neutron source. Moreover, Nicolussi et al. [14] suggested that neutron densities in most parent stars were lower than 10^7 cm^{-3}, based on the observation that the measured ^{88}Sr/^{86}Sr ratios in most SiC grains indicated the dominant β decay of short-lived ^{85}Kr. To better understand the physical parameters of nucleosynthesis in parent stars prior to formation of the Solar system, we focus here on the s-process parameters reflected in Eu isotopic compositions.

ANALYTICAL METHODS AND RESULTS

Presolar SiC grains were recovered from the Murchison meteorite using the procedure of Amari and others [15]. The grains remaining after acid digestion and density separation, corresponding to the KJ fraction [15], were mounted by pressing them onto a copper plate. They were analyzed for major elements and identified by the Electron Probe Micro Analyzer at Hiroshima University, then the three largest SiC grains (>5 µm) were selected for isotopic study.

First the Si and C isotopic compositions of individual grains were measured on the SHRIMP II at the Australian National University, using a Cs$^+$ primary ion beam. The Eu isotopic compositions of same grains were then measured on the SHRIMP II at Hiroshima University, using an O_2^- primary beam. Details of the experimental procedures and calibrations are given in [16].

The SiC isotopic compositions measured are listed in Table-1. Grains SiC35 and SiC67 have slightly lower ^{12}C/^{13}C than the terrestrial reference value of 89. They are also slightly enriched in ^{29}Si and ^{30}Si relative to ^{28}Si. These are the characteristics of Mainstream grains [17], considered to originate from low-mass AGB stars with a solar metallicity (abundance of all elements heavier than He). On the other hand, grain SiC39 is depleted in ^{12}C and has a large excess in ^{30}Si relative to ^{28}Si and ^{29}Si (^{12}C/^{13}C = 55, $\delta(^{30}$Si/^{28}Si) = 278 and $\delta(^{29}$Si/^{28}Si) = 12), indicating that it is a Z-grain (Zinner, 2004). It is proposed that such grains come from low-mass stars of much lower metallicity (approximately less than one-third solar) [17, 18].

The measured ^{153}Eu/^{151}Eu ratios in Mainstream grains SiC35 and SiC67 are 1.29 ± 0.24 and 1.33 ± 0.24, within error of the solar value of 1.09 [19], but significantly higher than the ratios of 0.85 and 0.71 predicted for products of the s-process by the stellar and classical models respectively [20]. In contrast, the Eu content of Z-grain SiC39 was so low that a precise isotopic analysis could not be obtained (0.97 ± 0.52). This might be concerned with the parent AGB star for this type of grain being a low-metallicity star. The discussion below is based on an average ^{153}Eu/^{151}Eu ratio of 1.3 ± 0.2 in the two Mainstream grains, which were considered to originate from low-mass AGB stars with a solar metallicity.

TABLE 1. Isotopic composition of single SiC grains

	^{12}C/^{13}C	δ ^{29}Si/^{28}Si (‰)	δ ^{30}Si/^{28}Si (‰)	^{153}Eu/^{151}Eu
SiC35	87.0 ± 0.2	27.0 ± 37.9	36.0 ± 27.0	1.29 ± 0.24
SiC39	55.4 ± 6.1	11.8 ± 13.2	278.0 ± 14.8	0.97 ± 0.52
SiC67	87.9 ± 0.3	31.4 ± 14.6	9.7 ± 14.5	1.33 ± 0.24
Solar System [19]	89	0	0	1.09

The error assigned to the isotopic ratios is 1 sigma

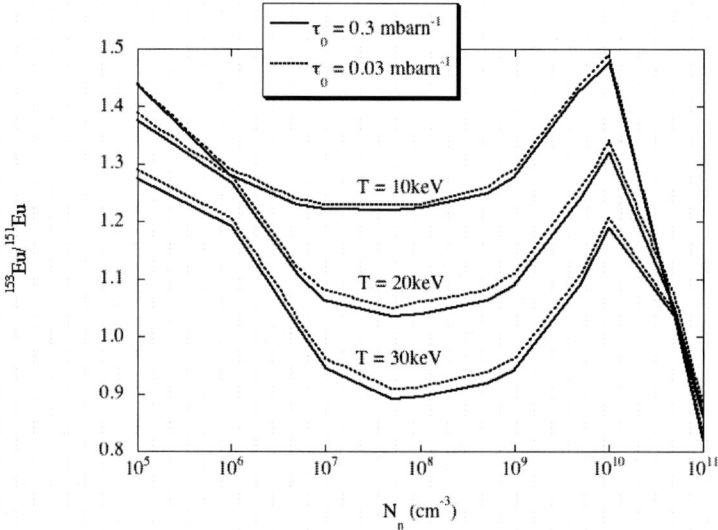

FIGURE 1. The dependence of ^{153}Eu/^{151}Eu ratios on temperature (T), neutron densitiy (N_n) and neutron exposure (τ_0).

DISCUSSION

To further investigate the s-process by using Eu isotopes, we have calculated expected Eu isotopic ratios as a function of nucleosynthetic conditions, in accordance with [21]. Here, in order to simply understand complicated nuclear flows (which are realized in the s-process of an AGB model) in wide ranges of temperature and neutron density, we adopted simple assumptions (constant temperature and neutron density conditions, in which recurrent neutron exposures are followed) for the usual s-process with the two major neutron sources. To follow the nuclear flow around the Nd-Pm-Sm-Eu-Gd region, we have used a nuclear network code in which the temporal variation of nuclides is solved with updated neutron capture rates [22], beta-decay rates and electron capture rates [23] under with various temperature and neutron density conditions. We used the recently reported neutron capture cross section for the unstable nucleus ^{151}Sm ([24], [25]). The new ^{151}Sm(n,γ) rate is a factor of 2 larger than previous theoretical predictions (e.g., [26]).

It is well known that the s-process in AGB stars depends on the temperature, neutron density and total neutron exposure in the He-layer. First we investigated the dependency of ^{153}Eu/^{151}Eu on these parameter (Figure-1). The predicted ^{153}Eu/^{151}Eu ratios are sensitive to temperature (T) and neutron density (N_n), so the observed ^{153}Eu/^{151}Eu ratio could be a good probe for the values of these s-process parameters in the He-layer. On the other hand, as Aoki et al. [23] previously pointed out in the range $0.2 \leq \tau_0 \leq 0.8$, the Eu isotopic ratio has almost no dependence on the mean neutron exposure (τ_0) for a fixed temperature in the range $0.03 \leq \tau_0 \leq 0.8$ under our present assumptions. As shown in Figure 1, the differences between ^{153}Eu/^{151}Eu ratios at $\tau_0 = 0.3$ and 0.03 mbarn^{-1} are less than 2%, indicating that the Eu isotopic ratio is not sensitive to the neutron exposure.

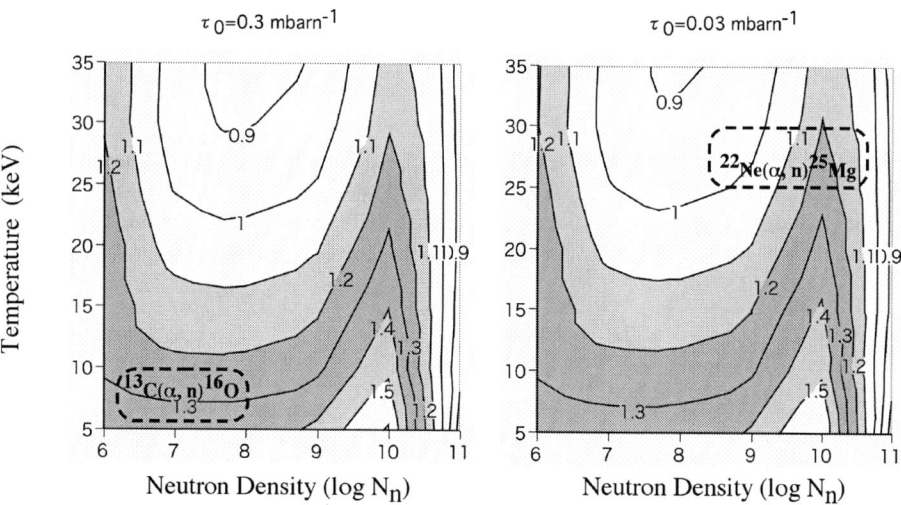

FIGURE 2. Contour maps of calculated ^{153}Eu/^{151}Eu ratios based on the thermally pulsed s-process model using the new ^{151}Sm(n,γ) reaction rate.

Figure 2 illustrates the changes in ^{153}Eu/^{151}Eu as a function of temperature and neutron density for different neutron exposures ($\tau_0 = 0.3$ mbarn^{-1} and $\tau_0 = 0.03$ mbarn^{-1}, respectively). The observed ^{153}Eu/^{151}Eu ratio of the Mainstream grains (average 1.3 ± 0.2) constrains the possible s-process temperature and neutron density conditions to the gray area in Figure 2; that is, (i) T < 15 keV for 10^7–10^9 cm^{-3}, and/or (ii) T > 10 keV for 10^{10} cm^{-3}. The plausible regions of ^{13}C(α, n)^{16}O and/or ^{22}Ne(α, n)^{25}Mg based on the recent stellar evolution models of thermally pulsing AGB stars [3-6], are also shown as dashed enclosures. It should be noted that ^{153}Eu/^{151}Eu values previously predicted by the stellar model (0.85) and the classical model (0.71) assuming that $N_n = 4.1 \times 10^8$ cm^{-3} [20] are out of range of not only the possible neutron sources of ^{13}C(α, n)^{16}O and ^{22}Ne(α, n)^{25}Mg, but also of the extended region of Figure 2 (5–35 keV and 10^6–10^{11} cm^{-3}). This discrepancy is possibly due to the different reaction rate previously used for ^{151}Sm(n,γ).

Both the recent theoretical studies [6] and the analysis of individual presolar grains [13] shows that s-process elements in the solar system are derived predominantly from ^{13}C(α, n)^{16}O sources. It should be noted that our data closely match the current ^{13}C-pocket model (<10 keV at about 10^7 cm^{-3}), but the neutron source of ^{22}Ne(α, n)^{25}Mg is still possibly about 10^{10} cm^{-3}. For tighter constraints on the s-process conditions, much higher precision analyses of ^{153}Eu/^{151}Eu ratios, coupled with analyses of isotopic compositions of other heavy elements sensitive to s-process branchings in the same grains, would be required.

Eu itotope ratios are also measured for a small number of metal-deficient stars (see [27] for a brief summary of isotope measurements for stars). Further observations of stellar isotope abundances and comparisons with measurements for pre-solar grains are desired for understanding the details of s-process nucleosynthesis.

ACKNOWLEDGMENTS

We cordially appreciate the instructive comments of Dr. S. Amari on the sample separation. We thank K. Itoh and Y. Shibata for EPMA analyses. This study was in part jointly supported by the Japan Society for the Promotion of Science and the Australian Academy of Science in 2000 and 2003, who funded the visits of K. Terada to the Australian National University. This contribution is an outcome of a joint project between the Hiroshima and RSES SHRIMP laboratories.

REFERENCES

1. E. M. Burbidge, G. R. Burbidge, W. A. Fowler and F. Hoyle, *Rev. Mod. Phys.* **29**, 547-650 (1957)
2. J. W. Truran, Jr. and A. Heger, "Origin of the Elements", in *Meteorites, Comets, and Planets*, edited by A. M. Davis, Vol. 1 *Treatise on Geochemistry, Oxford, Elsevier-Pergamon*, 2005, pp. 1-15.
3. R. Gallino, C. Arlandini, M. Busso, M. Lugaro, C. Travaglio, O. Straniero, A. Chieffi and M. Limongi, *Astrophysical Journal* **497**, 388-403 (1998).
4. M. Busso, R. Gallino and G. J. Wasserburg, *Annual Review of Astronomy and Astrophysics* **37**, 239-309 (1999).
5. S. Goriely and N. Mowlavi, *Astronomy and Astrophysics* **362**, 599-614 (2000).
6. J. C. Lattanzio and M. A. Lugaro, *Nuclear Physics A* **758**, 477c-484c (2005).

7. W. M. Howard, G. J. Mathews, K. Takahashi, and R. A. Ward, *Astrophysical Journal* **309**, 633-652 (1986).
8. K. Wisshak, K. Guber, F. Voss, F. Käppeler and G. Reffo, *Physical Review C* **48**, 1401-1419 (1993).
9. K. A. Toukan, K. Debus, F. Käppeler and G. Reffo, *Physical Review C* **51**, 1540-1550 (1995).
10. K. Wisshak, F. Voss, F. Käppeler, K. Guber, L. Kazakov, N. Kornilov, M. Uhl and G. Reffo, *Physical Review C* **52**, 2762-2779 (1995).
11. E. K. Zinner, "Presolar Grains", in *Meteorites, Comets, and Planets*, edited by A. M. Davis, Vol. 1 *Treatise on Geochemistry, Oxford, Elsevier-Pergamon, 2005, pp. 17-39.*
12. S. Amari (private communication)
13. G. K. Nicolussi, M. J. Pellin, R. S. Lewis, A. M. Davis, S. Amari and R. N. Clayton, *Geochimica et Cosmochimica Acta* **62**, 1093-1104 (1998)
14. G. K. Nicolussi, M. J. Pellin, R. S. Lewis, A. M. Davis, R. N. Clayton and S. Amari, *Physical Review Letters* **81**, 3583-3586 (1998)
15. S. Amari, R. S. Lewis and E. Anders, *Geochimica et Cosmochimica Acta* 58, 459-470 (1994).
16. Terada et al. New Astronomy Review submitted (2006).
17. P. Hoppe et al., *Astrophysical Journal Letter* **487**, 101-104 (1997).
18. M. Lugaro, E. K. Zinner, R. Gallino and S. Amari, *Astrophysical Journal Letter* 527, 369-394 (1999).
19. E. Anders and N. Grevesse, *Geochimica et Cosmochimica Acta* 53, 197-214 (1989).
20. C. Arlandini, F. Käppeler and K. Wisshak, *Astrophysical Journal* 525, 886-900 (1999).
21. W. Aoki et al. *Astrophysical Journal Letter* **592**, 67-70 (2003).
22. Z. Y. Bao, H. Beer, F. Käppeler, F. Voss, K. Wisshak and T. Rauscher, *At. Data Nucl. Data Tables* **76**, 70-154 (2000).
23. K. Takahashi and K. Yokoi, *At. Data Nucl. Data Tables* **36**, 375-409 (1987).
24. S. Marrone et al., *Nuclear Physics A* **758**, 533c- 536c (2005).
25. U. Abbondanno et al., *Physical Review Letter* **93**, 161103 (2004).
26. J. Best et al. *Physical Review C* **64**, 015801. (2001).
27. W. Aoki proc. of 'Origin of Matter and Evolution of Galaxies 2003' (World Scientific), 2004, pp429.

1.14 NUCLEAR ASTROPHYSICS and NUCLEOSYNTHESIS (II)

Composition of the Innermost Core Collapse Supernova Ejecta and the νp-Process

C. Fröhlich[*], M. Liebendörfer[*,†], G. Martínez-Pinedo[**,‡], F.-K. Thielemann[*], E. Bravo[§], N. T. Zinner[¶], W. R. Hix[‖], K. Langanke[**,¶], A. Mezzacappa[‖] and K. Nomoto[††]

[*]*Departement für Physik und Astronomie, Universität Basel, CH-4056 Basel, Switzerland*
[†]*CITA, University of Toronto, Toronto ON M5S 3H8, Canada*
[**]*Gesellschaft für Schwerionenforschung, D-64291 Darmstadt, Germany*
[‡]*ICREA and IEEC, Universitat Autònoma de Barcelona, E-08193 Barcelona, Spain*
[§]*Universitat Politècnica de Catalunya, E-08034 Barcelona, Spain*
[¶]*Institute of Physics and Astronomy, Aarhus University, Aarhus C, Denmark*
[‖]*Physics Division, Oak Ridge National Laboratory, Oak Ridge TN 37831-6374, USA*
[††]*Department of Astronomy, University of Tokyo, Tokyo 113-033, Japan*

Abstract. With presently known input physics and computer simulations in 1D, a self-consistent treatment of core collapse supernovae does not lead to explosions, while 2D models show some promise. Thus, there are strong indications that the delayed neutrino mechanism works combined with a multi-D convection treatment for unstable layers. On the other hand there is a need to provide correct nucleosynthesis abundances for the progressing field of galactic evolution and observations of low metallicity stars. The innermost ejecta is directly affected by the explosion mechanism, i.e. most strongly the yields of Fe-group nuclei for which an induced piston or thermal bomb treatment will not provide the correct yields because the effect of neutrino interactions is not included. We apply parameterized variations to the neutrino scattering cross sections and alternatively, parameterized variations to the neutrino absorption cross sections on nucleons in the "gain region". We find that both measures lead to similar results, causing explosions and a Ye larger than 0.5 in the innermost ejected layers, due to the combined effect of a short weak interaction time scale and a negligible electron degeneracy, unveiling the proton-neutron mass difference. The proton-rich environment results in enhanced abundances of ^{45}Sc, ^{49}Ti, and ^{64}Zn as requested by chemical evolution studies and observations of low metallicity stars. Moreover, antineutrino capture on the free protons allows for an appreciable production of nuclei in the mass range up to $A = 80$ by the νp-process.

Keywords: nucleosynthesis, supernovae
PACS: 26.30.+k, 97.60.Bw

INTRODUCTION

Core collapse supernovae are powered by the gravitational collapse of a stellar core. While the trajectory through core collapse determines the state of the cold nuclear matter inside the protoneutron star (PNS), the mass of the hot mantle around the PNS grows by continued accretion. The infalling matter is heated and dissociated by the impact at the fairly stationary accretion front and continues to drift inward. At first, it can still increase its entropy by the absorption of a small fraction of outstreaming neutrinos (heating region). Further in, when the matter settles on the surface of the PNS, neutrino emission dominates absorption and the electron fraction and entropy decrease significantly (cooling region). One half to two third of the neutrino luminosity in the heating region stems from the accreting matter in the cooling region, the smaller part diffuses out of the PNS. The accretion rate adjusts to the rate of neutrino absorptions in the heating region. And the neutrino luminosity adjusts to the accretion rate. This tight non-local feedback is well captured in computer simulations in spherical symmetry that accurately solve the Boltzmann neutrino transport equation for the three neutrino flavors [1, 2, 3, 4]. The latter two references implemented all equations in general relativity. One result of these calculations was that all progenitor stars between main sequence masses of 13 and 40 $M_\odot$ showed no explosions in simulations of the postbounce evolution phase (see Fig. 2). This indicates that the neutrino flux from the PNS has not the fundamental strength to blow off the surrounding layers for a vigorous explosion without the consideration of more details. Among the reasons for the failure are the strong deleptonization during collapse, the dissociation of heavy nuclei by the shock, a slow penetration of the neutrinos through layers with densities between $10^{12} - 10^{13}$ g/cm^3, efficient neutrino cooling at the surface of the PNS, and fast infall velocities through the gain region (see e.g. Ref. [5]).

Spherically symmetric simulations ignore fluid instabilities that are known to exist between the PNS surface and the stalled shock, as well as deep in the PNS [6]. However, first simulations in axisymmetry with energy-dependent neutrino transport do still not obtain vigorous explosions (see Ref. [7] and references therein). Neither did the originally suggested PNS convection [23] help because the crucial region around the neutrinospheres appears to remain convectively stable (see Ref. [8] and references therein). The reliable modeling of the coupling of the cooling and heating regions with the feedback between accretion flows and neutrino luminosity is much more involved in multiple dimensions than under the assumption of spherical symmetry. Although axisymmetric supernova models add an essential new dimension to the world of spherically symmetric models [7, 9, 10, 11], namely the possibility to accrete and expand matter at the same time in different locations, there are still important degrees of freedom missing. Hot and cool domains in the heating and cooling regions can only assume toroidal shapes around the axis of symmetry and convection will always imply the motion of the whole torus. Axisymmetric simulations cannot resolve small convective volumes or funnels that link the outer layers to the surface of the PNS. Three-dimensional simulations based on restricting neutrino physics approximations have been carried out in Ref. [12, 13]. In order to illustrate the emergence of non-trivial dynamics after bounce, Fig. 1 shows a snapshot at 40 ms after bounce from a three-dimensional simulation of collapse and bounce with magnetic fields based on Ref. [14].

MODELING CORE COLLAPSE SUPERNOVA NUCLEOSYNTHESIS

The complexity of neutrino transport and the frequent failure of self-consistent models of core collapse supernovae to produce explosions have generally divorced the modeling of core collapse supernova nucleosynthesis from the modeling of the central engine. Despite the difficulties with respect to the supernova mechanism, supernova nucleosynthesis predictions have a long tradition. All of these predictions rely on artificially intro-

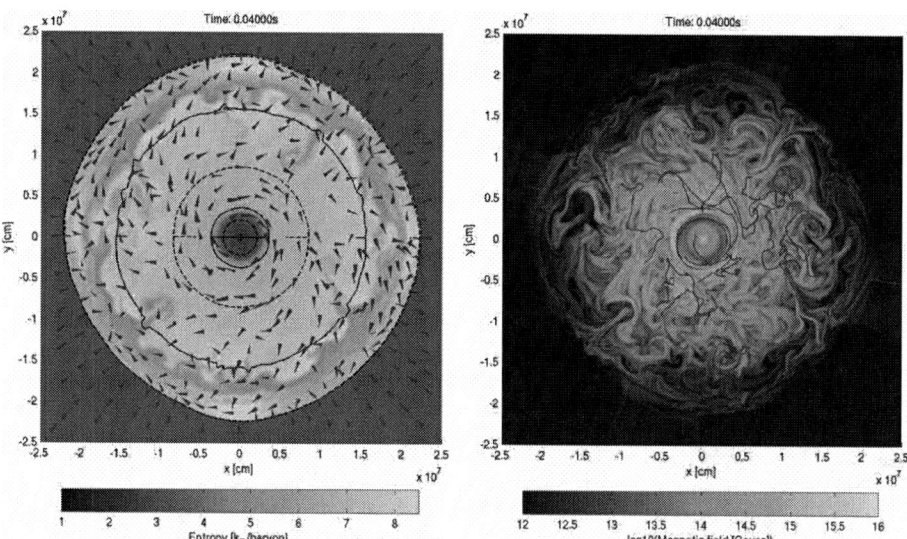

FIGURE 1. Snapshot of a three-dimensional simulation with rotation and magnetic fields at 40 ms after bounce. The simulation is based on Ref. [14] and spans a central region of 600 km cubed at a resolution of 1 km. Due to a more sophisticated parameterization of the deleptonization [15] and an effective GR gravitational potential [16] the simulation accurately reproduces the collapse phase and bounce, but becomes inaccurate in the postbounce phase due to yet missing neutrino physics. This therefore only illustrative figure has been obtained from the postbounce evolution of a 15 $M_\odot$ progenitor model [17] with a superimposed angular velocity of $\Omega = 31.4$ rad/s along the z-axis, and a quadratic cutoff at 100 km radius. A rather large initial poloidal magnetic field has been chosen ($\sim 10^{12}$ Gauss). The graph on the left hand side shows the matter entropy. The triangles indicate the direction of the velocity. Black solid lines at 10^{10} gcm^{-3} and 10^{12} gcm^{-3} show isodensity contours. The inner solid line at 10^{12} gcm^{-3} roughly coincides with the neutrinospheres at the surface of the PNS. The graph on the right hand side shows the magnetic field strength. The dark lines illustrate magnetic field lines wound up around the PNS by differential rotation. Variations in the shock strength cause entropy variations behind the expanding accretion front that induce fluid instabilities. The magnetic field lines then become entangled and more difficult to display.

duced explosions, replacing the central engine either with a parameterized kinetic energy piston [17, 18, 19] or a thermal bomb [20, 21]. The explosion energy and the placement of the mass cut (separating ejected matter from matter which is assumed to fall back onto the neutron star) are tuned to recover the observed explosion energy and ejected ^{56}Ni mass. Both approaches are largely compatible [22] and justifiable for the outer stellar regions. But most of the Fe-group nuclei are produced in the inner regions which are most affected by the details of the explosion mechanism. Nuclei in these layers interact with the large neutrino flux emerging from the PNS and the cooling region.

As this interaction is well accounted for in spherically symmetric models that are based on accurate neutrino transport, the question arises whether these models could be set into an exploding state in order to study the resulting nucleosynthesis. Having PNS convection [23] or convective turnover [6] in mind, we use two different methods to en-

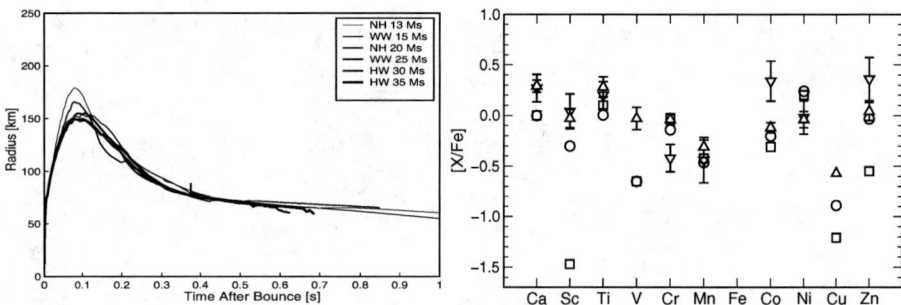

FIGURE 2. The graph on the left hand side shows the position of the accretion front as a function of time in simulations based on standard input physics, calculated with the code Agile-Boltztran [3]. The lines belong to different progenitor models ranging from 13 − 35 solar masses [25, 17, 26]. No explosion develops within the first second after bounce. The graph on the right hand side compares elemental overabundances in the mass range Ca to Zn. The triangles with error bars represent observational data. The triangles facing upward [27] originate from an analysis of stars with $-2.7 < [\text{Fe/H}] < -0.8$. The triangles facing downward [28] are data for a sample of extremely metal poor stars ($-4.1 < [\text{Fe/H}] < -2.7$). The circles are abundances of our recent calculation [24]. The squares are abundances of previous calculations in Ref. [20].

force explosions in otherwise non-explosive models. In a first approach, we parameterize the neutral current neutrino scattering opacities. This helps to artificially increase the diffusive fluxes from regions of high matter density, resulting in a faster deleptonization of the PNS and a boost of the neutrino luminosities. In a second approach, explosions are enforced by the multiplication of the reaction rates for forward and backward reactions in $\nu_e + n \rightleftharpoons p + e^-$ and $\bar{\nu}_e + p \rightleftharpoons n + e^+$ in the heating region by equal factors. This reduces the time scale for neutrino heating and leads to a more efficient deposition of neutrino energy behind the stalled shock. Both approaches allow successful explosions with a consistently emerging mass cut. Note that no external explosion energy is added and that the global energy in the simulation remains conserved.

The evolution through bounce is similar as with the unmodified models shown in Fig. 2: The shock stalls at few milliseconds after bounce and turns into an accretion front that hydrostatically expands to a radius ~ 150 km, before it starts to retract again. In the models where the input physics has been parameterized to enforce the explosion, a shock revival and explosion takes place ~ 200 ms later. See Ref. [24] for a description of the different parameter settings which we have explored.

NUCLEOSYNTHESIS IMPLICATIONS OF NEUTRINO INTERACTIONS

The electron fraction Y_e (the number of electrons per nucleon) is an indispensable quantity for the description of the explosive nucleosynthesis in the innermost ejecta. It is set by weak interactions in the explosively burning layers, i.e. electron and positron capture, beta-decays, and neutrino or anti-neutrino captures. We examined the effects of both electron and neutrino captures in the context of recent multi-group supernova simulations. These models are based on fully general relativistic, spherically symmetric Boltzmann neutrino transport [3]. Similar simulations using tracer particles from two-dimensional simulations [7] have been performed in Ref. [29]. In both cases, artificial adjustments to the simulations were needed to remedy the failure of the underlying models to produce self-consistent explosions. Also in both cases, the neutrino transport could not be run to later times and the simulations were mapped to a more simple model at later times. Despite these shortcomings, these simulations nevertheless reveal the significant impact of neutrino interactions on the composition of the ejecta.

Several phases can be identified in the evolution of the electron fraction of the matter that will become the innermost ejecta. At early times matter is degenerate and electron capture dominates. At the same time matter is being heated by neutrino energy deposition and subsequently, the degeneracy is lifted. While the ratio between electron captures and positron captures significantly decreases, neutrino absorption reactions start to dominate the change of Y_e. As the matter expands, the density decreases, and eventually the electron chemical potential drops below half the mass difference between the neutron and proton. At this energy scale, the proton is favored because of its slightly larger binding energy. With both neutrino absorption and emission processes favoring a higher electron fraction (charge neutrality links a higher proton abundance to a higher electron abundance), Y_e rises markedly in this phase, reaching values as high as 0.55. We found that all our simulations that lead to an explosion by neutrino heating developed a proton-rich environment around the mass cut with $Y_e > 0.5$ [24].

The nucleosynthesis in proton-rich ejecta has been investigated in Refs. [30, 31, 29]. The global effect of the proton-richness is the removal of previously documented overabundances of neutron rich iron peak nuclei [17, 20]. Production of ^{58,62}Ni is suppressed while ^{45}Sc and ^{49}Ti are enhanced. The results for the elemental abundances of scandium, cobalt, copper, and zinc are closer to those observed [24], as shown in Fig. 2.

However, in Ref. [24] the nucleosynthesis flow did not stop at waiting point nuclei with beta-decay lifetimes longer than the expansion time scale as in above investigations. Instead, the nucleosynthesis flow continued beyond ^{64}Ge to produce significant abundances in the range $A \sim 80$. Thus, the neutrino interactions are not only responsible for the proton-richness of the environment, they have a second important impact on the nucleosynthesis in the ejecta that has been included in these models: We found that the transformation of protons into neutrons by antineutrino capture caused (n,p)-reactions to substitute the slow β-decays in the waiting point nuclei, allowing significant flow to $A > 64$ within the expansion time scale. We termed this process the νp-process (see Ref. [32] for more details), due to the essential role of the neutrinos in producing light p-nuclei, and also, because one could interpret it as the capture of a $\bar{\nu}p$-pair in view of the long beta decay time scale of a waiting point nucleus. This process turns out to have a significant impact on the nucleosynthesis in the early neutrino wind [33] (see also Wanajo et al. in this volume), which establishes in the rarefied neutrino-irradiated region in the vicinity of the PNS after the ejection of the first layers. These results clearly illustrate the need to include the full effect of the supernova neutrino flux on the nucleosynthesis if we are to accurately calculate the nucleosynthesis from the inner layers of core collapse supernovae.

ACKNOWLEDGMENTS

This work is partly supported by the Swiss National Science Foundation grant 200020-105328 and PP002-106627, by the Spanish MCyT and European Union ERDF under contracts AYA2002-04094-C03-02 and AYA2003-06128. The work has been partly supported by the US National Science Foundation under contract PHY-0244783 and by a DoE ONP Scientific Discovery through Advanced Computing Program grant. Oak Ridge National Laboratory is managed by UT-Battelle, LLC, for the U.S. Department of Energy under contract DE-AC05-00OR22725. The supernova simulations have been performed at the Canadian Institute for Theoretical Astrophysics.

REFERENCES

1. M. Rampp, and H.-T. Janka, *A&A* **396**, 361–392 (2002).
2. T. A. Thompson, A. Burrows, and P. A. Pinto, *ApJ* **592**, 434–456 (2003).
3. M. Liebendörfer, O. E. B. Messer, A. Mezzacappa, S. W. Bruenn, C. Y. Cardall, and F.-K. Thielemann, *ApJS* **150**, 263–316 (2004).
4. K. Sumiyoshi, S. Yamada, H. Suzuki, Shen, Chiba, and Toki, (2005), astro-ph/0506620.
5. H.-T. Janka, *A&A* **368**, 527–560 (2001).
6. M. Herant, W. Benz, W. R. Hix, C. L. Fryer, and S. A. Colgate, *ApJ* **435**, 339–361 (1994).
7. R. Buras, M. Rampp, H.-T. Janka, and K. Kifonidis, *Phys. Rev. Lett.* **90** (2003).
8. S. W. Bruenn, E. A. Raley, and A. Mezzacappa (2004), astro-ph/0404099.
9. R. Walder, A. Burrows, C. D. Ott, E. Livne, I. Lichtenstadt, and M. Jarrah, *ApJ* **626**, 317–332 (2005).
10. F. D. Swesty, and E. S. Myra, *Journal of Physics Conference Series* **16**, 380–389 (2005).
11. K. Kotake, K. Sato, and K. Takahashi, (2005), astro-ph/0509456.
12. C. L. Fryer, and M. S. Warren, *ApJ* **601**, 391–404 (2004).
13. L. Scheck, T. Plewa, H.-T. Janka, K. Kifonidis, and E. Müller, *Phys. Rev. Lett.* **92**, 011103–+ (2004).
14. M. Liebendörfer, U. Pen, and C. Thompson, *Nucl. Phys. A* **758**, 59–62 (2005).
15. M. Liebendörfer, *ApJ* **633**, 1042–1051 (2005).
16. A. Marek, H. Dimmelmeier, H.-T. Janka, E. Müller, and R. Buras **445**, 273–289 (2006).
17. S. E. Woosley, and T. A. Weaver, *ApJS* **101**, 181 (1995).
18. T. Rauscher, A. Heger, R. D. Hoffman, and S. E. Woosley, *ApJ* **576**, 323–348 (2002).
19. A. Chieffi, and M. Limongi, *ApJ* **608**, 405–410 (2004).
20. F.-K. Thielemann, K. Nomoto, and M. Hashimoto, *ApJ* **460**, 408–436 (1996).
21. T. Nakamura, H. Umeda, K. Iwamoto, K. Nomoto, M.-a. Hashimoto, W. R. Hix, and F.-K. Thielemann, *ApJ* **555**, 880–899 (2001).
22. M. B. Aufderheide, E. Baron, and F.-K. Thielemann, *ApJ* **370**, 630–642 (1991).
23. J. R. Wilson, and R. W. Mayle, *Phys. Repts.* **227**, 97–111 (1993).
24. C. Fröhlich, P. Hauser, M. Liebendörfer, G. Martínez-Pinedo, F.-K. Thielemann, E. Bravo, N. T. Zinner, W. R. Hix, K. Langanke, A. Mezzacappa, and K. Nomoto (2004), astro-ph/0410208.
25. K. Nomoto, and M. Hashimoto, *Phys. Rep.* **163**, 13–36 (1988).
26. A. Heger, and S. E. Woosley, *ApJ* **567**, 532–543 (2002).
27. R. G. Gratton, and C. Sneden **241**, 501–525 (1991).
28. R. Cayrel, E. Depagne, M. Spite, V. Hill, F. Spite, P. François, B. Plez, T. Beers, F. Primas, J. Andersen, B. Barbuy, P. Bonifacio, P. Molaro, and B. Nordström, *A&A* **416**, 1117–1138 (2004).
29. J. Pruet, S. E. Woosley, R. Buras, H.-T. Janka, and R. D. Hoffman, *ApJ* **623**, 325–336 (2005).
30. F.-K. Thielemann, P. Hauser, E. Kolbe, G. Martinez-Pinedo, I. Panov, T. Rauscher, K.-L. Kratz, B. Pfeiffer, S. Rosswog, M. Liebendörfer, and A. Mezzacappa, *Space Sci. Rev.* **100**, 277–296 (2002).
31. H. Umeda, and K. Nomoto, *ApJ* **619**, 427–445 (2005).
32. C. Fröhlich, G. Martínez-Pinedo, M. Liebendörfer, F. . Thielemann, E. Bravo, W. R. Hix, K. Langanke, and N. T. Zinner, (2005), astro-ph/0511376.
33. J. Pruet, R. D. Hoffman, S. E. Woosley, H. . Janka, and R. Buras, (2005), astro-ph/0511194.

Universality of the p-process nucleosynthesis in supernova explosions and scaling laws for p- and s-process nuclei in the solar system abundances

T.Hayakawa[*,†], N.Iwamoto[**], T.Kajino[†,‡], T.Shizuma[*], H.Umeda[‡] and K.Nomoto[‡]

[*]*Japan Atomic Energy Agency, Kizu, Kyoto, 619-0215, Japan.*
[†]*National Astronomical Observatory, Osawa, Mitaka, Tokyo 181-8588, Japan.*
[**]*Japan Atomic Energy Agency, Toukai, Ibaraki 319-1195, Japan.*
[‡]*Department of Astronomy, School of Science, University of Tokyo, Tokyo 113-0033, Japan.*

Abstract. We find two empirical scaling laws in the solar system, which is evidence that the most probable origin of the p-nuclei is photodisintegration reactions in supernova explosions (p-process). We also find a novel concept of "the universality of the p-process". We discuss the mechanism of this universality in typical core-collapse supernova explosion models.

Keywords: γ-process; p-process; p-nuclei; photodisintegration nucleosynthesis
PACS: 26.30.+k; 98.80.Ft; 91.65.Dt

INTRODUCTION

Heavy elements in the Galaxy were dominantly synthesized by stellar nucleosynthesis process after the Galaxy formation. The solar system was formed from interstellar media (ISM), which composition was provided from different stellar nucleosynthesis episodes. The solar system abundance is, therefore, an important record of stellar nucleosynthesis and the galactic chemical evolution (GCE). About 99% of elements heavier than the iron group were produced by two different neutron-capture reaction chains, i.e. the s- and r-processes. The solar system abundance shows specific indication that the s- and r-processes have actually happened before the solar system formation. The first evidence is three pairs of two abundance peaks near neutron magic numbers, N = 50, 82, 126. These two peaks are corresponding to the s- and r-processes [1]. The second evidence for the s-process is an empirical relation, $N \cdot \sigma \sim$ constant, where N and σ are the solar abundance and the neutron capture cross-section [2, 3]. However, there are stable isotopes that cannot be synthesized by the neutron-capture reactions because they are located in neutron-deficient side of the β-stability line in the nuclear chart (see Fig. 1). These isotopes are called the "p-nuclei". They have a feature that their isotope abundances are very small (typically 0.1% $\sim$ 1 %). Their origin has long been discussed with many possible nuclear reactions, and their astrophysical sites have not been identified uniquely. The proposed nuclear processes are the rapid proton capture reactions in novae or Type I X-ray bursts in neutron stars (rp-process) [4], the proton-induced reactions by Galactic cosmic rays (cosmic-ray process) [5], the photodisintegration reactions in supernova (SN) explosions (γ-process or p-process) [6, 7, 8], and the neutrino-induced

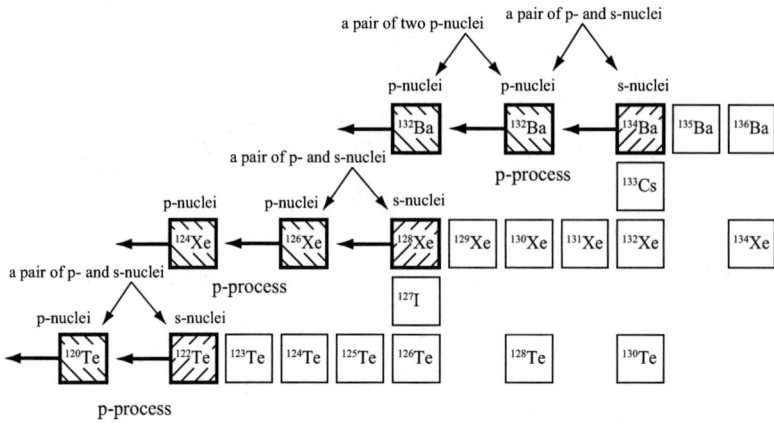

FIGURE 1. A partial nuclear chart. There are three pairs of a single p-nucleus and a single s-nucleus with the same atomic number and two pairs of two p-nuclei.

reactions in SN explosions (ν-process) [9, 10]. The origin of the p-nuclei is crucial to our understanding of how the solar-system material formed and evolved. We here present two empirical scaling laws obtained from a careful analysis of the solar system abundance, and discuss a new concept of "the universality of the p-process" [11]. We also calculate the p-process nucleosynthesis using a typical core-collapse SN model.

EVIDENCE OF THE P-PROCESS NUCLEOSYNTHESIS

Most of the p-nuclei are even-even nuclei, both the proton and neutron numbers of which are even. There are 22 pairs of a single p-nucleus and a single s-nucleus with the same atomic number. The s-nucleus is two neutron heavier than the p-nucleus in the individual pair. These s-nuclei are dominantly synthesized by the s-process and shielded by stable isobars against the β^--decay after the freezeout of the r-process. Figure 1 shows a partial nuclear chart in a Te-Xe-Ba region. A typical example is found in Ba isotopes: ^{132}Ba is a p-nucleus and ^{134}Ba is a pure s-nucleus shielded by an isobar ^{134}Xe against the β^--decay.

We here discuss the isotope abundance ratios of these two isotopes. Taking the abundance ratios of the s-nucleus to the p-nucleus, N(s)/N(p), where N(s) and N(p) are the solar abundances of the s- and p-nuclei, respectively, we find a clear correlation between them. The ratios N(s)/N(p) are almost constant (about 23) in a wide range of the atomic number with some exceptions exhibiting large deviations (see Fig. 2).

The scaling shows a strong correlation between p- and s-nuclei with the same atomic number. This indicates that the p-nuclei are synthesized from the s-nuclei, which is consistent with previous theoretical calculations that the p-nuclei are produced by the p-process (or γ-process) in SN explosions [12, 13]. In these models, the p-nuclei are synthesized by following sequence (see Fig. 3). Pre-existing nuclei in massive stars are affected by a weak s-process during the pre-supernova evolutionary stage in the massive

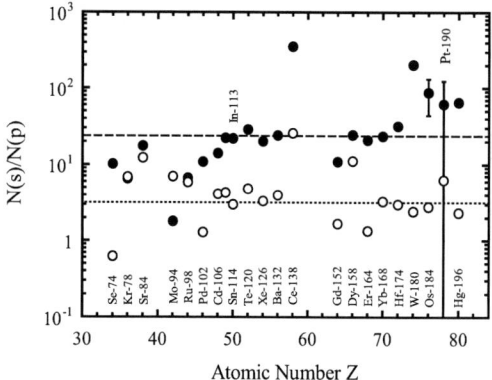

FIGURE 2. The abundance ratios of N(s)/N(p). The filled (open) circles are the observed ratios in the solar system (calculated ones). The dashed (dotted) line presents 23 (3.1).

stars. The p-nuclei are subsequently produced from the seed nuclei by photodisintegration reactions such as (γ,n) reactions in a huge photon bath at extremely high temperatures in SN explosions. The scaling can be explained by this nucleosynthesis mechanism. In contrast, the charged particle reactions in the rp-process [4] and the proton-induced reactions by the cosmic-rays process [5] change the proton number of the seed nuclei. The ν-process also changes the proton number because the charged current interaction has a contribution larger than the neutral current interaction in the ν-process [14, 15]. Therefore, the scaling is a piece of evidence that the p-process is the most promising origin of the p-nuclei.

We also find the second scaling. This is the correlation between two p-nuclei with the same atomic number. There exist nine such pairs of p-nuclei. A heavy p-nucleus, the first p-nucleus, is two neutrons heavier than the another p-nucleus, the second p-nucleus. The abundance ratios, N(2nd p)/N(1st p), are almost constant in a wide range [11].

UNIVERSALITY OF THE P-PROCESS

The material of the solar system originated from nucleosynthesis episodes in many different stars before the solar system formation. The abundance distribution in the solar system should be different from that of individual nucleosynthesis episode, because relevant conditions, such as the stellar mass, metallicity and explosion energy are different (see Fig. 3). However, recent astronomical observations for very metal-deficient stars reported a crucial discovery of the "universal" abundance distribution for the atomic number $Z > 56$, which are consistent with the abundance distribution of the r-process nuclides in the solar system [16, 17, 18, 19].

The first scaling observed in the solar system leads to a novel concept of "the universality of the p-process" that the N(s)/N(p) ratios produced by individual stellar p-process are constant in a wide range of the atomic number [11]. There are three possibilities for

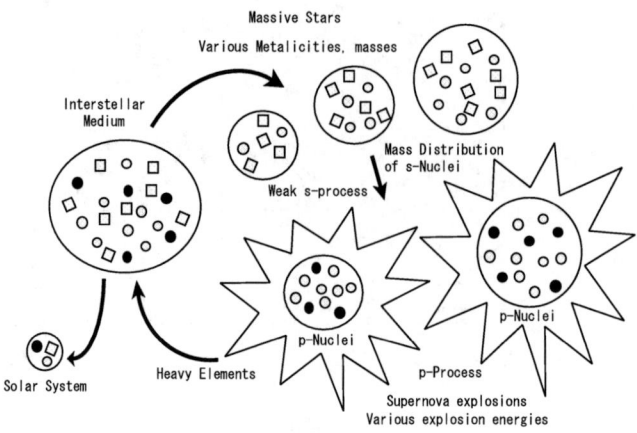

FIGURE 3. Stellar nucleosynthesis and the origin of the solar system.

the manifestation of this universality of the p-process. i) A single SN explosion strongly affects the composition of the solar system. However, this possibility is inconsistent with recent analyses of primitive meteorites that showed al least three SN explosions affect the solar system formation [24]. ii) Many SNe in the Galaxy occur and affect the solar system formation but the p-process nucleosynthesis can only occur under a specific SN. iii) The p-process episodes with various astrophysical conditions occur but the ratios are independent of the conditions. Thus, the understanding of the universality is crucial to constrain the astrophysical sites of the p-process.

We calculate nucleosynthesis of the *p*-process in oxygen-neon layers in core-collapse SN explosions [20]. We use a solar metallicity $Z = 0.02$ and a deficient metallicity $Z = 0.001$. The progenitor masses are 15, 25 and 40 solar masses ($M_\odot$). Massive stars evolve from main sequence to a core-collapse stage and explode with an explosion energy of 10^{51} ergs [21]. We use a solar abundances as the initial composition and calculate a weak s-process before SN explosions. The calculated N(s)/N(p) ratios for the various conditions are almost constant in the wide region, which is consistent with the observed ratios. Figure 2 shows a typical calculated result based on the metallicity-deficient model with $Z = 0.001$ and $M = 25$ $M_\odot$. These results suggest that even should the p-process occur under various conditions, the N(s)/N(p) ratios result in almost constant value independent of the conditions assumed.

Our calculated N(s)/N(p) ratios are fairly constant in the wide range but are an order of magnitude smaller than the observed ratios as shown in Fig. 2. We here would like to point out a fact that the s-nuclei in the solar system are consider to originate from the main s-process in asymptotic giant branch (AGB) stars [3, 22]. In contrast, the p-nuclei are synthesized by the p-process in SNe. Why is the abundances of the p-nuclei proportional to the s-nucleus abundance originated from the AGB stars ? This question can be understood by a weak s-process in a pre-supernova stage in massive stars. The heavy element seed in the massive stars are originally contaminations produced in early generation stars. Its mass distribution should be changed to that of the s-process nuclei.

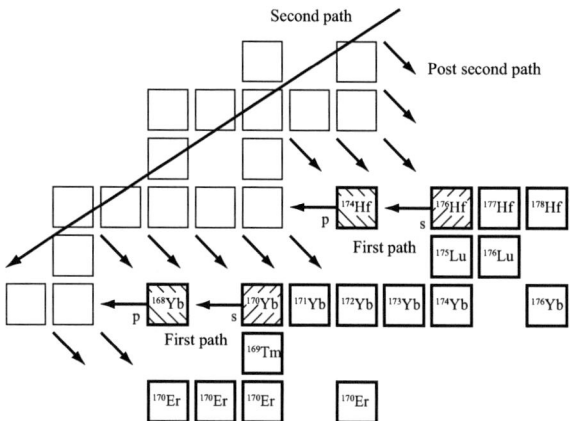

FIGURE 4. Two nucleosynthesis paths of the p-process. The neutron number of the second path flow depends on the neutron capture reaction rate.

To obtain the abundances at SN explosions, we adopt the solar abundances as the initial composition and subsequently calculate the weak s-process. The heavy elements are irradiated by neutrons provided mainly from the ^{22}Ne$(\alpha,n)^{25}$Mg reaction in the weak s-process during a core He burning stage.

The p-nuclei are produced by the two nuclear reaction paths as sketched in Fig. 4. The first is direct (γ,n) reactions from heavier isotopes with the same atomic number. Note that a part of the synthesized p-nuclei are also destroyed by the (γ,n) reactions. The second is β^--decay after the freezeout of the p process from neutron-deficient unstable nuclei via a flow on the neutron-deficient side. These nuclei are transmuted to neutron-deficient isotopes by successive (γ,n) reactions first, followed by transmutation to the lighter nuclei by nuclear reactions such as (γ,p), (γ,α), (γ,n) and (n,γ) reactions. The scaling shown in Fig. 2 indicates that the contribution of the direct reactions of the first path may be stronger than the population via the β^--decay. Woosley and Howard reported the anti-correlation between the photodisintegration reaction rates and the solar abundances of the p-nuclei [7]. This anti-correlation also suggests the large contribution of the direct reactions.

We find that the contribution of the second path is much smaller than that of the first path in our p-process nucleosynthesis calculation. The position of the second path in the nuclear chart is important for the contribution to the nucleosynthesis. Figure 4 shows a typical flow of the second path. Many nuclear reactions contribute to the nucleosynthesis in the second path but the position of the flow depends on the neutron reaction rate. The neutron capture reaction rate increases with decreasing the proton number of the nuclei and the flow of the second path shifts toward the β-stability line. A similar result was reported in a previous calculation [12]. In a region lighter than a neutron magic number, N = 82, the second path almost vanishes and the contribution of the first path becomes dominant. The neutron capture reaction rates on the p-nuclei have not been well established [25], the measurements are desired.

As the astrophysical sites of the p-process, deflagrating C/O white dwarfs [8] and accretion disks around neutron stars or black holes [26, 27] were also proposed. Whether these models can reproduce the universality of the p-process or not is a further subject.

SUMMARY

We find two empirical scaling laws in the solar system abundances. The abundance ratios of a single s-nucleus to a single p-nucleus in the same atomic number, N(s)/N(p), are almost constant in a wide region of the atomic number, where N is the isotope abundance. They are evidence that the most probable origin of the p-nuclei is photodisintegration reactions in supernova explosions (p-process). We propose a novel concept of "the universality of the p-process" that the N(s)/N(p) ratios for the elements produced by individual *p*-process are constant in a wide region of the atomic number. We calculate the p-process nucleosynthesis in typical core-collapse supernova explosion models and discuss the mechanism of the universality of the p-process.

We would like to thank M. Fujiwara, T.Komatsubara and Y.Aoki for valuable discussions and encouragement.

REFERENCES

1. E.M. Burbidge, G.R. Burbidge, W.A. Fowler, F. Hoyle, *Rev. Mod. Phys.* **29**, 548 (1957).
2. P.A. Seeger, W.A. Fowler, D.D. Clayton, Astrophys. J. **11**, 121 (1965).
3. R. Gallino, *et al.*,Astrophys. J. **497**, 388 (1998).
4. H. Schatz, *et al.*, Phys. Rev. Lett. **86**, 3471 (2001).
5. J. Audouze, Astron. Astrophys. **8**, 436 (1970).
6. M. Arnould, Astron. Astrophys. **46**, 117 (1976).
7. S.E. Woosley, W.M. Howard, Astrophys. J. Suppl. **36**, 285 (1978).
8. W.M. Howard, B.S. Meyer, S.E.Woosley, Astrophys. J. **373**, L5 (1991).
9. S. E. Woosley, D. H. Hartmann, R. D. Hoffman, W. C. Haxton, Astrophys. J. **356**, 272 (1990).
10. R.D. Hoffman, S.E. Woosley, G.M. Fuller, B.S. Meyer, Astrophys. J. **460**, 478 (1996).
11. T.Hayakawa, *et al.*, Phys. Rev. Lett. **93**, 161102 (2004).
12. M. Rayet, M. Prantzos and M. Arnould, Astron. Astrophys. 227, 271 (1990).
13. M. Rayet, *et al.*, Astron. Astrophys. **298**, 517 (1995).
14. S. Goriely, M. Arnould, I. Borzov, M. Rayet, Astron. Astrophys. **375**, L35 (2001).
15. A. Heger, *et al.*, Phys. Lett. B **606**, 258 (2005).
16. C. Sneden *et al.*, Astrophys. J. **496**, 235 (1998).
17. C. Sneden *et al.*, Astrophys. J. **533**, L139 (2000).
18. S. Honda *et al.*, Astrophys. J. **607**, 474 (2004).
19. K. Otsuki, G.J. Mathews, T. Kajino, New Astrom. **8**, 767 (2003).
20. N. Iwamoto, H. Umeda, K. Nomoto, International Symposium on Origin of Matter and Evolution of Galaxies, World Scientific, P.493 (2004).
21. K. Nomoto *et al.*, "Stellar Collapse" (Astrophysics and Space Science; Kluwer) ed. C. L. Fryer (2003)
22. M. Busso, R. Gallino, C.J. Wasserburg, Ann. Rev. Astron. Astrophys. **37**, 239 (1999).
23. T. Hayakawa *et al.*, Astriphys. J. in press.
24. Q. Yin, S.B. Jacobsen, K. Yamashita, Nature **415**, 881 (2002).
25. T.Nakagawa, S.Chiba, T.Hayakawa, T.Kajino, Atom. Data and Nucl. Data Tables, 91, 77 (2005).
26. S. Fujimoto, *et al.*, Astrophys. J. **585**, 418 (2003).
27. S. Fujimoto, *et al.*, Astrophys. J. **614**, 847 (2004).

The rp-Process in Core-collapse Supernovae

Shinya Wanajo

Research Center for the Early Universe, Graduate School of Science, University of Tokyo, Bunkyo-ku, Tokyo 113-0033, Japan

Abstract. Recent hydrodynamic simulations of core-collapse supernovae with accurate neutrino transport suggest that the bulk of the neutrino-heated ejecta is proton rich, in which the production of some interesting proton-rich nuclei is expected. However, there are a number of waiting point nuclei with the β^+-lives of a few minutes, which prevent the production of heavy proton-rich nuclei beyond iron in explosive events such as core-collapse supernovae. In this study, it is shown that the rapid proton-capture (rp) process takes place by bypassing these waiting points via neutron-capture reactions even in the proton-rich environment, if there is an intense neutrino flux as expected during the early phase of the neutrino-driven winds of core-collapse supernovae. The nucleosynthesis calculations imply that the neutrino-driven winds can be potentially the origin of light p-nuclei including ^{92,94}Mo and ^{96,98}Ru, which cannot be explained by other astrophysical sites.

Keywords: nuclear reactions, nucleosynthesis, abundances — supernovae: general
PACS: 26.30.+k; 26.50.+x; 97.10.Tk; 97.60.Bw; 98.35.Bd

INTRODUCTION

The rapid proton-capture (rp) process has been expected to take place in proton-rich compositions with sufficiently high temperature [1] such as in X-ray bursts [2], whose nucleosynthetic products might contribute to the Galactic chemical evolution of some proton-rich isotopes including p-nuclei. Recent hydrodynamic simulations of core-collapse supernovae with accurate neutrino transport show that the bulk of the neutrino-heated ejecta during the early phase is proton rich [3], in which the production of some proton-rich nuclei is expected. However, there are a number of waiting point nuclei with the β^+-lives of a few minutes on the rp-process path, which inhibit the production of heavy proton-rich nuclei beyond iron in core-collapse supernovae.

In this paper, it is shown that these waiting points are bypassed via (n,p) and (n,γ) reactions even in proton-rich environments, if there is an intense neutrino flux as expected during the early epoch of core-collapse supernovae (as also discussed in recent works [4, 5]). As a result, the rp-process takes place that leads to the production of proton-rich nuclei beyond $A \sim 100$. The nucleosynthesis calculations are performed in order to demonstrate this effect, with the semi-analytic neutrino-driven wind models [6, 7]. The results imply that this neutrino-induced rp-process in core-collapse supernovae can be potentially the origin of the light p-nuclei including ^{92,94}Mo and ^{96,98}Ru, which cannot be explained even by the most successful scenario (i.e., the O/Ne layers in core-collapse supernovae [8]) of the p-nuclei production. A more detailed discussion of the current results will be presented in a forthcoming paper [9].

NEUTRINO-DRIVEN WINDS

Assuming the spherical symmetry of the neutrino-driven winds from a nascent neutron star, the equations of baryon, momentum, and mass-energy conservation with the Schwarzschild metric can be solved numerically [6]. Once the neutron star mass (M), the neutrino sphere radius (R_ν), and the neutrino luminosity (L_ν, of one flavor, where each flavor is assumed to have the same luminosity) are specified along with the mass ejection rate ($\dot{M}$) as the boundary condition, the wind solution can be obtained. In this study, $\dot{M}$ is chosen so as the wind crosses the *sonic point* to be the transonic solution. The neutron star mass is taken to be $1.4 M_\odot$ and $2.0 M_\odot$.

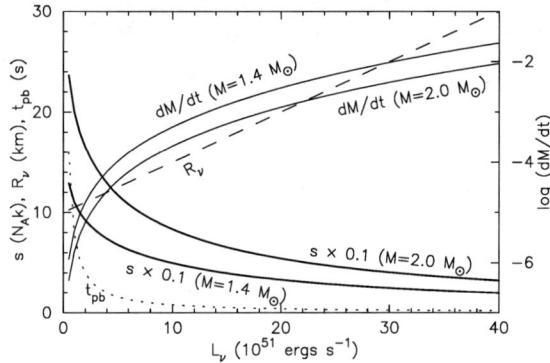

FIGURE 1. Model parameters (R_ν, t_{pb}, $\dot{M}$) are shown as functions of L_ν. Also denoted are the obtained entropies for $1.4 M_\odot$ and $2.0 M_\odot$ cases.

The time evolutions of L_ν and R_ν are assumed to be $L_\nu(t_{pb}) = L_{\nu 0}(t_{pb}/t_0)^{-1}$ and $R_\nu(t_{pb}) = (R_{\nu 0} - R_{\nu f})(t_{pb}/t_0)^{-1} + R_{\nu f}$, where t_{pb} is the post-bounce time, $t_0 = 0.2$ s, $L_{\nu 0} = 4 \times 10^{52}$ ergs s^{-1}, $R_{\nu 0} = 30$ km, and $R_{\nu f} = 10$ km, which mimic the hydrodynamic results of the neutrino-driven winds in [10]. The wind trajectories are calculated for 54 constant L_ν between 0.5 and 40 ergs s^{-1} (i.e., $0.2 < t_{pb} < 16$ s) with $R_\nu(L_\nu) = (R_{\nu 0} - R_{\nu f})(L_\nu/L_{\nu 0}) + R_{\nu f}$ deduced from the above equations. Figure 1 shows the obtained entropies in winds for $1.4 M_\odot$ and $2.0 M_\odot$ cases as functions of L_ν, along with the morel parameters $\dot{M}$, r_ν, and t_{pb}. For more details on these wind models, see [9].

NUCLEOSYNTHESIS

The nucleosynthetic yields for each wind trajectory are obtained by application of an extensive nuclear reaction network that consists of ~ 6000 species between the proton and neutron drip lines along with all relevant nuclear reaction and weak rates. The neutrino captures on free nucleons are also included. Each calculation is initiated ($t = 0$ s) when the temperature decreases to $T_9 = 9$ (where $T_9 \equiv T/10^9$ K).

Figure 2 shows the snapshot of nucleosynthesis at $t = 0.02$ s for the wind trajectory of $M = 2.0 M_\odot$, $L_\nu = 4.0 \times 10^{51}$ ergs s^{-1}, $t_{pb} = 2.0$ s, and $Y_{ei} = 0.600$, where the neutrino reactions are turned off (left panel) and on (right panel). Y_{ei} is the initial electron fraction (number of proton per baryon) at $T_9 = 9$ ($t = 0$ s). The entropy obtained for this trajectory

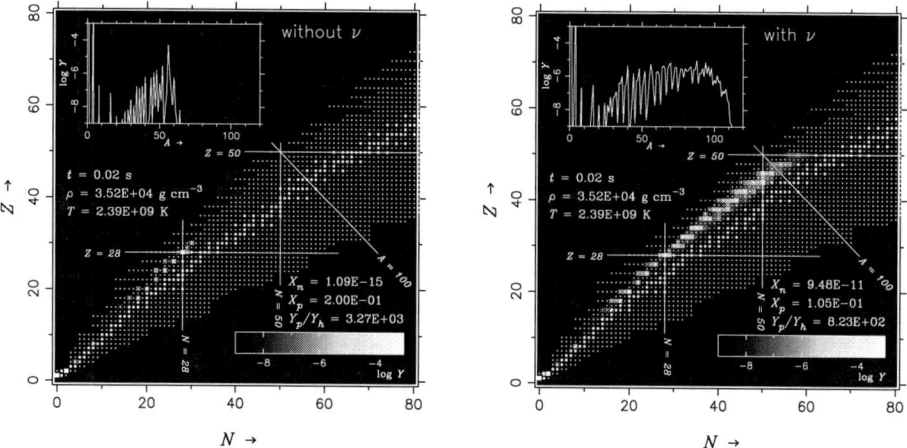

FIGURE 2. Snapshot of the nucleosynthesis at $t = 0.02$ s for the wind trajectory of $M = 2.0 M_\odot$, $L_\nu = 4.0 \times 10^{51}$ ergs s^{-1}, $t_{\rm pb} = 2.0$ s, and $Y_{ei} = 0.600$, where the neutrino reactions are turned off (left panel) and on (right panel). The abundances are shown by the image in the nuclide chart. The abundance curve as a function of mass number is shown in the upper left for each panel.

is $129 N_A k$. Without neutrino reactions, the abundances have a sharp peak at ^{56}Ni and the nuclear flow stops at ^{64}Ge owing to its long half life (1.062 min) as well as the small proton-separation energies around these species. In contrast, the inclusion of neutrino reactions leads to the nuclear flow reaching the $Z = 50$ proton-magic number, resulting in the production of proton-rich isotopes beyond $A = 100$. This *neutrino-induced rp-process* is a consequence of the presence of neutrons formed by the capture of anti-electron neutrinos on free protons. As a result, the waiting point nuclei are bypassed via neutron captures (n, p) and (n, γ), and the proton capture proceeds.

The time evolution of Y_{ei} that determines the initial composition for each wind trajectory is assumed as in the top-left panel of Figure 3, so as to mimic the hydrodynamic results in [10]. Y_{ei} is taken to be constant (Y_{e0}) for $t_0 < t_{\rm pb} \leq t_1$ and $Y_e(t_{\rm pb}) = (Y_{e0} - Y_{e1})(t_{\rm pb}/t_1)^{-1} + Y_{ef}$ for $t > t_1$, where $t_1 = 4.0$ s and $Y_{e1} = 0.100$. Y_{e0} is taken to be a free parameter, which varies from 0.460 to 0.630 with the interval of 0.005 (35 cases). This gives Y_{ei} for each wind such as $Y_e = Y_{e0}$ and $Y_e(L_\nu) = (Y_{e0} - Y_{e1})(L_\nu/L_{\nu 0})(t_1/t_0) + Y_{e1}$ for $L_\nu \geq 2.0 \times 10^{51}$ ergs s^{-1} and $L_\nu < 2.0 \times 10^{51}$ ergs s^{-1}, respectively. It should be noted that Y_e changes during the α-process phase ($T_9 \sim 7 - 4$), affected by the neutrino capture on free nucleons [11]. The bottom-left panel of Figure 3 shows the shifts ΔY_e at $t_{\rm pb} = 0.4, 1.0$, and 4.0 s ($L_\nu = 20, 8$, and 2×10^{51} ergs s^{-1}, respectively), when the temperature decreases to $T_9 = 3$ (at which the *rp*-process sets in). The left-top panel shows Y_{ei} (thin lines) and $Y_{ef} \equiv Y_{ei} + \Delta Y_e$ (thick lines) for $Y_{e0} = 0.600$ (solid lines) and 0.460 (dotted lines).

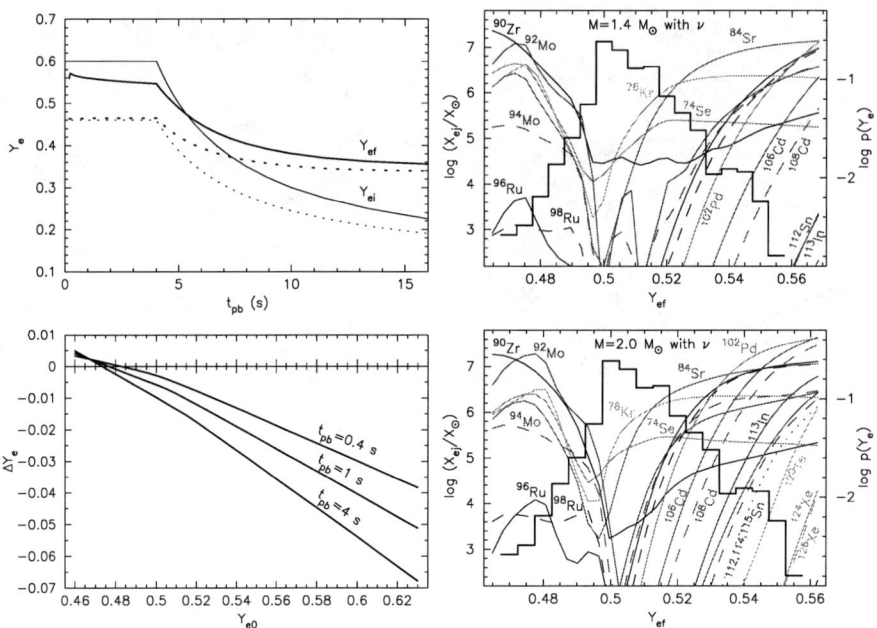

FIGURE 3. Left-top: time evolution of Y_{ei} (thin lines) and $Y_{ef} \equiv Y_{ei} + \Delta Y_e$ (thick lines) for $Y_{e0} = 0.600$ (solid lines) and 0.460 (dotted lines). Left-bottom: ΔY_e at $t_{pb} = 0.4$, 1.0, and 4.0 s ($L_\nu = 20$, 8, and 2×10^{51} ergs s^{-1}, respectively), when the temperature decreases to $T_9 = 3$ (at which the rp-process sets in). Right-top: mass-averaged abundances of p-nuclei (and ^{90}Zr for comparison purposes) with respect to their solar values for $M = 1.4 M_\odot$ case as functions of $Y_{ef}(t_{pb} = 4.0\,\text{s})$. Also shown is the histogram of the Y_e distribution $p(Y_e)$ of the neutrino-processed ejecta in a two-dimensional "exploding" core-collapse simulation by [3]. Right-bottom: same as the right-top panel, but for $M = 2.0 M_\odot$ case.

CONTRIBUTION TO THE GALACTIC CHEMICAL EVOLUTION

In order to examine the contribution of this neutrino-induced rp-process to the Galactic chemical evolution, the yields for each Y_{ei} model (35 cases) are mass-averaged over the 54 wind trajectories weighted by $\dot{M}(L_\nu)\Delta t_{pb}$. The mass-averaged abundances of p-nuclei with respect to their solar values for $M = 1.4 M_\odot$ (right-top) and $2.0 M_\odot$ (right-bottom) cases are shown in Figure 3, as functions of $Y_{ef}(t_{pb} = 4.0\,\text{s})$ (at $L_\nu = 2.0 \times 10^{51}$ ergs s^{-1}, as representative of different Y_{ef}s). Note that the production of p-nuclei during later phase ($t > 4.0$ s) are negligible owing to the neutron richness (Figure 3, left-top), where the neutron capture nucleosynthesis takes place. As can be seen, a variety of p-nuclei are produced with interesting amount for $Y_{ef} > 0.5$ models. The heavier p-nuclei (up to $A \approx 110 - 120$) appear for greater Y_{ef} models as well as for larger M ($= 2.0 M_\odot$) case (i.e., greater entropies, see Figure 1). It should be noted that ^{74}Se and ^{92}Mo are most enhanced in slightly neutron-rich compositions (at $Y_{ef} \approx 0.48$, see [9, 12]).

In reality, the neutrino-heated matter must have a certain distribution of Y_e as can be seen in a two-dimensional hydrodynamic simulation of core-collapse supernova [3].

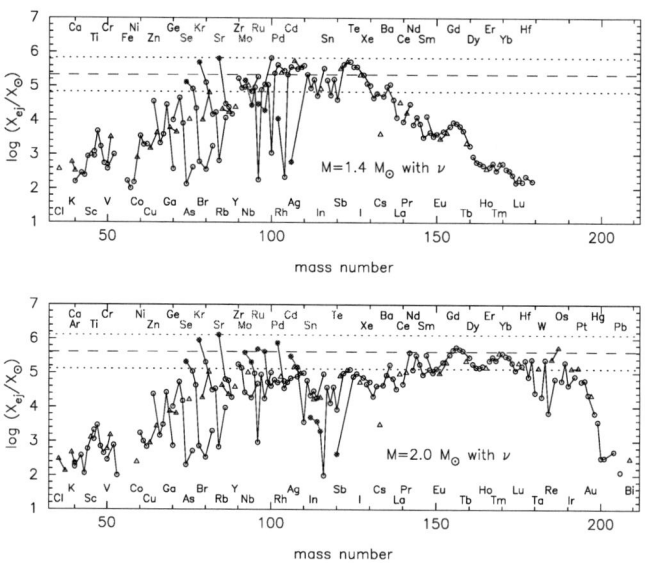

FIGURE 4. Mass-Y_e-averaged abundances with respect to their solar values for $M = 1.4M_\odot$ (top panel) and $2.0M_\odot$ (bottom panel) cases as functions of mass number. The isotopes (after decay) are denoted by circles (even-Z) and triangles (odd-Z). The p-nuclei are denoted with filled circles. The solid lines connect isotopes of a given element.

The Y_e distribution of the neutrino-processed ejecta during the first 468 ms after core bounce for a $15M_\odot$ progenitor star obtained by [3] is overlaid in Figure 3 (right panels), which has the maximum at $Y_e \approx 0.5$ and dominates in the proton-rich side. To test the contributions of the winds for $M = 1.4M_\odot$ and $2.0M_\odot$ cases, the mass-averaged yields for each Y_{ei} model are further Y_e-averaged ($54 \times 35 = 1890$ winds in total) with $p(Y_e)$ shown in Figure 3 (right panels), *assuming* this distribution to be representative of core-collapse supernovae. The resulting abundances with respect to their solar values are shown in Figure 4 for $M = 1.4M_\odot$ (top panel) and $2.0M_\odot$ (bottom panel) cases as functions of mass number. The p-nuclei are denoted with filled circles. The dotted horizontal lines indicate a "normalization band" between the largest production factor (^{84}Sr) and that by a factor of ten less than that, along with a median value (dashed line).

As can be seen in Figure 4, the p-nuclei up to ^{92}Mo and ^{108}Cd for $M = 1.4M_\odot$ and $2.0M_\odot$ cases, respectively, fall within the normalization band, which are regarded to be the dominant species produced by each event. Note that ^{74}Se and ^{92}Mo are mainly from the slightly neutron-rich ejecta as described above, while other p-nuclei are synthesized by the neutrino-induced rp-process in the proton-rich ejecta. The ejected masses by winds during the first 20 s are $2.8 \times 10^{-3} M_\odot$ and $1.1 \times 10^{-3} M_\odot$ for $M = 1.4M_\odot$ and $2.0M_\odot$ cases, respectively. Given that the progenitor mass for each case to be, e.g., $15M_\odot$ and $30M_\odot$, respectively, the overproduction factor is expressed as $\sim 10^{-4}(X_{\rm ej}/X_\odot)$. The requisite overproduction factor of a given isotope for the nucleosynthetic event to be the major source in the solar system is ~ 10 [10], assuming that *all* the core-collapse supernovae produce the same amount of the isotope. The overproduction factors of $\sim 10-100$ (see Figure 4) for the current models imply that the neutrino-driven winds can be potentially the major astrophysical site of these light p-nuclei.

The neutron-capture nuclei with $A > 100$ result from winds during the later phase ($t_{pb} > 4.0$ s, see Figure 3, left-top panel), where the initial compositions are presumed to be neutron rich. In fact, the high-entropy winds of $M = 2.0 M_\odot$ case are those proposed to be the astrophysical origin of the heavy r-process nuclei [6, 13], while that of $M = 1.4 M_\odot$ to be the origin of light r-process nuclei up to $A \sim 130$ [13]. It is interesting to note that no overproduction of the $A \approx 90$ nuclei [6, 10] appears owing to the neutron deficiency in the ejecta [13].

SUMMARY

It is shown that the interesting amounts of p-nuclei can be produced by the neutrino-induced rp-process in the proton-rich neutrino-driven winds of core-collapse supernovae. This is due to the presence of free neutrons by the anti-electron neutrino capture on free protons, which bypass the known waiting point nuclei along the nuclear path of the rp-process. The nucleosynthesis calculations imply that this neutrino-induced rp-process in core-collapse supernovae can be potentially the origin of light p-nuclei up to $A \sim 110$, which cannot be easily explained by other astrophysical scenarios.

ACKNOWLEDGMENTS

This work was supported in part by a Grant-in-Aid for Japan-France Integrated Action Program (SAKURA) from the Japan Society for the Promotion of Science, and Scientific Research (17740108) from the Ministry of Education, Culture, Sports, Science, and Technology of Japan.

REFERENCES

1. R. K. Wallace and S. E. Woosley, *Astrophys. J.* **45**, 389 (1981).
2. O. Koike, M. Hashimoto, K. Arai, and S. Wanajo, *Astron. Astrophys.*, **342**, 464 (1999).
3. R. Buras, M. Rampp, H. -Th Janka, and K. Kifonidis, *Astron. Astrophys.*, in press (2005, astro-ph/0507135).
4. J. Pruet, R. D. Hoffman, S. E. Woosley, R. Buras, and H. -Th. Janka, *Astrophys. J.*, submitted (2005, astro-ph/0511194).
5. C. Fröhlich, et al. *Astrophys. J.*, **637**, 415 (2006).
6. S. Wanajo, T. Kajino, G. J. Mathews, and K. Otsuki, *Astrophys. J.*, **554**, 578 (2001).
7. S. Wanajo, N. Itoh, Y. Ishimaru, S. Nozawa, and T. C. Beers, *Astrophys. J.*, **577**, 853 (2002).
8. M. Rayet, M. Arnould, M. Hashimoto, N. Prantzos, and K. Nomoto, *Astron. Astrophys.*, **298**, 517 (1995).
9. S. Wanajo, in preparation.
10. S. E. Woosley, J. R. Wilson, G. J. Mathews, R. D. Hoffman, and B. S. Meyer, *Astrophys. J.*, **433**, 229 (1994).
11. B. S. Meyer, G. C. McLaughlin, and G. M. Fuller, *Phys. Rev. C*, **58**, 3696 (1998).
12. R. D. Hoffman, S. E. Woosley, G. M. Fuller, and B. S. Meyer, *Astrophys. J.*, **460**, 478 (1996).
13. S. Wanajo and I. Ishimaru, *Nucl. Phys. A*, in press (2006, doi:10.1016/j.nuclphysa.2005.10.012)

Elastic α-scattering on proton rich nuclei at astrophysically relevant energies

Zs. Fülöp[*], D. Galaviz[1,†], Gy. Gyürky[*], G.G. Kiss[*], Z. Máté[*], P. Mohr[2,†], T. Rauscher[**], E. Somorjai[*] and A. Zilges[†]

[*]*ATOMKI, P.O. Box 51. H-4001 Debrecen, Hungary*
[†]*Technische Universität Darmstadt, D-64289 Darmstadt, Germany*
[**]*Universität Basel, CH-4056 Basel, Switzerland*

Abstract.
In order to improve the reliability of statistical model calculations in the region of heavy proton rich nuclei several elastic alpha scattering experiments have been carried out at low bombarding energies on various even-even and semi-magic nuclei. The extracted local optical potential parameters can be compared with the predictions of global alpha potentials. A study on ^{112,124}Sn$(\alpha,\alpha)^{112,124}$Sn has been made to test the global alpha potentials at both the proton and neutron rich sides of an isotopic chain. The present work describes the experimental challenges of high precision scattering experiments at low energy.

Keywords: Astrophysical p-process, Elastic scattering, Optical potential
PACS: 24.10.Ht, 25.55.-e, 25.55.Ci, 26.30.+k

INTRODUCTION

Although the nucleosynthesis of heavy elements is well described by neutron capture during the s- and r-processes, there are some isotopes not reachable by these processes. These stable neutron-deficient isotopes of the elements with charge number Z≥34 are classically referred to as p-nuclei.

The modeling of p-process nucleosynthesis requires a large network of nuclear reactions involving stable nuclei as well as unstable, proton-rich nuclides [1]. The relevant astrophysical reaction rates are inputs to this network, therefore their knowledge is essential for the p-process calculations. Still, there are very few charged-particle cross sections determined experimentally, despite big experimental efforts in recent years. Thus, the p-process reaction rates involving charged particles are still based mainly on theoretical cross sections obtained from Hauser-Feshbach statistical model calculations [2, 3]. Recently, Rapp et al. [4] studied the effect of cross section uncertainties on the p-nuclei overabundances for the relevant (γ,n), (γ,p) and (γ,α) reactions. The result underlines the importance of (γ,p) experiments at the lower mass region, while the (γ,α) reactions are dominating the higher mass region. Using the latest reaction rate compilations the p-process branching points can be reanalyzed as was done by Rauscher [5].

The primary aim of the present p-process experiments is the test of statistical model

---

[1] Present address: NSCL, MSU, 1 Cyclotron Lab, East Lansing, MI 48824-1321, USA
[2] Present address: Strahlentherapie, Diakoniekrankenhaus, D-74523 Schwäbisch Hall, Germany

calculations in the mass and energy range relevant to the astrophysical p-process. In general, the calculated (α,γ) and (γ,α) reaction cross sections are sensitive to the choice of the α–nucleus potential [6, 7, 8, 9, 10, 11]. Elastic α-scattering at low energies (close to the Coulomb barrier) should provide an additional test for the α–nucleus potentials considered in p-process network calculations. However, high precision data are needed for a clear determination of the optical potential properties at the measured energies.

In this work the cross sections for several elastic alpha scattering reactions [12, 13, 14] at energies above and below the Coulomb barrier are presented. The new experimental data provide a test for the global parameterizations considered in p-process network calculations. Furthermore, the study of both proton- and neutron-rich stable tin isotopes provides important information about the variation of α–nucleus potentials along an isotopic chain, since local α–nucleus potential is derived for both neutron–deficient (^{112}Sn) and neutron–rich (^{124}Sn) nuclei [13].

EXPERIMENTAL PROCEDURE AND RESULTS

The experiments have been performed at the cyclotron of ATOMKI in Debrecen, Hungary. Details about the scattering chamber can be found in [15]. A schematic view of the different elements inside the scattering chamber is shown in Fig. 1.

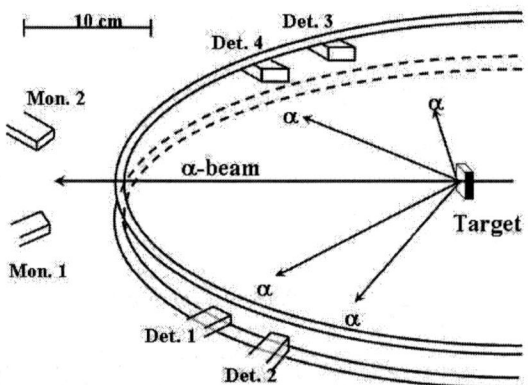

FIGURE 1. Scattering chamber at the cyclotron laboratory of ATOMKI.

The target holder was placed in the center of the scattering chamber, and could be rotated with respect to the incoming beam. The angular distribution of the elastically scattered particles is measured with two pairs of silicon barrier detectors (with an angular separation of 10° between each pair of detectors) each of which could be independently rotated around the target, thus covering almost the complete angular range between 20° and 170°. On the beam exit side of the scattering chamber two additional detectors are fixed at an angle of $\pm 15°$. These detectors are used to monitor the beam position during the experiment. Their spectra are also used to normalize the measured cross section relative to the Rutherford cross section. High energy resolution is desired for a clear identification of the elastically scattered particles. Therefore, it is necessary to ensure that the particles do not deposit too much energy in the target before and after their interaction with the target nucleus. This is achieved by using thin targets (between 200-300 μg/cm^2). The targets were produced by evaporation technique, and deposited onto thin carbon foils (approx. 20 μg/cm^2). The high enrichment of the targets (contamination of other isotopes less than 4%) helps to identify the peaks of the elastically scattered particles in the measured spectra.

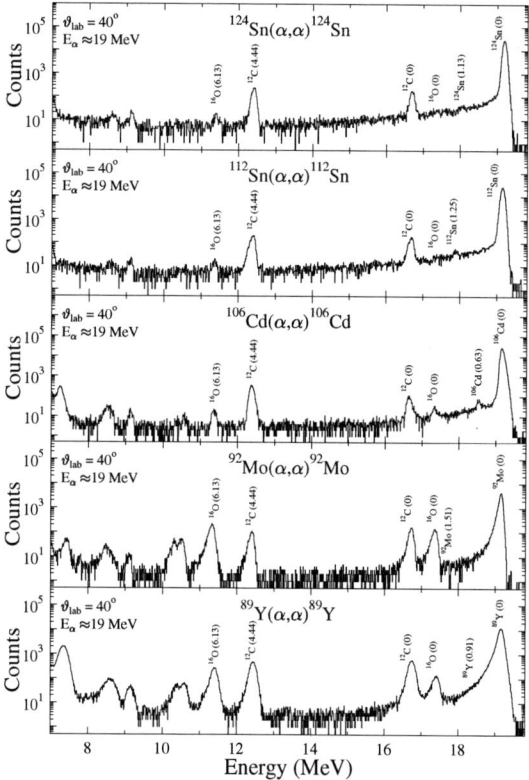

FIGURE 2. Measured elastic α-scattering spectra on the studied nuclei. In all spectra the peak corresponding to the elastically scattered particles can be separated from the inelastic scattering, the contaminations and carbon backing. For details see text.

Silicon surface barrier detectors, with thicknesses between 300 and 500 μm were used to measure the angular distributions. The distance from the detectors to the target was approximately 19.6 cm. The monitor detectors placed at the beam exit side of the scattering chamber were separated 35.1 cm from the target. In order to avoid very high counting rates and to achieve a high angular resolution, different collimators were placed in front of each detector. Another aspect which is necessary to achieve a good angular resolution is a well defined beam spot. For this purpose, two different apertures of dimensions 6×6 and 2×6 mm were placed at target position before each measurement. The incoming beam was tuned until no current was measured on both frames, which indicated that the beam spot had been adequately focused.

Complete angular distributions covering a range between 20° and 170° have been registered for all reactions. Typical spectra of these reactions can be seen in Fig. 2, where five spectra measured at $\theta_{lab} = 40°$ at energies of $E_\alpha \approx 19$ MeV are shown. In all cases, the peaks corresponding to the elastically scattered particles can be separated easily both from peaks belonging to inelastic events and elastic reactions on other target components (e.g. ^{12}C and ^{16}O). This clean separation is not always possible in the spectra corresponding to the angular distribution measurements of ^{92}Mo. Here, due to the high melting point of Molybdenum it was necessary to use oxide material to manufacture the targets. This fact leads to an increase of elastic scattering on ^{16}O, as well as the appearance of a low energy tail in the elastic peak (a consequence of the incremented target thickness). Such an effect, combined with a lower beam position stability in comparison to the ^{106}Cd or ^{112,124}Sn experiments, results in spectra with not so well defined peaks. Nevertheless, those effects could be corrected and it was possible to extract the necessary information from these spectra in order to derive the scattering cross section.

In each of the more than 4000 measured spectra (300 detector-spectra and 150 monitor-spectra for a complete angular distribution) the elastic alpha-peak was analyzed. Its integral is the observable needed for the determination of the experimental elastic cross section. A precise determination of the scattering cross section needs a proper knowledge of several factors. The number of incoming particles can be extracted from current measurements during the experiment with an uncertainty of approximately 3%. Though this error is comparable with the statistical error of the data acquired at backward angles, the much smaller relative errors at forward angles allows us to normalize the cross section based on the yields from the movable detectors relative to the yield measured by the monitor detectors fixed at forward angles ($\theta_{lab} = 15°$), described well by the Rutherford scattering formula.

Based on the high precision elastic scattering data obtained one can derive the angular distributions and —from its deviation from the Rutherford scattering— the corresponding optical potential parameters. Figure 3 shows the experimental cross sections normalized to the Rutherford cross section together with the results of our optical model analyses, which describe well the experimental data. As further test of the obtained optical potential parameters one can compare the present optical potentials derived from low energy scattering with the previous results of higher energy experiments. This can be seen in Fig. 4, where a reasonable reproduction of the diffraction pattern is obtained for almost all angular distributions. Here, a minor readjustment of the imaginary part of the potential improves the agreement between the experimental and calculated cross sections.

SUMMARY

We have measured with high precision the elastic alpha scattering cross section on several proton rich heavy nuclei at energies below $E_\alpha = 20$ MeV. The so far studied reactions are: ^{144}Sm$(\alpha,\alpha)^{144}$Sm [11], ^{92}Mo$(\alpha,\alpha)^{92}$Mo [12], ^{112,124}Sn$(\alpha,\alpha)^{112,124}$Sn [13], ^{106}Cd$(\alpha,\alpha)^{106}$Cd [14] and ^{89}Y$(\alpha,\alpha)^{89}$Y. The analysis performed within the optical model framework provided a remarkable reproduction of the measured angular distrib-

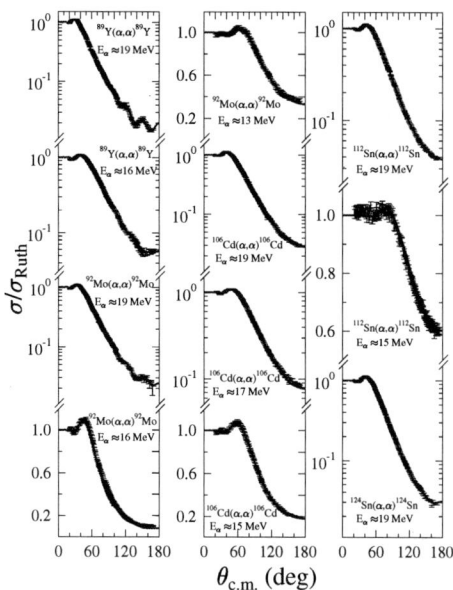

FIGURE 3. Comparison of the cross sections resulting from the optical model analysis (solid line) with the measured angular distributions (dots with error bars) for all the investigated nuclei. The solid lines are covered by the experimental data points and can hardly be seen.

utions. The results fit well with the systematic behavior of α–nucleus folding potentials [20, 21, 22]. The present data provide an excellent tool to test the behavior of global α–nucleus potentials proposed for astrophysical network calculations (see [23] and references therein). Additional elastic scattering experiments are planned in the p-process mass range at energies around the Coulomb barrier.

ACKNOWLEDGMENTS

This work is based on the Ph.D. thesis of D. Galaviz [23]. The experiments were supported by DFG (SFB634 and FOR 272/2-2) and OTKA (T049245, T042733, F043408, D048283). T. R. is supported by the Swiss NSF (grants 2024-067428.01, 2000-061031.02, 2000-105328). Zs. F. acknowledges support from a Bolyai grant.

REFERENCES

1. M. Arnould, and S. Goriely, *Phys. Rep.* **384**, 1. (2003)
2. T. Rauscher and F. K. Thielemann, *At. Data Nucl. Data Tables* **79**, 47. (2001) and *At. Data Nucl. Data Tables* **75**, 1. (2000)
3. S. Goriely, in Nuclei in the Cosmos V, Edition Frontieres Paris, p314. (1998)
4. W. Rapp et al, *Nucl. Phys.* **A758**, 545c. (2005)
5. T. Rauscher, *Phys. Rev.* **C73**, 015804. (2006)

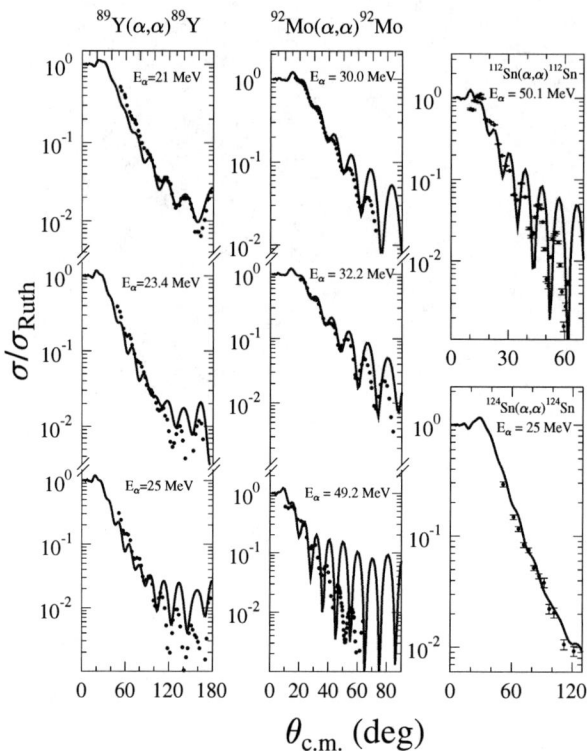

FIGURE 4. High energy elastic scattering cross sections (available in the literature) measured on the nuclei ^{89}Y at E_α = 21, 23.4 and 25 MeV [16]; ^{92}Mo at E_α = 30.0 MeV [17], 32.2 MeV [18] and 49.2 MeV [19]; ^{112}Sn at E_α = 50.1 MeV [19] and ^{124}Sn at E_α = 25 MeV. The solid line represents the resulting cross section predicted for the potentials derived from angular distributions of the present work at energies close to the Coulomb barrier.

6. Zs. Fülöp et al., *Z. Phys.* **A355**, 203. (1996)
7. E. Somorjai, et al., *Astron. Astrophys.* **333**, 1112. (1998)
8. W. Rapp et al., *Nucl. Phys. A* **688**, 427. (2001)
9. N. Özkan, et al., *Nucl. Phys.* **A710**, 469. (2002)
10. S. Harissopulos et al., *Nucl. Phys.* **A758**, 505. (2005)
11. P. Mohr, et al., *Phys. Rev.* **C55**, 1523. (1997)
12. Zs. Fülöp et al., *Phys. Rev.* **C64**, 065805. (2001)
13. D. Galaviz et al., *Phys. Rev.* **C71**, 065802. (2005)
14. G.G. Kiss et al., *Eur. Phys. J.* **A27**, (2006)
15. Z. Máté et al., *Acta Phys. Hung.* **65**, 287. (1989)
16. M. Wit et al., *Phys.Rev.* **C12**, 1447. (1975)
17. Y. Awaya et al., *J. Phys. Soc. Jap.* **33**, 881. (1972)
18. S.J. Burger et al., *Nucl.Phys.* **A243**, 143. (1975)
19. N.T. Burtebaev et al., *Sov. J. Nucl.Phys.* **51**, 827. (1990)
20. U. Atzrott et al., *Phys.Rev.* **C53**, 1336. (1996)
21. P. Demetriou et al., *Nucl. Phys.* **A707**, 253. (2002)
22. M. Avrigeanu et al., *Nucl. Phys.* **A764**, 246. (2006)
23. D. Galaviz, Ph. D. thesis TU Darmstadt (2004)

2. POSTERS

R-matrix and Potential Model Extrapolations for NACRE Update and Extension Project

Masayuki Aikawa[*], Koji Arai[†,*], Masahiko Katsuma[*], Kohji Takahashi[*], Marcel Arnould[*] and Hiroaki Utsunomiya[**]

[*] *Institut d'Astronomie et d'Astrophysique, Université Libre de Bruxelles, Campus de la Plaine, CP226, 1050 Brussels, Belgium*
[†] *Division of General Education, Nagaoka National College of Technology, 888 Nishikatakai, Nagaoka, Niigata 940-8532, Japan*
[**] *Department of Physics, Konan University, Okamoto 8-9-1, Higashinada, Kobe 658-8501, Japan*

Abstract. NACRE, the 'nuclear astrophysics compilation of reaction rates', has been widely utilized in stellar evolution and nucleosynthesis studies. Its update and extension programme started within a Konan-Université Libre de Bruxelles (ULB) collaboration. At the present moment, experimental data in refereed journals have been collected, and their theoretical extrapolations are being performed using the R-matrix or potential models. For the ^{3}H(d,n)^{4}He and ^{2}H(p,γ)^{3}He reactions, we present preliminary results that could well reproduce the experimental data.

Keywords: Nuclear Reactions
PACS: 95.30.-k

The elemental composition in the Universe are continuously changed by nuclear reactions. Detailed knowledge about nuclear reactions is thus important to investigate stellar evolution and so on. Fowler and his collaborators ([1], and references therein) provided astrophysicists with reaction rate compilations. More recently, a new compilation referred to as NACRE [2] has been published. Since its publication in 1999, lots of nuclear reaction experiments related to these studies have been performed. These experiments, therefore, should be taken into account in the forthcoming compilation. In order to achieve this requirement, the work programme started within a Konan-Université Libre de Bruxelles (ULB) collaboration.

At the present moment, the collection of experimental data in refereed journals have been finished, and theoretical extrapolations of the data are being performed using the R-matrix or potential models. In view of the large variety of reactions to be dealt with, the development of an automatic parameter search is of substantial interest. Computer codes developed for this purpose have been applied on the ^{3}H(d,n)^{4}He and ^{2}H(p,γ)^{3}He reactions, which have been already analyzed by [3].

Applying the R-matrix code to the former reaction, the minimum of the reduced χ^2 value with regard to the experimental data is reached, and the resulting S-factor is shown in Fig. 1. This is very close to the analysis of [3] in the energy region lower than around 1 MeV. The discrepancy between two results in the higher energy region is caused by the choice of the experimental data and the treatment of resonances at around 3 MeV. That is, however, small enough even at the Gamow window at $T = 10^{10}$ K, which is also shown in Fig. 1.

As an example for the potential model, its analysis for the ^{2}H(p,γ)^{3}He reaction is

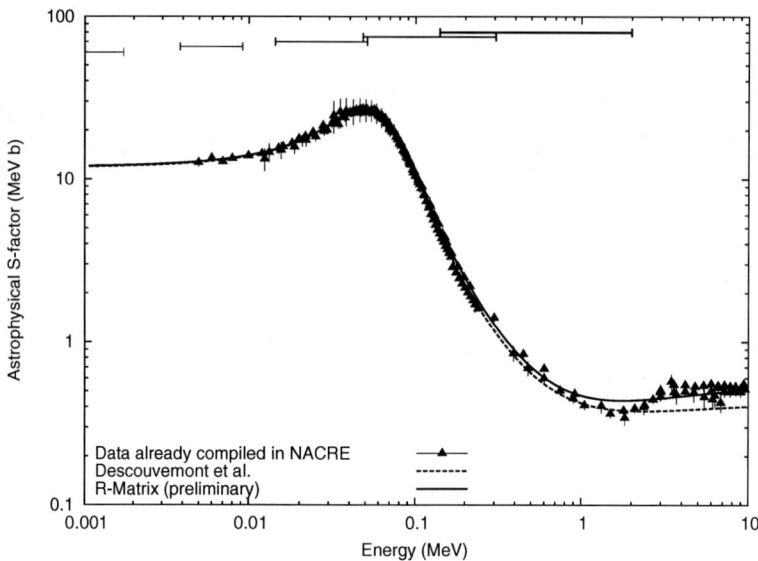

FIGURE 1. Astrophysical S-factor for ^{3}H(d,n)^{4}He. The experimental data are from [2]. Line segments on top of the figure are Gamow windows for temperatures (from left to right) of 10^6, 10^7, 10^8, 10^9, and 10^{10} K

shown in Fig. 2. The result of the potential model can well reproduce the experimental data and is very similar to that of the R-matrix method. We emphasize that the different codes could reach the similar results individually. This agreement is considered to be very important when applying these codes to other reactions.

In summary, the update and extension of NACRE is proceeding in line with the programme. Experimental data that have been published for the reactions already included in NACRE have been gathered, and their theoretical treatment in order to sort out reaction rates has been started within the R-matrix or potential models. The computer codes have been developed and applied on the ^{3}H(d,n)^{4}He and ^{2}H(p,γ)^{3}He reactions. The results could well reproduce the experimental data and we are analyzing other reactions using these codes. The proper handling of the uncertainties remains a matter to be investigated further. The extension to reactions not included in NACRE will start soon.

With this new version of NACRE, it is hoped to provide astrophysicists with a quality and reliable nuclear database. It comes in complement to other theoretical nuclear data provided by the BRUSLIB library [5] that is also continuously improved and extended.

ACKNOWLEDGMENTS

We thank P. Descouvemont, S. Goriely, A. Jorissen and P. Leleux for helpful discussions. The first author acknowledges the financial support of the FNRS (Belgium). This work is supported in part by the Interuniversity Attraction Pole IAP 5/07 of the Belgian Federal

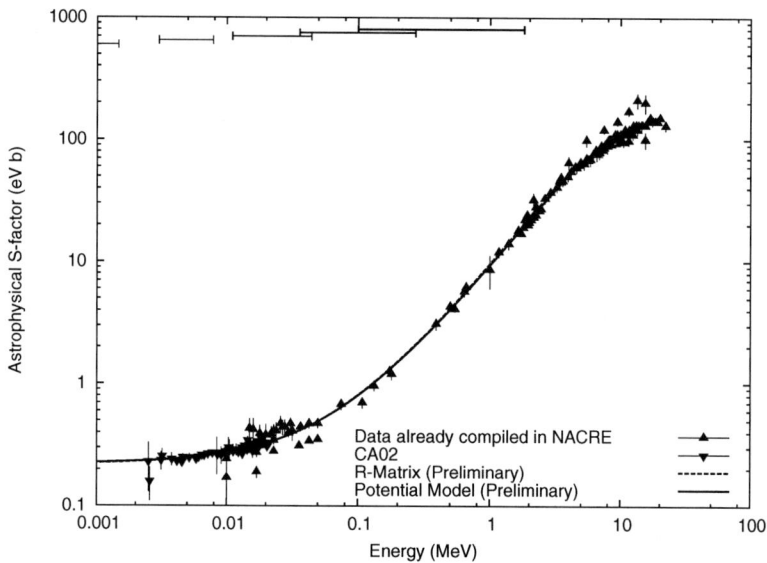

FIGURE 2. Astrophysical S-factor for ^{2}H(p,γ)^{3}He. The experimental data are from [2] and [4]. Line segments on top of the figure are the same, but for the ^{2}H + p system, as those in Fig. 1.

Science Policy and by the Konan University - Université Libre de Bruxelles Convention 'Construction of an Extended Nuclear Database for Astrophysics'. In addition to the compilation work, this convention also concerns the direct experimental and theoretical study of photonuclear reactions of astrophysical interest. This activity and the first results are described in [6, 7, 8].

REFERENCES

1. G. R. Caughlan, and W. A. Fowler, *At. Data Nucl. Data Tables*, **40**, 283 (1988).
2. C. Angulo, M. Arnould, M. Rayet *et al.*, *Nucl. Phys.*, **A656**, 3 (1999). [NACRE]
3. P. Descouvemont, A. Adahchour, C. Angulo *et al.*, *At. Data Nucl. Data Tables*, **88**, 203 (2004).
4. C. Casella, H. Costantini, A. Lemut *et al.*, *Nucl. Phys.*, **A706**, 203 (2002). [CA02]
5. M. Aikawa, M. Arnould, S. Goriely, A. Jorissen, and K. Takahashi, *Astron. Astrophys.*, **441**, 1195 (2005).
6. H. Utsunomiya, P. Mohr, A. Zilges, and M. Rayet, *Nucl. Phys.*, **A**, in press; nucl-ex/0502011.
7. H. Utsunomiya, S. Goko, H. Toyokawa *et al.*, Proc. of the Nuclear Physics in Astrophysics II, Debrecen Hangary, May 16-20, 2005; to be published in European Physical Journal.
8. H. Utsunomiya, S. Goko, K. Soutome *et al.*, *Nucl. Instr. Meth.*, **A538**, 225 (2005).

Study of inelastic contribution in the ^{7}Be + p scattering experiment at CRIB

G. Amadio[*], H. Yamaguchi[*], J.J. He[*], A. Saito[*], Y. Wakabayashi[*], H. Fujikawa[*], S. Kubono[*], L.H. Khiem[*], Y.K. Kwon[*], T. Teranishi[†], S. Nishimura[**], M. Niikura[‡], Y. Togano[**], N. Iwasa[§] and S. Inafuku[§]

[*]*CNS, University of Tokyo, Japan, Saitama, Wako, Hirosawa, 2-1, Zip Code 351-0198*
[†]*Kyushu University, Japan*
[**]*RIKEN, Japan*
[‡]*Kyoto University, Japan*
[§]*Tohoku University, Japan*

Abstract.
The ^{7}Be(p,γ)^{8}B reaction is undoubtfully important to the understanding of the solar model. As a step in the direction of improving the accuracy of the S_{17} astrophysical factor, a study of the ^{7}Be + p scattering was performed with the thick target method at the CRIB facility. In addition to its astrophysical significance, this reaction is also useful to clarify the nuclear structure of ^{8}B. A primary beam of ^{7}Li and a hydrogen gas target were used to produce a ^{7}Be secondary beam at 7.69 MeV/u. This was the first time in which γ-rays were measured in coincidence with protons, and we have successfully measured the inelastic contribution to the scattering cross section.

Keywords: Solar neutrinos, Proton resonant scattering, nucleosynthesis
PACS: 26.56.+t, 25.45.De, 25.40.Ep

INTRODUCTION

The solution of the solar neutrino problem depends on the knowledge of several experimental inputs that are used in the calculation of the neutrino production rates[1]. The astrophysical factor for the ^{7}Be(p,γ)^{8}B reaction ($S_{17}(0)$), despite its importance in the prediction of the high-energy neutrino production rates, still remains one of the most uncertain[2] parameters in the solar model. Although it has been measured many times in the last decades[3], the challenges and difficulties in performing the measurement have limited the accuracy of existing experimental values, preventing it to reach the desirable accuracy of ∼5%. In addition, recent experimental studies [4, 5, 6] have opened to discussion what kind of influence the suggested low-lying 2$^-$ broad state in ^{8}B might produce in the extrapolations of $S_{17}(0)$.

Having all this in mind, and the ^{7}Be(p,γ)^{8}B measurement as a goal, a study of the ^{7}Be and p scattering was performed to both help to improve the current setup, as to make it possible to measure the capture reaction cross section - including the production of a high-current pure ^{7}Be beam - and to improve the current knowledge of ^{8}B nuclear structure. This was also the first time in which γ-rays from inelastic scattering were simultaneously measured, providing precious extra information.

EXPERIMENT

The experiment was performed at the CRIB facility, of University of Tokyo. We have used the inverse kinematics technique by bombarding a thick polyethylene target with a ^{7}Be beam at a laboratory energy of 7.69 MeV/u. The secondary beam was produced through the reaction ^{7}Li(p,n)^{7}Be by bombarding an 8 cm thick, 1 atm pressure hydrogen gas target with a ^{7}Li primary beam from RIKEN's AVF cyclotron, at 8.76 MeV/u. The energy spread of the secondary beam was set to 0.8%. After the production target, the secondary beam passed through a double achromatic system and a velocity filter, leading to a virtually pure beam at the end. The setup at the scattering chamber (F3 chamber) consisted of two PPACs that measured the beam position and angle of incidence into the target, as well as time of flight between them, a set of four silicon detector telescopes, composed by a position-sensitive 70 μm detector backed by two or three 1.5 mm detectors, that measured the recoiled protons and an array of ten NaI detectors that measured the γ-rays from inelastic scattering. The array of NaI detectors was set at a distance of 7 cm above the target position. An aluminum plate of 2 mm thickness separated these detectors from vacuum, to avoid absorption. The angular range covered by the silicon telescopes was $0 \leq \theta_{lab} \leq \pi/4$, with a solid angle of about 60 msr per detector. The NaI detectors covered a total solid angle of $\Delta\Omega \sim 4\pi/5$, with a resolution of 61 keV at 429 keV (FWHM).

PRELIMINARY ANALYSIS

In the thick target method, that was adopted in this experiment, the beam has its energy set to the maximum value to be measured and the target is chosen thick enough to cover the energy range of interest. Because of the energy loss of the beam inside the target, it is possible to measure the excitation function all at once. Notwithstanding, an adequate treatment using the kinematics and energy loss of the recoiled proton inside the target is required in the analysis. The development of the code to treat the data with all these considerations is now under development. In order to be able to subtract the background contribution due to carbon in the polyethylene target, carbon foil runs were also performed. These data also require similar treatment, since the target thickness seen by the beam is different, according to the reaction point and density of the target. The inelastic contribution can be distinguished by taking coincidence between the proton and the γ-ray (spectrum shown in Fig. 1) from the deexcitation of ^{7}Be to its ground state ($E_\gamma = 429$ keV). The background contribution from the β-decay of the implanted ^{7}Be to ^{7}Li*, that decays to the ground state emmiting a γ-ray ($E_\gamma = 470$ keV) was found to be very low (less than 7%), as expected. The similarities between the raw proton spectra (left of Fig. 2) and the raw proton spectra in coincidence with a γ-ray gated around the 429 keV peak and the timing peak (right of Fig. 2) indicate that some of the resonances in ^{8}B might have a significant branching to the first excited state in ^{7}Be as well. In addition, we may be able to confirm the now uncertain existence of a 2^- state at around 3 MeV[4, 5], since our data covers a wide range up to $E_{cm} = 7$ MeV. A detailed analysis is though necessary before we make a conclusion.

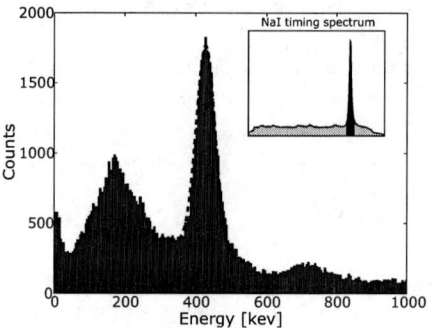

FIGURE 1. The γ-ray spectrum for the polyethylene target runs. The inset shows the timing spectrum for the detectors and the region where the gate is set.

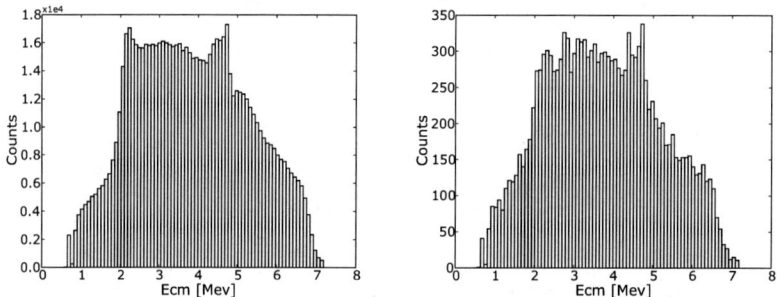

FIGURE 2. Raw proton spectra in coincidence with γ-rays measured by the NaI detector array (right) and events not in coincidence (left). The timing gate is shown in the inset of Fig. 1, the energy gate covers the region of the 429 keV peak.

REFERENCES

1. J. N. Bahcall and C. Peña-Garay, *New J. Phys.* **6**, 63 (2004).
2. E. G. Adelberger, *Rev. Mod. Phys.* **70**, 1265–1291 (1998).
3. A. R. Junghans et al., *Phys. Rev. C* **68**, 065803 (2003).
4. V. Z. Gol'dberg et al., *JETP Lett.* **67**, 1013 (1998).
5. G. V. Rogachev et al., *Phys. Rev. C* **64**, 061601 (2001).
6. C. Angulo et al., *Nucl. Phys. A* **716**, 211–229 (2003).

Neutrino Flavor Changing Neutral Currents and Stellar Collapse

Philip S. Amanik* and George M. Fuller*

Department of Physics - 0319, University of California - San Diego, La Jolla, CA 92093

Abstract. Flavor changing neutral current interactions are predicted in some extensions of the Standard Model. We discuss a process by which a neutrino can change flavor by scattering with a quark. We show how the stellar collapse environment is sensitive to neutrino flavor changing scattering on heavy nuclei. In particular, when this interaction is included in the stellar collapse model, the dynamics of the core's evolution and outcome of collapse are drastically changed.

Keywords: non-standard neutrino interactions, core collapse supernovae, neutrino-nucleus scattering
PACS: 97.60.Bw, 13.15.+g, 26.50.+x, 25.30.Pt

INTRODUCTION

Neutrino flavor changing can occur due to the fact that neutrinos have mass and their flavor eigenstates do not coincide with their mass eigenstates. Neutrino flavor changing is a result of the quantum mechanical effect of flavor states oscillating. Another possible way neutrinos could change flavor is by flavor violating interactions. Such interactions are predicted by some extensions of the Standard Model but have yet to be found by experiments. We have investigated the effects of these proposed flavor changing interactions on the core collapse supernova model [1, 2]. We have found that physical effects due to flavor changing scattering processes are qualitatively different from the effects due to flavor oscillations, and furthermore that this type of flavor changing significantly alters the standard core collapse supernova model.

INTERACTIONS AND CROSS SECTIONS

There are several models that predict flavor violating interactions of different types. We will consider a general model for neutrino flavor changing scattering with a quark. A low energy effective Lagrangian describing this theory is

$$\mathscr{L} = G_F \bar{\nu}^i \gamma^\mu (1-\gamma_5) \nu^j \bar{q} \gamma_\mu (\varepsilon^q_{V_{ij}} + \varepsilon^q_{A_{ij}} \gamma_5) q. \tag{1}$$

This Lagrangian has the form of a neutral current. The indices on the neutrino fields indicate flavor. Interactions of this form are known as flavor changing neutral currents (FCNCs). The epsilons in the quark currents are dimensionless parameters which quantify the strength of the interaction relative to the Fermi constant.

The Lagragian describes the interaction of a neutrino and quark. However, the core of a collapsing star contains nuclear matter, not free quarks. We need cross sections for neutrinos interacting with the matter in the core. It is believed that different phases of nuclear matter arise in the core throughout collapse, but this topic is an area of active research. For our calculations we have considered the simplest opacity source. It is known that nuclei exist in the core for at least some portion of the collapse. We therefore consider neutrino flavor changing elastic scattering with a spin zero nucleus in the $q^2 = 0$ limit. The cross section [1] for this process is

$$\sigma_{ij} \approx \frac{2G_F^2}{\pi} \left| \varepsilon_{V_{ij}}^d (2N+Z) + \varepsilon_{V_{ij}}^u (2Z+N) \right|^2 E_\nu^2. \quad (2)$$

For comparison, the cross section for Standard Model neutral current elastic scattering with a spin zero nucleus is

$$\sigma \approx \frac{2G_F^2}{\pi} (Z+N)^2 E_\nu^2. \quad (3)$$

Notice that both these cross sections have a coherent amplification: the factors of $(2N+Z)^2$ and $(2Z+N)^2$ in the flavor changing case, and the factor $(Z+N)^2 = A^2$ in the Standard Model case. The core of a collapsing star has nuclei with mass number of order $A = 100$. The coherent scattering cross sections for these nuclei are roughly four orders of magnitude larger than cross sections for neutrino scattering on free nucleons. The FCNCs we are considering are weaker [3] than SM interactions, but because of coherent amplifications the flavor changing cross sections are appreciable and enough neutrino flavor changing occurs in the core to effect the standard stellar collapse model.

EFFECTS OF NEUTRINO FLAVOR CHANGING

After neutrino trapping, when the Fermi sea of electron neutrinos reaches a maximum level, net electron capture becomes blocked. Electron capture and beta decay reactions are still taking place in the core, but they are in equilibrium. Once the neutrino sea reaches its maximum level, the electron fraction Y_e no longer changes. However, if electron neutrinos could change flavor to mu and tau neutrinos, then phase space would open to allow electron capture to occur and cause net reduction in Y_e.

The neutrino FCNCs we are considering are weaker than Standard Model interactions and in particular, the cross sections for neutrino flavor changing coherent scattering with nuclei are smaller than cross sections for electron capture. We therefore restrict our discussion to the following limiting case. When neutrino flavor changing scattering is operating in the core, the number of electron neutrinos remains the same: every time a hole is opened in the electron neutrino sea, it is filled by a neutrino created from electron capture. By means of neutrino flavor changing then, electron fraction is converted to mu and tau neutrino fraction:

$$\Delta Y_e = -(\Delta Y_{\nu_\mu} + \Delta Y_{\nu_\tau}). \quad (4)$$

The change in electron fraction can be used to quantify dynamical changes to the core collapse model.

The homologous core mass and initial shock energy depend on the core's electron fraction at bounce. Lowering the core's electron fraction on infall results in a smaller homologous core and therefore more outer core material for the shock to photodissociate. The shock therefore loses more energy before reaching the outer envelopes of the star. A lower electron fraction also results in a weaker initial shock. Both of these effects disfavor getting an explosion.

We have performed a one-zone collapse simulation (details can be found in Ref. [2]) to calculate scattering rates and reduction in Y_e due to the neutrino flavor changing. We assumed a maximum trapped neutrino fraction of $Y_\nu^{\text{trap}} = 0.05$ so that according to Eq. (4), the maximum reduction in Y_e possible is $\Delta Y_e = 0.1$. We have found from our simulation that this maximum reduction in Y_e occurs for our neutrino FCNCs up to their current level of experimental constraint. We note here that a change in Y_e as low as 0.02 is dynamically significant [4].

Neutrino FCNCs also have consequences unrelated to the core's electron fraction. Neutrinos may aid the explosion by depositing energy behind the stalled shock. Flavor changing interactions result in many more neutrinos to participate in shock revival, thus positively effecting the explosion. An observable feature of neutrino FCNCs is the existence of mu and tau neutrinos in the supernova spectrum, with the same distribution for each of the flavors. This result is different from what would occur solely due to neutrino oscillations. For example, neutrino oscillations could cause certain portions of the neutrino distributions to be exchanged.

CONCLUSIONS

We have examined the effects of neutrino flavor changing neutral current interactions on the core collapse supernova model. We have seen that neutrino-quark interactions are enhanced because of coherent scattering with nuclear matter, and that these interactions cause qualitatively different flavor transformation that matter enhanced oscillations. In particular, neutrino flavor changing scattering processes during infall have both negative and positive effects regarding the explosion. They cause the dynamics of collapse, and the expected neutrino spectrum to be significantly changed from the predictions of the current supernova model.

If new physics of this type is discovered experimentally, it must be included in the supernova model. Conversely, detection of supernova neutrinos could give evidence for discovery or constraint of this type of interaction. A complete supernova simulation that includes these interactions is needed to quantify how they would effect the explosion mechanism and neutrino spectra, and also determine other effects they would have on the model.

REFERENCES

1. P. S. Amanik, G. M. Fuller, and B. Grinstein, *Astroparticle Phys.*, **24**, 160-182 (2005).
2. P. S. Amanik, and G. M. Fuller, in preparation.
3. S. Davidson, C. Pena-Garay, N. Rius, and A. Santamaria, *JHEP*, **0303**, 011 (2003).
4. W. R. Hix et al., *Phys. Rev. Letters*, **91**, 201102 (2003).

Low-Metallicity Lead Stars: Comparison between Theory and Observations

S. Bisterzo*, R. Gallino*,[†], O. Straniero**, W. Aoki[‡], S. Ryan[§] and T. C. Beers[¶]

Dipartimento di Fisica Generale, Universitá di Torino, 10125 (To) Italy
[†]*Center for Stellar and Planetary Astrophysics, School of Mathematical Sciences, PO Box 28M, Monash University, 3800 Victoria, Australia*
**Osservatorio Astronomico di Collurania, Teramo, 64100 Italy*
[‡]*National Astronomical Observatory, Tokyo (Japan)*
[§]*Open University, Milton Keynes, (UK)*
[¶]*Dept. of Physics and Astronomy, Michigan State University, East Lansing (MI, USA)*

Abstract. We compare AGB theoretical models with spectroscopic abundances of a sample of very metal-poor, C-rich, s-rich and lead-rich stars observed at high-resolution spectroscopy. Fits are obtained for AGB models with different ^{13}C-pocket efficiencies and initial masses. The two intrinsic indicators, [hs/ls] and [Pb/hs] versus [Fe/H], are analyzed. An extended analysis of all the observed elements is made, outlining apparent discrepancies for a few elements. The analysis of C and N abundances strengthen the need of a strong cool bottom process occurring during the AGB. A significant number of these stars are both s-enriched and r-enriched. For them, the envelope abundances are predicted by mass transfer from the more massive AGB companion in a binary system from a parental cloud already enriched in r-elements.

Keywords: C stars, S stars, metal poor stars
PACS: 97.30, 97.20

We have analysed a sample of very metal poor C-rich, s-rich and lead rich stars, using AGB models of different ^{13}C-pocket efficiencies, initial masses, metallicities, and different initial r-enrichment. In the sample there are stars showing both s and r enrichment. Since AGB stars do not synthesize the r elements, to reproduce these abundances we make different choice of initial Eu abundance in the progenitor cloud, assuming that the binary system formed from a parental cloud already enriched in r elements. In Gallino et al. (this Symposium), the effect of pre r-enrichment in s-enhanced stars is shown for different choices of the ^{13}C-pocket. In Fig. 1, left panel, we report the star CS 29526-110 [2] with [Fe/H] = −2.38, fitted by an AGB model with a ^{13}C-pocket of ST/6, where ST is the standard case of [1]. In the same plot we compare the s+r process prediction (thick line), with the model without r-enrichment (thin line) where all the heavy element abundances prediction are from the s-process. In the case s+r, the pre r-enrichment is normalised to [Eu/Fe] = 2.0 in the parental cloud. The choice of the initial r-rich isotopes abundances normalised to Eu was made considering the r-process solar prediction from [3]. With this model it is possible to fit stars with a ratio of [La/Eu] ≈ 0, while [La/Eu]$_s$ ≈ 1 dex. For high to moderate ^{13}C-pocket efficiencies the r-enrichment does not affect the abundances of three peaks, ls, hs, and Pb. Instead, for very low ^{13}C-pocket (e.g. ST/30 for [Fe/H] = −2.6), a high r-enhancement influences the s-process distribution. Another example is the star CS 22898-027 [2] with [Fe/H] =

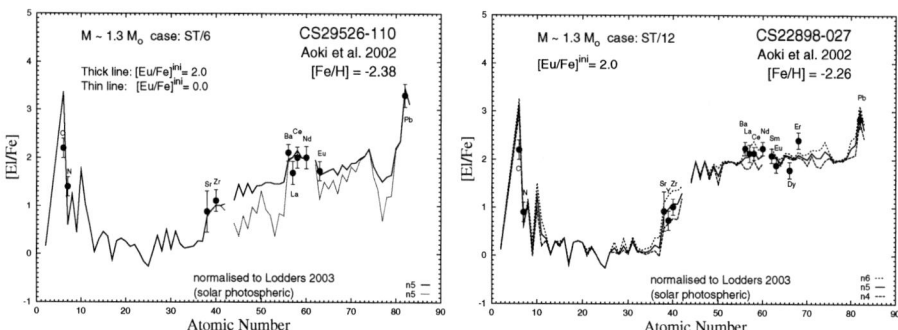

FIGURE 1. Fits of two stars with updated AGB model predictions. *Left panel:* the CS 29526-110 [2], fitted by s+r process (thick line) and by s-process only (thin line). *Right panel:* the CS 22898-027 [2] with our s+r process predictions.

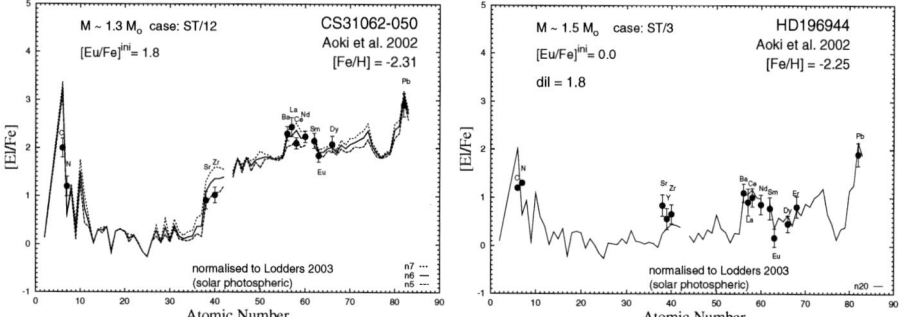

FIGURE 2. Fits of CS 31062-050 [2] (*Left panel*), and HD 196944 [2] (*Right panel*) by updated AGB model predictions.

−2.26 (Fig. 1, right panel), where a best fit is obtained for an AGB model of initial M = 1.3 $M_\odot$, a ^{13}C-pocket ST/12, and r-enrichment [Eu/Fe] = 2.0. In general, we used three lines to fit the data, that correspond to an uncertainty in the mass range of M = 1.3 ± 0.05 $M_\odot$. The discrepancy in the analysis of C and N abundances strengthen the need of a strong cool bottom process occurring during the AGB phase. In Fig. 2, left panel, the star CS 31062-050 [2], with [Fe/H] = −2.31, is fitted with an initial [Eu/Fe] = 1.8 and ST/12. In the right panel, the giant HD 196944 [2] is fitted with an AGB model of 1.5 $M_\odot$, ST/3, [Eu/Fe] = 0.0 and a high dilution factor (dil = 1.8 dex); this dilution factor represents the ratio of the accreted mass from the primary AGB by stellar winds over the envelope of the observed companion. In Table 1 we collect the sample of lead stars, and its correspondent best fits with models are reported. For HE 0338-3945 [13] we give a prediction of [Pb/Fe]. In columns 4 and 5 the two intrinsic index, [hs/ls] and [Pb/hs], are reported. The [Pb/hs] is indicative of the s-process efficiency: for a given [Fe/H], higher lead implies a more efficient ^{13}C-pocket (Gallino et al. this Symposium).

TABLE 1. Sample of s-rich and Pb-rich stars considered

Lead Stars	[Fe/H]	[Pb/Fe]	[hs/ls]	[Pb/hs]	M	Pocket	dil	[Eu/Fe]	Ref
CS 22183-015	-3.12	3.17	1.24	1.39	1.3	ST/15	0.00	1.5*	4
CS 22880-074	-1.93	1.90	1.03	0.71	1.3	ST/6	0.85	0.0*	1
CS 22898-027	-2.26	2.84	1.30	0.67	1.3	ST/12	0.00	2.0*	1
CS 22942-019	-2.64	≤1.6	-0.13	≤0.09	1.5	ST/75	0.00	0.5*	1
CS 29497-030	-2.70	3.55	0.97	1.51	1.3	ST*2	0.40	0.5*	5
CS 29497-030	-2.57	3.65	1.04	1.42	1.3	ST*1.3	0.00	2.0*	6
CS 29526-110	-2.38	3.30	0.83	1.36	1.3	ST/6	0.00	1.5*	1
CS 30301-015	-2.64	1.70	0.81	0.60	1.3	ST/24	1.00	0.0*	1
CS 31062-012	-2.55	2.40	1.34	0.47	1.3	ST/30	0.00	1.5*	1
CS 31062-050	-2.31	2.90	1.26	0.62	1.3	ST/12	0.00	1.8*	1
CS 31062-050	-2.42	2.81	1.55	0.59	1.3	ST/12	0.00	1.8*	7
LP 625-44	-2.70	2.60	1.07	0.27	1.3	ST/30	0.00	1.8*	8
HD 196944	-2.25	1.90	0.30	0.99	1.5	ST/3	1.80	0.0*	1
HD 196944	-2.40	2.10	0.17	1.33	1.5	ST/3	1.3	0.5	9
HD 26	-1.25	2.00	0.73	0.37	1.5	ST/2	0.80	0.5	9
HD 187861	-2.30	3.30	0.6	1.33	1.5	ST/2	0.25	0.5	9
HD 189711	-1.80	0.90	0.60	-0.70	1.3	ST/24	0.40	0.5	9
HD 198269	-2.20	2.40	0.93	1.07	1.2	ST/9	0.40	0.5	9
HD 201626	-2.10	2.60	0.70	1.00	1.5	ST/3	0.70	0.5	9
HD 224959	-2.20	3.10	1.07	1.03	1.5	ST/2	0.35	0.5	9
V-Ari	-2.40	1.20	0.50	-0.40	1.2	ST/30	0.05	0.5	9
HE 0024-2523	-2.70	3.30	0.56	1.67	1.5	ST/2	1.00	0.0*	10
HE 2148-1247	-2.30	3.12	1.10	0.87	1.3	ST/12	0.00	2.0*	11
CS 22948-27	-2.47	2.72	1.27	0.45	1.3	ST/24	0.25	1.5*	3
CS 29497-34	-2.90	2.95	0.98	0.87	1.3	ST/30	0.00	1.5*	3
HE 0338-3945	-2.41	3[a]	1.52	0.75[a]	1.3	ST/12	0.00	2.0*	12

[a] Prediction of Pb
* [Eu/Fe] measured

Acknowledgments. Italian FIRB Project Astrophysical Origin of the Heavy Elements beyond Fe.

REFERENCES

1. R. Gallino, et al. *Astrophysical Journal*, **497**, pp. 338–403 (1998).
2. W. Aoki, et al. *Astrophysical Journal*, **580**, pp. 1149–1158 (2002).
3. C. Arlandini, et al., *Astrophysical Journal*, **525**, pp. 886–900 (1999).
4. B. Barbuy, et al., *Astronomy & Astrophysics*, **429**, pp. 1031–1042 (2005).
5. J. A. Johnson, and M. Bolte, *Astrophysical Journal*, **579**, pp. L87–L90 (2002).
6. T. Sivarani, et al. *Astronomy & Astrophysics*, **413**, pp. 1073–1085 (2004).
7. I. Ivans, et al., *Astrophysical Journal*, **627**, pp. 145–148 (2005).
8. J. A. Johnson, and M. Bolte, *Astrophysical Journal*, **605**, pp. 462–471 (2004).
9. W. Aoki, et al. *Publ. Astron. Soc. Japan*, **54**, pp. 427–449 (2002).
10. S. Van Eck, S. Goriely, A. Jorissen, B. Plez, *Astronomy & Astrophysics*, **404**, pp. 291–299 (2003).
11. S. Lucatello, et al., *Astronomical Journal*, **125**, 875 (2003).
12. J. G. Cohen, et al., *Astrophysical Journal*, **588**, pp. 1082–1098 (2003).
13. P. S. Barklem, et al., *Astronomy & Astrophysics*, **439**, 129–151 (2005).

Cosmic history of star formation and metal production

Francesco Calura and Francesca Matteucci

Dipartimento di Astronomia, Universitá di Trieste, via G.B. Tiepolo 11 34131 Trieste,

Abstract. By means of detailed galactic chemical and photometric evolution models, we study the star formation and metal production in galaxies of different morphpological types, namely elliptical, spirals and dwarf irregulars. We analyze the cosmic evolution of the galaxy luminosity density in various bands, comparing the predictions by our pure-luminosity evolution (PLE) scenario with the ones coming from a semi-analythical hierarchical model (SAM), predicting strong galaxy density evolution. Once corrections for dust extinction are taken into account, the PLE models can reproduce the UV luminosity density observed at $z \geq 3$, whereas the hierarchical SAM underestimates it. We study the evolution of the cosmic type Ia and II Supernova rate, showing that our pure-luminosity evolution picture accounts for most of the current observations. Finally, we estimate the total metal density at redshift $z = 0$, showing that a Salpeter IMF allows us to reproduce the local metal inventory.

Keywords: Document processing, Class file writing, LaTeX 2_ε
PACS: 43.35.Ei, 78.60.Mq

INTRODUCTION

We compute the cosmic star formation history by means of detailed chemical evolution models for galaxies of different morphological types, namely ellipticals, spirals and irregular galaxies. All the details concerning these models can be found in [2], [3]. Elliptical galaxies are assumed to form by means of a rapid collapse of a gas cloud occurring at high redshift [15]. Spirals are formed through a double-Infall process [6]. Irregulars galaxies assemble through continuos infall and form stars at a continuous and very low ($\sim 0.01 M_\odot/yr$) star formation rate. The adopted initial stellar mass function is the one by [21], assumed to be constant in both space and time. To calculate galaxy spectra, we use the spectro-photometric codes by [12] and [1], by taking into account also dust extinction effects. The number densities of the various galactic morphological types are normalized according to the local B-band luminosity function [14]. We assume a scenario of pure-luminosity evolution (PLE) and no evolution in number. We assume that all galaxies started forming stars at redshift $z_f = 5$, and adopt a Lambda Cold dark Matter cosmology with $h = 0.65$, $\Omega_m = 0.3$, $\Omega_\Lambda = 0.7$.

THE COSMIC EVOLUTION OF THE GALAXY LUMINOSITY DENSITY AND SN RATE

In figure 1 (left side), we show our predictions for the evolution of the UV and B luminosity density and we compare our results with the predictions calculated with a

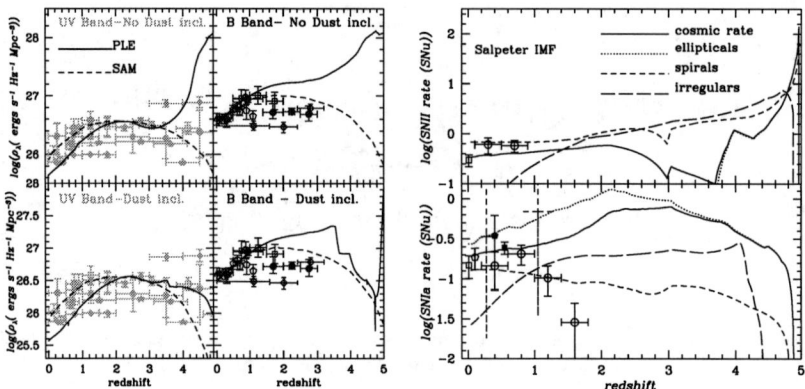

FIGURE 1. **Left:** Observed evolution of the UV (left panels) and B (right panels) luminosty densities as observed by various authors (points with error bars, see [4] and references therein) and as predicted by a pure luministy evolution model (solid lines) and a hierarchical semi-analytic model (dashed lines). In the upper (lower) panels, dust extinction is not (is) included in the predictions. **Right:** Evolution of the type Ia (lower panel) and II (upper panel) Supernova rate as predicted by the PLE model for various galactic morphological types and as observed by various authors. Solid square: [20]; open squares: [5]; open penthagon: [11]; solid penthagon: [19]; open circles: [7]; dashed crosses: [10].

semi-analythical model (SAM) of galaxy formation [16]. The SAM follows the merging histories of the DM haloes hosting the galaxies and, at variance with the PLE model, predicts a strong galaxy density evolution. The observational data are from various authors (see [4] and references therein). We note that, once dust corrections are applied to the models (lower panels), the PLE scenario reproduces the UV luminosity density observed at redshift $z \geq 3$, whereas the hierarchical SAM underestimates these data [4]. In figure 1 (right side), we show the evolution of the type Ia and II cosmic Supernova (SN) rates, expressed in SNu ($1SNu = 1SN/century/10^{10}L_{B,\odot}$), as predicted by the PLE model for various galactic morhological types and as observed by various authors. The data up to redshift $z \sim 1$ are well reproduced by our models. On the other hand, the data at higher redshifts by [7] are overestimated by our predictions. These data are very difficult to reproduce by means of standard SN rate models, unless a significant delay time ($\tau \geq 2$ Gyr) between the epoch of star formation and the explosion of type Ia SN is assumed. These data are the first collected at redshift > 1 for the type Ia SN rate and certainly need to be confirmed by future surveys. Finally, in Table 1, we show the present-day metal density as predicted by our models by assuming a standard Salpeter IMF and a Top-heavy IMF in spheroids, compared to different estimates by various authors. Our results indicate that with the Salpeter IMF, the local metal budget is reproduced with good accuracy. The adoption of a Top-Heavy IMF in spheroids leads to an overestimation of the local metal density by a factor of ~ 4.

TABLE 1. Predicted total metal density in the local universe, by assuming different IMFs and compared to various estimates by other authors.

Author	Ω_Z	$\rho_Z (M_\odot/Mpc^3)$
[13]	$7.8 \cdot 10^{-5}$	$5.4 \cdot 10^6$
[17]	$2. \cdot 10^{-4}$	$1.4 \cdot 10^7$
[8]	$9.6 \cdot 10^{-5} - 1.9 \cdot 10^{-4}$	$6.7 - 13.3 \cdot 10^6$
[18]	$2.5 \cdot 10^{-4}$	$1.7 \cdot 10^7$
[9]	$9.9 \cdot 10^{-5} - 2.00 \cdot 10^{-4}$	$6.9 - 13.8 \cdot 10^6$
present work (Salpeter IMF)	$1.35 \cdot 10^{-4}$	$9.37 \cdot 10^6$
present work (Top-Heavy IMF in spheroids)	$5.9 \cdot 10^{-4}$	$4.12 \cdot 10^7$

REFERENCES

1. Bruzual, A. G., Charlot, S., 2003, MNRAS, 344, 1000
2. Calura, F., Matteucci, F., 2003, ApJ, 596, 734
3. Calura, F., Matteucci, F., 2004, MNRAS, 350, 351
4. Calura F., Matteucci F., Menci N., 2004, MNRAS, 353, 500
5. Cappellaro, E., Evans, R., Turatto, M., 1999, A&A, 351, 459
6. Chiappini, C., Matteucci, F., Gratton, R. 1997, ApJ, 477, 765
7. Dahlen, T., et al., 2004, 613, 189
8. Dunne, L., Eales, S. A., Edmunds, M. G., 2003, MNRAS, 341, 589
9. Finoguenov, A., Burkert, A., Boehringer, H., 2003, ApJ, 594, 136
10. Gal-Yam, A., Maoz, D., Sharon, K., 2002, MNRAS, 332, 37
11. Hardin, D., et al., 2000, A&A, 362, 419
12. Jimenez, R., et al., 1998, MNRAS, 299, 123
13. Madau, P., Ferguson, H. C., Dickinson, M. E., et al., 1996, MNRAS, 283, 1388
14. Marzke, R. O., et al., 1998, ApJ, 503, 617
15. Matteucci, F., 1994, A&A, 288, 57
16. Menci, N., Cavaliere, A., Fontana, A., Giallongo, E., Poli, F., 2002, ApJ, 575, 18
17. Mushotszky, R.F., Loewenstein, M., 1997, ApJ, 481, L63
18. Pagel, B., 2002, in "Chemical Enrichment of Intracluster and Intergalactic Medium", ASP Conference Proceedings, Edited by R. Fusco-Femiano and F. Matteucci, 253, 489
19. Pain R., et al., 1996, ApJ, 473, 356
20. Pain R., et al., 2002, ApJ, 577, 120
21. Salpeter, E. E., 1955, ApJ, 121, 161

A New Measurement of the ^{8}Li(α,n)^{11}B Reaction for Astrophysical Interest

Suranjan K. Das [*], T. Fukuda[*], Y. Mizoi[*], H. Ishiyama[†], H. Miyatake[†],
Y. X. Watanabe[†], Y. Hirayama[†], M. H. Tanaka[†], N. Yoshikawa[†],
S. C. Jeong[†], Y. Fuchi[†], I. Katayama[†], T. Nomura[†], T. Ishikawa[**],
K. Nakai[**], T. Hashimoto[‡], S. Mitsuoka[‡], K. Nishio[‡], Pranab K. Saha[‡],
M. Matsuda[‡], S. Ichikawa[‡], H. Ikezoe[‡], T. Furukawa[§], H. Izumi[§],
T. Shimoda[§] and T. Sasaqui[¶]

[*]*Osaka Electro-Communication University*
[†]*Institute of particle and Nuclear Studies, High Energy Accelerator Research Organaization (KEK)*
[**]*Tokyo University of Science*
[‡]*Japan Atomic Energy Research Institute (JAERI)*
[§]*Osaka University*
[¶]*University of Tokyo*

Abstract. The ^{8}Li(α,n)^{11}B reaction has been measured directly and exclusively in the energy region of E_{cm}=0.45-1.75 MeV by using highly efficient detector system covering E_{cm}= 0.56 MeV, which corresponds to the Gamow window at T_9=1. This experiment has been performed in the condition of inverse kinematics by using low-energy radioactive ^{8}Li beam at the Tandem accelerator facility of Japan Atomic Energy Research Institute. The reaction cross section obtained in the present measurement is consistent with that of the previous exclusive measurements within the errors in an overlapping energy region, but is less than half of that of the inclusive measurements, in particular for lower energy region.

Keywords: r-process, Big Bang nucleosynthesis, radioactive nuclear beam (RNB)
PACS: 26.30.+k

INTRODUCTION

It has been discussed that the α n reactions of light neutron-rich radioactive nuclei play important roles in r-process nucleosynthesis as well as in Big Bang nucleosynthesis. In particular, the ^{8}Li(α,n)^{11}B reaction is one of the key reactions going to heavier elements across the stability gap of A=8. We have measured the ^{8}Li(α,n)^{11}B reaction by using highly efficient detector system and high-purity low energy ^{8}Li beam. In this paper, a new result with E_{cm}=0.45-1.75 MeV is presented.

EXPERIMENT

The present experiment was carried out at tandem facility in Japan Atomic Energy Research Institute (JAERI). The ^{8}Li beam was produced via the reaction ^{9}Be(^{7}Li,^{8}Li). The recoil mass separator (RMS) was used as an in-flight secondary beam separator. The detector system consists of a Micro-Channel Plate (MCP), a Parallel-Plate Avalanche

Counter (PPAC), a Multiple-Sampling and Tracking Proportional Chamber (MSTPC), and a plastic scintillator array. See also the Ref. [1] for the details. The absolute energy of ^{8}Li beam was determined by the time-of-flight information between the MCP and PPAC event by event. Then ^{8}Li beam was injected into the MSTPC, which was filled with ^{4}He(90%)+CO_2(10%) gas. The efficiency of the MSTPC is very high because the chamber gas, ^{4}He, acts both as a target and as a counter gas. In this experiment the gas pressure was set to be 140 Torr so as to cover the energy region of E_{cm}= 0.14 - 1.75 MeV. The MSTPC identified the ^{8}Li$(\alpha,n)^{11}$B reaction by observing the sudden change of the energy loss along the beam direction, and also determined the reaction point and reaction energy. The neutron was detected by the plastic scintillator array. The reaction event thus selected was checked with its kinematical condition by using all information on its reaction energy, scattering angle and the energy of ejected nuclei and neutron.

RESULTS

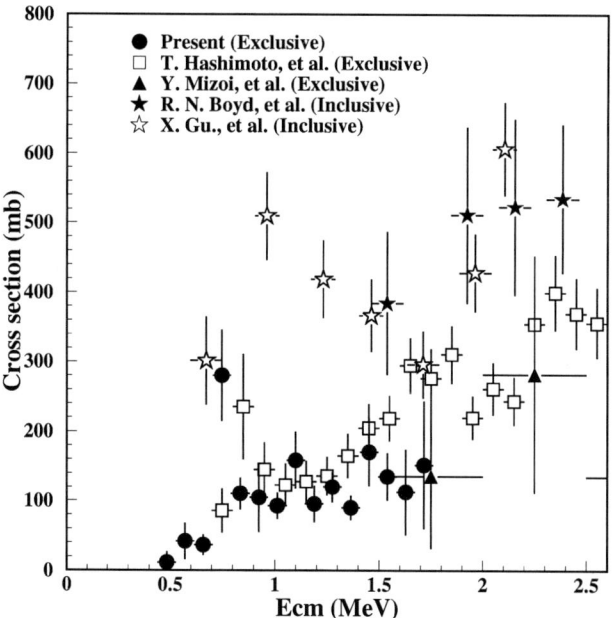

FIGURE 1. Total cross section of ^{11}B reaction with earlier works.

The FIG.1 shows the present result of the cross section. The present result is consistent with that of the previous exclusive measurements reported by Y. Mizoi et al. [2] and T. Hashimoto et al. [3] within the errors in an overlapping energy region. However, the present result is less than half of that of the inclusive measurements reported by R. N. Boyd et al. [4] and X. Gu et al. [5], in particular for lower energy region. In the result by

T. Hashimoto et al., there is a resonance like structure located around 0.85 MeV. In the present result, we could see a similar structure around 0.8 MeV. The FIG. 2 shows the cross sections to the ground state of ^{11}B as a function of the center-of-mass energy. The present result is consistent with that of the previous exclusive measurement performed by T. Hashimoto et al. within the error bars. This result also agrees with that of the T. Paradellis et al. except for a resonant-like structure around the energy region of 1.3 MeV.

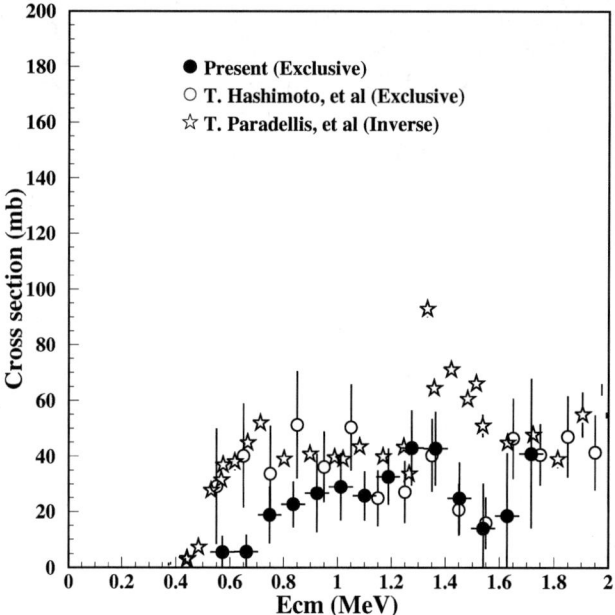

FIGURE 2. Ground state cross section of ^{8}Li$(\alpha,n)^{11}$B reaction.

REFERENCES

1. Ishiyama, H. et al., *in this Proceedings.*
2. Mizoi, Y. et al., *Phys. Rev.*, **C 62** (2000) 065801.
3. Hashimoto, T. et al., *to be submitted.*
4. Boyd, R. N et al., *Phys. Rev. Lett.*, **68** (1992) 1283.
5. Gu, X. et al., *Phys. Lett.*, **B 343** (1995) 31.
6. Paradellis, T. et al., *Z. Phys.*, **A 337** (1990) 211.

Oxygen in very metal-poor stars

É. Depagne, F. Primas
ESO

1 Introduction

Oxygen is the third most abundant element in the Universe. But abundance determination in very metal poor stars is still, even if things are getting better thanks to large amount of data gathered at the largest telescopes, an heavily debated subject.

One major problem enountered is that up to now, abundance determination was depending on the Oxygen lines chosen. And since oxygen abundance is ofcrucial importance for many astrophysical topics let's briefly mention the following :How old are the oldest stars?How can we interpret abundance patterns in terms of Galactic Chemical Evolution? How does spallation really work? How do Massive stars and their lives? it is needed that we can rely ont whichever line system used.

2 How to measure Oxygen?

There are four main system used to measure oxygen abundance.

- the UV OH lines,

- the forbidden visible line

- the IR triplet

- the OH IR lines

The forbidden visible line is the best indicator for oxygen abundance, as it is very little sensitive to models errors, but this line is very faint in metal poor stars.

3 What have been measured ?

Various studies have been conducted on oxygen abundances and all the systems mentionned have been used. The results, presented in the following figures show some discrepancy.

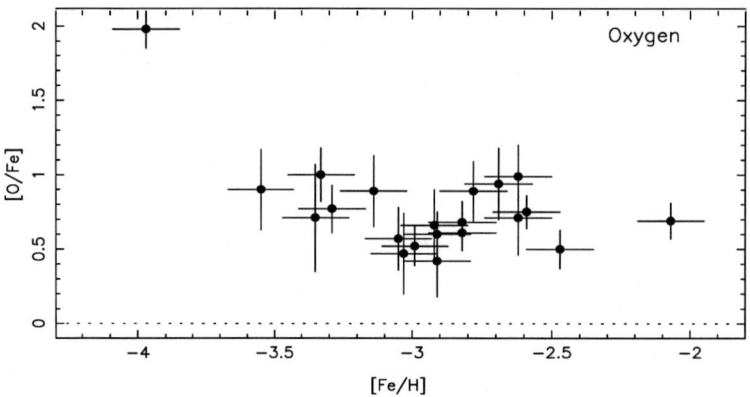

Figure 1: Oxygen abundances measured using the [OI] line by Cayrel et al 2004, A&A 416, 1117

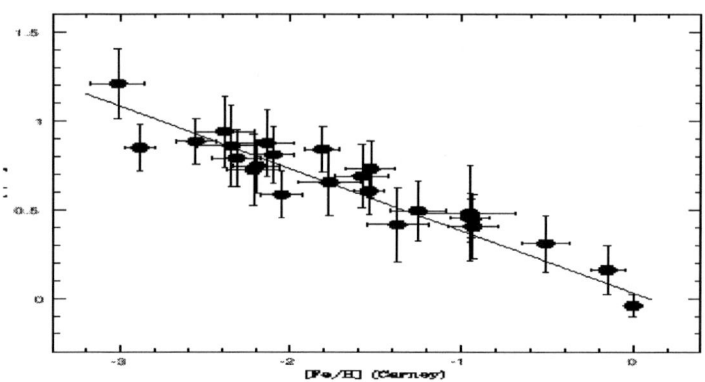

Figure 2: Oxygen abundances measured using the OH IR lines by Boesgaard et al 2001, ApJ, 117 492

The apparent discrepancy comes from modelling. The lines are not formed in the same physical conditions, and thus, errors on the models lead to errors in abundance determination.

4 How to solve this?

We have measured the IR triplet (771-4nm) and we have used some 3D models with NLTE corrections. Observations were conducted using the VLT/UVES spectrograph giving us spectra of an unprecedented quality. With these data, we have obtained the results shown figure 3.

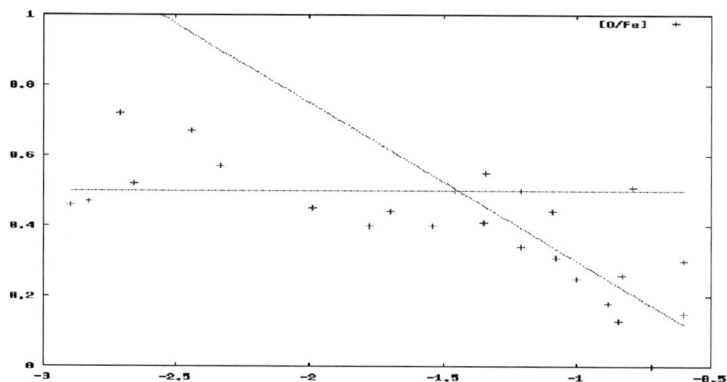

Figure 3: Oxygen abundances measured in our sample

5 Interpretation

The results we have obtained show that when the metallicity decreases, the oxygen abundance reaches a plateau at a value of 0.5. The behaviour is exactly the one we expect from an α element. This is now a proved fact, which was not the case when comparing to other studies using the same lines.

These results show that the IR triplet can now be used safely to determine O abundance in combination with 3D models.

Finite-size effects on the hadron-quark mixed phase

Tomoki Endo*, Toshiki Maruyama[†], Satoshi Chiba[†] and Toshitaka Tatsumi*

*Department of Physics, Kyoto University, Kyoto 606-8502, Japan
[†]Japan Atomic Energy Agency, Tokai, Ibaraki 319-1195, Japan

Abstract. We show that the hadron-quark mixed phase is restricted to narrow range of baryon chemical potential by the charge screening effect. Accordingly the mixed phase expected in hadron-quark hybrid stars should be narrow. Although the screening would not have large effect in bulk properties of the star such as mass or radius, it change the internal structure of the star very much, which may be tested by the cooling curve, glitch phenomena or gravitational waves.

Keywords: screening effect, mixed phase, hybrid star
PACS: 21.65.+f, 25.75.Nq, 26.60+c, 97.60.Jd

INTRODUCTION

Recently it seems to be widely believed that the *structured mixed phase* (SMP) would appear in the wide density range during the first order phase transitions with many (≥ 2) chemical potentials [1, 2, 3]. Applying the Gibbs conditions to get the equation of state (EOS), one may see non-uniform structures of not only baryon density but also charge density distribution for the mixed phase, and no constant pressure region in EOS. If this is the case, the Maxwell construction (MC) should become meaningless, where the phase equilibrium between two bulk matters with local charge neutrality is assumed. The appearance of such SMP in the hadron-quark deconfinement transition has been expected inside neutron stars and its implications have been discussed [3].

The importance of the *finite-size effects* in the mixed phase, which has not been fully included in the previous calculations, has been emphasized in recent papers: especially, the charge rearrangement effect induced by the Coulomb interaction should be carefully taken into account. Actually it has been demonstrated [4, 5, 6] that the Debye screening effect should greatly modify the description of the SMP.

To study non-uniform structure, we solve the coupled equations of motion based on the density functional theory [7], using the Wigner-Seitz approximation. In our calculation we can derive the density profiles of all the particle species and determine the configuration of the Coulomb potential *exactly* without recourse to any approximations included in the recent study.

NUMERICAL RESULTS

We present the thermodynamic potential density $\omega \equiv \Omega/V$ of each uniform matter in Fig. 1 (a) as a function of the baryon number chemical potential μ_B: the hadron phase (H) is thermodynamically favorable for μ_B below 1225 MeV and the quark phase (Q) above it. We also depict the one denoted by "bulk Gibbs" for comparison, where the Gibbs conditions are applied but the finite-size effects are completely discarded. MC can be represented as a point in this figure. We can immediately see that "bulk Gibbs" smoothly connects H and Q, and the mixed phase appears in this wide interval. We plot $\delta\omega$, the difference of the thermodynamic potential density between the mixed phase and each uniform matter, in Fig. 1 (b): the curve denoted by "screening" is the result of self-consistent calculation, while the one denoted by "bulk Gibbs" corresponds to that in Fig. 1 (a). We also depict another curve denoted by "no screening" to elucidate the charge screening effect, which is given by a perturbative treatment [2, 3] of the Coulomb interaction. Then we can see that the large reduction of the thermodynamical potential from "bulk Gibbs" is mainly given the effect of the surface tension, while the screening effect further reduces it. $\delta\omega$ given by MC appears as a point denoted by a circle in Fig.

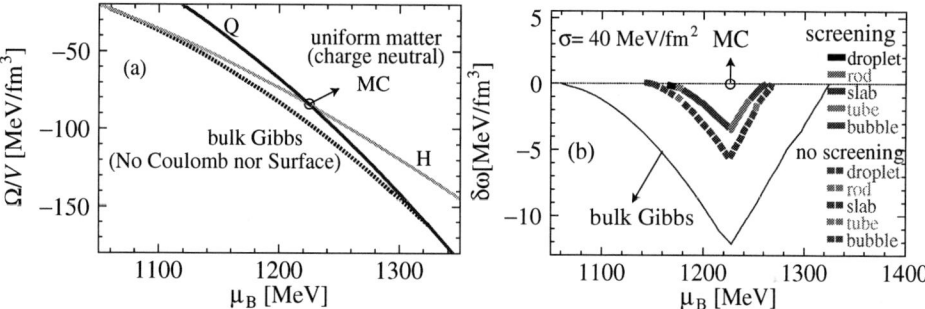

FIGURE 1. Thermodynamic potential density. (a) shows the results of each uniform matter and "bulk Gibbs". (b) shows the difference between the mixed phases and the uniform matter.

1 (b) where two conditions, $P^Q = P^H$ and $\mu_B^Q = \mu_B^H$, are satisfied. On the other hand the mixed phase derived from "bulk Gibbs" appears in a wide region of μ_B. Therefore, if the region of the mixed phase becomes narrower, it signals that the properties of the mixed phase become close to those given by MC. One may clearly see that the thermodynamic potential becomes close to that given by MC due to the finite-size effects.

Figures. 2 (a) and (b) show EOS in the cases with and without the screening effect. The pressure becomes more similar to that given by MC by the finite-size effects. Moreover, in Fig. 2 (b), it becomes more close to MC by the charge screening effect, which shows a larger pressure in the beginning and weaker one in the end of the phase transition.

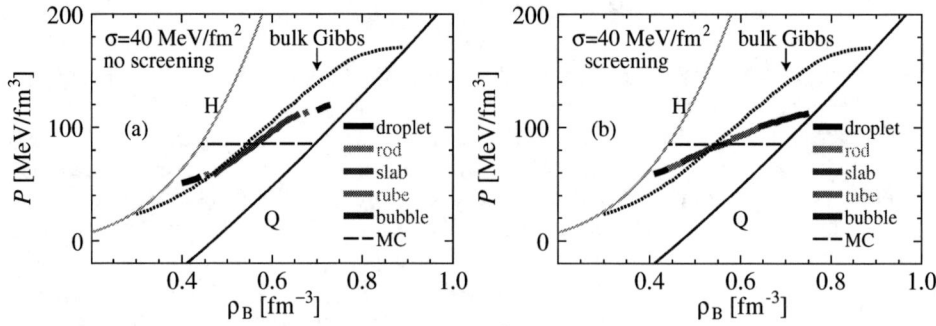

FIGURE 2. Pressure as a function of baryon-number density. (a) is the result of "no screening" and (b) "screening". The results given by "bulk Gibbs" and MC are also presented for comparison.

SUMMARY

We have seen that the finite-size effects changes the properties of the hadron-quark mixed phase which is expected in hybrid stars. In particular, the region in the baryon-number chemical potential is restricted by the charge screening effect. We have seen that EOS becomes close to that with MC by the finite-size effects; EOS becomes more similar to that with MC by the charge screening effect.

Let us briefly consider some implication of these our results for compact star phenomena. Glendenning and Pei [3] suggested many SMPs appear in the core region by using "bulk Gibbs": the mixed phase should appear for several kilometers. However we can say that the region of SMP should be narrow in the μ_B space and EOS is more similar to that of MC due to the finite-size effects. These results seem to be consistent with those given by other studies. Bejger et al. [8] have examined the relation between the mixed phase and glitch phenomena, and shown that the mixed phase should be narrow if the glitch is generated by the mixed phase in the inner core. On the other hand the gravitational wave asks for density discontinuity in the core region [9]. It is very interesting to study the relation between these phenomena and our results in more detail.

REFERENCES

1. N. K. Glendenning, Phys. Rev. **D46** (1992) 1274.
2. H. Heiselberg, C. J. Pethick and E. F. Staubo, Phys. Rev. Lett. **70** (1993) 1355.
3. N. K. Glendenning and S. Pei, Phys. Rev. **C52** (1995) 2250.
4. D. N. Voskresensky, M. Yasuhira and T. Tatsumi, Nucl. Phys. **A723** (2003) 291.
5. T. Endo, Toshiki Maruyama, S. Chiba and T. Tatsumi, Nucl. Phys.**A749** (2005) 333c.
6. T. Endo, Toshiki Maruyama, S. Chiba and T. Tatsumi, hep-ph/0502216.
7. T. Endo, Toshiki Maruyama, S. Chiba and T. Tatsumi, Prog. Theor. Phys. in press; hep-ph/0510279.
8. M. Bejger, P. Haensel and J. L. Zdunik, Mon. Not. Roy. Astron. Soc. **359** (2005) 699.
9. G. Miniutti, J. A. Pons, E. Berti, L. Gualtieri, V. Ferrari, Mon. Not. Roy. Astron. Soc. **338** (2003) 389.

Search for α-enhanced stars in the spectroscopic SDSS stellar database

M. Franchini[1], C. Morossi[1], P. Di Marcantonio[1], M.L. Malagnini[2] and M. Chavez[3]

[1] INAF - Osservatorio Astronomico di Trieste, Via G.B. Tiepolo, 11, I-34131 Trieste, Italy
[2] Dip. di Astronomia, Università degli Studi di Trieste, Via G.B. Tiepolo, 11, I-34131 Trieste Italy
[3] Instituto Nacional de Astrofisica, Optica y Electronica, A.P. 51 y 216, 72000 Puebla, Mexico

Abstract. We analyze a sample of about 8000 stars extracted from the Sloan Digital Sky Survey with T_{eff} in the range 4500-7000 K and log g greater than 2.0 dex as estimated by comparing measured and synthetic Lick indices. We present preliminary results on the dependence of the α-enhancement phenomenon on the stellar metallicity and on the galactic position.

Keywords: Stars – abundances, chemical composition; stellar content and populations.
PACS: 97.10.Tk, 98.35.Ln

INTRODUCTION

In the study of the formation and structure of the Milky Way (MW) the abundance and kinematical analysis of stars characterized by the so-called α-enhancement phenomenon is of great relevance to get insights into the role of SN I and SN II in the chemical enrichment of different stellar populations. In particular, Franchini et al. [1], [2], [3], on the basis of synthetic spectra [4] computed with Solar Scaled Abundances (SSA) and enhanced [α/Fe] ratios (NSSA), introduced four Lick/IDS [5] index-index diagrams characterized by the presence of two different loci representative of SSA and NSSA points. Therefore, they were able to mark an observed star either as "NSSA" or "SSA" by looking at its position in these diagrams regardless to the knowledge of its main atmospheric parameters (i.e. T_{eff}, log g and [Fe/H]) and by using as dividing line the lower boundary of the theoretical SSA locus.
In this contribution we use the combinations of indices introduced in [1] to study the behavior of a sample of several thousand cool stars from the Sloan Digital Sky Survey (SDSS-DR4, [6]). Preliminary results on the dependency of the α-enhancement phenomenon on the stellar metallicity and on the galactic position are presented.

PRELIMINARY RESULTS

Since we were interested in finding mainly late F, G and early K stars belonging to the disk or to the halo, we selected among all the spectra classified as STARs in the SDSS DR4 [6] more than 20,000 spectra having 0.7<(g-r)<1.3 mag and (g+r)/2<21 mag. A check on the consistency of the calculated SDSS spectrophotometric color indices $(g-r)$ and $(r-i)$ with the photometric values and the rejection of spectra that show hints of possible non-stellar objects wrongly classified as stars reduced the number of object to slightly less than 17,000.

We present preliminary results on a sub-sample of 7766 stars with T_{eff} in the range 4500-7000 K and log g greater than 2.0 dex as estimated by comparing measured and synthetic Lick indices. The measured indices are calibrated in the Lick/IDS system as described in [7] and the synthetic ones are derived from synthetic spectra computed by means of the SPECTRUM v.2.56 code [8] starting from NEWODF Atlas9 models [9]. Figure 1 shows the g versus $(g-r)$ diagram (left panel) of our sample and the distributions in metallicity of two sub-samples representing stars at low ($|Z_{gal}|<3000$ pc, empty area) and high ($|Z_{gal}|>5000$ pc, dashed area) distances from the Galactic plane. Colors were dereddened by using the extinction maps by [10]. Stellar distances were computed by using evolutionary tracks. Our results agree with those found also by Allende Prieto et al. [11] (see their figure 12 panel d): the average [Fe/H] decreases as $|Z_{gal}|$ increases with peaks in distributions at about -0.7 and below -1.0 dex for $|Z_{gal}|<3$ Kpc and $|Z_{gal}|>5$ Kpc, respectively.

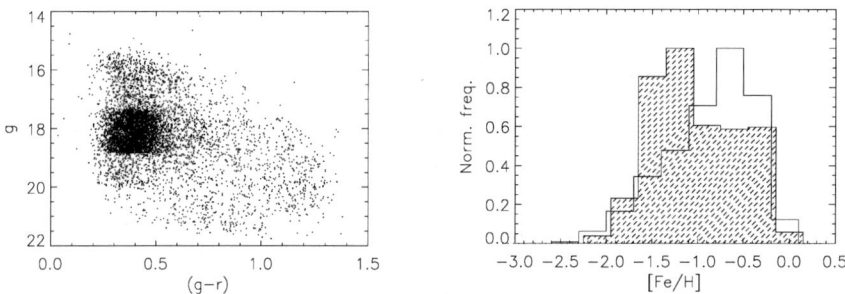

FIGURE 1. g versus $(g-r)$ diagram (left panel) and the distributions in metallicity of two sub-samples representing stars at low ($|Z_{gal}| < 3000$ pc, empty area) and high ($|Z_{gal}| > 5000$ pc, dashed area) distances from the Galactic plane (see text).

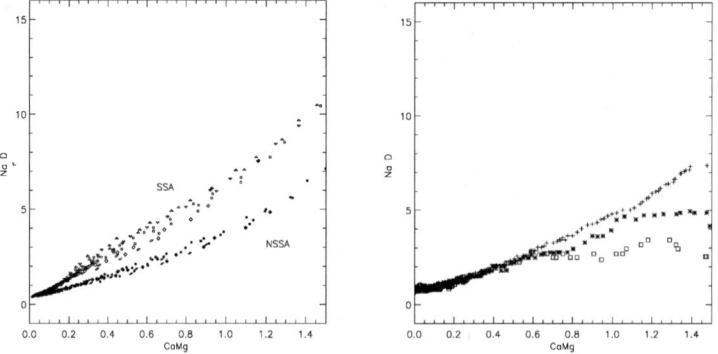

FIGURE 2. Search for α-enhanced stars via Lick/IDS NaD-CaMg diagram; left panel: theoretical predictions for α-enhanced stars (empty symbols) and Solar Scaled Abundance Stars (filled symbols); rigth panel: average observational trends for different $|Z_{gal}|$ (see text).

In [1] we introduced four Lick/IDS index-index diagrams (i.e. NaD vs Ca4227, NaD vs Mg_2, NaD vs Mgb, and NaD vs $CaMg=0.125*Ca4227+Mg_2$), which allow us to identify different loci representative of SSA and α-enhanced stars irrespectively of their T_{eff}, log g and [Fe/H]. In Figure 2 the median trend of three sub-samples of our SDSS stars is plotted in one of the four diagrams (i.e. NaD versus CaMg) as an example; the three sub-samples contain stars with $|Z_{gal}|<500$ pc (plus symbols), $1800< |Z_{gal}|<3000$ pc (asterisks), and $|Z_{gal}|>4000$ pc (squares), respectively. As can be seen at greater $|Z_{gal}|$ correspond lower Na D and higher Ca4227, Mgb, Mg_2, and CaMg index values. If we take into account the theoretical predictions shown in the left panel, we can conclude that the observations clearly indicate an increase in the percentage of α-enhanced stars with increasing $|Z_{gal}|$. Work is in progress to extend this analysis to a larger sample of SDSS stars, to study in details their kinematical and metallicity properties and to achieve quantitative estimates of individual elemental abundances.

ACKNOWLEDGMENTS

This work received partial financial support from the Mexican CONACyT via grant 36547-E from MIUR COFIN-2003028039.

Funding for the creation and distribution of the SDSS Archive has been provided by the Alfred P. Sloan Foundation, the Participating Institutions, the National Aeronautics and Space Administration, the National Science Foundation, the U.S. Department of Energy, the Japanese Monbukagakusho, and the Max Planck Society. The SDSS Web site is http://www.sdss.org/. The SDSS is managed by the Astroph. Research Consortium (ARC) for the Participating Institutions. The Participating Institutions are The University of Chicago, Fermilab, the Institute for Advanced Study, the Japan Participation Group, The Johns Hopkins University, the Korean Scientist Group, Los Alamos National Laboratory, the Max-Planck-Institute for Astronomy (MPIA), the Max-Planck-Institute for Astrophysics (MPA), New Mexico State University, University of Pittsburgh, University of Portsmouth, Princeton University, the United States Naval Observatory, and the University of Washington.

REFERENCES

1. Franchini, M., Morossi, C., Di Marcantonio, et al., 2004°, *ApJ*, **601**, 485
2. Franchini, M., Morossi, C., Di Marcantonio, et al., 2004b, *ApJ*, **613**, 312
3. Franchini, M., Morossi, C., Di Marcantonio, et al., 2005, *ApJ*, **634**, 1319-1335
4. Malagnini, M. L., Franchini, M., Morossi, C., et al. 2005, ESA **SP-576**, 595
5. Worthey, G., Faber, S. M., Gonzalez, J. J., et al., 1994, *ApJ*, **94**, 687
6. Adelman-McCarthy, J. K. et al 2005, astro-ph/0507711
7. Morossi, C., Franchini, M., Di Marcantonio, et al., 2005, Poster No. 36, this Symposium
8. Gray, R. O., Corbally, C. J., 1994, *AJ*, **107**, 742
9. Castelli, F., & Kurucz, R.L. 2003, in IAU Symp. 210, poster A20 on enclosed CD-ROM(astro-ph/0405087).
10. Schlegel, D. J., Finkbeiner, D. P., Davis, M. 1998, *ApJ*, **500**, 525
11. Allende Prieto, C. et al. 2005, *ApJ* in press, astro-ph/0509812.

Nucleosynthesis inside Magnetically-Driven Jets in A Gamma-Ray Burst

Shin-ichirou Fujimoto*, Masa-aki Hashimoto[†], Kei Kotake** and Shoichi Yamada**

*Kumamoto National College of Technology
[†]Department of Physics, Kyushu University
**Science and Engineering, Waseda University

Abstract. We investigate nucleosynthesis inside jets in a gamma ray bursts (GRB). We have calculated detailed composition of magnetically driven jets related to GRBs with post-processing calculation, which is based on long-term, magneto-hydrodynamic simulations of a rapidly rotating massive star of $40M_\odot$ during core collapse. We follow abundance evolution of about 4000 nuclides inside the jets from the collapse phase to the ejection phase through the jet generation phase with a large nuclear reaction network. We find that the r-process successfully operates inside the jets, so that U and Th are synthesized abundantly. Abundance pattern inside the jets is similar compared to that of r-elements in the solar system. Heavy neutron-rich nuclei $\sim 0.001 M_\odot$ can be ejected through the jets. Furthermore, we find that p-nuclei are produced without seed nuclei: not only light p-nuclei, such as ^{74}Se, ^{78}Kr, ^{84}Sr, and ^{92}Mo, but also heavy p-nuclei, ^{113}In, ^{115}Sn, and ^{138}La, can be abundantly synthesized in the jets. The amounts of p-nuclei in the ejecta are much greater than those in core collapse supernovae (SNe). In particular, ^{92}Mo, ^{113}In, ^{115}Sn, and ^{138}La deficient in core collapse SNe, are significantly produced in the ejecta.

Keywords: stars: nuclear reactions, nucleosynthesis, abundances - stars: evolution - stars: magnetic fields - supernova: general - gamma rays: bursts
PACS: 97.60.Bw

MAGNETICALLY DRIVEN JETS IN A GAMMA-RAY BURST

During the collapse of a massive star greater than 35-40 $M_\odot$, stellar core is considered to promptly collapse to a black hole. Stellar material can fall onto the hole at extremely high accretion rates. If the star has sufficiently high angular momentum before the collapse, an accretion disk is likely to be formed around the hole. Jets are suggested to be launched from the inner region of the disk near the hole through magnetic and/or neutrino processes. Gamma-ray bursts (GRBs) are expected to be driven by relativistic jets launched from the accretion disk around the black hole. This scenario of GRBs is called a collapsar model [1]. The disk is so dense and hot that a sequence of nuclear burnings proceed explosively. Using one dimensional models of (horizontal) disks and (vertical) outflows, r-nuclei [2] as well as p-nuclei [3] have been shown to be ejected through the outflows. However, the composition of the ejecta highly depends on the degree of neutronization of the ejecta and thus the dynamics of the disk and the outflows. Therefore, multi dimensional effects are important for nucleosynthesis inside the outflows. In the present work, we investigate nucleosynthesis inside magnetically-driven jets from GRB, based on two dimensional magneto-hydrodynamic (MHD) simulations of the jets.

We have firstly carried out two dimensional axisymmetric Newtonian MHD calculation of the collapse of a rotating massive star of $40M_\odot$ in light of the collapsar model of GRBs (see Fujimoto et al. [4] in detail). The numerical code for the MHD calculation employed in the present paper is based on the ZEUS-2D code with a realistic equation of state [5]. We consider neutrino cooling processes but ignore resistive heating, whose properties are highly uncertain. Fluid is freely absorbed through the inner boundary of 50km, which mimics a surface of the black hole, whose mass is continuously increased by the mass of the infalling gas through the inner boundary. We adopt an analytical form of the angular velocity $\Omega(r) = 10\,\mathrm{rad\,s^{-1}}/[1+(r/1000\mathrm{km})^2]$ of the star before the collapse. Initial magnetic field, B_0, is assumed to be uniform and parallel to the rotational axis. We consider case with $B_0 = 10^{12}$ G.

We find that jets can be magnetically driven from the central region of the star along the rotational axis; After material reaches to the black hole with high angular momentum, a disk is formed inside a surface of weak shock, which is appeared near the hole due to the centrifugal force and propagates outward slowly. The magnetic fields, which are dominated by the toroidal component, are chiefly amplified due to the wrapping of the field inside the disk and propagate to the polar region along the inner boundary near the black hole through the Alfvén wave. Eventually, the jets can be driven by the tangled-up magnetic fields at the polar region near the hole.

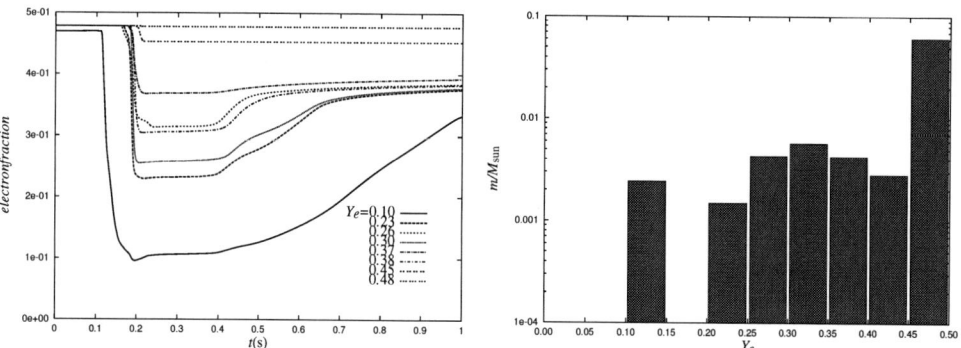

FIGURE 1. Evolution of the electron fraction for neutronized ejecta (left panel) and mass distribution with respect to Y_e inside ejecta (right panel), where Y_e is the electron fraction of the ejecta at 9×10^9 K.

The jets are tightly collimated and so dense and hot that electron capture on protons proceeds efficiently. Consequently, the electron fraction, Y_e, of ejecta at 9×10^9 K decreases enough for r-process to operate successfully (left panel in Figure 1). The ejected mass is found to be $0.081M_\odot$ through the jets. Note that masses of neutronized material of $Y_e \leq 0.1$ and $Y_e \leq 0.4$ are $2.4\times 10^{-3}M_\odot$ and $1.8\times 10^{-2}M_\odot$, respectively.

NUCLEOSYNTHESIS INSIDE THE JETS

Next we follow abundance evolution of the ejecta through the jets using a large nuclear reaction network (NETWORK B in Nishimura et al. [6]). The network contains about 4000 nuclides up to atomic number $Z = 100$, in which spontaneous and β-delayed

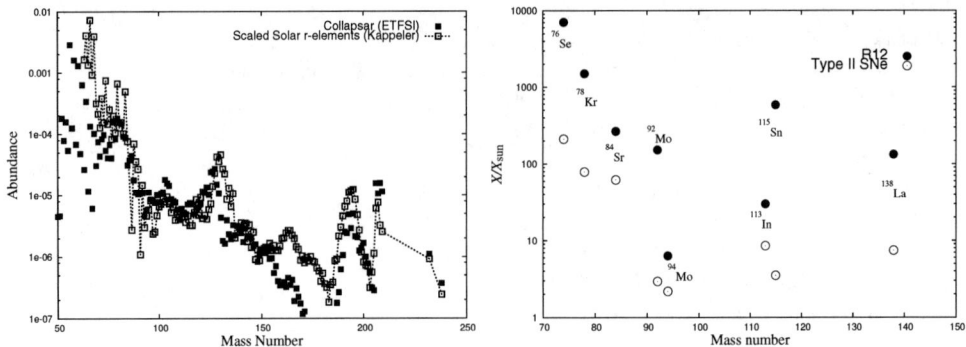

FIGURE 2. Integrated abundances in ejecta with respect to mass number, A, (left panel) and mass fractions of p-nuclei abundantly produced in the ejecta (right panel). Eu-scaled abundances of solar r-elements are shown in left panel. In right panel, the fractions are normalized by those in the solar system. IMF averaged abundances of p-nuclei from Type II supernovae (SNe) [7] are shown with open circles.

fission is taken into account. We find that r-process efficiently operates inside the highly neutronized ejecta of $Y_e \sim 0.1$. The abundances in the jets well reproduce the solar pattern of r-elements (left panel in Figure 2). However, nuclei with $160 < A < 180$ are underproduced compared with those in the solar system because of the lack of ejecta with $Y_e = 0.15$-0.20 (right panel in Figure 1). Nuclei with $A < 70$ are also less abundant due to low Y_e of the ejecta. It should be noted that U and Th are comparably produced as those in the solar system.

Moreover, we find that p-nuclei are significantly synthesized in the ejecta in spite of neutron-richness of the ejecta (right panel in Figure 2). Not only light p-nuclei, such as ^{74}Se, ^{78}Kr, ^{84}Sr, and ^{92}Mo, but also heavy p-nuclei, ^{113}In, ^{115}Sn, and ^{138}La, are produced in the ejecta abundantly The abundances of p-nuclei in the ejecta are much greater than those in Type II SNe [7]. It should be emphasized that ^{92}Mo, ^{113}In, ^{115}Sn, and ^{138}La, which are deficient in Type II SNe [7], are significantly produced in the ejecta. We also find that the light p-nuclei are produced via sequences of proton capture while the heavy p-nuclei can be produced through fission processes only in ejecta of $Y_e \sim 0.1$.

REFERENCES

1. Woosley, S.E. 1993, ApJ 405, 273
2. Fujimoto, S., Hashimoto, M., Arai, K., & Matsuba, R. 2004, ApJ, 614, 817
3. Fujimoto, S., Hashimoto, M., Arai, K., & Matsuba, R. 2005, Nucl. Phys. A758, 47
4. Fujimoto, S., Kotake, K., Yamada, S., Hashimoto, M., & Sato, K. 2005, (ApJ submitted)
5. Kotake, K., Sawai, H., Yamada, S., & Sato, K. 2004, ApJ, 608, 391
6. Nishimura, S., Kotake, K., Hashimoto, M., Yamada, S., Nishimura, N., Fujimoto, S., & Sato, K. 2005, preprint, Astro-ph/0504100 (to appear in ApJ)
7. Rayet, M., Arnould, M., Hashimoto, M., Prantzos, N., & Nomoto, K. 1995, A&A, 298, 517

Radioactive elements in stellar atmospheres

Vira Gopka*, Alexander Yushchenko[†,*], Stephane Goriely**, Angelina Shavrina[‡] and Young Woon Kang[†]

*Astronomical observatory, Odessa National University, Odessa, Ukraine
[†]Astrophysical Research center for the Structure and Evolution of the Cosmos, Sejong University, Seoul, Korea
**Institut d'Astronomie et d'Astrophysique, Universite Libre de Bruxelles, Brussels, Belgium
[‡]Main Astronomical observatory, National Academy of Sciences of Ukraine, Kiev, Ukraine

Abstract. The identification of lines of radioactive elements (Tc, Pm and elements with 83<Z<100) in the spectra of chemically peculiar stars HD101065, HR465, HD965 is made. Three possible explanations are proposed: natural radioactive decay of Th and U in the upper levels of stellar atmospheres, contamination of stellar atmosphere by recent SN explosion, and spallation reactions.

Keywords: atomic data, nucleosynthesis, stars: abundances, stars: chemically peculiar, stars: individual (HD101065, HD965, HR465)
PACS: 26.20+k 26.50+x 95.30.Ky 95.75.Fg 97.10.Ex 97.30.Fi

INTRODUCTION

Nucleosynthesis at stellar surfaces was discussed from the fifties to the late seventies of the former century as a possibility to explain the observed stellar abundances. Later, atomic diffusion in stellar atmospheres was successfully proposed as an explanation of the overabundances of r- and s-process elements in peculiar B-F stars (for a recent review see [4]). In particular, Proffitt & Michaud [5] pointed out that one out of 500 peculiar stars can be expected to have surface abundance anomalies due to accretion from a binary companion that exploded as a supernova. Radiative diffusion is expected to make these peculiarities undetectable after a few millions years.

60 chemical elements (including Th and U) were investigated in the photosphere of Przybylski's star (HD101065) which is after the sun the most spectroscopically studied star. However, near half of the strong lines in the spectrum of HD101065 still remain unidentified. The possibility to relate these lines to radioactive elements was discussed in the seventies. Cowley et al. [2] made a review of earlier investigations and found the lines of Pm and Tc in the spectra of HD101065 and HD965. Gopka et al. [3] confirmed the identifications of Tc & Pm and identified in HD101065 the lines of radioactive elements with atomic numbers 83<Z<100, except At and Fr. These results were confirmed by [1]. Hereafter, we give some results on the identification of such radioactive elements in the atmospheres of the chemically peculiar stars HD101065, HR465, and HD965 and discuss several scenarios which can explain the observational data.

RESULTS OF OBSERVATIONS

Observations of HR465 and HD965 were obtained at the 1.8-meter telescope of Bohyunsan astronomical observatory (Korea) with the spectral resolution R=80000 and signal to noise ratio S/N=150. The wavelength interval is 3780-9500 Å. For HD101065 we used observations from archive of the 8-meter telescope of the European Southern Observatory (with R=110000, S/N>300 and wavelength interval of 3040-10400 Å). Atomic data for lines of radioactive elements were taken from the NIST database. The details of the calculations can be found in [3, 6, 8]. More details about the lines identified in the spectrum of HD101065, and most particularly for the Po, Rn, Ra, Ac, Pa, Np, Pu, Am, Cm, Bk, Cf and Es elements, can be found in [3]. In the spectra of HR465 and HD965 we identified the lines of Tc (12 & 17 lines in the spectra of HR465 and HD965 respectively), Pm (26 & 12), Rn (4 & 2), Ra (4), Ac (4 & 1), Pa (12), Np (2), Pu (4 & 3), Am (4 & 2), Cm (3 & 2), Bk (18 & 6), Cf (2). Only the line identification was performed because of lack of atomic data for these investigated elements. For example, in the case of Es (Z=99), only wavelengths are available in NIST (even the ionization stages are not marked). Only some oscillator strengths for Ra and Am are published.

INTERPRETATION

We can propose three hypotheses to explain the existence of radioactive elements in the stellar atmospheres. The first is the interplay of atomic diffusion (overabundance of elements with long decay times in the upper layers of the atmosphere) and natural radioactive decay of these elements [3]. The second is the contamination of stellar atmosphere by a recent SN explosion and the third is nuclear reactions involving energetic particles at the stellar surface.

The diffusion theory [4] predicts the overabundances of heavy elements in the certain layers of stellar atmosphere. Observations show that the overabundances of the order of 6-7 dex (with respect to the Sun) are possible. So the abundances of Th and U can reach $N/N_{total}=10^{-5}$ or higher. On earth, Th and U concentrations near $N/N_{total}=10^{-2}$ are found in the radioactive ores which are considered as the best ore deposits available for industry. All elements from Z=84 to 99 are found in these uranium ores. Z<92 elements can be explained by the natural decay of Th and U. Z>92 elements can be produced by reactions on lighter elements with neutrons and α-particles produced in the decay chains. The usual equations of equilibrium show that the concentration of the products of radioactive decays are proportional to decay times, i.e the longer the decay times the higher the concentration. The above mentioned stratification of Th and U in stellar atmospheres could explain the existence of long-lived isotopes of other elements (with decay timescale of the order of several months).

The atmosphere of HD101065 could also have been contaminated by a recent SN explosion of the binary component. The surface abundances of HD101065 follow rather well the solar system pattern of r-elements [2], though no lines of Pb and Bi are observed [3]. It can be considered as a sign of a recent r-process event. The solar abundances were produced in r-process events several billions years ago. This time is sufficient for natural decay of all short-lived isotopes of elements with Z>83 to Pb and Bi. This scenario of the

SN contamination should also be accompanied with an overabundance of some specific light elements, and most particular oxygen. Furthemore, as pointed out by [5], diffusion in stellar atmospheres is believed to destroy the relative pollution a few millions years after the contamination.

Finally, spallation reactions at the stellar surface can also be at the origin of the production of radioactive elements. The large magnetic field observed in Ap stars can be the origin of a significant acceleration of charged-particles, mainly protons and α-particles, that in turn can by interaction with the stellar material modify the surface content. Due to the unknown characteristics of the accelerated particles, a purely parametric approach was followed, taken as free parameters the proton and α-particle flux amplitude and energy distribution, and the time of irradiation. To estimate the resulting nucleosynthesis, a nuclear reaction network including all nuclei heavier than oxygen up to $Z = 102$ and located between the proton drip line and the neutron-rich side of the valley of stability is used. Proton, α and neutron captures, as well as α-, β- and spontaneous fission decays are considered. This includes some 240000 reactions on 3940 different species. The initial abundance distribution is assumed to be the solar one. Our calculation shows that the many aspects of the composition of HD101065 and other stars can be explained by spallation events. In particular, a significant production of $Z > 30$ heavy elements can be explained through secondary neutron captures. Another very attractive feature of the spallation process is the systematic production of Tc and Pm, and the possible production of actinides and subactinides, as suggested by observations [1,3]. More details on these calculations will be published somewhere else.

All the three scenarios described here still need to be investigated in more details. The combination of different physical mechanisms, like nuclear reaction and diffusion, may also have contributed to the presently observed stellar surface of HD101065.

REFERENCES

1. W. P. Bidelman, "Tc and Other Unstable Elements in Przybylski's Star" in *Cosmic Abundances as Records of Stellar Evolution and Nucleosynthesis in honor of David L. Lambert*, Edited by T. G. Barnes III, and F. N. Bash, Astronomical Society of the Pacific Conference Series 336, San Francisco, 2005, pp. 309–312.
2. C. R. Cowley, W. P. Bidelman, S. Hubrig, G. Mathys, and D. J. Bord, *Astron. & Astrophys*, **419**, 1087–1093 (2004).
3. V. F. Gopka, A. V. Yushchenko, A. V. Shavrina, D. E. Mkrtichian, A. P. Hatzes, S. M. Andrievsky, and L. V. Chernysheva, "On the radioactive shells in peculiar main sequence stars: the phenomenon of Przybylski's star" in *The A-Star Puzzle*, Edited by J. Zverko, J. Ziznovsky, S. J. Adelman, and W. W. Weiss IAU Symp. 210, Cambridge University Press, Cambridge, UK, 2004, pp. 734-742.
4. G. Michaud, "Atomic diffusion in stellar surfaces and interiors" in *The A-Star Puzzle*, Edited by J. Zverko, J. Ziznovsky, S. J. Adelman, and W. W. Weiss IAU Symp. 210, Cambridge University Press, Cambridge, UK, 2004, pp. 173–183.
5. C. R. Proffitt, and G. Michaud, *Astrophys. J.*, **345**, 998–1007 (1989).
6. A. V. Shavrina, N. S. Polosukhina, Ya. V. Pavlenko, A. V.Yushchenko, P. Quinet, M. Hack, P. North, V. F. Gopka, J. Zverko, J. Zhiznovsky, and A. Veles, *Astron. & Astrophys*, **409**, 707-713 (2003).
7. D. O'Sullivan, A. Thompson, J. Donnelly, L. O'C. Drury, and K.-P. Wenzel, *Adv. Space Res.*, **27**, 785–789 (2001).
8. A. Yushchenko, V. Gopka, S. Goriely, F. Musaev, A. Shavrina, C. Kim, Y. Woon Kang, J. Kuznietsova, and V. Yushchenko, *Astron. & Astrophys*, **430**, 255–262 (2005).

Missing Mass in Galaxies in Dynamic Universe Model of Cosmology (Part 3)

SNP. Gupta

Bhilai Steel Plant, 1b / 57 / sector 8, 490008, Bhilai, India, snp.gupta@gmail.com; snp.gupta@indiatimes.com

Abstract. In this present work SITA simulations were used to find out Theoretical star circular velocity curves in a Galaxy (star circular velocity verses star distance from the center of galaxy), depends on various initial conditions and are never half bell shaped curves as predicted by Bigbang cosmologies. Here we are presenting four main cases. In the first case A Galaxy with a huge central mass with star like masses in presence of external galaxies were taken. Theoretical prediction of circular velocities were matching with the observed velocities. In the later cases either Huge central mass was absent or external galaxies were absent or both were absent, the theoretical circular velocities did not match the observations. Hence the question of missing mass does not arise.

Large-scale structures of universe could not be explained by Big bang based theories using additional repulsive forces like "Einstein's λ", as it requires isotropy and homogeneity. Our universe is neither isotropic nor homogeneous. It is LUMPY. And there is no gravitational repulsive force found in the universe even after almost a century after publication of General theory of Relativity. We find that for all fringe effects, Special theory of Relativity is sufficient. Things can be explained by Newtonian gravitation. This proves Galaxy disk formation require some external forces other than self-gravitation of Galaxy it self. Here the there is universal gravitational effect at that position and time are calculated due to ALL the bodies present in the universe. This forms a repulsive force. And this force varies with time, position, structure, masses, their distances, their dynamic movement etc.

SITA (Simulation of Inter-intra-Galaxy Tautness and Attraction forces) was successful in the formation of Dynamic universe model where Blue shifted Galaxies were also present (Paper presented by SNP. Gupta, GR17, Dublin, 2004 & Presented in ICR 2005 International Conference on Relativity), at Amravati University, India, Jan 11- 14, 2005. Testing of model and its behavior at micro sec, 1 sec, 1 month, I year, 10 year done. The pictures show a non- collapsing mass distributions and formations of orbits due to mutual gravitational attraction forces. (paper presented by SNP. Gupta, Brit Grav 4, Oxford, 2004).

Keywords: DYNAMIC UNIVERSE MODEL, MODELS OF COSMOLOGY, DARK MATTER, DISK FORMATION IN GALAXIES, EQUILIBRIUM OF DISKS, LARGE SCALE STRUCTURE OF UNIVERSE, SINGULARITY-FREE COSMOLOGY, SITA SIMULATIONS, DYNAMIC BALANCING STRUCTURES, TAUTNESS AND GRAVITATION ATTRACTION
PACS: 95.35.+d, 98.62.Ai, 98.62.Dm, 98.80.-k

BASIS : DYNAMIC UNIVERSE MODEL OF COSMOLOGY

In SITA Non uniform mass distribution and 3-D positioning done and Newtonian attraction forces and Special Relativistic effects were used. No long distance repulsive (λ) General relativistic effects. No space-time continuum. SITA results are encouraging. The masses do not collapse into lump but they are orbiting each other. Large-scale structures of universe could not be explained by Big bang based theories without using additional repulsive forces like "Einstein's λ", as it requires isotropy and homogeneity. Our universe is neither isotropic nor homogeneous. It is LUMPY. And it is not empty. Special theory of Relativity is sufficient. Things can be explained by Newtonian gravitation. This proves Galaxy disk formation require some external forces other than self-gravitation of Galaxy it self. <u>Here the there is universal gravitational effect at that position and time are calculated due to ALL the bodies present in the universe. This forms a repulsive force</u>. This force varies with time, position, structure, masses, their distances, their dynamic movement etc. Our universe is not a Newtonian static universe.

Theoretical star circular velocities in a Galaxy, in Bigbang cosmologies are predicted as shown in the left in the Picture. The observed rotation curves are shown on the right side. Is this means that the mass of the Galaxy increases with increasing distance from the center? The observed rotation curves are shown on the right side of Pic 1 Is this means that the mass of the Galaxy increases with increasing distance from the center? Here we have predicted them using Dynamic universe model cosmology in FIVE different cases and present them in the following table.

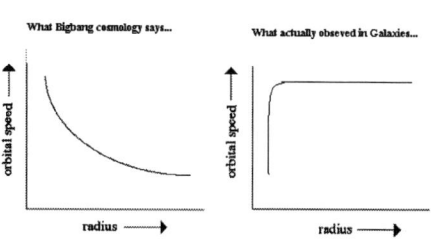

Pic1: Rotation Curve of the Galaxy

RESULT : DYNAMIC UNIVERSE MODEL OF COSMOLOGY

The following table shows results SITA simulations. The first column gives the explanations for the row. The second column gives 'xy' projections for the 'xyz' coordinates of the masses in the simulation at the START. Third column gives the 'xy' projections of the masses after 100 iterations. All these graphs show the theoretical Galaxy Circular Velocities vs radius. Row 1 gives the Case 1. *showing disk formation and velocities achieved graph. This case is with a Huge central mass at the center of galaxy, sun like stars with different mases and external galaxies.* Lets see Case 2. Here disk formation is not seen. This case is *without a Huge central mass at the center* of galaxy, sun like stars and external galaxies xy. Last column shows the horizontal portion only. Now see for Case 3. Here disk formation is seen. This is *with a Huge central mass at the center* of galaxy, sun like stars and *no external galaxies*. Radius vs. velocity curve shows only vertical portion. Go for Case 4. *no disk formation* This is *without a Huge central mass at the center* of galaxy, sun like stars and *no external galaxies* A total distribution of velocities seen here. As a theoretical example lets see Case 5. A gravitationally stabilized system of masses after forming a galaxy disk, and after it's stability analysis was done by giving perturbations and jeans swindle test. Cir Vel vs. Radius curve is similar to Case 1. We can see clearly external Galaxies and Central mass in Galaxy are required as to form a Galaxy disk. Only then radius vs. velocity curve is near to actual observational results for galaxies. Now lets ask our selves do the Galaxies are to be assumed to have some missing mass? Is that required?

DYNAMIC UNIVERSE MODEL: ACKNOWLEDGEMENTS

Almighty Goddess VAK continuously guided this work. I sincerely thank Her

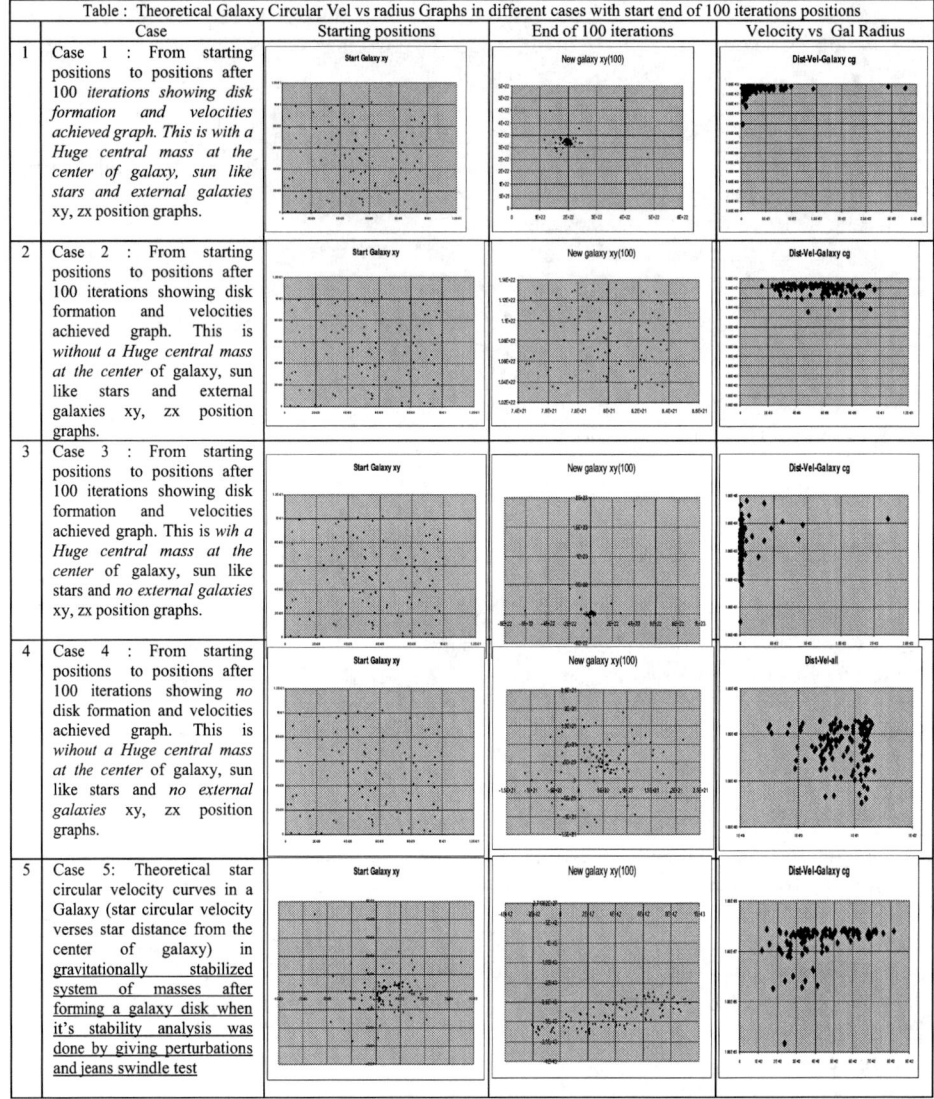

REFERENCES

1. SNP.GUPTA, "DYNAMIC UNIVERSE MODEL of cosmology: *Missing mass* in Galaxy" Presented in 7th Astronomical conf by HEL.A.S,. Kefallinia, Greece 8-11,Sept, 2005.

2. SNP.GUPTA, "*Missing mass* in Galaxy using regression analysis in DYNAMIC UNIVERSE MODEL of cosmology" Presented at PHYSTAT05 Conference on 'Statistical Problems in Particle Physics, Astrophysics and Cosmology" held in Oxford, UK on Sept 12th to 15th, 2005.

Weak-coupling structure of proton resonant states in ^{23}Al studied with RI beam at CNS

J. J. HE, S. KUBONO, T. TERANISHI[1], M. NOTANI[2], S. MICHIMASA[3], H. BABA[3]

Center of Nuclear Study (CNS), University of Tokyo, RIKEN Campus, Wako, Saitama 351-0198, Japan

S. NISHIMURA, M. NISHIMURA, Y. YANAGISAWA

The Institute of Physical and Chemical Research (RIKEN), 2-1 Hirosawa, Wako, Saitama 351-0198, Japan

N. HOKOIWA, M. KIBE, Y. GONO

Department of Physics, Kyushu University, 6-10-1, Hakozaki, Fukuoka 812-8581, Japan

J. Y. MOON, J. H. LEE, C. S. LEE

Department of Physics, Chung-Ang University, Seol 156-756, Korea

H. IWASAKI

Department of Physics, University of Tokyo, 7-3-1 Hongo, Bunkyo-ku, Tokyo 113-0033, Japan

S. KATO

Department of Physics, Yamagata University, Yamagata 990-8560, Japan

Abstract. Proton resonances in ^{23}Al have been investigated for the first time by the resonant elastic and inelastic scattering of ^{22}Mg+p by using a 4.38 MeV/nucleon ^{22}Mg beam bombarding a thick Hydrogen target. The low-energy ^{22}Mg beam was separated by the CNS radioactive ion beam separator (CRIB). A new resonant state due to elastic scattering was observed at $E_x = 3.00$ MeV with a $J^\pi = (3/2^+)$ assignment. Other three excited states due to resonant inelastic scattering at 3.14, 3.26 and 3.95 MeV were identified and all mainly decay to the first excited state in ^{22}Mg by the proton emissions. The newly observed 3.95-MeV state probably has a spin-parity of $J^\pi = (7/2^+)$. The resonant properties were determined from an R-matrix analysis of the excitation functions. The weak-coupling structure in ^{23}Al is discussed in conjunction with a shell-model calculation.

Keywords: Weak-coupled states; Resonance states; Inelastic scattering; R-matrix.
PACS: 23.50.+z; 21.10.Jx; 25.60.-t; 27.30.+t

INTRODUCTION

By so far, the nuclear structure data for ^{23}Al are very limited as compared with those of the mirror nucleus ^{23}Ne. The excited states in ^{23}Al were studied previously via the two-body ^{24}Mg(^{7}Li, ^{8}He)^{23}Al reaction [1], the β-delayed proton-decays of ^{23}Si [2],

---

[1] Present address: Department of Physics, Kyushu University, 6-10-1 Hakozaki, Fukuoka 812-8581, Japan.
[2] Present address: Argonne National Laboratory, 9700 S. Cass Avenue Argonne, Illinois 60439, USA.
[3] Present address: RIKEN (The Institute of Physical and Chemical Research), 2-1 Hirosawa, Wako, Saitama 351-0198, Japan.

and the Coulomb dissociation of ^{23}Al [3]. The proton resonant states in ^{23}Al cannot be necessarily excited by the (^{7}Li, ^{8}He) because of different reaction mechanism; The populated states in β-decay study of ^{23}Si are restricted by the selection rules; In the Coulomb-dissociation study only the first and the second excited states in ^{23}Al was studied. Therefore, although several excited states were reported before, most of the spin-parity assignments have not been established yet.

In the present study, proton resonances in ^{23}Al have been investigated firstly by the resonant elastic and inelastic scattering of a ^{22}Mg RI beam on a thick hydrogen target. By using the thick target method [4], a wide energy-range of excitation functions for ^{22}Mg+p were obtained simultaneously by measuring the energies of recoiled protons. The resonance energy, width and spin-parity information have been deduced by an R-matrix analysis of the excitation functions. Thereby, the weak-coupling structure in ^{23}Al has been discussed together with a shell-model calculation.

EXPERIMENT

The experiment was performed using the CNS Radioactive Ion Beam separator (CRIB) [5], which was installed by the Center for Nuclear Study (CNS), University of Tokyo. An 8.11-MeV/nucleon, 200-pnA ^{20}Ne^{8+} beam from a K=70 AVF cyclotron at RIKEN, bombarded a water-cooled ^{3}He gas target, and a ^{22}Mg beam was produced via the (^{3}He, n) reaction and separated by CRIB. The 4.38 MeV/nucleon (4% in FWHM) with a typical intensity of 4.4 kpps ^{22}Mg beam, bombarded a 7.9-mg/cm^2 polyethylene (CH$_2$)$_n$ target, in which the beam was stopped. In inverse kinematics, the excitation-energy range from 3.4 MeV to 0.8 MeV above proton threshold in ^{23}Al was thereby scanned in one single step. The emitted particles were measured at 4°, 17° and 23° in the laboratory system by silicon ΔE-E telescopes. The energy calibrations of the detector system were made using the secondary proton beams separated by CRIB at several energy points. Additionally a carbon target with equivalent thickness to that of the (CH$_2$)$_n$ target was also used in a separate run to evaluate the background contributions. The experiment setups and preliminary results have been reported in details [6, 7], and a full description of the experiment will be published elsewhere [8].

The identification of excited states in ^{23}Al based on the elastic and inelastic scattering events were discussed in Ref. [7]. As a conclusion, the 3.00-MeV state decays to the ground state in ^{22}Mg, while all other states mainly decay to the first excited state in ^{22}Mg. A new level scheme of ^{23}Al is proposed in Fig. 1. The uncertainties of energy determination, which include both the systematic errors and the ones associated with the fitting, are indicated in the unit of keV in the parentheses. The 3.00-MeV and 3.95-MeV states are newly observed, and the states at 3.14 and 3.26 MeV could correspond to a broad peak at E_x=3.204 MeV reported previously [1]. The comparison is shown in Fig. 1.

DISCUSSION

The experimental differential elastic and inelastic cross sections were analyzed by an R-matrix code SAMMY-M6-BETA [9], which enables multi-level R-Matrix fits to neutron and charged-particle cross-section data using Bayes' equations [10]. The resonant properties, energies, spin-parities and proton partial widths to the ground state (Γ_p) and to the first excited state ($\Gamma_{p'}$) in ^{22}Mg, have been deduced as listed in Table 1. Note that the present determination of J^π values (the second column of Table 1) is based on the results of R-matrix analysis and of the shell model calculation discussed later.

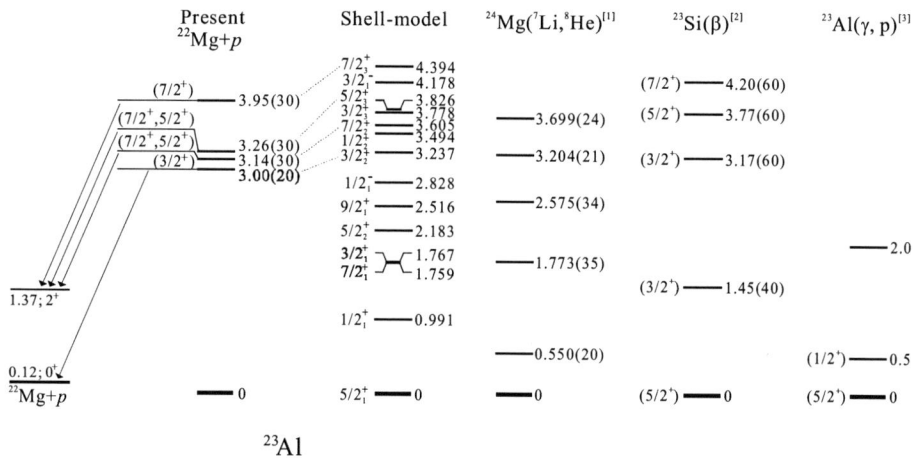

FIGURE 1. A new level scheme in ^{23}Al proposed in this work. The calculated levels together with the previously observed levels are shown for comparison.

The shell-model calculations have been performed with an OXBASH [11] code. The calculations were carried out in a complete model space *spsdpf* using an isospin-conserving *WBT* interaction.

The calculated levels are shown in Fig. 1. The calculated energies of positive-parity states agree very well with those observed in the mirror ^{23}Ne nucleus, and the calculated C^2S factors [8] are also in good agreement with those determined from the (d, p) reaction. To study the weak-coupling structure in ^{23}Al, the spectroscopic factors of the excited states in A=23 nucleus for the single-particle structure with the A=22 core in its ground (or the first excited) state were calculated. Furthermore, the proton partial widths calculated [12] by using the calculated spectroscopic factors are listed in Table 1 as well. The experimental values are in reasonable agreement with the calculated ones.

TABLE 1. Resonant properties determined in this work. The widths are for the inelastic channels for the last three states. The predicted J_n^π indicates the *n*th level of the same J^π. (widths in keV)

E_x (MeV)	$2J^\pi$	$2J_n^\pi$	$\Gamma_{p,(p')}^{exp}$	$\Gamma_{p,(p')}^{cal}$	Coupling
3.00	(3^+)	3_2^+	32 (5)	44	$0^+ \otimes 1d_{3/2}$

3.14	$(7^+, 5^+)$	7_2^+	30 (20)	6	$2^+ \otimes 1d_{3/2}$
3.26	$(7^+, 5^+)$	5_3^+	30 (20)	10	Mixed
3.95	(7^+)	7_3^+	30 (20)	18	$2^+ \otimes 1d_{3/2}$

A correspondence assignment is tentatively made in Fig. 1. We suggest that the 3.00-MeV state corresponds to the calculated 3.237-MeV state. The predicted Γ_p value is 44 keV (see Table 1), which agrees well with the experimental value of $\Gamma_p = 32\pm5$ keV (with $J^\pi=3/2^+$). The main configuration of this state may be written as $0^+ \otimes 1d_{3/2}$, where 0^+ indicates ^{22}Mg core in the ground state, and the proton occupies the $1d_{3/2}$ orbit in ^{23}Al nucleus. It is mainly of a $1d_{3/2}$ single-particle character. In comparison, the energies of $1d_{3/2}$ single-particle state are 3.43 MeV in ^{23}Ne, and 3.36 MeV (=11.25-7.89) in ^{23}Na (for $T=3/2$) [13], respectively.

According to the shell-model calculations, the 3.14-MeV state mainly decays to the first excited state in ^{22}Mg by a $1d_{3/2}$ proton emission, and this result is consistent with the experimental observations. The predicted proton width $\Gamma_{p'}$ value is 6 keV, which is not very far from the experimental value $\Gamma_{p'}=30\pm20$ keV. Its main configuration can be written as $2^+ \otimes 1d_{3/2}$, where 2^+ implies the first excited state in ^{22}Mg, and the proton occupies the $1d_{3/2}$ orbit in ^{23}Al nucleus. The 3.26-MeV state also mainly decays to the first excited state in ^{22}Mg, and the predicted $\Gamma_{p'}$ values are 10 keV and 1 keV for the $2s_{1/2}$ and $1d_{3/2}$ proton emissions, respectively. The main configuration is possibly a mixture of $2^+ \otimes 2s_{1/2}$ and $2^+ \otimes 1d_{3/2}$. The 3.95-MeV state has most probably an assignment of $(7/2^+)$. It mainly decays to the first excited state in ^{22}Mg by a $1d_{3/2}$ proton emission. The predicted $\Gamma_{p'}$ value is 18 keV, which agrees well with that deduced from the R-matrix analysis of $\Gamma_p = 30\pm20$ keV. The main configuration can be written as $2^+ \otimes 1d_{3/2}$. Overall, the nuclear structure in ^{23}Al can be well explained within the shell model.

ACKNOWLEDGMENTS

We wish to thank the CNS/RIKEN staff for their operation of the ion source and the AVF cyclotron. This work is partially supported by Grant-in-Aid for Science Research from the Japanese Ministry of Education, Culture, Sports, and Technology under the contract number 13440071, 14740156. It is also supported by the Grants-in-Aids for JSPS fellows (1604055), and the Korea Research Foundation Grant (KRF-2002-070-C00031).

REFERENCES

1. J. A. Caggiano et al., *Phy. Rev. C* **64**, 025802 (2001).
2. B. Blank et al., *Z. Phys. A* **357**, 247-254 (1997).
3. T. Gomi et al., *Nucl. Phys. A* **718**, 508c-509c (2003).
4. S. Kubono et al., *Nucl. Phys. A* **693**, 221-248 (2001).
5. Y. Yanagisawa et al., *Nucl. Instr. and Meth. A* **539**, 74-83 (2005).

6. J. J. He et al., *CNS Annual Report* **2003**, 34-35 (2004).
7. J. J. He et al., *CNS Annual Report* **2004**, 3-4 (2005).
8. J. J. He et al., under preparation.
9. N.M. Larson, *A Code System for Multilevel R-Matrix Fits to Neutron Data Using Bayes' Equations*, ORNL/TM-9179/R **5**, Oak ridge, USA, (2000).
10. N.M. Larson, *User's Guide for BAYES: A General-Purpose Computer Code for Fitting a Functional Form to Experimental Data*, ORNL/TM-8185, ENDF-323, Oak ridge, USA, (1982).
11. B. A. Brown et al., *MSU-NSCL* report number 1289.
12. C. Iliadis: *Nucl. Phys. A* **618**, 166-175 (1997).
13. P.M. Endt and C. Van Der Leun, *Nucl. Phys. A* **310**, 1 (1978).

Cosmological solutions to the discrepancy among the light elements abundances and WMAP

Kazuhide Ichikawa

Institute for Cosmic Ray Research, University of Tokyo, Kashiwa 277-8582, Japan

Abstract. Within the standard big bang nucleosynthesis (BBN) and cosmic microwave background (CMB) framework, the baryon density measured by the Wilkinson Microwave Anisotropy Probe (WMAP) or the primordial D abundance is much higher than the one measured by the ^{7}Li abundance. We propose two non-standard cosmological scenarios to solve the discrepancy. In the first scenario, we consider BBN with non-standard values of the fine structure constant and/or the cosmic expansion rate. We show that the discrepancy is not solved by considering a varying fine structure constant or a non-standard expansion rate alone but solutions are found by their simultaneous existence. In the second scenario, we consider additional baryons which appear after BBN. We show that simply adding the baryons can not be a solution but the existence of a large lepton asymmetry before BBN makes the scenario successful. These extra baryons and leptons, in addition to the initial baryons which exist before the BBN, can be all produced from Q-balls.

Keywords: Big Bang nucleosynthesis; CMBR; Primordial element abundances
PACS: 26.35.+c,98.62.Bj,98.80.Cq,98.80.Ft

INTRODUCTION

Recently, following the precise measurement of the CMB by the Wilkinson Microwave Anisotropy Probe (WMAP), its concordance with the BBN and the light elements observations has been investigated in Refs [1, 2, 3, 4]. The baryon density measured from the WMAP data is $\omega_b \equiv \Omega_b h^2 = 0.024 \pm 0.001$ (with the power-law ΛCDM model) [5], where Ω_b is the baryon energy density divided by the critical energy density today and h is the Hubble constant in units of 100 km s^{-1} Mpc^{-1}. The uncertainty is very small because the WMAP has detected the first and second peaks accurately in the temperature angular spectrum [6]. This corresponds to $\eta \equiv n_b/n_\gamma = (6.6 \pm 0.3) \times 10^{-10}$, where n_b and n_γ are baryon and photon number densities, via the relation $\eta = \omega_b/(3.65 \times 10^7)$. Refs [1, 2, 3, 4] take this well-determined WMAP ω_b as the BBN input and calculate the light elements abundances and their theoretical errors using improved evaluations of nuclear reaction rates and uncertainties. The results are compared with the received measurements of the primordial abundances of three light elements, D (D/H=$2.78^{+0.44}_{-0.38} \times 10^{-5}$ [7]), ^{4}He ($Y_p = 0.238 \pm 0.002 \pm 0.005$ [8] or 0.2421 ± 0.0021 [9]) and ^{7}Li (^{7}Li/H = $1.23^{+0.68}_{-0.32} \times 10^{-10}$ (95%) [10] or $2.19^{+0.46}_{-0.38} \times 10^{-10}$[11]). Although there are small differences concerning their adopted reaction rates or observation data, their conclusions agree: from the WMAP baryon density, the predicted abundances are highly consistent with the observed D but not with ^{4}He or ^{7}Li. They are produced more than observed. Especially, the ^{7}Li-WMAP discrepancy is severer and it may require an explanation.

For ^{4}He, the discrepancy may not exist since recent analysis by Refs. [12, 13] found higher value and larger error estimate 0.249 ± 0.009 by treating stellar absorption in a different manner. For ^{7}Li, most favored and discussed solution has been a depletion in halo stars. However, since recent measurements by Ref. [14] reported detection of significant amount of more fragile isotope ^{6}Li along with ^{7}Li in metal-poor halo stars, the solution by stellar depletion is not likely. Therefore, we here propose two cosmological solutions to ^{7}Li discrepancy. One is based on the idea of varying coupling constants and another considers baryon number production after BBN.

VARYING COUPLING CONSTANTS

Coupling constants may vary in some frameworks of unified theory. Also, Ref. [15] has reported possible evidence for varying fine structure constant α from QSO absorption lines, $\Delta\alpha/\alpha = (-0.573 \pm 0.113) \times 10^{-5}$. We propose a scenario to solve ^{7}Li discrepancy using this idea [16].

The key point is that, around the baryon density measured by WMAP, ^{7}Li is mainly produced by the reaction ^{4}He(^{3}He,γ)^{7}Be (this ^{7}Be turns into ^{7}Li through the electron capture afterward) and this process is suppressed by increasing α because of the large Coulomb barrier between incident He nuclei. However, when α is increased, ^{4}He abundance also increases somewhat beyond the observed values. This is because the neutron-proton mass difference Δm decreases following $\Delta m = -0.76(1 + \Delta\alpha/\alpha) + 2.05$ MeV and more neutrons survive (most of which in turn are synthesized into ^{4}He) at the freeze-out of the neutron-proton interconversion. We can cancel this ^{4}He excess without affecting ^{7}Li abundance by introducing nonstandard (slower) expansion rate of the universe during BBN. In terms of the effective number of neutrino species N_ν, we need, precise values depending on the value of $\Delta\alpha$, $N_\nu < 1$ to adjust ^{4}He within the observation error bars. Such N_ν, less than standard value of three, may arise from a varying (smaller at BBN) gravitational constant. In that case, possible connection with the varying α is an interesting issue to be investigated.

Other scenarios to solve ^{7}Li problem by varying coupling constants are discussed by Refs. [17, 18, 19, 20].

BARYON PRODUCTION AFTER BBN

Roughly speaking, the discrepancy exists because the WMAP data needs more baryons than those required to account for the primordial light elements abundances, especially ^{7}Li. Then a naive solution would be allowing to increase the baryons after the BBN to the amount required to explain the WMAP data before the physics which form the acoustic peaks takes place.

However, such increase in the baryons (in the form of protons *i.e.* H nucleus) considerably affects the observation of the light elements abundances because they are always measured in the ratio to H. Since the numerator does not change and the denominator increases, the abundances decrease in general. To obtain a solution to reconcile three elements abundance measurements simultaneously, we first choose to make D and ^{7}Li consistent by the adding baryons $\eta_{add} \approx 1.5 \times 10^{-10}$. Then we introduce a (negative) lepton asymmetry before the onset of the BBN in order to enhance ^{4}He abundance (about 1.5 times) while retaining D and ^{7}Li abundances. In terms of the degeneracy parameter $\xi_e \equiv \mu_e/T$ where μ_e is the electron-type neutrino chemical potential and T is the temperature, the lepton asymmetry needed is calculated to be $\xi_e = -0.2 \sim -0.8$.

We also showed a model that explains the large lepton asymmetry, the initial baryon asymmetry before BBN, and additional baryon appearance after the BBN all together. This is accomplished within the framework of the Affleck-Dine (AD) baryogenesis mechanism and the subsequent Q-ball formation in the gauge-mediated supersymmetry breaking model. Details are described in Ref. [21].

CONCLUSION

The observed ^{7}Li abundance is known to be in conflict with D abundance and WMAP. This discrepancy is sharpened by the recent detection of ^{6}Li in metal-poor halo stars because it limits the degree of stellar ^{7}Li depletion. Although systematic errors of the lithium measurements need to be further investigated, it might be a signal of new physics. Thus, we presented two possible cosmological solutions to the discrepancy. Other solutions considering decaying particles during BBN are investigated by Refs. [22, 23, 24, 25, 26]. Future observations will hopefully tell us what the solution is.

REFERENCES

1. R. H. Cyburt, B. D. Fields and K. A. Olive, Phys. Lett. B **567**, 227 (2003).
2. A. Coc *et al.*, Astrophys. J. **600**, 544 (2004).
3. R. H. Cyburt, Phys. Rev. D **70**, 023505 (2004).
4. P. D. Serpico, S. Esposito, F. Iocco, G. Mangano, G. Miele and O. Pisanti, JCAP **0412**, 010 (2004).
5. D. N. Spergel *et al.*, Astrophys. J. Suppl. **148**, 175 (2003).
6. L. Page *et al.*, Astrophys. J. Suppl. **148**, 233 (2003).
7. D. Kirkman *et al.*, Astrophys. J., Suppl. Ser. **149**, 1 (2003).
8. B. D. Fields and K. A. Olive, Astrophys. J. **506**, 177 (1998).
9. Y. I. Izotov and T. X. Thuan, Astrophys. J. **602**, 200 (2004).
10. S. G. Ryan *et at.*, Astrophys. J. Lett., **530**, L57 (2000).
11. P. Bonifacio *et al.*, Astron. Astrophys. **390**, 91 (2002).
12. K. A. Olive and E. D. Skillman, Astrophys. J. **617**, 29 (2004).
13. R. H. Cyburt, B. D. Fields, K. A. Olive and E. Skillman, Astropart. Phys. **23**, 313 (2005).
14. M. Asplund, D. L. Lambert, P. E. Nissen, F. Primas and V. V. Smith, arXiv:astro-ph/0510636.
15. M. T. Murphy, J. K. Webb and V. V. Flambaum, Mon. Not. Roy. Astron. Soc. **345**, 609 (2003).
16. K. Ichikawa and M. Kawasaki, Phys. Rev. D **69**, 123506 (2004).
17. V. F. Dmitriev, V. V. Flambaum and J. K. Webb, Phys. Rev. D **69**, 063506 (2004).
18. B. Li and M. C. Chu, arXiv:hep-ph/0511013.
19. B. Li and M. C. Chu, arXiv:astro-ph/0511642.
20. J. Larena, J. M. Alimi and A. Serna, arXiv:astro-ph/0511693.
21. K. Ichikawa, M. Kawasaki and F. Takahashi, Phys. Lett. B **597**, 1 (2004).
22. J. L. Feng, A. Rajaraman and F. Takayama, Phys. Rev. D **68**, 063504 (2003).
23. K. Jedamzik, Phys. Rev. D **70**, 063524 (2004).
24. F. Wang and J. M. Yang, Nucl. Phys. B **709**, 409 (2005).
25. K. Kohri, T. Moroi and A. Yotsuyanagi, arXiv:hep-ph/0507245.
26. K. Jedamzik, K. Y. Choi, L. Roszkowski and R. R. de Austri, arXiv:hep-ph/0512044.

Origin of Cosmological Magnetic Fields

Kiyotomo Ichiki*,†, Keitaro Takahashi**, Hiroshi Ohno‡, Hidekazu Hanayama* and Naoshi Sugiyama*

*National Astronomical Observatory, Mitaka, Tokyo 181-8588, JAPAN
†Research Fellows of Japan Society for the Promotion of Science
**Department of Physics, Princeton University, Princeton, NJ 08544
‡Corporate Research and DevelopmentCenter, Toshiba Corporation, Kawasaki 212-8582, Japan

Abstract. Astronomical observations indicate there exist substantial magnetic fields in galaxies and on even larger scales such as in cluster of galaxies. The origin of such magnetic fields with large coherent length, however, is still a mystery. Several models have been proposed to explain the origin, which are often involved with inflation in the early universe, or astrophysical activities during structure formation of the universe. Here we show that cosmological density perturbations in photons and baryons, especially the anisotropic stress in photons can source the cosmological magnetic fields. Our study is based upon the cosmological perturbation theory, which is quite successful in a sense that it can explain the large scale structure of the universe and the anisotropies in the cosmic microwave background. Thus our results are quite robust and give the precise amount of magnetic fields which should inevitably exist at cosmological scales.

Keywords: primordial magnetic fields, cosmological perturbation
PACS: 91.25.Cw, 98.80.-k

Conventional models for generation of large scale magnetic fields are mostly categorized into two parts. The one is astrophysical mechanism, which is often involving stellar activities or biermann battery in the non-adiabatic processes [1, 2, 3]. It may explain the strength and total amount of magnetic fields with help from dynamo mechanism. Their coherence scales are, however, much smaller than those of intergalactic magnetic fields and thus magnetic fields generated from these mechanisms could not be directly the origin of large scale magnetic fields.

The other is the cosmological mechanism, which is often related with inflation, have no difficulty in accounting for the length of coherence; accelerating expansion of the universe stretches enough the small scale structures, even up to super horizon scales. The problem here which is well known is that we need to break conformal invariance of electro magnetic interaction at inflationary epoch. Therefore, the nature of generated magnetic fields, amplitude or its spectrum for example, is highly depend on the models beyond the standard cosmology [4, 5, 6].

In this contribution, we show an numerical result of magnetic-field generation from cosmological perturbations after inflation. In the early universe before cosmological recombination, Compton and Coulomb scatterings are efficient enough that photons, protons and electrons can be approximately treated as a tightly coupled fluid. If these three kind of fluids moved in exactly the same way, magnetic field could not be generated. However, since photons scatter off electrons preferentially compared with protons, small difference in velocity between photons and electrons generates that between electrons and protons [7, 8]. Moreover, we show that anisotropic pressure of photons pushes the electrons in different way from protons. Thus the electric current is indeed induced and this leads the generation of magnetic field.

Let us first consider the collision term in the Euler equation of photons. The collision term representing Compton scattering with a non-relativistic approximation for electrons is given by

$$C_{\gamma e}^{(T)i}[f(p_i)] = \int \frac{d^3 p}{(2\pi)^3} p^i C_{\gamma e}^{(T)}[f(p_i)]$$
$$= \frac{4\sigma_T \rho_\gamma n_e}{3}\left[(v_e^i - v_\gamma^i) + \frac{1}{8}v_{ej}\Pi_\gamma^{ij}\right], \tag{1}$$

where f is distribution function of photons, σ_T is the cross section of the Thomson scattering, ρ_γ, v_γ, and Π_γ^{ij} are energy density, bulk velocity, and anisotropic stress of photons, n_e and v_e are number density and velocity of electrons. It should be noted that the collision term (1) was obtained non-perturbatively with respect to the cosmological perturbation [9]. This collision term also pushes electrons via momentum conservation. From the equations of motion for protons and electrons, we obtain a generalized Ohm's law, which, combined with the Maxwell equations, leads to an evolution equation for magnetic field,

$$\dot{B}^i \sim \frac{4\sigma_T \overset{(0)}{\rho_\gamma}}{3e}\varepsilon^{ijk}\left[\frac{\overset{(1)}{\rho_{\gamma,k}}}{\overset{(0)}{\rho_\gamma}}\left(\overset{(1)}{v_{ej}} - \overset{(1)}{v_{\gamma j}}\right) + \left(\overset{(2)}{v_{ej,k}} - \overset{(2)}{v_{\gamma j,k}}\right)\right.$$
$$\left. + \frac{1}{8}\left(\overset{(1)}{v_{el,k}}\overset{(1)}{\Pi_{\gamma j}^l} + \overset{(1)}{v_{el}}\overset{(1)}{\Pi_{\gamma j,k}^l}\right)\right], \tag{2}$$

where the dot denotes a derivative with respect to the cosmic time, and superscript (i) denote the order of expansion.

The best advantage of this study is that it is based on the successful theory of cosmological perturbations. Recent observations of CMB temperature and polarization anisotropies have already revealed that the linear-order quantities are dominated by scalar type components and we can evaluate the first and third terms in the source term in eq.(2) exactly.

Generated spectrum of magnetic fields is obtained by a complicated non-linear convolution of these perturbation variables. In figure 1, we depict the spectrum of magnetic field, $\left\langle B^*(\vec{k})B(\vec{k}')\right\rangle = S(k)\delta^{(3)}(\vec{k}-\vec{k}')$, produced from cosmological perturbations. For this illustration all cosmological parameters were fixed to the ΛCDM values, i.e., $(h, n_s, \Omega_b h^2, \Omega_\Lambda, A) = (0.71, 0.99, 0.022, 0.74, 0.83)$, where h is Hubble parameter, n_s is power spectrum index, A is overall amplitude, Ω_b and Ω_Λ are energy densities of baryon and cosmological constant in critical density units. The field has the spectrum $S(k) \propto k^{-4}$ at super-horizon scales at recombination and $S(k) \propto k$ at smaller scales.

At even smaller scales, one may have a cut-off of the spectrum due to the diffusion of the magnetic field, and thus the energy density in the magnetic field does not diverge. The effect of diffusion of magnetic field can be token into consideration by evaluating Coulomb scattering correctly, which have been omitted above. If one keeps the Coulomb

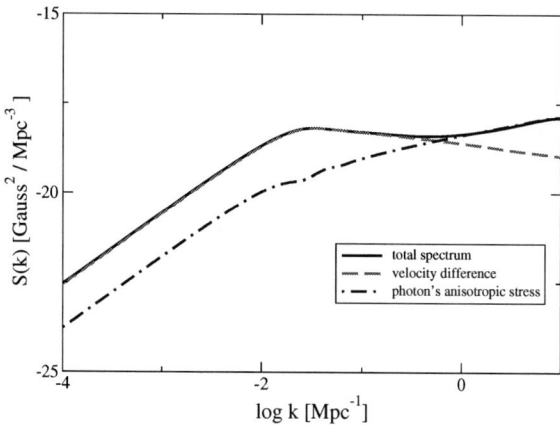

FIGURE 1. Spectrum of magnetic fields S(k) at $z = 1100$ generated from cosmological perturbations. Contributions from the velocity difference between baryon and photon, and photon's anisotropic stress are shown in the figure as indicated.

scattering term in the formulation, one finds the diffusion term in the induction equation, $\dot{B} \sim \eta \nabla^2 B$, with η being resistivity of the plasma. This is usually neglected at cosmological scales in the previous studies on generating magnetic fields due to the high conductivity of the primordial plasma in the early universe. However, it should be correctly included to estimate the diffusion scale. Estimating the scale in which the diffusion can not be neglected against adiabatic decay due to the cosmic expansion, we have $k \sim 10^{12}$ Mpc^{-1}.

From the spectra we can estimate the field strength of generated magnetic field at the recombination epoch as $B \approx k^3 S(k) \approx 10^{-16.5}$G at 1 Mpc comoving scale. If the magnetic field decays adiabatically according to the cosmic expansion, the current magnetic field is about $10^{-22.5}$G.

REFERENCES

1. Biermann, L., & Schlüter, A. , 1951, Physical Review 82, 863-868.
2. Gnedin, N. Y., Ferrara, A., & Zweibel, E. G. , 2000, Astrophys. J. 539, 505-516.
3. Hanayama, H. et al. , 2005, Astrophys. J., in press.
4. Ratra, B. , 1992, Astrophys. J. 391, L1-L4.
5. Lemoine, D. & Lemoine, M. , 1995, Phys. Rev. D 52, 1955.
6. Turner, M. S. & Widrow L. M. , 1988, Phys. Rev. D 37, 2743.
7. Harrison, E. R. , 1970, Mon. Not. Roy. Astron. Soc. 147, 279-286.
8. Lesch, H. & Chiba, M. , 1995, Astron. Astrophys. 297, 305-310.
9. Takahashi, K., Ichiki, K., Ohno, H. & Hanayama, H. , 2005, Phys. Rev. Lett. 95, 121301.; Ichiki, K., Takahashi, K., Ohno, H., Hanayama, H., & Sugiyama, N. , 2006, Science, in press.

Neutrino Emission from Type Ia Supernovae

K. Iwamoto and T. Kunugise

Department of Physics, Faculty of Science and Technology,
Nihon University
1-8-14 KandaSurugadai, Chiyoda-ku, Tokyo 101-8308, Japan

Abstract. Type Ia supernovae(SNe) produce a burst of neutrino emission due to the electron capture on free protons and nuclei in the hot and dense matter processed by explosive nucleosynthesis. We calculated its luminosity and energy spectrum for a standard SN Ia model and with updated weak reaction rates. We found the peak neutrino (v_e) luminosity of $\sim 10^{50}$ erg s^{-1} and the duration of the burst ~ 1 sec. We also calculated the spectrum of the neutrino emission.

Keywords: Supernova, Neutrino, Nucleosynthesis
PACS: 43.35.Ei, 78.60.Mq

INTRODUCTION

Type Ia SNe are thought to be thermonuclear explosions of accreting white dwarfs. Neutrinos are mainly produced by electron captres on free protons and nuclei.

$$e^- + (A,Z) \longrightarrow (A,Z-1) + v_e$$

The neutrino luminosity curve and the energy spectrum would provide clues to the understanding of the explosion mechanism of SNe Ia.

We use the result of a hydrodynamical simulation of deflagration models of SN Ia by Nomoto, Thielemann, & Yokoi (1984) [1] and the weak reaction rates by Langanke & Martínez-Pinedo (2001) [3].

NUCLEOSYNTHESIS AND WEAK PROCESSES

To calculate the neutrino emission rate, we have to determine the abundances in the inner region of SN Ia ejecta. The nucleosynthesis are calculated with the assumption of NSE(Nuclear Statistical Equilibrium). Under the NSE,

$$(A,Z) \leftrightarrow Z\,p + (A-Z)\,n\,,$$

nuclear abundances depend only on the density ρ, the temperature T, and the electron mole fraction Y_e. In the NSE region, nuclei and nucleons are supposed to be an ideal non-relativistic Maxwell-Boltzmann gas. Then, the nuclear abundances are determined by the following Saha-like equations.

$$\frac{X(A,Z)}{X_p^Z X_n^{A-Z}} \frac{A g(A,Z)}{g_p^Z g_n^{A-Z}} \left(\frac{\rho}{M_\mu}\right)^{A-1} \left(\frac{2\pi\hbar^2}{kT}\right)^{3(A-1)/2} \left[\frac{M(A,Z)}{M_p^Z M_n^{A-Z}}\right]^{3/2} \exp(Q/kT)$$

$X's$ are normalized such that
$$\sum_{(A,Z)} X(A,Z) = 1$$
and $g(A,Z)$ is the partition function of the nucleus (A,Z),
$$g(A,Z) = \sum_r (2I_r + 1)\exp(-E_r/kT),$$
where I_r and E_r are the spin and the energy level of the r th excited state, respectively. Q is the binding energy of the nucleus (A,Z),
$$Q = c^2\{ZM_p + (A-Z)M_n - M(A,Z)\},$$
The abundances are uniquely determined with an additional condition of charge neutrality,
$$\sum_{(A,Z)} \frac{Z}{A} X(A,Z) = Y_e.$$
Electron captures reduce the Y_e (electron fraction) at a rate
$$\frac{d}{dt} Y_e = -\sum_i \lambda_i^{ec} \frac{X_i}{A_i},$$
where λ_i^{ec} is the electron capture rate of the i th isotope.

The following weak interaction processes are also considered.

β^- decay
$$(A,Z) \longrightarrow (A,Z+1) + e^- + \bar{\nu}_e$$

β^+ decay
$$(A,Z) \longrightarrow (A,Z-1) + e^+ + \nu_e$$

positron capture
$$e^+ + (A,Z) \longrightarrow (A,Z+1) + \bar{\nu}_e$$

The neutrino energy spectrum via an electron capture are approximately given by
$$n(E_\nu) = E_\nu^3 (E_\nu - q)^2 \frac{N}{1 + \exp\{(E_\nu - q - \mu_e)/kT\}},$$
where q is a parameter that defines the effective Q-value averaged over allowed channels and N is a normalization constant, μ_e is the chemical potential of electrons [5]. Summation of the contribution to the spectrum for all the relevant isotopes over the whole SN ejecta provides a neutrino spectrum from the SN Ia explosion.

RESULTS AND DISCUSSION

The neutrino luminosity reaches a maximum of $\sim 10^{50}$erg/s at about 0.65 sec since the start of explosion. The duration of the ν_e burst is ~ 1 sec. The energy spectra have peaks at around 4 MeV. The peak neutrino energy decreases slowly with time. The peak of the time integrated energy spectrum is located at ~ 4 MeV.

We will study the effect of neutrino oscillations in matter (MSW effect) on the emergent energy spectra of neutrinos. Contributions of the neutrinos from SNe Ia to the neutrino background is also under investigation.

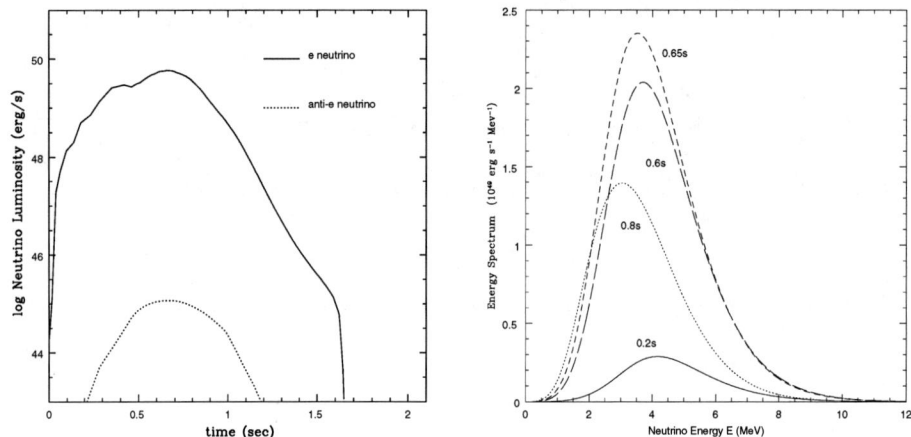

FIGURE 1. The luminosity(*left*) and energy spectrum(*right*) of a SN Ia v_e burst.

REFERENCES

1. Nomoto, K., Thielemann, F.-K., Yokoi, K., 1984, Astrophys.J. **286**, 644-658
2. Thielemann, F.-K., Nomoto, K., Yokoi, K., 1986, Astron. Astrophys. **158**, 17-33
3. Langanke, K., Martínez-Pinedo, G., 2001, Atomic Data and Nuclear Data Tables, **79**, 1-46
4. Fuller, G.M., Fowler, W.A., Newman, M.J., 1982, Astrophys.J.Suppl.Ser. **42**, 447
5. Langanke, K., Martínez-Pinedo, G., Sampaio, J.M., 2001, Phys.Rev.C **64**, 055801

Explosive Nucleosynthesis in Different Y_e Conditions

Nobuyuki Iwamoto*, Hideyuki Umeda[†], Ken'ichi Nomoto[†], Nozomu Tominaga[†], Friedrich-Karl Thielemann** and W. Raphael Hix[‡]

*Japan Atomic Energy Agency, Tokai, Ibaraki 319-1195, Japan
[†]Department of Astronomy, School of Science, University of Tokyo, Hongo, Bunkyo-ku, Tokyo 113-0033, Japan
**Departement für Physik and Astronomie, Universität Basel, CH-4056 Basel, Switzerland
[‡]Department of physics and Astronomy, University of Tennessee, Knoxville, TN and Physics Division, Oak Ridge National Laboratory, Oak Ridge, TN, USA

Abstract. The influence of a large variation of Y_e on explosive yield is investigated. We calculate nucleosynthesis with the initial electron fraction Y_e ranging from 0.48 to 0.58 in explosive Si burning region in Population III, 25 $M_\odot$ supernovae. We obtain the significant overproduction of odd elements, K and Sc. In the $Y_e < 0.5$ cases light p-process nuclei are enhanced. We find that the abundance pattern taken from arbitrary mixture of each nucleosynthesis yield in various values of Y_e can reasonably explain that in observed extremely metal-poor stars.

Keywords: nucleosynthesis, supernovae, nuclear reactions, abundances
PACS: 26.30.+k; 26.50.+x; 97.60.Bw;

INTRODUCTION

The observations of metal-poor stars with metallicity [Fe/H] < -2.5 have been extensively performed [1]. The abundance patterns of these stars are well explained by supernova models with extensive material mixing during the explosion and fallback onto the central remnant [2, 3, 4]. Nevertheless significant underproduction of odd elements (in particular K and Sc) are present in most investigations.

Recent simulations of core collapse supernovae with accurate neutrino transport show that the innermost ejecta is exposed to large amounts of neutrino flux [5] and that interactions of neutrinos with the ejecta change electron fraction to $Y_e > 0.5$ [6, 7, 8, 9]. They found that nucleosynthesis in the proton-rich environments allows Sc and even Zn (which is less produced in normal supernova models without neutrino interactions) to be enriched in low mass supernovae.

INITIAL COMPOSITIONS AND MODELS

The initial abundances are taken from a Population III, 25 $M_\odot$ progenitor model which is exploded with explosion energies of $E_{51} = 1$ and 20 (in units of 10^{51} ergs). The Y_e of the initial composition is changed to values ranging from 0.48 to 0.58 in a region where explosive Si burning occurs. The large variations of Y_e are obtained by modifying neutron or proton abundance. Detailed nucleosynthesis is then performed by a nuclear reaction

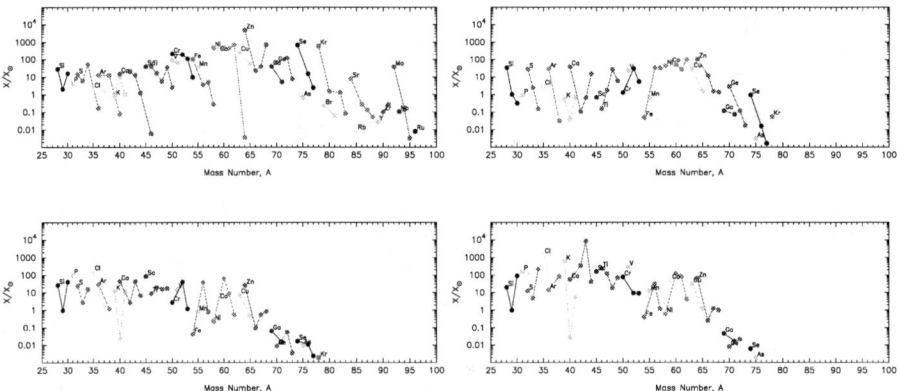

FIGURE 1. Production factors of nuclei in the yields of a 25 $M_\odot$ hypernova ($E_{51} = 20$) model with $Y_e = 0.48$ (upper left), original (upper right), 0.505 (lower left) and 0.56 (lower right) in the explosive Si burning region. Isotopes of a given element are connected by solid lines.

network with 1305 nuclei, but neutrino interactions during the explosive burning are not taken into account in this investigation.

RESULTS

Figure 1 shows production factors of each nuclide in the ejecta of a 25 $M_\odot$ hypernova ($E_{51} = 20$) model. Nuclei lighter than ^{56}Fe (including ^{39}K and ^{45}Sc) are enhanced with increasing Y_e in both explosion models. Heavier nuclei are also enhanced with increasing Y_e in the normal supernova model, but in the hypernova model the production of heavier nuclei is suppressed. This difference is attributable to the difference in the timescale of temperature decrease after heating by the passage of shock wave. In $Y_e > 0.5$ the production of nuclei heavier than ^{56}Ni can be expected, if there has sufficient time to decay and capture protons. In normal supernova model large amounts of heavy nuclei ($A \geq 64$) are produced by proton captures on ^{56}Ni and heavier nuclei owing to the slow decline of temperature. In contrast, hypernova model experiences rapid decrease in temperature due to fast expansion. Therefore, in the short expansion timescale the proton captures on nuclei with high Coulomb barrier and the decay of waiting point nuclei are too slow to induce the further production of heavy nuclei.

In contrast to the above results in the hypernova model, large amounts of heavy nuclei with $A \geq 64$ are synthesized in the $Y_e = 0.48$ case [10] as shown in Figure 2 which also gives the results in $Y_e = 0.56$ and original cases. This is because $Y_e < 0.5$ makes nuclear path to be more neutron-rich and thus the main nuclear flow bypasses β-decay waiting-point nuclei (e.g., ^{64}Ge) located at the neutron-deficient side of the β-stability line. Therefore, light p-process nuclei (^{74}Se, ^{78}Kr, ^{84}Sr and ^{92}Mo) are largely enhanced in the $Y_e = 0.48$ case of Figure 1.

We compare the average abundance pattern of observed metal-poor stars with $-4.2 < $ [Fe/H] $ < -3.5$ [1] with the explosive yield of the hypernova model. The yield is

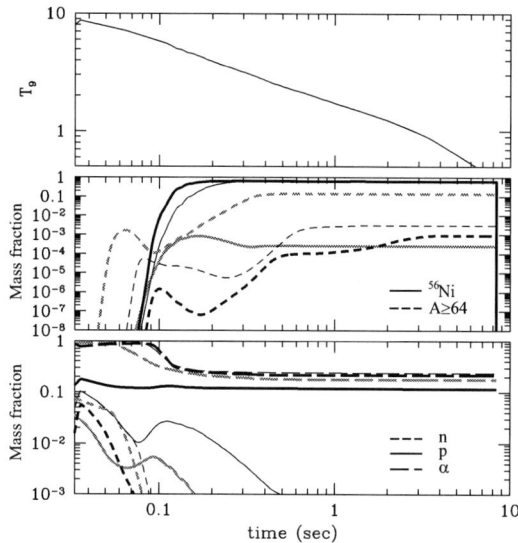

FIGURE 2. Time variations of temperature in 10^9 K (top), abundances of ^{56}Ni and heavy nuclei with $A \geq 64$ (middle), and neutron, proton and ^{4}He abundances (bottom) in a shock-heated layer with a peak temperature $T_9 \sim 8.7$ in the hypernova model. The results for the initial values of $Y_e = 0.48$ (gray thick lines), original (black thin lines), and 0.56 (black thick lines) are shown.

obtained by applying the mixing-fallback process [2, 3]. In this process the mixing and fallback region is set to be the explosive Si burning region. We assume arbitrary mixture of nucleosynthesis results for the initial values of $Y_e = 0.48 - 0.58$ as shown in [7]. We find that the resulting abundance pattern improves the fitting to K and Sc and well reproduces the observation.

REFERENCES

1. R. Cayrel, *et al.*, *Astronomy and Astrophysics*, **416**, 1117–1138 (2004).
2. H. Umeda, and K. Nomoto, *Nature*, **422**, 871–873 (2003).
3. H. Umeda, and K. Nomoto, *Astrophysical Journal*, **619**, 427–445 (2005).
4. N. Iwamoto, H. Umeda, N. Tominaga, K. Nomoto, and K. Maeda, *Science*, **309**, 451–453 (2005).
5. R. Buras, M. Rampp, H. -Th. Janka, K. Kifonidis, *Astronomy and Astrophysics*, submitted (astro-ph/0507135), (2005)
6. C. Fröhlich,, et al., *Astrophysical Journal*, in press (astro-ph/0410208), (2006).
7. J. Pruet, S. E. Woosley, R. Buras, H. -Th. Janka, and R. D. Hoffman, *Astrophysical Journal*, **623**, 325–336 (2005).
8. C. Fröhlich, G. Martínez-Pinedo, M. Liebendörfer, F. -K. Thielemann, E. Bravo, W. R. Hix, K. Langanke, and N. T. Zinner, *Physical Review Letters*, submitted (astro-ph/0511376), (2005).
9. J. Pruet, R. D. Hoffman, S. E. Woosley, H. -Th. Janka, and R. Buras, astro-ph/0511194 (2005).
10. R. D. Hoffman, S. E. Woosley, G. M. Fuller, and B. S. Meyer, *Astrophysical Journal*, **460**, 478–488 (1996).

Solar Neutrino Fluxes Using The Exponential S-Factor

Hasan Abu Kassim, Ithnin Abdul Jalil and Norhasliza Yusof

Department of Physics, University of Malaya, 50603 Kuala Lumpur, Malaysia

Abstract. Recently we propose an exponential form for the astrophysical S-factor. This form produces about 20% more solar ^{3}He production through the ^{3}He-^{3}He reaction. In this note, we investigate the effects on the ^{7}Be and ^{8}B neutrino productions since the neutrino fluxes depend on the ^{3}He abundance.

Keywords: Solar neutrinos, solar physics.
PACS: 26.65+t ; 96.60-j.

INTRODUCTION

The sun is like a huge thermonuclear reactor. Nuclear reaction schemes i.e. the proton-proton chain and carbon-nitrogen-oxygen cycle in the sun in principle could be tested if the neutrinos produced by these reactions are detected by terrestrial neutrino detectors.

SOLAR NEUTRINOS

There are seven sources of electron neutrinos produced by the proton-proton chain and carbon-nitrogen-oxygen cycle in the sun. The production of neutrino luminosity L_ν is described by

$$\frac{dL_\nu}{dM} = \varepsilon_\nu \qquad (1)$$

where M is the solar mass, ε_ν is the rate of the nuclear energy carried away by the neutrinos per unit mass of the solar material and is given by

$$\varepsilon_\nu = \frac{n_i n_j \langle \sigma v \rangle_{ij}}{\rho} \langle q_\nu \rangle. \qquad (2)$$

Here $\langle\sigma v\rangle_{ij}$ is the reaction rate per pair of particles i and j that produces the neutrino with number densities n_i and n_j respectively. The average energy of the neutrino calculated from the decay rate is $\langle q_v \rangle$.

The astrophysical S-factor is taken to be dependent on the exponential of some function in energy:

$$S \propto e^{f(E)}. \qquad (3)$$

Using the known experimental cross section of the $^3He(^3He,2p)^4He$,

$$\ell n S = 1.7092 - 0.86E + 0.59E^2 - 0.11E^3. \qquad (4)$$

The neutrino fluxes of the ^7Be and ^8B arriving on the Earth surface are computed by integrating the neutrino luminosity over the solar model,

$$\phi_v = \int_0^{M_{sun}} L_v dM. \qquad (5)$$

We have evolved a standard solar model in which the temperature, density and chemical abundance profiles are used to calculate $\langle\sigma v\rangle$ and the neutrino fluxes. The details of the solar model will be described elsewhere. Analytical formulas for $\langle\sigma v\rangle$ are taken from the NACRE compilation and the ^7Be and ^8B neutrinos produced in the proton-proton chain are determined. The $\langle\sigma v\rangle$ calculated using the exponential S-factor given by equation (4) is then used to compute the neutrino fluxes and compare with the ones calculated using NACRE compilation. Table 1 shows the ^7Be and ^8B neutrino fluxes.

TABLE 1. Integrated Neutrino Flux on the Earth $cm^{-2}s^{-1}$.

Neutrino	NACRE	This Work	Differences
^7Be	4.981×10^9	7.731×10^9	55%
^8Be	5.572×10^6	8.170×10^6	47%

CONCLUSIONS

The exponential S-factor gives 50% more neutrinos than the neutrinos from the standard solar model. This would translate to a higher capture in solar neutrino detectors.

ACKNOWLEDGMENTS

This work is funded by the Intensification of Research in Priority Areas (IRPA) grant 09-02-03-0149 from the Malaysian Government and a short term research grant F0204/2004D from University of Malaya. The authors would like to thank Raphael Hirshi for useful discussion on the solar model.

REFERENCES

1. D.Clayton, *Pinciples of Stellar Evolution and Nucleosynthesis*, Chicago:The University of Chicago Press, 1983, pp. 296-309.
2. C.E. Rolfs and W.S. Rodney, *Cauldrons in the Cosmos : Nuclear Astrophysics*, The University of Chicago Press, Chicago, 1988
3. C. Angulo et al., *Nucl. Phys. A* **656**, 3 (1999).
4. Norhasliza Yusof, Ithnin Abdul Jalil and Hasan Abu Kassim., "The Exponential S-Factor in the PP Chain" presented in *Origin of Matter and Evolution of Galaxies 2005*.

MSW Effect in Supernova-Shock Propagation

Shiou Kawagoe*, T. Kajino*,†, K.Yoshihara**, H. Suzuki**, K. Sumiyoshi*,‡ and S. Yamada§

*Department of Astronomical Science, School of Physical Sciences, The Graduate University for Advanced Studies(SOKENDAI), National Astronomical Observatory of Japan,2-21-1 Osawa, Mitaka-shi, Tokyo 181-8588, Japan
†Department of Astronomy, Graduate School of Science, University of Tokyo, 7-3-1 Hongo, Bunkyo-ku, Tokyo 113-0033,Japan
**Department of Physics, Faculty of Science and Technology, Tokyo University of Science, 2641 Yamazaki, Noda-shi, Chiba 278-8510, Japan
‡Numazu College of Technology, 3600 Ooka, Numazu-shi, Shizuoka 410-8501, Japan
§Department of Physics, Faculty of Science and Engineering, Waseda University, 3-4-1 Okubo, Shinjuku-ku, Tokyo 169-8555, Japan

Abstract. It is pointed out that shock wave propagation has influences on the supernova neutrino oscillation by changing density profile and neutrino survival probability (MSW effect). Using an implicit Lagrangian code for general relativistic spherical hydrodynamics, we succeeded in calculating propagation of shock waves which are generated by adiabatic collapse of iron cores and pass into the stellar envelopes. We apply our model to the neutrino oscillation and calculate neutrino energy spectra of three light-neutrino families. We examined how the influence of the shock wave appear in the neutrino spectra, and found that the influences of the shock wave in the neutrino spectra move from low-energy side to high-energy side according to the shock propagation.

Keywords: Supernova, Shock wave, Supernova neutrinos
PACS: 97.60.BW

INTRODUCTION

The supernova neutrinos have very interesting nature such that the average energies of three different species are different from one another. These differences make some very interesting and important effects on neutrino oscillations. The supernova neutrinos generated in the core, propagate through the envelope. Therefore, it is necessary to consider the matter effect of the neutrino oscillations. It is known that the shock wave influences the resonance area of neutrino oscillations[1]. In this article we discuss how the energy spectrum of each neutrino species can change due to the MSW effect. This effect of supernova neutrino is one of the foci of recent neutrino astrophysics. It has still been an open question how the shock wave propagation affects the neutrino oscillation.

NUMERICAL METHOD AND RESULT

We use an implicit Lagrangian code for general relativistic spherical hydrodynamics[2]. As the first step, we perform simplified calculations of core collapse and bounce by following adiabatic collapse with fixed electron fraction, because we intend to construct an approximate models of prompt explosion. We adopt the presupernova model of $15 M_\odot$

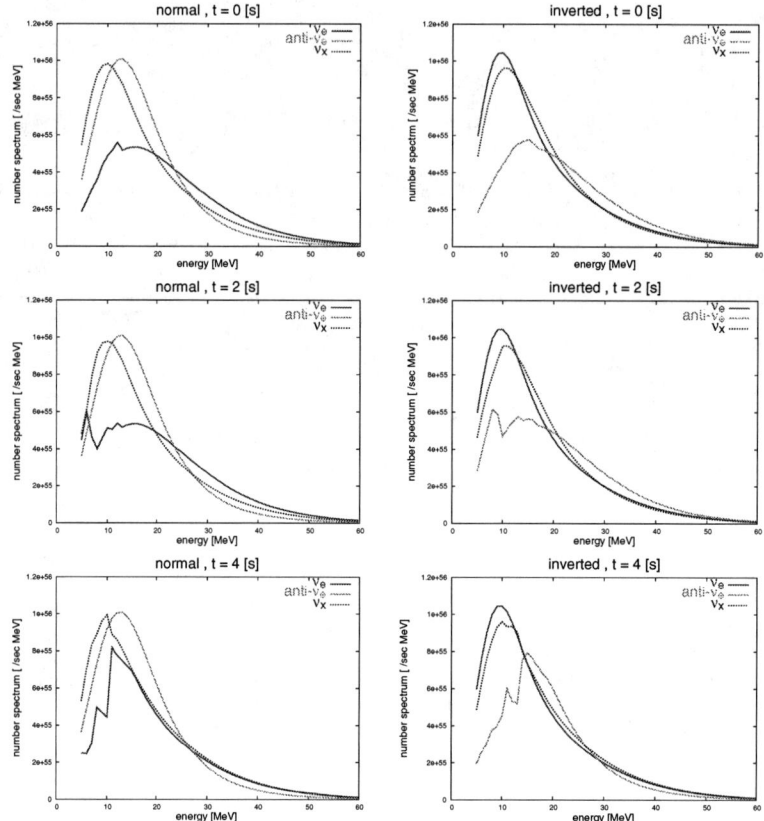

FIGURE 1. The neutrino spectra as functions of neutrino energy at different times 0, 2 and 4 s after the core bounce, respectively. Left column is for normal mass hierarchy, and right column for inverted mass hierarchy. Spectra for ν_e, $\bar{\nu}_e$ and ν_x are displayed by red, green and blue curves, respectively. We set $\sin^2 2\theta_{13} = 6 \times 10^{-4}$ in these calculations. See the text for the other oscillation parameters.

star provided by Woosley and Weaver (WW95) [4] and calculate the region of the iron core ($\sim 10^{15}$g/cm^3) and the stellar envelope ($\sim$1g/cm^3) simultaneously. We succeeded in the calculation of propagation of shock wave which is generated by adiabatic collapse of iron core passing through the stellar envelope consistently.

Then we investigate how the shock propagation affects the neutrino oscillation for the supernova neutrinos. We solved numerically the time evolution equation of the neutrino wave function along the density profile of our supernova calculation[5]. We calculated the neutrino survival probabilities and neutrino energy spectra for the normal mass hierarchy and for the inverted mass hierarchy, respectively. The neutrino oscillation parameters are taken from the best fit values in various analyses of the observations, except θ_{13} : $\sin^2 2\theta_{12} = 0.84$, $\sin^2 2\theta_{23} = 1.00$, $\Delta m_{12}^2 = 7.3 \times 10^{-5}$eV^2, and $\Delta m_{13}^2 = 2.5 \times 10^{-3}$eV^2. If θ_{13} is relatively large even within the observed conservative constant

$\sin^2 2\theta_{13} < 0.1$, neutrino oscillation might be affected by the shock wave. The influence appears in neutrino spectra for normal mass hierarchy, and in anti-neutrino spectra for inverted mass hierarchy [6].

Fig.1 depicts the spectra of three flavor neutrinos as functions of the neutrino energy at three different times of 0, 2 and 4 s after the core bounce. Left and right column are the spectra for the normal mass hierarchy and for the inverted mass hierarchy, respectively. We use the energy spectra of $v_e, \bar{v}_e$ and v_x of the numerical supernova model[7]. Compared with t=0 s (top), only a low energy parts of v_e for the normal mass hierarchy and of $\bar{v}_e$ for the inverted mass hierarchy increase slightly at t=2 s (middle). At later times (bottom), the effects move toward higher energy.

DISCUSSIONS AND CONCLUSION

MSW effect appears from low-energy side and moves toward high-energy side according to the shock wave propagation. This is because, as shock wave propagates outward, the density at the shock front decreases and the resonance condition is satisfied for higher energy neutrinos. The electron number density at the resonance point is

$$n_{e,res} \equiv \frac{1}{2\sqrt{2}G_F} \frac{\Delta m^2}{E} \cos 2\theta, \qquad (1)$$

where G_F is Fermi coupling constant, Δm^2 is the mass squared difference, θ is the mixing angle, and E is the neutrino energy. Δm^2 and θ correspond to Δm_{13}^2 and θ_{13} at H-resonance and to Δm_{12}^2 and θ_{12} at L-resonance, respectively.

The neutrino spectra of the normal mass hierarchy (left column of Fig.1) and of the inverted hierarchy (right column) are different. Therefor, we expect that the neutrino mass hierarchy could be determined from observation of the supernova neutrino spectra.

To summarize, we succeeded in calculating propagation of shock wave which is generated by adiabatic collapse of iron core and passes into the stellar envelope in single simulation. There is a possibility of finding the movement of the shock wave inside the star as such supernova neutrino signal. However, in our simulations, the density behind the shock wave hardly decreases because we neglected neutrino cooling of the protoneutron star. We need several improvements. Detailed studies of θ_{13} dependence and the cooling effect of the protoneutron star are underway.

REFERENCES

1. M. Fukugita and T. Yanagida, *Physics of Neutrinos and Applications to Astrophysics*, edited by R. Balian et al., Springer, Berlin, 2003, pp 348–363.
2. S. Yamada, *Astrophys. J*, **475**, 720 (1997)
3. H. Shen, H. Toki, K. Oyamatsu and K. Sumiyoshi, *Nucl. Phys. A*, **637**, 435 (1998)
4. S. E. Woosley and T. Weaver, *Astrophys. J. Suppl*, **101**, 181 (1995).
5. K. Takahashi and K. Sato, *Prog. Theor. Phys.*, **109**, 919 (2003).
6. R. Tomàs, M. Kachelrie, G. Raffelt, A. Dighe, H. T. Janka and L. Scheck, *JCAP*, **09**, 015 (2004).
7. K. Takahashi, M. Watanabe and K. Sato, *Phys. Letters* B, **510**, 189 (2001).

Neutrino-Induced Hydrogen Burning

Chad T. Kishimoto* and George M. Fuller*

Department of Physics, University of California, San Diego, La Jolla, CA 92093-0319

Abstract.
The principal hydrogen burning mechanisms that take place in stars have been elucidated and explored for many decades. However, the introduction of a prodigious flux of electron anti-neutrinos would significantly accelerate these mechanisms and change the path toward the production of an α particle. We discuss the nature of such changes in the hydrogen burning mechanisms, and the side effects spawned from such alterations.

Keywords: neutrinos; nuclear reactions
PACS: 25.60.Pj (fusion reactions); 13.15.+g (neutrino interactions)

INTRODUCTION

Hydrogen burning involves the conversion of four protons into an alpha particle, liberating energy in photons and neutrinos, through fast radiative proton captures and slow weak interactions. The weak interaction is necessary to transform protons into neutrons during hydrogen burning and provides the bottleneck reactions in the hydrogen burning sequence. In the proton-proton chain (pp-chain), the weak reaction $p(p,e^+\nu_e)d$ is the rate limiting step, while the CNO cycle relies on the positron decay of isotopes of oxygen. We investigate the repercussions of introducing a prodigious flux of neutrinos with large energies into the mix.

NEUTRINO-INDUCED HYDROGEN BURNING MECHANISMS

The weak reaction $p(p,e^+\nu_e)d$ provides the bottleneck in the pp-chain. A significant flux of electron anti-neutrinos allows an alternate mechanism to be favored, where anti-neutrino capture on a proton creates a neutron and positron, followed by fast radiative proton capture to form a deuteron, $p(\bar{\nu}_e,e^+)n(p,\gamma)d$. If a large $\bar{\nu}_e$-flux with high energies is introduced ($\phi_{\bar{\nu}_e} \gtrsim 10^{40}$ cm^{-2} s^{-1}, $\langle E_{\bar{\nu}_e}\rangle \gtrsim$ a few MeV), this alternate reaction path is favored in astrophysical environments where hydrogen burning would be relevant. This provides not only a new mechanism for hydrogen burning, but increases the energy generation rate by several orders of magnitude.

The positron decay of ^{14}O and ^{15}O are the rate limiting steps in the β-limited CNO cycle, with half lives of 71 s and 122 s respectively. This decay can be facilitated by anti-neutrino capture on these nuclei. Figure 1 illustrates the acceleration of the relevant weak rates as a function of total electron anti-neutrino flux. For a large enough flux, the reaction rates are proportional to the flux of electron anti-neutrinos, while at low fluxes, the decay rates of ^{14,15}O asymptote to their laboratory decay values, meaning there is no discernible effect at those levels.

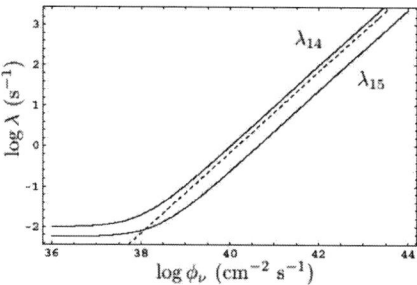

FIGURE 1. Key weak decay rates as a function of electron anti-neutrino flux, assuming a Fermi-Dirac $\bar{v}_e$-energy spectrum with zero chemical potential and $\langle E_{\bar{v}_e}\rangle = 25$ MeV. The solid lines labeled $\lambda_{14,15}$ are the decay rates of ^{14,15}O. The dashed line is the conversion of a proton to a neutron through $\bar{v}_e$-capture.

OTHER EFFECTS

The rationale for introducing a large $\bar{v}_e$-flux was to accelerate the rate limiting steps in hydrogen burning. However, a number of side effects are possible as this flux will accelerate the rates of other positron decays.

The principal mechanism for break out into the rp-process is ^{15}O$(\alpha,\gamma)^{19}$Ne$(p,\gamma)^{20}$Na. Break out into the rp-process occurs when proton capture on ^{19}Ne competes favorably with the decay of ^{19}Ne. For densities and temperatures that satisfy the inequality

$$\rho X \lambda_{p\gamma}(^{19}\text{Ne}) > \lambda_{e^+}(^{19}\text{Ne}), \qquad (1)$$

where ρ is the density in g cm^{-3}, X is the hydrogen mass fraction, $\lambda_{e^+}(^{19}$Ne$)$ is the decay rate of ^{19}Ne, and $\lambda_{p\gamma} = N_A\langle\sigma v\rangle_{p\gamma}$ taken from Caughlin and Fowler [1], break out into the rp-process will occur. [4]

A large $\bar{v}_e$-flux would upset the inequality (1) by increasing the decay rate of ^{19}Ne with anti-neutrino capture. This would demand higher temperatures (increasing $\lambda_{p\gamma}$) at a given density for break out into the rp-process to occur. Figure 2a shows how the conditions for break out into the rp-process are altered in the presence of a strong $\bar{v}_e$ flux.

Another relevant question is if the conditions are favorable to break out into the rp-process, what are the effects of a prodigious electron anti-neutrino flux on the subsequent nucleosynthesis? The competition between the slow capture of an α particle on ^{15}O and the faster decay of ^{15}O becomes even more lopsided in favor of nuclear material remaining in the CNO cycle rather than escaping into the rp-process. As a result, the total yield in rp-process elements will decrease with increasing neutrino flux and energy. Figure 2b shows the effects of large electron anti-neutrino fluxes on the total yield in rp-process elements. Notice that at large $\bar{v}_e$-fluxes, break out into the rp-process is nearly completely suppressed.

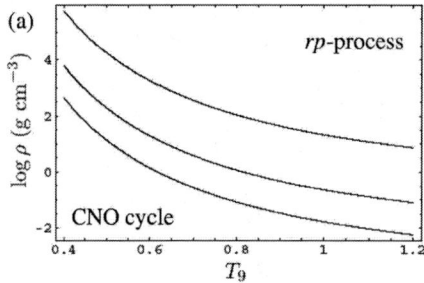
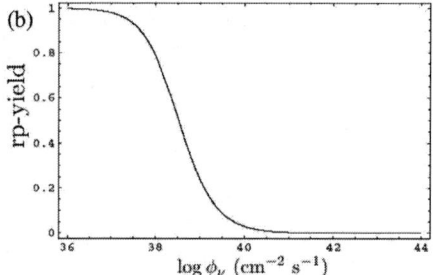

FIGURE 2. (a) Conditions required for break out into the rp-process. The plotted contours are for, in ascending order, $\bar{\nu}_e$-fluxes of $10^{38,40,42}$ cm^{-2} s^{-1}. Zero flux is indistinguishable from the lowest contour. (b) Ratio of the yield of rp-process elements with neutrinos to without neutrinos. Both figures are made assuming a Fermi-Dirac $\bar{\nu}_e$-energy spectrum with zero chemical potential and $\langle E_{\bar{\nu}_e}\rangle = 25$ MeV

DISCUSSION

We have examined the effects of a prodigious flux of electron anti-neutrinos on hydrogen burning. Large fluxes and high energies allow anti-neutrino capture to accelerate the rate limiting steps in hydrogen burning, causing a significant increase in the energy generation rates due to the pp-chain and β-limited CNO cycle. This large flux of electron anti-neutrinos would also lead to a change in the conditions necessary for break out into the rp-process, and suppress the expected yield in rp-process elements.

The next step is to find an environment where this effect may be relevant. A high entropy electron-positron plasma is an efficient engine for the production of neutrinos and anti-neutrinos of all flavors, and could provide the $\bar{\nu}_e$-flux necessary to drive this effect. Possible environments that would merit future investigations into the effects of anti-neutrino capture on hydrogen burning include high mass accretion disks and collapsing supermassive stars.

In [3], we explore these effects in supermassive stars, where in the final stages of collapse, a prodigious flux of neutrinos is emitted from the collapsing homologous core. We find that changes from current simulations may occur on the lower end of the mass spectrum. Further investigation using computer simulations would be necessary to see if this effect could be relevant and leave a possible observational signature.

REFERENCES

1. G. R. Caughlan and W. A. Fowler 1988, ADNDT, 40, 283
2. R. B. Firestone, *Table of Isotopes*, 8th ed., edited by V. S. Shirley et al., John Wiley, New York, 1996.
3. C. T. Kishimoto and G. M. Fuller, in preparation
4. R. K. Wallace and S. E. Woosley 1981, ApJS, 45, 389

Multigroup Flux-limited Diffusion Neutrino Transport Simulations for Magnetized and Rotating Core-Collapse Supernovae

Kei Kotake[*,†], Naofumi Ohnishi[**], Shoichi Yamada[‡] and Katsuhiko Sato[§]

[*] *Science & Engineering, Waseda University, 3-4-1 Okubo, Shinjuku, Tokyo, 169-8555, Japan*
[†] *<kkotake@heap.phys.waseda.ac.jp>*
[**] *Department of Aerospace Engineering, Tohoku University, 6-6-01 Aramaki-Aza-Aoba, Aoba-ku, Sendai, 980-8579, Japan*
[‡] *Advanced Research Institute for Science and Engineering, Waseda University, 3-4-1 Okubo, Shinjuku, Tokyo, 169-8555, Japan*
[§] *Department of Physics, School of Science, the University of Tokyo, 7-3-1 Hongo, Bunkyo-ku, Tokyo 113-0033, Japan*

Abstract. We report a current status of our radiation-magnetohydrodynamic code for the study of core-collapse supernovae. In this contribution, we discuss the accuracy of our newly developed numerical code by presenting the test problem in a static background model. We also present the application to the spherically symmetric core-collapse simulations. Since close comparison with the previously published models is made, we are now applying it for the study of magnetorotational core-collapse supernovae.

Keywords: supernovae: general–neutrinos–radiative transfer–magnetohydrodynamics–methods: numerical
PACS: 97.60.Bw

INTRODUCTION

A leap beyond the spherical models in core-collapse supernovae seems indeed meaningful, because asphericities in the supernova core should have influence on the nucleosynthesis, neutrino and gravitational-wave emissions [1]. Before the advent of the observations using neutrinos and gravitational waves as new eyes, one hopes to clarify the explosion mechanism of aspherical supernovae.

So far, many physical ingredients to produce aspherical explosion have been investigated, such as convection, possible density inhomogeneities formed prior to core-collapse, rotation and magnetic fields (see [1] for a collective reference), more recently standing shock instability (e.g., [2]) and the excitations of g-modes in the protoneutron star [3]. Whatsoever the origin of the asphericity, the neutrino heating mechanism should play a key role to drive explosions. Thus multidimensional neutrino transport simulations are indispensable. In this contribution, we report the current status of our radiation transport calculations.

BASIC EQUATIONS OF RADIATION-MAGNETOHYDRODYNAMICS

As for the neutrino reactions, we include the so-called standard set denoted in table 1 in [4], plus nucleon bremsstrahlung [5]. Since fully angle-dependent solutions in multidimensional models (even in 2D) are computationally prohibitive, we employ the flux limited diffusion approximation to relate the first and zeroth angular moment of the neutrino distribution function,

$$\psi_1 = \begin{matrix} \psi_{1,r} \\ \psi_{1,\theta} \end{matrix} = -\Lambda \cdot \left[\nabla \psi_0 - A_1 \psi_0 - C_1 \right], \quad (1)$$

where ψ_0 and ψ_1 is the zeroth and first angular moment of the neutrino distribution function, respectively. See equation (A25) in [4] for the definition of A_1 and C_1. Here Λ is a flux limiter generalized to 2D, namely, $(\Lambda_r, \Lambda_\theta)$ whose components are given by $\Lambda_r = (3\lambda^t)/(3 + \lambda^t |\nabla_r \psi_0|/\psi_0)$, and $\Lambda_\theta = (3\lambda^t)/(3 + \lambda^t |\nabla_\theta \psi_0|/\psi_0)$, where λ^t is the transport mean free path (see Equation (A26) in [4]), ∇_r and ∇_θ represents $\partial/\partial r$ and $\partial/(r\partial\theta)$, respectively. With equation (1), one can determine ψ_0 by solving the following zeroth angular moment equation of Boltzmann equation,

$$\frac{1}{c}\frac{d}{dt}\psi_0 + \frac{1}{3}\nabla \cdot \psi_1 + \frac{1}{c}\nabla \cdot v\, \omega\, \frac{\partial}{\partial \omega}\psi_0 = j(1-\psi_0) - \frac{\psi_0}{\lambda^a} + A_0 \psi_0 + B_0 \cdot \psi_1 + C_0, \quad (2)$$

which is often referred to as a multi-group flux limited diffusion (MGFLD) equation for neutrinos (again see appendix A in [4] for the definition of unmentioned variables). For the time evolution of transport, we solve the neutrino-matter coupling implicitly with performing the Newton Raphson iteration until $\delta\psi_0$, δY_e, and $\delta(e/\rho)$ converges to a certain value with e being the specific internal energy. More details for the numerical implementation of our code will be presented in the forthcoming paper.

CODE TEST

Before we apply the newly developed code to the magnetized and rotating supernova simulations, we need to verify the accuracy by the test calculations. Taking the profiles of the density, temperature, and the electron fraction from Bruenn's 1D model after core bounce [6] as a background, we calculate the neutrino distribution functions of each species ($v_e, \bar{v}_e, v_x, \bar{v}_x$) by our 2D computations and compare them with the ones obtained in 1D simulation [6]. Using 16 energy mesh points which is logarithmically uniform and covers 0.95 - 255 MeV, we saw the good agreement with each other within the errors of $O(10)\%$ near the surface of the neutrinospheres. This may be mainly due to the difference of the numerical implementation of the neutrino transport, and of the employed equation of state. Since close comparison is made in the static background computation, we move on to apply the code to the spherically symmetric core-collapse simulations of 15 $M_\odot$ progenitor star (Woosley & Weaver 1995).

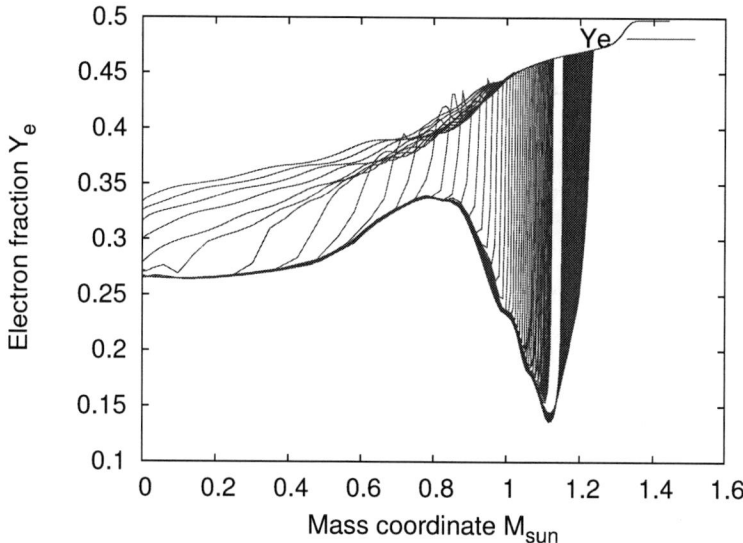

FIGURE 1. Time evolution of Y_e up to ~40 msec after bounce (small blank is due to the data loss).

In Figure 1, the evolution of Y_e up to ~ 40 msec after bounce is shown. The trough of Y_e represents the neutronization occuring outside the neutrinospheres. The global profiles are well in good agreement with the previous spherically symmetric calculations (e.g., [7]). More details about the tests of our newly developed code will be presented soon in elsewhere, with its applications to the magnetorotational core-collapse simulations of massive stars.

This work was supported in part by the Japan Society for Promotion of Science (JSPS) Research Fellowships (K.K) and a Grant-in-Aid for Scientific Research from the Ministry of Education, Science and Culture of Japan through No.S 14102004, No. 14079202, and No. 14740166.

REFERENCES

1. Kotake K *et al* 2006 *Reports on Progress in Physics* in press
2. Ohnishi N, Kotake K, and Yamada S 2006 *Astrophys. J.* in press
3. Burrows A *et al* 2005 submitted to *Astrophys. J.*
4. Bruenn S W 1985 *Astrophys. J. Suppl.* **58** 771
5. Hannestad S and Raffelt G 1998 *Astrophys. J.* **507** 339
6. Bruenn, S. W. 1992, in Nuclear Physics in the Universe (Philadelphia: Institute of Physics Pub.)
7. Rampp M and Janka H T 2000 *Astrophys. J.* **539** L33., Liebendörfer M *et al* 2004 *Astrophys. J. Suppl.* **150** 263

The p-Process in the Carbon Deflagration Model for Type Ia Supernovae and Chronology of the Solar System Formation

Motohiko Kusakabe[*,†], Nobuyuki Iwamoto[**] and Ken'ichi Nomoto[*]

[*]*Department of Astronomy, School of Science, University of Tokyo,*
7-3-1 Hongo, Bunkyo-ku, Tokyo 113-0033, Japan
[†]*National Astronomical Observatory of Japan, Mitaka, Tokyo 181-8588, Japan*
[**]*Japan Atomic Energy Agency, Tokai, Ibaraki 319-1195, Japan*

Abstract. We study nucleosynthesis of p-nuclei in the carbon deflagration model for Type Ia supernovae (SNe Ia) by assuming that seed nuclei are produced by the s-process in accreting layers on a carbon-oxygen white dwarf during mass accretion from a binary companion. We find that about 50 % of the p-nuclides are synthesized in proportion to the solar abundance and that p-isotopes of Mo and Ru which are significantly underproduced in Type II supernovae (SNe II) are produced up to a level close to other p-nuclei. Comparing the yields of iron and p-nuclei in SNe Ia we find that SNe Ia can contribute to the galactic evolution of the p-nuclei. Next, we consider nucleochronology of the solar system formation by using four radioactive nuclides and apply the result of the p-process nucleosynthesis to simple galactic chemical evolution models. We find that when assumed three phases of interstellar medium are mixed by the interdiffusion with the timescale of about 40 Myr ^{53}Mn/^{55}Mn value in the early solar system is consistent with a meteoritic value. In addition, we put constraints to a scenario that SNe Ia induce the core collapse of the molecular cloud, which leads to the formation of the solar system.

Keywords: nucleosynthesis, abundances, solar system, supernovae
PACS: 91.65.Dt, 91.65.Sn, 96.12.Bc, 97.60.Bw

INTRODUCTION

The stable isotopes with proton number $Z \geq 34$ on the neutron-deficient side of β-stability line are classified as p-nuclei, and the nucleosynthesis processes which produce the p-nuclei are called p-process. The nuclear photodisintegration reactions on pre-existing nuclei are considered to be important to make the bulk of p-isotopes. One of the sites thought to contribute significantly to the production of the p-nuclei is the oxygen/neon layers of Type II supernovae (SNe II) [1, 2] where the solar p-abundance pattern is reproduced for $\sim 60\%$ of the p-nuclei. However, ^{92,94}Mo and ^{96,98}Ru are severely underproduced in comparison with the other p-nuclides. In addition, the ejected yield of p-nuclei is a factor of 4 smaller than that of oxygen which is the main product of SNe II [2]. This result may indicate that the other sites play an important role in providing the p-nuclei.

In this paper we investigate the p-process nucleosynthesis in the outermost layers of an exploding CO white dwarf as Type Ia supernovae (SNe Ia) [3] and estimate their contribution to the galactic p-nuclei. Then applying this result to galactic chemical evolution models of short-lived radioactive nuclei, we discuss the solar system formation.

TABLE 1. Production ratios of chronometers

Radionuclide	Halflife (Myr)	Chronometer	Yield ratio
^{53}Mn	3.74	^{53}Mn/^{55}Mn	0.032*
^{92}Nb	34.7	^{92}Nb/^{92}Mo	8.7×10^{-3}
^{97}Tc	2.6	^{97}Tc/^{92}Mo	0.018
^{146}Sm	103.	^{146}Sm/^{144}Sm	0.41

* Data taken from W7 model in [7]

THE p-PROCESS NUCLEOSYNTHESIS IN SNe Ia

We use a Chandrasekhar-mass, carbon-deflagrating white dwarf model for Type Ia supernova (W7 model) [4]. We assume that the s-process occurs during a mass accretion phase of presupernova and enhances the abundances of heavy nuclei. The initial seed abundances for heavy nuclei are calculated by a thermally pulsed model [5] with updated neutron capture rates and revised treatments for many branchings.

Our calculation shows that p-isotopes of Mo and Ru are more produced than in SNe II. We investigate the influences of an unknown initial abundance on the yields of p-nuclei. The large variations of C/O ratio in a white dwarf has a negligible impact on the p-process, whereas the abundances of ^{22}Ne and seed s-nuclei have large influences on the p-process yield. The enhanced yields of p-nuclei are obtained in high abundances of s-nuclei and low abundance of ^{22}Ne.

By comparing the net yields of ^{56}Fe and the p-nuclei with those in SNe II we find that SNe Ia could effectively contribute to the galactic evolution of the p-nuclei. This result, however, strongly depends on the unknown s-process history at an accreting stage of presupernova. For details on the above results, see [6].

CHRONOLOGY OF THE SOLAR SYSTEM FORMATION

Several short-lived radioactive nuclides of halflives up to a hundred Myr are found to have existed in the early solar system from meteoritic analyses of their isotopic anomalies. Especially four radionuclides listed in Table 1 are produced at a certain level in our model of SNe Ia. Table 1 shows the halflife in a million years (Myr), a pair of unstable and stable nuclei, and the production ratio of the pair for each radionuclides. We use two models in order to constrain the formation process of the solar system.

First we consider a three-phase mixing model of the interstellar medium (ISM) [8], supplemented by an analytic chemical evolution model [9]. The three phases of ISM are considered as the process to the birth of a star which is formed in molecular cloud (MC) core. They are (a) warm HI clouds to which the hot ISM quickly changes, (b) large HI clouds which are formed by the accumulation of warm HI clouds and (c) molecular clouds which are formed by the condensation of the large HI cloud. In this model the SN ejecta is injected into the hot phase of the ISM. The matter exchanges between adjacent phases only occur with the mixing timescales of T_1 and T_2 ($= T_1$ in this investigation). We found that if a typical lifetime of the galactic disk until the birth of the solar system is assumed to be 5 Gyr, the mixing timescale is about $T_1 = 40$ Myr from meteoritic data of ^{53}Mn/^{55}Mn [10]. This result is almost consistent with the formation cycle of the MC [11] and large HI cloud [8]. However, relatively long mixing timescales of 200 and 320 Myr are obtained for ^{92}Nb [12] and ^{146}Sm [13], respectively. This indicates that the two p-process radionuclides might be overproduced. However, the ratios in the bulk ISM are still uncertain due to unknown s-process seed distribution for the p-process and uncertain amounts of stable isotopes previously comprised in the ISM. If the ratios for two p-process nuclei were a factor of 4 smaller than used in this investigation the mixing timescales would be consistent for the above three chronometers. ^{97}Tc isotopic ratios measured in meteorites are so uncertain that the constraint on the timescale is not convincing yet [14, 15].

Secondly, we consider the triggered solar formation model using the chronology formalism in [16]. The model deals with a situation where a constant-rate supply of p-nuclei to a gas cloud is followed by the last injection by the last SN, then the solar system forms after some decay interval. The halflives of ^{53}Mn and ^{97}Tc are short so that the steady-state model above might not be good. If so, the second model would indicate more precise information. We assumed that the solar material is a marginally stable Bonnor-Ebert sphere with 1 $M_\odot$ in mass [17], use the momentum conservation of the SN shock, and consider the geometry between the position of the SN and that of the solar material. In this model if about $(10^{-3}\text{-}0.1)$ % of p-nuclei mass came from the last supernova and the solar system condensed within a few Myr after the SN explosion, then the abundance is consistent with meteoritic data for the four nuclides. More meteoritic data are necessary to know more about the environment of the solar formation.

REFERENCES

1. N. Prantzos, M. Hashimoto, M. Rayet and M. Arnould, *Astron. Astrophys.*, **238**, 455–461 (1990).
2. M. Rayet, M. Arnould, M. Hashimoto, N. Prantzos and K. Nomoto, *Astron. Astrophys.*, **298**, 517–527 (1995).
3. W. M. Howard, B. S. Meyer and S. Woosley, *Astrophys. J.*, **373**, L5–L8 (1991).
4. K. Nomoto, F. K. Thielemann and K. Yokoi, *Astrophys. J.*, **286**, 644–658 (1984).
5. W. M. Howard, G. J. Mathews, K. Takahashi and R. A. Ward, *Astrophys. J.*, **309**, 633–652 (1986).
6. M. Kusakabe, N. Iwamoto and K. Nomoto, *Nucl. Phys.*, **A758**, 459–462 (2005).
7. K. Iwamoto, F. Brachwitz, K. Nomoto, N. Kishimoto, H. Umeda, W. R. Hix and F. K. Thielemann, *Astrophys. J. Suppl.*, **125**, 439–462 (1999).
8. D. D. Clayton, *Astrophys. J.*, **268**, 381–384 (1983).
9. D. D. Clayton, "Galactic Chemical Evolution and Nucleocosmochronology: A Standard Model", in *Nucleosynthesis : Challenges and New Developments*, edited by W. David Arnett and James W. Truran, University of Chicago Press, Chicago, 1985, pp. 65–88.
10. G. W. Lugmair and A. Shukolyukov, *Geochim. Cosmochim. Acta*, **62**, 2863–1886 (1998).
11. L. Hartmann, J. Ballesteros-Paredes and E. A. Bergin, *Astrophys. J.*, **562**, 852–868 (2001).
12. C. L. Harper Jr., *Astrophys. J.*, **466**, 437–456 (1996).
13. L. E. Nyquist, B. Bansal, H. Wiesmann and C. Y. Shih, *Meteoritics*, **29**, 872–885 (1994).
14. Q. Z. Yin and S. B. Jacobsen, *Lunar Planet. Sci.*, **XXIX**, 1802 (1998).
15. N. Dauphas, B. Marty and L. Reisberg, *Astrophys. J.*, **565**, 640–644 (2002).
16. D. N. Schramm and G. J. Wasserburg, *Astrophys. J.*, **162**, 57–69 (1970).
17. H. A. T. Vanhala and A. P. Boss, *Astrophys. J.*, **575**, 1144–1150 (2002).

Current progress of nuclear astrophysics experiments at CIAE

Weiping Liu, Zhihong Li, Jun Su, Xixiang Bai, Youbao Wang, Gang Lian, Bing Guo, Sheng Zeng, Shengquan Yan, Baoxiang Wang, Nengchuan Shu, and Yongshou Chen

China Institute of Atomic Energy, P.O.Box 275(46), Beijing 102413, P.R.China

Abstract. This paper described current progress of nuclear astrophysical studies using the unstable ion beam facility GIRAFFE. We measured the angular distributions for some low energy reactions, such as $^{11}C(d,n)^{12}N$, $^{8}Li(d,p)^{9}Li$ and $^{17}F(d,n)^{18}Ne$ in inverse kinematics, and indirectly derived the astrophysical S-factors or reaction rates of $^{11}C(p,\gamma)^{12}N$, $^{8}Li(n,\gamma)^{9}Li$, $^{8}B(p,\gamma)^{9}C$ at astrophysically relevant energies.

Keywords: Secondary beam facility, ANC, DWBA, S-factor, Reaction rate
PACS: 25.60.Je, 21.10.Jx, 25.40.Lw, 26.35.+c

INTRODUCTION

Aiming at the studies of nuclear astrophysics, the secondary beam facility (GIRAFFE) [1, 2] for producing and utilizing low energy beams of unstable nuclei has been constructed at the HI-13 tandem laboratory in 1993. It comprises a primary reaction chamber, a dipole-quadrupole doublet (D-Q-Q), a wien filter and a secondary reaction chamber. Up to now, the ion beams of ^{6}He, ^{7}Be, ^{8}Li, ^{10}C, ^{11}C, ^{13}N, ^{15}O and ^{17}F have been delivered, as summarized in Table 1. Recently, we have carried out some measurements of astrophysical interest, including $^{11}C(d,n)^{12}N$, $^{8}Li(d,p)^{9}Li$ and $^{17}F(d,n)^{18}Ne$, and derived the astrophysical S-factors or reaction rates of $^{11}C(p,\gamma)^{12}N$, $^{8}Li(n,\gamma)^{9}Li$, $^{8}B(p,\gamma)^{9}C$ at astrophysically relevant energies.

TABLE 1. Summary of the produced unstable ion beams at GIRAFFE

RNB	Reaction	Energy (MeV)	FWHM (MeV)	Purity (%)	Beam intensity (pps)*
^{6}He	$^{2}H(^{7}Li, ^{6}He)^{3}He$	35.3	0.5	90	500
^{7}Be	$^{1}H(^{7}Li, ^{7}Be)n$	23.0	1.3	99	1000
^{8}Li	$^{2}H(^{7}Li, ^{8}Li)^{1}H$	39.0	0.5	90	500
^{10}C	$^{1}H(^{10}B, ^{10}C)n$	55.9	3.5	96	200
^{11}C	$^{1}H(^{11}B, ^{11}C)n$	38.2	2.7	85	1000
^{13}N	$^{2}H(^{12}C, ^{13}N)n$	57.8	2.1	86	500
^{15}O	$^{2}H(^{14}N, ^{15}O)n$	66.0	3.6	91	800
^{17}F	$^{2}H(^{16}O, ^{17}F)n$	76.1	3.7	90	2000

* With 2 mm diameter collimator and primary beam intensity 100-700 enA.

EXPERIMENTS AND THEORETICAL ANALYSIS

One of the key reactions in the hot pp chains is $^{11}C(p,\gamma)^{12}N$ which is believed to play a pivotal role in the evolution of Pop III stars. The angular distribution of $^{11}C(d,n)^{12}N$ at E_{cm} = 9.8 MeV was measured with the secondary ^{11}C beam. The experimental data were analyzed with distorted wave Born approximation (DWBA) and thereby the square of asymptotic normalization coefficient (ANC)^2 was extracted to be 2.86 ± 0.91 fm^{-1} for the virtual decay $^{12}N \to {^{11}C} + p$. The astrophysical S-factors of $^{11}C(p,\gamma)^{12}N$ were then derived, as shown in Fig. 1 (a). The temperature dependence of the direct capture, resonant capture and total reaction rates for $^{11}C(p,\gamma)^{12}N$ were derived [3]. This work shows that the direct capture dominates the $^{11}C(p,\gamma)^{12}N$ in the wide energy range of astrophysical interest except the ranges corresponding to two resonances.

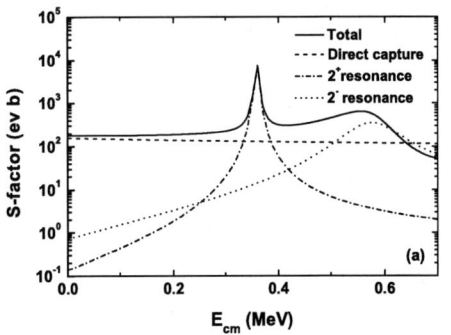 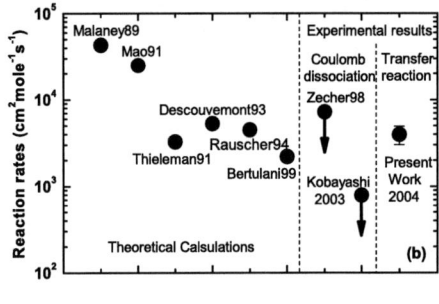

FIGURE 1. (a) E_{cm} dependence of the direct capture, resonant capture and total astrophysical S-factors for $^{11}C(p,\gamma)^{12}N$. (b) The $^8Li(n,\gamma)^9Li$ reaction rates for the direct capture derived from theoretical calculations and experiments.

The $^8Li(n,\gamma)^9Li$ reaction plays an important role in both the r-process nucleosynthesis and the inhomogeneous big bang models. We have measured the angular distribution of the $^8Li(d,p)^9Li_{g.s.}$ reaction at E_{cm} = 7.8 MeV in inverse kinematics using coincidence detection of ^9Li and recoil proton, for the first time. The single particle spectroscopic factor for the ground state of $^9Li={^8Li} \otimes n$ was derived to be 0.68 ± 0.14, and then used to calculate the direct capture cross sections for the $^8Li(n,\gamma)^9Li$ reactoin at energies of astrophysical interest. The astrophysical $^8Li(n,\gamma)^9Li$ reaction rate for the direct capture was found to be 3970 ± 950 cm^{3}mole^{-1}s^{-1} at T_9 = 1 [4]. The $^8Li(n,\gamma)^9Li$ reaction rates for the direct capture derived from theoretical calculations and experiments are shown in Fig. 1 (b).

The angular distribution of $^8Li(d,p)^9Li_{g.s.}$ was also used to extract the ANC for the virtual decay $^9Li \to {^8Li} + n$. According to charge symmetry, the ANC for $^9C \to {^8B} + p$ was derived to be 1.14 ± 0.29 fm^{-1}. The astrophysical $S_{18}(0)$ factor for the direct capture of the $^8B(p,\gamma)^9C$ reaction was found to be 44 ± 11 eV b. Fig. 2 (a) shows the astrophysical $S_{18}(0)$ factor extracted from theoretical calculations and experiments. We have also deduced the astrophysical reaction rates for direct capture in $^8B(p,\gamma)^9C$ at energies of astrophysical relevance [5].

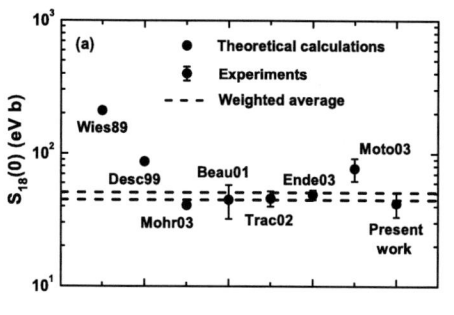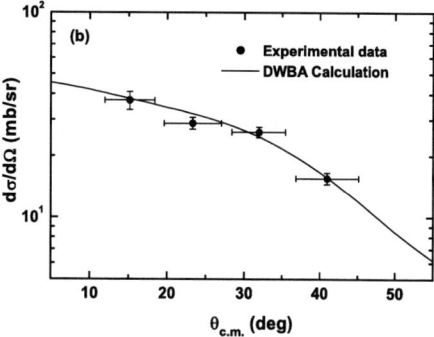

FIGURE 2. (a) Astrophysical $S_{18}(0)$ factors of $^8\text{B}(p,\gamma)^9\text{C}$ extracted from theoretical calculations and experiments. (b) Measured angular distribution of $^{17}\text{F}(d,n)^{18}\text{Ne}$ at $E_{cm} = 7.0$ MeV, together with DWBA calculation.

Very recently we have measured the $^{17}\text{F}(d,n)^{18}\text{Ne}$ reaction angular distribution at $E_{cm} = 7.0$ MeV. Fig. 2 (b) shows the experiment result, together with preliminary DWBA analysis. This reaction will be used to determine the direct capture contribution of $^{17}\text{F}(p,\gamma)^{18}\text{Ne}$, which can provide an alternate path from hot CNO cycle to rp process.

SUMMARY

In summary, GIRAFFE, a tandem based one stage unstable beam facility proved to be effective to produce beams suitable for the study of nuclear astrophysics reactions. Angular distribution measurement of transfer reactions in inverse kinematics, together with DWBA/ANC theoretical approach have been used to study the astrophysical reactions indirectly. The astrophysical S-factors and/or reaction rates for $^{11}\text{C}(p,\gamma)^{12}\text{N}$, $^8\text{Li}(n,\gamma)^9\text{Li}$, $^8\text{B}(p,\gamma)^9\text{C}$ were deduced by using the measurements of the $^{11}\text{C}(d,n)^{12}\text{N}$, $^8\text{Li}(d,p)^9\text{Li}$ reactions at the energies of astrophysical interest.

ACKNOWLEDGEMENT

The above research programs were supported by the Major State Basic Research Development Program under Grant Nos. G200077400 and 2003CB716704, the National Natural Science Foundation of China under Grant Nos. 19935030, 10025524 and 10375096.

REFERENCES

1. X. X. Bai, W. P. Liu, J. C. Qin et al., Nucl. Phys. A 588 (1995) 273c.
2. W. P. Liu, Z. H. Li, X. X. Bai et al., Nucl. Instr. Meth. B 204 (2003) 62.
3. W. P. Liu, Z. H. Li, X. X. Bai et al., Nucl. Phys. A 728 (2003) 275.
4. Z. H. Li, W. P. Liu, X. X. Bai et al., Phys. Rev. C 71(2005)052801(R)
5. B. Guo, Z. H. Li, W. P. Liu et al., Nucl. Phys. A 761(2005)162

Two-Dimensional Simulation of Core-Collapse Supernovae: Role of Anisotropic Neutrino Radiation on Explosion Dynamics

Hideki Madokoro, Tetsuya Shimizu and Yuko Motizuki

RIKEN, Hirosawa 2-1, Wako, Saitama 351-0198, Japan

Abstract. We perform two-dimensional numerical simulations of core-collapse supernovae. Starting from a rotating progenitor, we obtain a deformed neutrino sphere in a oblate shape in which the ratio of the polar axis to the equatorial one is $\sim$ 5%. This implies anisotropy of $\sim$ 5% in the emerging neutrino flux. Combined with our previous results, this anisotropy in neutrino radiation is sufficient to increase the explosion energy compared with that of spherical models, and can lead to a successful explosion.

Keywords: hydrodynamics—shock waves—stars:neutron—supernovae:general
PACS: 97.60.Bw

ANISOTROPIC NEUTRINO RADIATION

In multidimensional simulations of core-collapse supernovae, the neutrino flux from a nascent neutron star is expected to become anisotropic because of multidimensional effects such as convection and rotation. Janka and Mönchmeyer (1989)[1, 2] first pointed out that the neutrino flux along the polar axis can be about three times larger than that on the equatorial plane. Inspired by this work, Shimizu et al. (2001)[3] investigated the effects of globally anisotropic neutrino radiation on the explosion dynamics. In our subsequent studies[4, 5], we performed a series of 2-D simulations with various profiles of neutrino anisotropy. We have shown that globally anisotropic neutrino radiation is the most effective, and anisotropy of only $\sim 5-10$ % can greatly increase the explosion energies when the total neutrino luminosity is fixed. In these studies, however, the profile of anisotropy in the neutrino flux was assumed because we did not solve the region inside the proto-neutron star. It is expected that anisotropy will naturally appear as a result of multidimensional effects in the course of core-collapse, bounce, and shock propagation. Therefore, as the next step, we have performed two-dimensional core-collapse simulations starting from pre-collapse models.

CORE-COLLAPSE SIMULATION

We solve 2-D hydrodynamic equations in spherical coordinate. Our simulations start from the progenitor models developed by Heger et al. (2000) [6]. Both non-rotating progenitors and rotating ones are used. Table 1 lists our models investigated in this study. As for neutrino transport, either a (simple) diffusion approximation (inside the neutrino sphere) or the light-bulb approximation (outside the neutrino sphere) is adopted. This simplified treatment is sufficient for our purpose in this paper. Our equation of state (EOS) for lower ($\rho < 10^{11}$ g cm^{-3}) densities includes contributions from radiation fields, gas of nucleons and nuclei, and electrons. When the density becomes higher than 10^{11} g cm^{-3}, we switch to the EOS by Lattimer and Swesty[7].

ESTIMATE OF ANISOTROPY

Figures 1 and 2 show contour plots of entropy distribution with velocity fields at $t = 20$ ms after bounce. It is seen that the shock wave is stalled at $r \sim 100$ km. The radius of the neutrino sphere is $r \sim 55$ km for spherical models (sph15 and sph20), while it is located at $r \sim 52$ km (on the pole) and ~ 55 km (on the equatorial plane) for rotating ones (models rot15 and rot20). The neutrino sphere is, as expected, deformed in a oblate shape due to rotation. We find only minor differences between models with $M = 15 M_\odot$ and those with $M = 20 M_\odot$.

Now we can estimate the degree of anisotropy in the neutrino flux as follows[3]: Let us assume that neutrinos are emitted isotropically from each point on the surface of a protoneutron star. If a deformed neutrino sphere is regarded as a Maclaurin spheroid, the ratio of the neutrino flux along the polar axis (l_z) to that on the equatorial plane (l_x) then becomes $l_z/l_x = R_x/R_z$, where R_x and R_z represent the equatorial and polar radius of the neutrino sphere, respectively. Our calculations give $R_x/R_z \sim 1.05$. This implies that the neutrino flux has an anisotropy of $\sim 5\%$. Although this value of anisotropy is a bit smaller than we had expected, our previous studies in which the anisotropic neutrino flux was assumed [3, 4, 5] showed that anisotropy of $\sim 5\%$ is sufficient to increase the explosion energy compared with that of the spherical model.

CONCLUDING REMARKS

We performed 2-D simulations of core-collapse supernovae. Due to centrifugal forces, the neutrino sphere is deformed in a oblate shape, which causes anisotropic neutrino radiation from the surface of protoneutron star. We estimated the degree of anisotropy and obtained anisotropy of $\sim 5\%$. Combined with our previous studies, we found that this value is sufficient to increase the explosion energy. In future, we should improve our method of neutrino transport to a more sophisticated one, and continue our simulations to later times to see whether the shock revival does occur in our simulations.

TABLE 1. Simulated models.

model	progenitor mass	equatorial velocity of progenitors
sph15	$15 M_\odot$	0 km s^{-1}
sph20	$20 M_\odot$	0 km s^{-1}
rot15	$15 M_\odot$	200 km s^{-1}
rot20	$20 M_\odot$	200 km s^{-1}

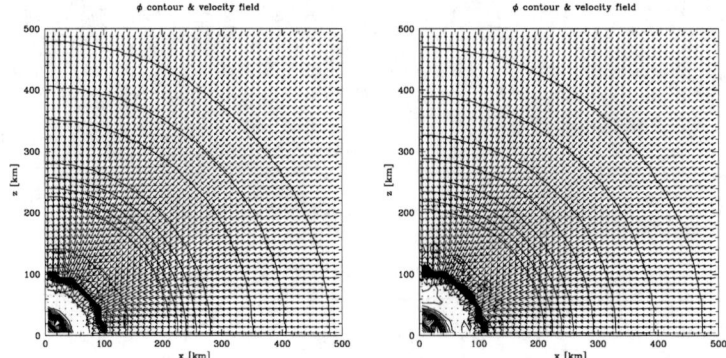

FIGURE 1. Contour plot of entropy distribution with velocity fields at $t = 20$ ms after bounce for a non-rotating and a rotating model, Left: model sph15, Right: model rot15. The progenitor mass is $15M_\odot$. The stalled shock wave is represented by crowded contour lines at $r \sim 100$ km. The deformed neutrino sphere is located at $r \sim 52-55$ km.

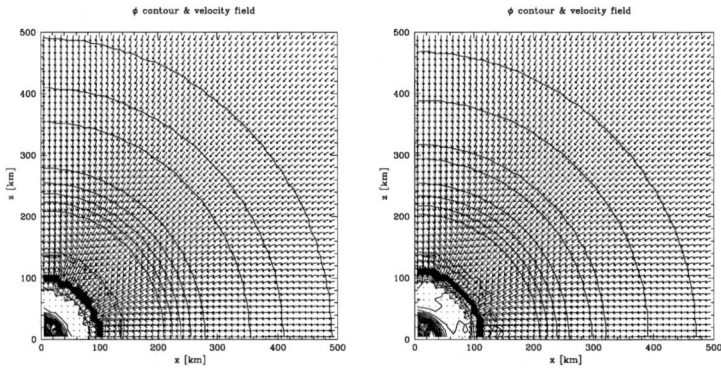

FIGURE 2. Same as Fig.1 but for the models with the progenitor mass of $20M_\odot$, Left: model sph20, Right: model rot20.

REFERENCES

1. H. -T. Janka, & R. Mönchmeyer, *A& A* **209**, L5–L8 (1989).
2. H. -T. Janka, & R. Mönchmeyer, *A& A* **226**, 69–87 (1989).
3. T. M. Shimizu, T. Ebisuzaki, K. Sato, & S. Yamada, *ApJ* **552**, 756–781 (2001).
4. H. Madokoro, T. Shimizu, & Y. Motizuki: *ApJ* **592**, 1035–1041 (2003).
5. H. Madokoro, T. Shimizu, & Y. Motizuki:, *Publ. Astron. Soc. J* **56**, 663–669 (2004).
6. A. Heger, N. Langer, & S. E. Woosley, *ApJ* **528**, 368–396 (2000).
7. J. M. Lattimer, & F. D. Swesty, *Nucl. Phys.* **A535**, 331–376 (1991)

Time-Variable Complex Metal Absorption Lines in the Quasar HS1603+3820

Toru Misawa*, Michael Eracleous*, Jane C. Charlton* and Akito Tajitsu[†]

Department of Astronomy and Astrophysics, Pennsylvania State University, 525 Davey Laboratory, University Park, PA 16802
[†]*Subaru Telescope, National Astronomical Observatory of Japan, Hilo, HI 96720*

Abstract. We present five spectra of the quasar HS1603+3820 (z_{em}=2.542) taken over intervals of 0.2–1.6 years (0.7–5.4 months in the quasar rest frame) with the High Dispersion Spectrograph on the Subaru telescope, for the purpose of studying its intrinsic narrow absorption lines (NALs). This quasar shows a rich complex of C IV NALs near the emission redshift ($dN/dz \sim 12$). We perform time variability analysis as well as covering factor analysis to separate intrinsic NALs, which are physically related to the quasar, from intervening NALs in 8 C IV systems. Only one of them, at $z_{abs} \sim 2.43$, shows both partial coverage and large variation in line strength, width, and position. Assuming that a change in the ionization state of the absorber causes the variability, a lower limit can be placed on the electron density ($n_e > 3.2 \times 10^4$ cm^{-3}) and an upper limit on the distance from the continuum source ($r < 6$ kpc). On the other hand, if motion of clumpy gas causes the variability, the crossing velocity and the distance from the continuum source are estimated to be $v_{cross} > 8,000$ km s^{-1} and $r < 3$ pc, assuming that the observed shift velocity does not exceed the escape velocity at that radius. If we adopt the dynamical model of Murray et al. (1995) [1], we can obtain a much more strict constraint on the radius of the gas parcel, $r < 0.2$ pc. We are planning to monitor this quasar for several years so as to use intrinsic absorber's behavior as a direct check of wind models.

Keywords: quasars, absorption lines, accretion disks
PACS: 98.54.Aj

Introduction, Data, and Analysis

The *broad* absorption lines observed in the UV spectra of quasars (BALs; FWHM $\geq$ 2000 km s^{-1}) can only be plausibly produced by absorbers in the accretion disk winds of quasars (i.e., intrinsic absorbers). In contrast, the *narrow* absorption lines found in these spectra (NALs; FWHM $\leq$ 500 km s^{-1}) can arise in intervening absorbers (e.g., cosmologically intervening galaxies) as well as intrinsic absorbers. Intrinsic NALs, by virtue of their narrow profiles, lend themselves as probes of the physical conditions in quasar accretion disks winds, unlike BALs. With this goal in mind, we have been monitoring the variations of a NAL complex (a "mini-BAL") in the spectrum of the bright, high-z quasar HS1603+3820 for 3 years. Using our spectra we have been able to separate intrinsic NALs from intervening NALs by using covering factor analysis. Moreover, by using the dramatic variations of one intrinsic NAL, we have been able to constrain its location and velocity relative to the quasar accretion disk, thereby testing dynamical models for the accretion disk wind.

We have at least two methods of determining that given NALs are intrinsic to the quasars: (1) covering factor analysis and (2) time variability analysis.

Covering factor analysis — The covering factor, C_f, represents the fraction of photons

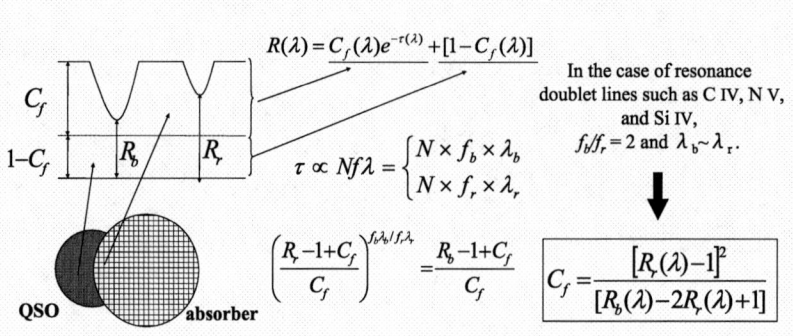

FIGURE 1. Absorption profile of resonance doublet produced by intrinsic absorber, where C_f is the covering factor, $R_{b,r}$, $f_{b,r}$ and $\lambda_{b,r}$ are the residual flux, oscillator strength, and rest-frame wavelength, respectively. The final formula shows how the observed quantities can be combined to obtain the effective covering factor of the continuum source by the absorber.

from the background source(s) that pass through the absorber [2]. For resonant doublets such as C IV, N V, and Si IV, we can evaluate C_f, by using the residual fluxes of the blue and red members of the doublets (R_b and R_r) in the normalized spectrum as illustrated in detail in Figure 1.

Time variability analysis — The detection of variability of the strength, profile, or position of absorption lines is also a reliable indicator of intrinsic NALs. There are at least two causes of variability. *(1) Change of ionization state* [3]: supposing that the gas is in ionization equilibrium, we can calculate the electron density of the absorber (n_e) from the recombination time (i.e., variability time) given by $t_{i-1}^{recom} \sim (n_e \, \alpha_{i-1})^{-1}$, where α_{i-1} is the rate coefficient for recombination from the stage i to the next lower stage $i-1$. If we know the ionization parameter, it is also possible to estimate the distance of the absorber (r) from the continuum source [4]. *(2) Motion of the absorption gas across the line of sight* [5]: assuming a clumpy gas crossing the background source, we can set the following lower limit on the crossing velocity $v_{cross} > (R/t_{var})\,[C_f(2) - C_f(1)]$, where R is the size of the background source, and $C_f(1)$ and $C_f(2)$ are the covering factors in the observed spectra at two different epochs, and t_{var} is the time scale for the variation.

Results and Discussion

In our analysis of the spectrum of HS1603+3820, we found 48 C IV, 15 Si IV, and 6 N V doublets, as well as 54 single metal lines in 8 absorption systems. Among them, only one system at $z_{abs} = 2.42 - 2.45$ ($v_{ej} = 8,300 - 10,600$ km s^{-1}) shows partial coverage as well as time variability (Figure 2, *left*). This system has many broad C IV components, which makes the line profile very difficult to decompose into kinematic components. Therefore, we concentrate on the region at 5290-5315Å which can be decomposed relatively simply. We have so far collected five high-quality spectra ($S/N \sim 70$ per resolution element; $R = 45,000$) with *Subaru*+HDS. The observed equivalent width of

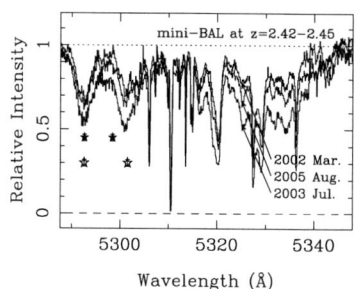 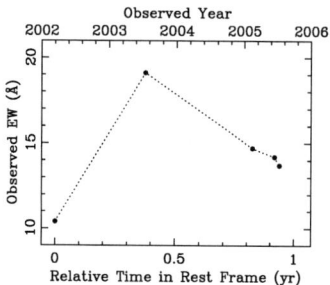

FIGURE 2. *Left*: Three of our five spectra of HS1603+3820 showing C IV NALs at z = 2.42–2.45. Open and filled stars denote the covering factors in 2002 March and 2003 July with 1σ errors. *Right*: The equivalent width first increased by a factor of two (2002 March → 2003 July), and then decreased again (→ 2005 February–August).

the C IV components in this complex system has varied dramatically over the course of our observations, as depicted in Figure 2 (*right*).

We have carried out a detailed analysis of the line profiles for the first two epochs. We found a significant change in the covering factors, from 0.30 to 0.45, between the 1st (March 2002) and 2nd (July 2003) spectra. The results of this analysis lead to the following conclusions [5]:

1. If the observed equivalent width variation is a result of a change in the ionization state of the absorber, we can estimate the electron density ($n_e > 3.2 \times 10^4$ cm^{-3}) and the absorber's distance from the quasar ($r < 6$ kpc).

2. If, on the other hand, the observed variation is a result of gas motion across the background UV source, we can estimate the crossing velocity ($v_{cross} > 8,000$ km s^{-1}) and the distance from the continuum source ($r < 0.2$ pc). These estimates rely on assuming the dynamical model for the outflow by Murray et al. (1995) [1]. This distance is larger than the size of the continuum source, $R_{cont} \sim 0.02$ pc but smaller than that of the broad-emission line region, $R_{BELR} \sim 3$ pc, of this quasar.

We are planning to continue monitoring this quasar in order to follow the variability of this absorption line system, and derive the physical conditions in the absorbing gas. The constraints that we obtain from our observations will be very helpful in refining models of accretion disk winds.

REFERENCES

1. N. Murray, J. Chiang, S. A. Grossman, and G. M. Voit, Astrophysical Journal, **451**, 498–509 (1995)
2. E. J. Wampler, N. N. Chugai, and P. Petitjean, Astrophysical Journal, **443**, 586–605 (1995)
3. F. Hamann, T. A. Barlow, V. Junkkarinen, and E. M. Burbidge, Astrophysical Journal, **478**, 80–86 (1997).
4. D. Narayanan, F. Hamann, T. Barlow, E. M. Burbidge, R. D. Cohen, V. Junkkarinen, and R. Lyons, Astrophysical Journal, **601**, 715–722 (2004)
5. T. Misawa, M. Eracleous, J. C. Charlton, and A. Tajitsu, Astrophysical Journal, **629**, 115–130 (2005)

Calibration of Lick indices from SDSS spectra

C. Morossi[*], M. Franchini[*], P. Di Marcantonio[*], M.L. Malagnini[*,**]
and M. Chavez[***]

[*] INAF - Osservatorio Astronomico di Trieste, Via G.B. Tiepolo, 11, I-34131 Trieste, Italy
[**] Dip. di Astronomia, Università degli Studi di Trieste, Via G.B. Tiepolo, 11, I-34131 Trieste Italy
[***] Instituto Nacional de Astrofisica, Optica y Electronica, A.P. 51 y 216, 72000 Puebla, Mexico

Abstract. A method to calibrate the Lick indices directly measured from the Sloan Digital Sky Survey spectra into the Lick/IDS standard system is presented. The soundness of the derived calibration coefficients is checked by comparing indices from the Worthey Catalogue with those obtained from SDSS spectra.

Keywords: main sequence stars, late-type stars
PACS: 97.20.Jg

INTRODUCTION

One of the most helpful constraints on the models of the Milky Way formation and structure is given by the observations of abundance ratios and abundance gradients in the different stellar galactic components. Observations in extended zones of the Galaxy are now possible by using the huge amount of low and intermediate resolution spectra from the Sloan Digital Sky Survey (SDSS, [1]). To this aim, the use of the indices in the Lick/IDS system [2] seems very promising. In fact, these indices have been particularly successful in interpreting both low resolution spectra of cool stars and integrated spectra of galaxies. Unfortunately these indices were defined in spectra that were not flux calibrated and with a resolution three times lower than that achieved in the SDSS. In fact, even if the SDSS data processing provides indices measured in the Lick bands it is clearly stated that: "*The Lick standards are too bright to be observed by SDSS and therefore these indices are not on the Lick system in the classical sense, nor is any attempt made to smooth to Lick resolution*" (http://www.sdss.org/dr4/dm/flatFiles/spSpec.html#lineindex). Therefore, the indices measured from the SDSS spectra and provided in the fourth SDSS Data Release (DR4, [3]) cannot be used as-they-are. Actually, not only the SDSS spectra must first be degraded to the specific wavelength resolution of the Lick/IDS system but also a further calibration of the measured indices is needed to remove the differences between the flux calibrated SDSS spectra and the Lick ones.

In this paper we present our solution to the calibration problem. We adopt a two-steps strategy: *i)* we look for SDSS spectra similar to those in a flux calibrated database containing stars in common with [2]; *ii)* the Lick/SDSS indices of the above-selected stars are compared with the Lick/IDS ones of the similar stars.

THE CALIBRATION OF LICK/SDSS INDICES

Three observational data-sets are used to derive the set of calibration coefficients: the Worthey catalog [2], the low resolution ELODIE library (V3) [4], and the SDSS dataset. The first contains the standard stars of the Lick/IDS system and their measured indices. The second one is a collection of flux calibrated spectra from which spectral indices can be measured and transformed into the Lick/IDS system [5]. The SDSS data-set is extracted from [3], the fourth major data release containing 102,714 spectra classified as STARs. Since we were interested in finding mainly late F, G and early K stars belonging to the disk and to the halo, we selected among all the spectra classified as STARs those having $0.7 < (g-r) < 1.3$ mag and $(g+r)/2 < 21$ mag. The ELODIE library and the SDSS spectra have a spectral resolution higher than the Lick/IDS spectra used to compute the indices in the Worthey catalogue, namely $R_E=10,000$, $R_{SDSS}=1800$, and $R_{Worthey}=630$, respectively. Therefore, before computing indices in the Lick bands from the ELODIE library and from the SDSS data-set, we have corrected the original spectra for radial velocity and degraded them to the resolution of the Lick/IDS spectra.

At this point the derivation of the linear transformation coefficients between the ELODIE library indices and the Lick/IDS system ones is straightforward by means of the stars in common between the two data-sets. Unfortunately, the same approach cannot be taken as far as the SDSS indices are concerned since there are no stars in common between our SDSS sample and the Worthey catalogue. To transform SDSS indices into the Lick/IDS system we take advantage of the fact that both ELODIE and SDSS spectra are flux calibrated. In fact, we can look for pairs of SDSS and ELODIE stars formed by very similar objects ("twin stars") by using ELODIE stars present in the Worthey Catalog. Then, the actual calibration can be done by associating the measured indices of the SDSS twin stars to those given in the Worthey Catalog for the ELODIE twin stars. Out of the 1391 ELODIE stars we found 43 stars which have Worthey Lick indices and twin SDSS objects. Figure 1 shows an example of the similarity in the spectra of twin stars in a broad wavelength range and in the wavelength regions of the G band, Hβ, and Mg indices. Figure 2 illustrates the linear regression fit used to calibrate the four indices used in [5], [6], [7] to discriminate between the Solar Scaled composition stars and the α-enhanced ones. A check on the validity of the calibration has been performed by comparing the overall behavior of the calibrated Lick/SDSS indices with that of the Lick/IDS ones. In order to have comparable samples we use only those SDSS objects having a distance lower than 3500 pc, an absolute height from the galactic plane less than 700 pc, and $0.52<(B-V)_0<1.15$ mag. The good agreement between the Lick/SDSS and Lick/IDS loci makes us confident about the soundness of our calibration procedure. In the near future, we have already planned observations of few bright SDSS stars and of Worthey standards at the G. Haro Observatory in Cananea (Mexico) in order to achieve a more quantitative estimate of the goodness of our results. An example of the possible use of these calibrated Lick/SDSS indices to infer the properties of different components of our Galaxy is shown in the contribution presented at this Symposium by Franchini et al. [8].

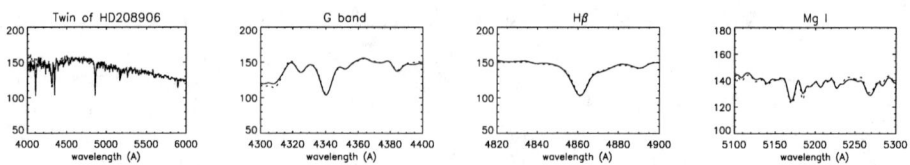

FIGURE 1. Comparison of SDSS (solid line) and ELODIE (dotted line) spectra of twin stars.

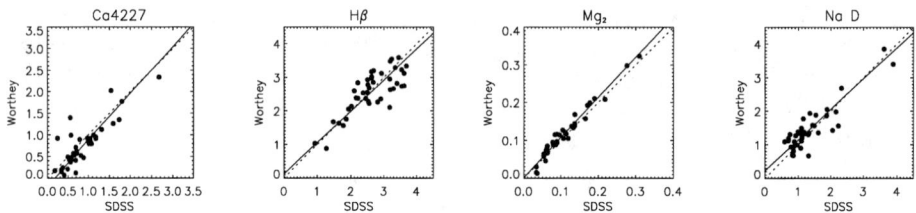

FIGURE 2. Comparison of Lick/SDSS and Lick/IDS index values of pairs of SDSS and ELODIE twin stars. Regression (solid line) and 45 degree lines (dashed line) are superimposed.

ACKNOWLEDGMENTS

This work received partial financial support from the Mexican CONACyT via grant 36547-E from MIUR COFIN-2003028039.

Funding for the creation and distribution of the SDSS Archive has been provided by the Alfred P. Sloan Foundation, the Participating Institutions, the National Aeronautics and Space Administration, the National Science Foundation, the U.S. Department of Energy, the Japanese Monbukagakusho, and the Max Planck Society. The SDSS Web site is http://www.sdss.org/. The SDSS is managed by the Astroph. Research Consortium (ARC) for the Participating Institutions. The Participating Institutions are The University of Chicago, Fermilab, the Institute for Advanced Study, the Japan Participation Group, The Johns Hopkins University, the Korean Scientist Group, Los Alamos National Laboratory, the Max-Planck-Institute for Astronomy (MPIA), the Max-Planck-Institute for Astrophysics (MPA), New Mexico State University, University of Pittsburgh, University of Portsmouth, Princeton University, the United States Naval Observatory, and the University of Washington.

REFERENCES

1. D. G. York et al, 2000 AJ,120, 1579
2. G.Worthey, S.M. Faber, J. J. Gonzalez, & D. Burstein 1994, *ApJ*, **94**, 687
3. J. K. Adelman-McCarthy, et al., 2005, astro-ph/0507711
4. J. Moultaka, S. A. Ilovaisky, P. Prugniel, & C. Soubiran 2004, PASP 116, 693
5. M. Franchini, C. Morossi, P. Di Marcantonio, et al., 2005, *ApJ*, **634**, 1319-1335
6. M. Franchini, C. Morossi, P. Di Marcantonio, et al. 2004a, *ApJ*, **601**, 485
7. M. Franchini, C. Morossi, P. Di Marcantonio, et al. 2004b, *ApJ*, **613**, 312
8. M. Franchini, C. Morossi, P. Di Marcantonio, et al., 2005b, Poster No. 9, this Symposium

Self-bound object with kaon condensates as a baryonic dark matter

Takumi Muto

Department of Physics, Chiba Institute of Technology
2-1-1 Shibazono, Narashino, Chiba 275-0023, Japan

Abstract. A self-bound object, which is composed of hyperon-mixed neutron-star matter with kaon condensates, is discussed as a candidate of baryonic dark matter. A kaon-condensed nucleus is obtained in a liquid-drop picture, and its relation to deeply bound kaonic nuclei is mentioned.

Keywords: kaon-baryon interactions, kaon condensation, neutron stars, baryonic dark matter
PACS: 05.30.Jp, 13.75.Jz, 26.60.+c, 95.35.+d

INTRODUCTION

Kaon condensation as a possible new hadronic phase in neutron stars has been extensively studied. Recently, we have considered interplay of kaon condensation with *hyperonic matter*, where hyperons (Λ, Σ^-) are mixed in addition to the neutrons, protons and leptons in the ground state of neutron-star matter[1]. It has been shown that the equation of state (EOS) of the kaon-condensed phase becomes considerably soft owing to the (anti) kaon-baryon attractions in addition to the softening effect from mixing of hyperons. On the other hand, the attractive energy from kaon condensation has been shown to be saturated at high densities. As a result, there appears a local energy minimum at a high baryon density (a density isomer). Based on the EOS, we have obtained the structure of the kaon-condensed neutron star as a self-bound star and its implications for observations of mass-radius relations[1]. Such a self-bound object may exist in any scale in nature ranging from atomic nuclei to neutron stars. In this paper, we discuss a possible existence of a self-bound object with kaon condensates with a mass of the order of nuclear mass as a candidate of a baryonic dark matter.

EQUATION OF STATE

We take into account the *p*-wave kaon-baryon interactions as well as the s-wave ones as driving forces of kaon condensation by the use of $SU(3)_L \times SU(3)_R$ effective chiral Lagrangian. The classical kaon field is naturally assumed to be a plane wave type as $\langle K^- \rangle = (f/\sqrt{2})\theta e^{-i(\mu t - \mathbf{k}\cdot\mathbf{r})}$, where f (=93 MeV) is the meson decay constant, θ the chiral angle (the amplitude of the condensate), μ the kaon chemical potential which is equal to be the charge chemical potential owing to chemical equilibrium, $n \rightleftharpoons pK^-$, $n \rightleftharpoons pe^-(\bar{\nu}_e)$, and $\mathbf{k}$ the spatial momentum of the condensate. At the *KNY* vertex, a form factor $F = (\Lambda^2 - m_K^2)/(\Lambda^2 + \mathbf{k}^2)$ is introduced with a cut-off factor Λ (=1.2 GeV). After diagonalizing the baryonic part of the effective Hamiltonian, one obtains the quasibaryonic states which are given by superposition of the p and Λ states or the n and Σ^- states. We also take into account the nonrelativistic effective baryon-baryon interactions $\mathscr{E}_{pot}$[2], the parameters of which are determined from the saturation properties of nuclear matter and recent information from hypernuclear experiments. We introduce the baryon potentials V_i ($i = p, n, \Lambda, \Sigma^-$) such that $V_i = \partial \mathscr{E}_{pot}/\partial \rho_i$, where ρ_i is the number density of each baryon (i). The kaon-condensed ground state is obtained by occupation of quasiparticles $\tilde{p}$, $\tilde{\Lambda}$, $\tilde{n}$, and $\tilde{\Sigma}^-$ over each Fermi sea, with the charge neutrality condition.

The results for the EOS are summarized as follows: (In Fig. 1, the energy per baryon of the kaon-condensed phase in hyperonic matter, which corresponds to a case $A \to \infty$, is shown as a function of the baryon number density ρ_B by the dashed line for the Kn sigma term Σ_{Kn}=305 MeV.) Kaon condensation appears at $\rho_B \sim 0.60$ fm^{-3} as a p-wave condensate. The EOS with kaon condensates becomes very soft, which mainly stems from the s-wave attractive kaon-baryon interactions in addition to softening by mixing of hyperons. It should be noted that there appears a local energy minimum at $\rho_B = 1.206$ fm$^{-3} \sim 7.5\rho_0$ with ρ_0(=0.16 fm^{-3}) being the nuclear saturation density.

KAON-CONDENSED NUCLEUS

With the EOS, we obtain the kaon-condensed nucleus (KCN) with a baryon number A and uniform baryon number density ρ_B^0 inside the nucleus. We take a spherical liquid-drop picture. The free energy of the KCN at zero temperature is written as $F(\rho_B^0, A) = A[f_{vol}(\rho_B^0) - f_{vol}(\rho_B^0 = 0)] + F_{surf}(R_A, A)$, where $f_{vol}(\rho_B^0)$ is the volume part of the free energy per baryon, and $F_{surf}(R_A, A)$ (=$4\pi R_A^2 \sigma$) is the surface energy of the whole nucleus with radius R_A. The surface tension coefficient σ is defined as the difference between the total energy per unit plane with a density distribution $\rho_B(z)$ and that of the reference system inside which the density distribution is uniform (ρ_B^0) and vanishes at R_A :

$$\sigma(\rho_B^0) \equiv \int_{-\infty}^{\infty} dz f_{vol}(\rho_B(z))\rho_B(z) - \int_{-\infty}^{R_A} dz f_{vol}(\rho_B^0)\rho_B^0 . \quad (1)$$

For simplicity, the form of the density distribution $\rho_B(z)$ is assumed to be of the Fermi distribution function with a diffuseness parameter b= 0.2 fm.

In Fig. 1, we show the total free energies per baryon, F/A (the solid lines), and the surface energies per baryon, F_{surf}/A (the dotted lines), as functions of ρ_B^0 for A =10, 100, 1000. The volume part of the free energy per baryon (F/A in the limit of $A \to \infty$) is shown as a function of ρ_B^0 by the dashed line. There is a local minimum with the baryon number density $\rho_{B,min}^0(A) \sim 1.206$ fm^{-3}. The total free energy per baryon at $\rho_{B,min}^0(A)$ is mostly determined by the volume part f_{vol} only, as long as A is large enough ($A \gtrsim 100$). We summarize the properties of the KCN in Table 1. One finds that the KCN is a highly dense and compact object with a small radius R_A and a huge incompressibility K and that the saturation properties are ensured. The number of negative strangeness for the KCN is of $O(A)$, so that the KCN is expected to have a long lifetime (larger than the age of the universe), only decaying through the multiple weak processes. Towards further study, one should look into a collisionless property of the KCN with other particles.

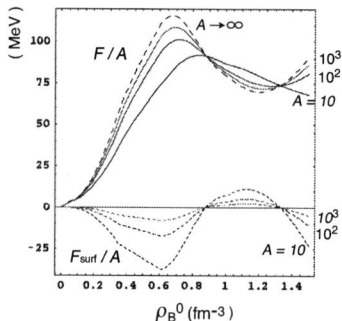

FIGURE 1. Total free energies per baryon, F/A (the solid lines) and the surface energies per baryon F_{surf}/A (the dotted lines), as functions of the uniform density ρ_B^0 inside the nucleus.

TABLE 1. Profile of a kaon-condensed nucleus.

A	ρ_B^0 (fm^{-3})	R_A (fm)	K ($\times 10^3$ MeV)	$|S|/A$
20	1.46	1.5	2.9	0.91
100	1.28	2.7	4.3	0.89
500	1.24	4.6	4.8	0.88
1000	1.23	5.8	5.9	0.88
∞	1.21	∞	6.9	0.88

SUMMARY AND CONCLUDING REMARKS

With regard to the structure of neutron stars, the EOS with a local energy minimum leads to a metastable (substantially stable) configuration of self-bound kaon-condensed star in addition to a usual kaon-condensed star with a two-phase structure obtained by the Maxwell's construction[1]. The KCN may be closely connected with the deeply bound kaonic nuclei which have been predicted theoretically and have been recently detected by some experiments [3, 4]. In order to create the KCN in a laboratory where the strangeness of the system is conserved, one should relax the chemical equilibrium conditions for the weak processes which have been imposed throughout this work.

ACKNOWLEDGMENTS

This work is supported in part by funds provided by Chiba Institute of Technology.

REFERENCES

1. T. Muto, Nucl. Phys. **A 697**, 225 (2002) ;*ibid* **A 754**, 350c (2005).
2. S. Balberg and A. Gal, Nucl. Phys. **A625**, 435 (1997).
3. Y. Akaishi and T. Yamazaki, Phys. Rev. **C 65**, 044005 (2002);
   A. Dote et al., Phys. Lett. **B 590**, 51 (2004).
4. T. Kishimoto, Nucl. Phys. **A 754**, 383c (2005). M. Iwasaki et al., nucl-ex/0310018. T. Suzuki et al., Phys. Lett. **B 597**, 167 (2004).

The metal enrichment of galaxies and galaxy clusters in the cold dark matter universe

M. Nagashima*, C.G. Lacey[†], T. Okamoto**, C.M. Baugh[†], C.S. Frenk[†] and S. Cole[†]

*Department of Physics, Kyoto University, Sakyo-ku, Kyoto 606-8502, Japan
[†]Department of Physics, Durham University, South Road, Durham DH1 3LE, England
**National Astronomical Observatory, Mitaka, Tokyo 181-8588, Japan

Abstract. We investigate the metal enrichment due to type II and Ia supernovae using semi-analytic models of galaxy formation based on the cold dark matter model of the Universe.

Keywords: Metal Enrichment, Formation of Galaxies, Large-scale Structure of the Universe
PACS: 98.62.Ai, 98.62.Bj, 98.65.-r

Recent observational and theoretical studies have strongly supported the cold dark matter (CDM) model, and its natural consequence is the hierarchical growth of cosmic structure, different from the monolithic collapse scenario. Therefore realistic models of formation and evolution of galaxies must be based on the CDM model. We have developed such hierarchical models of galaxy formation, which are often called as semi-analytic (SA) models. By extending the models to include metal enrichment due to type Ia supernovae (SNe Ia), we investigate the metal enrichment of the intracluster medium (ICM) and galaxies.

Firstly, we studied the metallicity distribution function (MDF) and [O/Fe] of disk stars in Milky Way-like galaxies and compared them with those for solar-neighborhood stars by extending a SA model developed by Nagashima & Yoshii[1]. To compare our model with a monolithic collapse model by Yoshii, Tsujimoto & Nomoto[2], we used the same chemical yields as those used by them, assuming Salpeter-type initial mass function (IMF) of stars. As shown in Fig.1, our SA model well reproduces the observed MDF and [O/Fe] as a function of [Fe/H][3].

Next, by extending a SA model developed by Cole et al.[4], we investigated the metal enrichment of the ICM and elliptical galaxies. We found that, as shown in Fig.2, if all stars form with an IMF similar to that found in the solar neighborhood (Kennicutt's IMF), then the metallicities of O, Mg, Si and Fe in the ICM are predicted to be 2-3 times lower than observed values. In contrast, a model in which stars formed in bursts triggered by galaxy mergers have a top-heavy IMF reproduces the observed ICM abundances of O, Mg, Si and Fe. The same model predicts ratios of ICM mass to total stellar luminosity in clusters which agree well with observations. According to our model, the bulk of the metals in clusters are produced by L_* and brighter galaxies[5].

We also found that the α-element abundance in elliptical galaxies is consistent with observed values only if the top-heavy IMF is used, as shown in Fig.3. This result is consistent with our previous study on the metal enrichment of the ICM. We also discuss the abundance ratio of α elements to iron as a function of velocity dispersion and

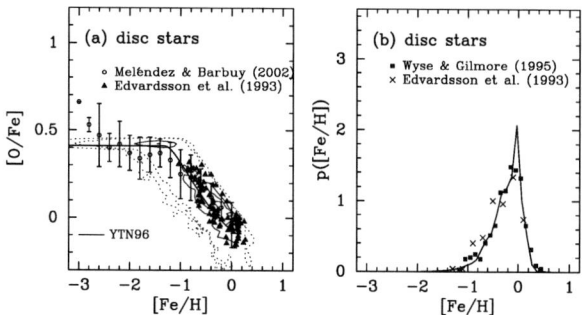

FIGURE 1. (a) [O/Fe] distribution against [Fe/H] for disk stars. The levels of contours drawn by the thin solid and dashed lines indicate 0.5, 0.4, 0.3, 0.2 and 0.1, and 0.02 and 0.005 times the largest number of stars in grids, respectively. The solid curve indicates a chemical enrichment model based on a monolithic cloud collapse model [2]. The circles with error bars and filled triangles denote observations. (b) Iron MDF for disk stars. The solid line indicates the model prediction for disk stars. The filled squares and crosses denote observations.

metallicity. We find that models with a top-heavy IMF match the α/Fe ratios observed in typical L_* ellipticals, but none of the models reproduce the observed increase of α/Fe with velocity dispersion.[6]

While top-heavy IMFs for elliptical galaxies have been suggested for a long time, for the first time, we showed that such an IMF is required to explain observations within the framework of the CDM cosmology in a statistical sense.

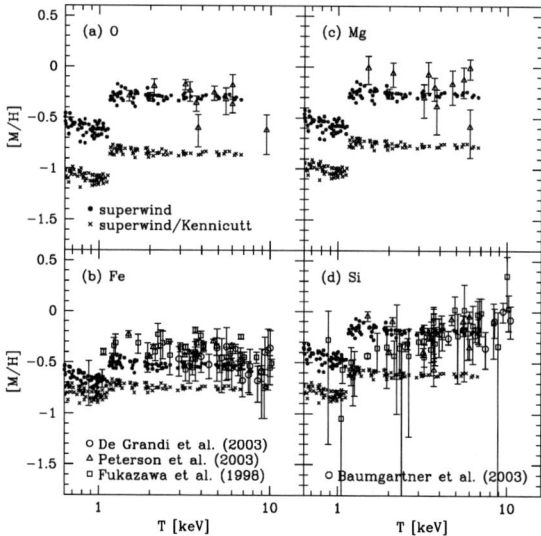

FIGURE 2. The metal abundances of the ICM. Each panel shows a different element: (a) [O/H] (b) [Fe/H] (c) [Mg/H] (d) [Si/H]. The predictions of the superwind model are shown by dots in the standard case where starbursts have a top-heavy IMF, and by crosses for the variant model in which all star formation takes place with a Kennicutt IMF. Open symbols denote observational data.

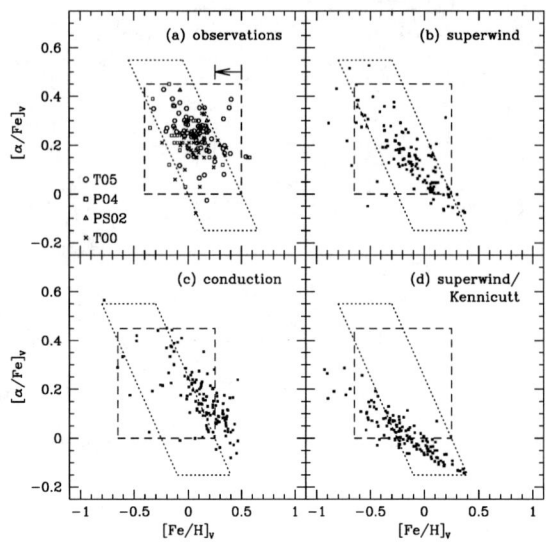

FIGURE 3. The α/Fe abundance ratio vs. Fe-abundance for stars in elliptical galaxies. (a) Observational data. Panels (b), (c) and (d) show predictions for the three different models. For reference, dashed boxes surrounding most of the observational data are plotted in all panels. In the panels showing the models, these boxes are shifted by the estimated central-to-global correction for metallicity gradients.

REFERENCES

1. M. Nagashima, and Y. Yoshii, *Astrophys. J.* **610**, 23–44 (2004).
2. Y. Yoshii, T. Tsujimoto, and K. Nomoto, *Astrophys. J.* **462**, 266–275 (1996).
3. M. Nagashima, and T. Okamoto, submitted to *Astrophys. J.*, astro-ph/0404486.
4. S. Cole, C. G. Lacey, C. M. Baugh, and C. S. Frenk, *Mon. Not. R. Astr. Soc.* **319**, 168–204 (2000).
5. M. Nagashima, C. G. Lacey, C. M. Baugh, C. S. Frenk, and S. Cole, *Mon. Not. R. Astr. Soc.* **358**, 1247–1266 (2005).
6. M. Nagashima, C. G. Lacey, T. Okamoto, C. M. Baugh, C. S. Frenk, and S. Cole, *Mon. Not. R. Astr. Soc. Letters* **363**, L31–L35 (2005).

Explosive nucleosynthesis inside/outside of the jet launched by a collapsar

S. Nagataki*, A. Mizuta[†] and K. Sato**

*Yukawa Institute for Theoretical Physics, Kyoto University, Oiwake-cho Kitashirakawa Sakyo-ku, Kyoto 606-8502, Japan
[†]Max-Planck-Institute für Astrophysik, Karl-Schwarzschild-Str. 1, 85741 Garching, Germany
**Department of Physics, The University of Tokyo, Bunkyo-ku, Tokyo 113-0033, Japan

Abstract. Two-dimensional hydrodynamic simulations are performed to investigate explosive nucleosynthesis in a collapsar using the model of MacFadyen and Woosley (1999). It is shown that ^{56}Ni is not produced in the jet of the collapsar sufficiently to explain the observed amount of a hypernova when the duration of the explosion is ~ 10 sec, which is considered to be the typical timescale of explosion in the collapsar model. Even though considerable amount of ^{56}Ni is synthesized if all explosion energy is deposited initially, the opening angles of the jets become too wide to realize highly relativistic outflows and gamma-ray bursts in such a case. Moreover, the synthesized ^{56}Ni can not be used to brighten the supernova, since most of ^{56}Ni exist in the jet component rather than in the supernova ejecta, and the jet becomes optically thin before considerable amount of ^{56}Ni decays. From these results, it is concluded that the origin of ^{56}Ni in hypernovae is not the explosive nucleosynthesis in the jet. We consider that the idea that the origin is the explosive nucleosynthesis in the accretion disk is more promising. We also show that the explosion becomes bi-polar naturally due to the effect of the deformed progenitor. This fact suggests that the ^{56}Ni synthesized in the accretion disk and conveyed as outflows are blown along to the rotation axis, which will explain the line features of SN 1998bw and double peaked line features of SN 2003jd. This feature will help the idea of the accretion disk mentioned above. We predict that some fraction of ^{56}Ni synthesized in the jet may show Lorentz boosted line profiles. That is, highly blue shifted (or red shifted) broad line features might be observed in the future. We show that abundance of nuclei whose mass number ~ 40 in the ejecta depends sensitively on the energy deposition rate, which is a result of active incomplete silicon burning and alpha-rich freezeout. So it may be determined by observations of chemical composition in metal poor stars which model is the proper one as a model of a gamma-ray burst accompanied by a hypernova.

Keywords: gamma rays: bursts — accretion, accretion disks — black hole physics — nuclear reactions, nucleosynthesis, abundances — supernovae: general — galaxy: halo
PACS: 97.60.Bw

REFERENCES

. A. I. MacFadyen, S. E. Woosley, *ApJ*, 1999, 524, 262
. S. Nagataki, S., *ApJS*, 2000, 127, 141

Light Elements Produced by Nitrogen-rich Type Ic Supernovae

Ko Nakamura*, Susumu Inoue†, Shinya Wanajo**, Takeru Suzuki‡ and Toshikazu Shigeyama*,**

Department of Astronomy, Graduate School of Science, University of Tokyo, Japan
†*National Astronomical Observatory of Japan, Mitaka, Tokyo, Japan*
**Research Center for the Early Universe, Graduate School of Science, University of Tokyo, Japan*
‡*Department of Physics, Kyoto University, Kyoto, Japan*

Abstract. We investigate energetic type Ic supernovae as a production site for ^{6}Li and other light elements in the early stage of the Milky Way.

Keywords: nuclear reactions, nucleosynthesis, abundances — supernovae; general
PACS: 26.30.+k, 26.40.+r, 97.20.Tr, 97.60.Bw, 98.35.Bd

INTRODUCTION

The amounts of light elements ^{6,7}Li, ^{9}Be, and ^{10,11}B (LiBeB) at present are thought to be the sum of products of the big bang nucleosynthesis [1] and subsequent processes. Among them cosmic-ray spallation reactions are thought to play an important role in ^{6}Li and ^{9}Be productions. In particular, recent observations of extremely metal-poor stars suggest that the primary spallation process dominated light-element nucleosynthesis in the early Galaxy [2]. Theoretically Fields et al. [3, 4], and later Nakamura & Shigeyama [5], investigated explosions of Type Ic supernovae (SNe Ic) as a site for this primary process. They considered the interactions between ejecta and the interstellar matter (ISM), assuming that ISM mainly consists of H (90%) and partly of He (10%) and implicitly that the interaction with circumstellar matter (CSM) could be ignored.

However, spallation reactions ^{12}C,^{16}O + ^{1}H,^{4}He → LiBeB+ fragment(s) are not the only source of ^{6}Li. We investigate a fusion reaction ^{4}He +^{4}He → ^{6}Li +··· in energetic SNe Ic as an alternative channel to ^{6}Li. This reaction will be significant in a conceivable case that there may remain a small amount of helium in the surface layers of progenitor stars and a remarkable fraction of CSM may be dominated by helium stripped from progenitor stars. Furthermore, Meynet & Maeder [6] and Meynet et al. [7] calculated evolutions of metal-poor massive stars taking account of effects of rotation-induced mixing and found a large amount of mass loss even in an extremely metal-poor star and the enhancement of nitrogen near the surface. If this nitrogen is accelerated after the shock breakout of SN explosion, a significant amount of beryllium will be produced through a spallation reaction ^{14}N +^{4}He → ^{9}Be +··· because of its low threshold and a high cross section at peak compared with other spallation reactions producing ^{9}Be (see Fig. 1a). Indeed, recent observations indicate that abundances of nitrogen and beryllium seem to be enhanced in some very metal-poor stars from which ^{6}Li is detected.

SUMMARY OF OBSERVATIONS

Some detections of ^{6}Li in metal-poor stars were reported in the last several years. Asplund et al. [8] observed 24 metal-poor halo stars and acquired Li isotopic abundances, in particular, ^{6}Li abundances in nine of their 24 stars at the $\geq 2\sigma$ significance level. A subdwarf LP 815-43 was the most metal-poor star ([Fe/H] = -2.74) with ^{6}Li detection in their sample. The most metal-poor star with ^{6}Li detection to date might be G 64-12 ([Fe/H]= -3.17) reported by Aoki et al. (in preparation). Primas et al. [9, 10] had already analyzed the Be abundances of these two stars and determined that $\log\varepsilon(\text{Be}) = -1.09 \pm 0.20$ for LP 815-43 and -1.10 ± 0.15 for G 64-12. The Be abundance of G 64-12 is considerably higher than that expected from previous measurements of Be in stars with similar metallicities, and it might be the case for LP 815-43. Israelian et al. [11] analyzed nitrogen abundances in 31 metal-poor stars and found that both LP 815-43 and G 64-12 are N rich. We adopt the values obtained by 1-D LTE analyses to ensure consistency.

CALCULATIONS AND RESULTS

We consider stars that have lost all of their H-rich layer and most of their He/N layer before explosion. Here we use an explosion model of a $\sim 15\,M_\odot$ star, originated from a main-sequence star with the mass $M_{\text{ms}} \sim 40\,M_\odot$ [12]. The explosion energy is assumed to be 3×10^{52} ergs corresponding to SN 1998bw. The mass of ejecta becomes $13\,M_\odot$, containing $10\,M_\odot$ oxygen. Accelerated ejecta consisting of ^{4}He, ^{12}C, ^{14}N, and ^{16}O (HeCNO) will collide with the circumstellar He and N stripped from the progenitor star and produce LiBeB through spallation reactions and He-He fusion (HeCNO + ^{4}He,^{14}N $\to$ LiBeB + $\cdots$). As an initial condition of the transfer equation for ejecta we use the energy distribution of C/O ejecta calculated by Nakamura & Shigeyama [5], which is shown in Figure 1a with the cross sections for some spallation reactions [13, 14]. The "thick target" approximation is used to estimate energy loss of accelerated ejecta assuming that circumstellar He is so thick that ejecta lose energy mainly by Coulomb collisions with free electrons and most of light elements are produced within CSM. Then the yields from nuclear reactions are calculated using the cross sections for spallation and fusion reactions given by Read & Viola [13] and Mercer et al. [14].

We target on the abundances of a very metal-poor star LP 815-43 from which ^{6}Li is detected with enhanced beryllium and nitrogen. It can be derived from observational data that $X_{^6\text{Li}}/X_\text{O} \sim 6.88^{+3.08}_{-3.22} \times 10^{-7}$ and $X_{^9\text{Be}}/X_\text{O} \sim 1.32^{+0.77}_{-0.49} \times 10^{-8}$ for this star. The mass $M_{\text{He,N}}$ of He/N layer and its mass fraction X_N of nitrogen are parameters changed to reproduce the observed ratios. Figure 1bc show our results. The yield of ^{6}Li increases until $M_{\text{He,N}} \sim 0.01\,M_\odot$, which corresponds to the threshold energy of the ^{4}He $+^4$He $\to$ ^{6}Li $+\cdots$ reaction (~ 11 MeV/A, see Fig.1a), and then becomes flat. It is insensitive to X_N as long as X_N is small because most of ^{6}Li is produced through He + He reaction. Only the line corresponding to the case of $X_\text{N} = 0.005$ is shown in Figure 1b. On the other hand, the yield of ^{9}Be, which is mainly produced through ^{12}C,^{16}O $+^4$He $\to$ ^{9}Be $+\cdots$ reaction when He/N envelope is deficient, strongly depends on X_N (Fig. 1c). Without nitrogen ($X_\text{N} = 0$) ^{9}Be yield rapidly decreases with $M_{\text{He,N}}$ since He + He reaction produces no

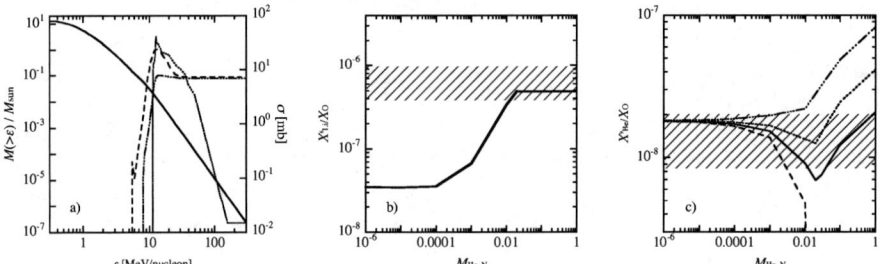

FIGURE 1. *a)* Energy distribution of ejecta (solid line) and the cross sections σ of reactions ^{4}He + ^{4}He → ^{6}Li (dotted), ^{14}N + ^{4}He → ^{9}Be (dashed), and ^{16}O + ^{4}He → ^{9}Be (dash-dotted). *b)* Mass ratio of ^{6}Li to O as functions of $M_{\text{He,N}}$ when $X_{\text{N}} = 0.005$. Shaded regions represent the observed ratios for LP 815-43 [8, 9] including the error. *c)* Same as *b)* but the cases for ^{9}Be with $X_{\text{N}} = 0.005$ (solid line), $X_{\text{N}} = 0.01$ (dotted), $X_{\text{N}} = 0.02$ (dash-dotted), and without N ($X_{\text{N}} = 0$; dashed) are shown.

^{9}Be. For small X_{N}, ^{9}Be yield once decreases because of the above reason, and then turns to increase since the cross section of ^{14}N + ^{4}He → ^{9}Be + ⋯ reaction has a peak around ∼ 13 MeV/A, which corresponds to ∼ 0.013 $M_\odot$ integrated from outside for SN 1998bw model (see Fig. 1a). Both $X_{^6\text{Li}}/X_{\text{O}}$ and $X_{^9\text{Be}}/X_{\text{O}}$ show good agreement with observational data of LP 815-43 when $M_{\text{He,N}} \sim 0.01 - 0.1\ M_\odot$ and $X_{\text{N}} \sim 0.005 - 0.01$. These are consistent with simulations of evolution for a metal-poor massive star with rotation: $X_{\text{N}} \sim 0.008$ for $M_{\text{ms}} = 40\ M_\odot$ (Hirschi et al., in preparation) and $X_{\text{N}} \sim 0.01$ for $M_{\text{ms}} = 60\ M_\odot$ [7]. We also consider G 64-12, though the abundance of ^{6}Li in it by Aoki et al. (in preparation) is a tentative one. For G 64-12 $X_{^6\text{Li}}/X_{\text{O}} \sim 1.89^{+1.40}_{-1.54} \times 10^{-6}$ and $X_{^9\text{Be}}/X_{\text{O}} \sim 2.04^{+0.84}_{-0.59} \times 10^{-8}$. We can reproduce the observed ratios with parameters analogous to those for LP 815-43.

We are grateful to Georges Meynet and Raphael Hirschi for valuable discussions. This work has been partially supported by the grant in aid (16540213, 17740108) of the Ministry of Education, Science, Culture, and Sports in Japan.

REFERENCES

1. Spite, F., & Spite, M. 1982, A&A, 115, 357
2. Duncan, D. K., Lambert, D. L., & Lemke, M. 1992, ApJ, 401, 584
3. Fields, B. D., Casse, M., Vangioni-Flam, E., & Nomoto, K. 1996, ApJ, 462, 276
4. Fields, B. D., Daigne, F., Cassé, M., & Vangioni-Flam, E. 2002, ApJ, 581, 389
5. Nakamura, K. & Shigeyama, T. 2004, ApJ, 610, 888
6. Meynet, G., & Maeder, A. 2002, A&A, 390, 561
7. Meynet, G., Ekstrom, S., & Maeder, A. 2005, preprint (astro-ph/0510560)
8. Asplund, M. et al. 2005, preprint (astro-ph/0510636)
9. Primas, F., Molaro, P., Bonifacio, P., & Hill, V. 2000a, A&A, 362, 666
10. Primas, F., Asplund, M., Nissen, P. E., & Hill, V. 2000b, A&A, 364, L42
11. Israelian, G. et al. 2004, A&A, 421, 649
12. Nakamura, T., Mazzali, P. A., Nomoto, K., & Iwamoto, K. 2001, ApJ, 550, 991
13. Read, S. M., & Viola, V. E. 1984, Atomic Data and Nuclear Data Tables, 31, 359
14. Mercer, D. J., et al. 2001, Phys. Rev. C, 63, 065805

Constraints on Brans-Dicke cosmology with a varying Λ term due to the big-bang nucleosynthesis and WMAP

Riou Nakamura*, Masa-aki Hashimoto* and Kenzo Arai[†]

*Department of Physics, Kyushu University, Fukuoka 810-8560, Japan
[†]Department of Physics, Kumamoto University, Kumamoto 860-8555, Japan

Abstract. We investigate the big-bang nucleosynthesis (BBN) in a Brans-Dicke model with a varying Λ term. We find that the cosmic expansion rate differs appreciably from that of the Friedmann model during the BBN epoch. The produced abundances of ^{4}He, D, and ^{7}Li are consistent with both the observed ones and the baryon density obtained from WMAP.

Keywords: nucleosynthesis; cosmology; isotope relative abundance; cosmic background radiation

INTRODUCTION

The big-bang nucleosynthesis (BBN) has succeeded in explaining the origin of the light elements ^{4}He, D, and ^{7}Li. Although the baryon density has been derived from the observations of the Wilkinson Microwave Anisotropy Probe (WMAP), the value seems to be inconsistent with the results of the standard BBN. Therefore, non-standard models of BBN has been proposed with the Friedmann model modified (e.g. decaying cosmological term [1], neutrino degeneracy [2], quintessence [3]).

For non-standard models, scalar-tensor theories have been investigated (e.g. [4]). For a simple model with a scalar field ϕ, it is shown that a Brans-Dicke generalization of gravity with torsion includes the low-energy limit string effective field theory. Related to the *cosmological constant problem*, a Brans-Dicke model with a varying $\Lambda(\phi)$ term (BDΛ) has been presented.

On the other hands, BBN has been studied in the BDΛ [5]. Therefore, it is worthwhile to check the validity of BDΛ related to the recent observations. In the present paper, we investigate the parameters in BDΛ how BBN can be reconciled with η from WMAP. A precise analysis of this paper have been performed in ref. [6].

BRANS-DICKE COSMOLOGY WITH A Λ TERM

In BDΛ model, the field equations are written as follows [5, 6]:

$$\left(\frac{\dot{x}}{x}\right)^2 = \frac{8\pi}{3\phi}\left(\rho_m + \rho_\gamma + \rho_{e^\pm}\right) - \frac{k}{x^2} + \frac{\Lambda}{3} + \frac{\omega}{6}\left(\frac{\dot{\phi}}{\phi}\right)^2 + \frac{\dot{x}\dot{\phi}}{x\phi} \tag{1}$$

$$\ddot{\phi} = \frac{1}{x^3}\left[\frac{8\pi\mu}{2\omega+3}\left(\rho_{m0}t + \int(\rho_e - 3p_e)dt\right) + B\right], \tag{2}$$

where x, k, ρ and p are the scale factor, the spatial curvature, the energy density and the pressure, respectively. Subscripts m, γ and $e^{\pm}$ means matter, photon+neutrino and electron+positron, respectively. Here the energy density of matter varies as $\rho_m = \rho_{m0} x^{-3}$ and the radiation density is $\rho_\gamma = \rho_{\gamma 0} x^{-4}$. B is the constant [5, 6]. Hereafter we use the normalized values $B^* = B/(10^{-24} \text{ g s cm}^{-3})$.

In BDΛ model, evolution of Λ-term and the gravitational "constant" G are described as

$$\Lambda = \frac{2\pi(\mu - 1)}{\phi} \rho_{m0} x^{-3}, \quad G = \frac{1}{2}\left(3 - \frac{2\omega + 1}{2\omega + 3}\mu\right)\frac{1}{\phi}. \quad (3)$$

It is noted that we have $G\phi < 0$ if $\mu > 3$ and $\omega \gg 1$.

The coupled equations (1)–(3) can be solved numerically with the physical parameters $\omega = 500$, $G_0 = 6.672 \times 10^{-8}$ dyn cm^2 g^{-2} and $H_0 = 71$ km s^{-1} Mpc^{-1} [7].

Figure 1 illustrates the evolution of the Hubble parameter ($\dot{x}/x$) in BDΛ and the Friedmann model in the early universe. During the epoch of $T > 10^9$ K, if $|B^*|$ increases, the expansion rate increases. However, around $T = 10^9$ K, the expansion rate in BDΛ crosses that of the Friedmann model, which will have noticeable effects on BBN.

BIG-BANG NUCLEOSYNTHESIS

We have performed the calculations of the primordial nucleosynthesis. For the BBN calculation, we adopt the observational abundances of ^{4}He [8], D/H [9] and ^{7}Li/H [10] as follows:

$$Y_p = 0.2391 \pm 0.0020, \quad \text{D/H} = 2.78^{+0.44}_{-0.38} \times 10^{-5}, \quad ^7\text{Li/H} = (2.19 \pm 0.28) \times 10^{-10}. \quad (4)$$

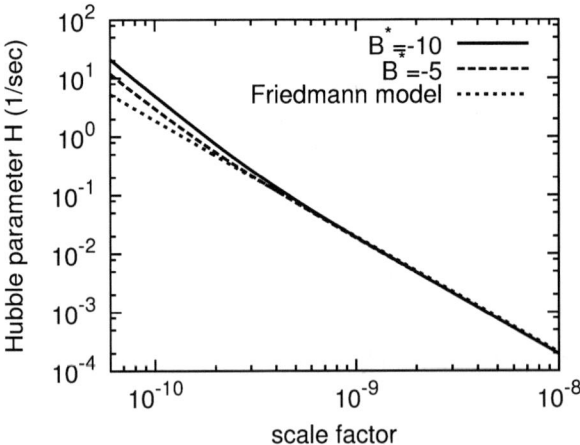

FIGURE 1. Expansion rates of the universe in BDΛ ($\mu = 0.7, \eta_{10} = 6.1$) and Friedmann model.

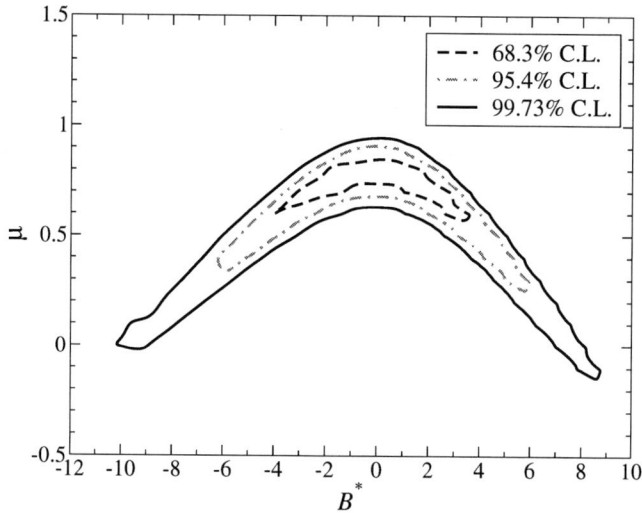

FIGURE 2. 1, 2 and 3σ confidence regions on the $\mu - B^*$ plane constrained from both BBN and η by WMAP

We find that the ^{4}He abundance is very sensitive to both B^* and μ. This is because changes of the cosmic expansion rate compared to the Friedmann model affect the primordial nucleosynthesis; the neutron-to-proton ratio is sensitive to the expansion rate.

Considering uncertainties in observational data, we calculate the χ^2-fitting to the observational abundances (4) and $\eta_{10} = 6.1^{+0.3}_{-0.2}$ obtained by WMAP [7]. Figure 2 shows $1\sigma, 2\sigma$ and 3σ confidence regions in the $\mu - B$ plane.

As the results, the 2σ confidence limits on the $\mu - B^*$ plane from BBN and η obtained by WMAP are $0.4 \leq \mu \leq 0.9$ and $-5.0 \leq B^* \leq 5.0$. While bounds on μ comes from ^{4}He, constraints on B^* comes from ^{7}Li. In our analysis, the Brans-Dicke model (for $\mu = 1$) seems to be inconsistent with the observational ^{4}He abundances and η by WMAP.

REFERENCES

1. M. Hashimoto, T. Kamikawa, and K. Arai, *Astrophys. J.*, **598**, 13–19 (2003)
2. M. Orito, T. Kajino, G. J. Mathews and Y. Wang, *Phys. Rev. D* **65**, 123504 (2002)
3. M. Yahiro, G. J. Mathews, K. Ichiki, T. Kajino and M. Orito, *Phys. Rev. D* **65**, 063502 (2002)
4. T. Fukui, K. Arai and M. Hashimoto, *Class. Quant. Grav.* **18**, 2087–2096 (2001).
5. K. Arai, M. Hashimoto & T. Fukui, *Astron. & Astrophys.* **179**, 17–22 (1987): T. Etoh, M. Hashimoto, K. Arai, & S. Fujimoto, *Astron. & Astrophys.* **325**, 893–897 (1997)
6. R. Nakamura, M. Hashimoto, S. Gamow and K. Arai , *Astron. & Astrophys.* **448**, 23–27 (2006)
7. C. L. Bennett et al., *Astrophys. J. Suppl.* **148**, 1–27 (2003)
8. V. Luridiana, A. Peimbert, M. Peimbert and M. Cervino, *Astrophys. J.* **592**, 846–865 (2003)
9. D. Kirkman, D. Tytler, N. Suzuki, J. M. O'Meara and D. Lubin, *Astrophys. J. Suppl.* **149**, 1–28 (2003)
10. P. Bonifacio et al., *Astron. & Astrophys.* **390**, 91–101 (2002)

Heavy Element Nucleosynthesis in the MHD Jet Explosions of Core-Collapse Supernovae

Nobuya Nishimura*, Masa-aki Hahimoto*, Shin-ichirou Fujimoto[†], Kei Kotake** and Shoichi Yamada**

Department of Physics, Kyushu University
[†]*Kumamoto National Collage of Technology*
**Science and Engineering, Waseda University*

Abstract. For the first time heavy element nucleosynthesis in the magneto-hydrodynamical (MHD) explosions of core-collapse supernovae are investigated using a massive star of $13 M_\odot$ in a main sequence stage. Contrary to the case of the spherical explosion, jet-like explosion due to the combined effects of the rotation and magnetic field lowers the electron fraction significantly inside the layers above the Fe-core. Then, the anisotropic shock waves pass through the oxygen rich layers. As a consequence, we find that the nucleosynthesis of the r- and p-process proceeds appreciably compared to the models previously considered.

Keywords: stars: nuclear reactions, nucleosynthesis, abundances - stars: evolution - stars: magnetic fields - supernova: general
PACS: 97.60.Bw

SUPERNOVA EXPLOSION MODEL

The presupernova model has been constructed from the evolution of He-core of $3.3 M_\odot$ that corresponds to $13 M_\odot$ in the main sequence stage [1]. We incorporate the effects of the rotation and the magnetic field by the following parameterized forms.

$$\Omega(X,Z) = \Omega_0 \times \frac{X_0^2}{X^2 + X_0^2} \cdot \frac{Z_0^4}{Z^4 + Z_0^4}, \quad B_\phi(X,Z) = B_0 \times \frac{X_0^2}{X^2 + X_0^2} \cdot \frac{Z_0^4}{Z^4 + Z_0^4} \quad (1)$$

where X and Z are the distances from the rotational axis and the equatorial plane with X_0 and Z_0 being model parameters. Under the condition of $T/|W| = 0.5\%$ and $E_m/|W| = 0.1\%$, we choose $X_0 = 100$ km and $Z_0 = 1000$ km as a strong differential rotation model. We perform the calculations of the collapse, bounce, and the propagation of the shock wave with use of ZEUS-2D in which the realistic equation of state has been implemented by Kotake et al. [2].

R-PROCESS NUCLEOSYNTHESIS

We use the full nuclear reaction network consists of about 4000 nuclear species up to $Z = 100$. This network includes two body reactions, i.e., (n,γ), (p,γ), (α,γ), (p,n), (α,p), (α,n), and their inverses and contains specific reactions such as three body reactions, heavy ion reactions and weak interactions. For nuclear masses, the experimental data

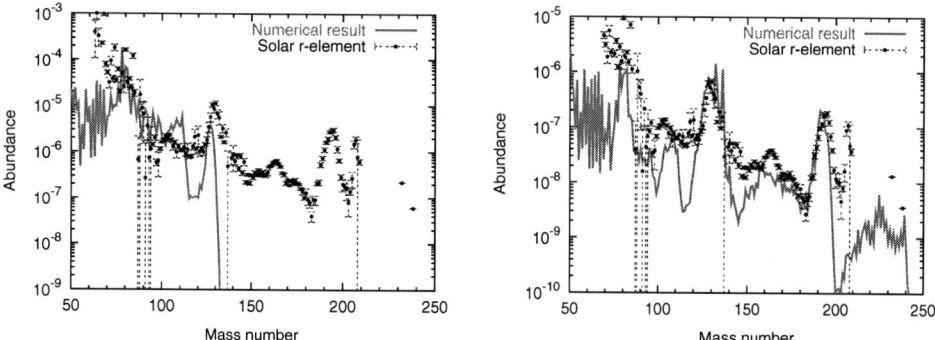

FIGURE 1. Abundances obtained from spherical explosion (left), and jet-like explosion (right).

is used if available; otherwise, the theoretical data by the mass formula FRDM is adopted. Most reaction rates are taken from the compilation (REACLIB) that includes experimental and theoretical data for the reaction rates and partition functions. The rates on decay channels, α-, $\beta^{\pm}$-decay, and β-delayed neutron emission, are taken from JAERI, which that includes experimental and theoretical decay rates of nuclei near the stability line. On the β-decay rates not available in JAERI are taken from FRDM. Fission is taken into consideration.

We calculate the r-process nucleosynthesis of ejected masses. For the high temperature area, NSE approximation is used. The calculation is executed by using a detailed nuclear reaction network when the temperature lowers. Low values of the electron fraction result in heavy element production. Figure 1 shows the comparison of the solar r-process abundances with obtained abundances. More details of the results are presented in our preprint [4].

JET PROPAGATION TO THE OUTER LAYER: P-PROCESS NUCLEOSYNTHESIS

It has been considered that the p-process occurs by the explosion in the layers where the peak temperature becomes $2 - 3 \times 10^9$ K. When the shock wave passes through the oxygen rich layer, the temperatures behind the shock reach these values. The nal products depend on the peak temperature. The MHD jet models are affected by the stellar model, and the directions of the jet determine the peak temperatures. Thus, p-process products must be different from those produced in the spherical explosions depending on the direction.

We use nuclear reaction network including about 2000 nuclides for the p-process calculations [5]. Figure 2 shows the overproduction factor relative to the solar abundances for two directions.

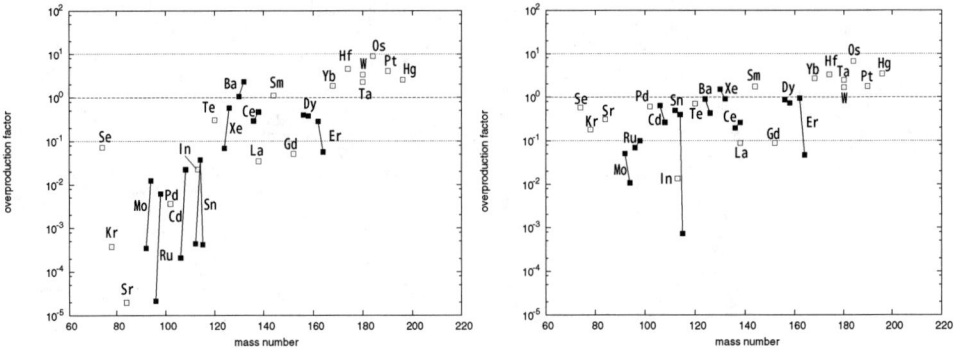

FIGURE 2. Overproduction factors from two jet MHD explotsion models. Polar direction (left) and 15 dgrees form the rotational axis (right).

CONCLUSION

We have investigated heavy element nucleosynthesis of both the r- and p-process during the magneto-hydrodynamical (MHD) explosions of supernovae in a massive star of 13 $M_\odot$ without neutrino effect. Contrary to the case of the spherical explosion, jet-like explosion due to the combined effects of the rotation and magnetic eld lowers the electron fraction signi cantly inside the Si-rich layers above the Fe-core. The ejected material of low electron fraction responsible for the r-process comes out from the Si-rich layer of the presupernova mode. This leads to the production up to the third peak in the solar r-process element.

Furthermore during the MHD shock wave propagation in the oxygen-neon layers, another kind of nucleosynthesis of the p-process occur inside the MHD jets. Until now, the p-process has been studied with use of spherical symmetric explosion models, where signi cant de ciencies in some *p*-elements compared to solar ones have been found. We have presented rst non-spherical effects on the production of p-elements using the MHD explosion model. We nd that the production of the *p*-elements depends on the jet direction signi cantly. We point out that there are variations in the heavy element nucleosynthesis such as the r- and p-process when the MHD effects play an important role in the supernova explosion.

REFERENCES

1. M. Hashimoto, *Prog. Theor. Phys*, 1995, 94, 663–736.
2. K. Kotake, S. Yamada, and K Sato, *ApJ*, 2003, 603, 242.
3. M. Rayet, M Arnould, M. Hashimoto, N. Prantzos and K. Nomoto, *A&A* **298**, 517–527 (1995).
4. S. Nishimura, K. Kotake, M. Hashimoto, S. Yamada, N. Nishimura, S. Fujimoto and K. Sato, *ApJ*, in press (astro-ph/0504100).
5. S. Fujimoto, M Hashimoto, O. Koike, K. Arai and R. Matsuba, *ApJ* **585**, 418–428 (2003).

Neutron-capture Nucleosynthesis in the He-Flash Convective Zone in Extremely Metal-Poor Stars

Takanori Nishimura*, Nobuyuki Iwamoto[†], Takuma Suda**,
Masayuki Aikawa[‡], Masayuki Y. Fujimoto* and Icko Iben Jr.[§]

*Graduate School of Science, Hokkaido University, Kita-ku, Sapporo 060-0810, Japan
[†]Nuclear Data Center, Japan Atomic Energy Agency, Tokai, Ibaraki 319-1195, Japan
**Meme Media Laboratory, Hokkaido University, Kita-ku, Sapporo 060-0810, Japan
[‡]Institut d'Astronomie et d'Astrophysique, Université Libre de Bruxelles, Campus de la Plaine, CP226, 1050 Brussels, Belgium
[§]Departments of Astronomy and of Physics, University of Illinois at Urbana-Champaign, Urbana, Illinois 61801 USA

Abstract. We investigate the nucleosynthesis in the helium flash convective zone, triggered by the hydrogen mixing, for extremely metal-poor stars of low and intermediate mass. Mixed hydrogen is converted into neutron through $^{12}C(p,\gamma)^{13}N(e^+v)^{13}C(\alpha,n)^{16}O$ and the doubly neutron-recycling reactions $^{12}C(n,\gamma)^{13}C(\alpha,n)^{16}O(n,\gamma)^{17}O(\alpha,n)^{20}Ne$ operate. In addition to oxygen and neon, not only light elements from sodium through phosphorus but also the s-process elements, heavier than iron, are synthesized via successive neutron captures with ^{20}Ne as seeds even in the stars originally devoid of metals. We follow the both the doubly neutron-recycling reactions and the s-process nucleosynthesis up to Pb and Bi by varying model parameters such as the amount of mixed ^{13}C. The resultant abundance patterns is shown to reproduce the observed enhancement not only of oxygen, the light elements but also Sr observed from HE 0107-5240 and HE 1327-2326.

Keywords: stellar evolution, nucleosynthesis, population III stars
PACS: 97.10.Cv;97.20.Wt

INTRODUCTION

During a past decade, known members of extremely metal-poor (EMP) stars in Galactic Halo have greatly increased in number and their surface abundances have been attracting wide interest in relation to the chemical evolution in early Universe and to the formation of galaxies. In particular, recently discovered HE 0107-5240 and HE 1327-2326 attract lots of attention because of the iron-poorest iron abundances of Fe H] ≤ -5, which are by an order of magnitude smaller than other halo stars observed to date. If they belong to the first generation stars, they should carry information about the physical conditions when they were born. At present they display a peculiar surface abundance pattern, which suggests that their surface abundances have suffered modifications due to mechanism(s) different from those in the stars of younger populations.

It is known that EMP stars of Fe H] $\leq -2\,5$ exhibit a unique feature in their evolution during the core helium flash at the tip of RGB and during helium shell flashes in early phase of AGB; the convection driven by helium flash extends into the hydrogen containing layer. Mixed hydrogen is carried inward by convection and captured by ^{12}C on the half way, and ^{13}C thus formed is carried further inward to produce neutrons

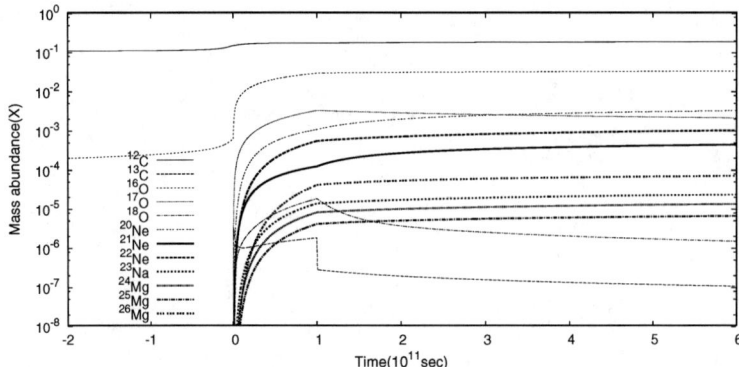

FIGURE 1. Time variations of the light elements abundances for the mixing amount $^{13}C/^{12}C = 0.02$ and the mixing interval 10^{11} seconds. Horizontal axis is the elapsed time from the peak of helium shell flash.

through the reaction $^{13}C(\alpha,n)^{16}O$. We investigate the nucleosynthesis, triggered by this hydrogen mixing, under the condition that metals are initially absent in order to explore the possibility that the characteristic abundance patterns, observed from the recently discovered iron-poorest stars, can be explained by the nucleosynthesis in the helium flash convective zone in population III stars.

Method and Approximations

After reaching the AGB, the helium shell burning becomes unstable and gives rise to recurrent shell flashes. As the strength of shell flashes grows, convection, driven by the shell flashes, extends outward to reach the hydrogen-containing layer. During the recurrence, the amount of hydrogen engulfed into the helium convective zone increases, and eventually, becomes large enough to cause the split of convective zone into two, the upper zone driven by burning of mixed hydrogen and the lower by helium burning. Then, the upper zone occupied by hydrogen convection expands greatly during the decay phase, into which the surface convective zone penetrates to bring out the carbon and other nuclear products. Once this helium-flash deep mixing (He-FDDM) enriches the surface with CNO elements, hydrogen mixing no longer occurs.

We compute the nucleosynthesis, triggered by hydrogen mixing, during helium shell flashes preceding He-FDDM. In our computations, we adopt an analytical approximation to the progress of helium shell flashes. The nucleosynthesis is followed by two nuclear networks; the synthesis of lighter nuclei from hydrogen and neutron through ^{35}S is computed by a nuclear network of 65 nuclei via proton-, neutron-, and alpha-captures, beta decays and their inverse reactions, and by using the production rate of ^{34}S and time variation of neutron density, the synthesis of heavier elements via neutron captures is pursued by another network of 926 nuclei up to Bi. We deal with the mixing of ^{13}C, instead of hydrogen, at a given rate and for a given interval from the peak of helium

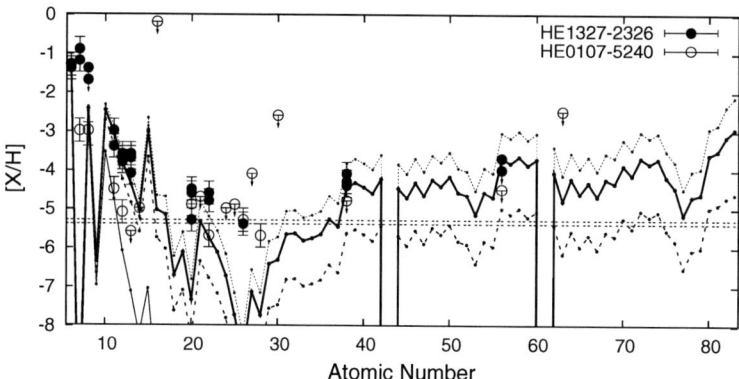

FIGURE 2. Resultant abundances of nucleosynthesis in the helium flash convection, triggered by hydrogen mixing, and comparison with observed abundances of HE 1327-2326 and HE 0107-5240 (horizontal lines denote their [Fe/H] values). Lines plot the results for models of $^{13}C/^{12}C = 0.03$ (dotted), 0,02 (thick solid), 0.01 (broken), and 0.001 (thin solid) from top to bottom; the abundances are normalized with respect to the observed carbon abundance of HE 1327-2326.

shell flash; the amount of mixed ^{13}C is taken to be $^{13}C/^{12}C = 0.0001$ to 0.07 relative to the abundance of ^{12}C in the helium convective zone.

Results and Discussion

Under extremely metal-poor conditions and at high temperatures achieved in the helium-flash convection, neutrons are captured by the exceedingly most abundant carbon and enter into the doubly neutron-recycling reactions; first into $^{12}C(n,\gamma)^{13}C(\alpha,n)^{16}O$, and then, into $^{16}O(n,\gamma)^{17}O(\alpha,n)^{20}Ne$. Figure 1 illustrates the time variations of light elements abundances for $^{13}C/^{12}C = 0.02$. We see that ^{16}O is produces much more than mixed ^{13}C, indicative of efficient recycling (about 10 times) of the first reactions. Also seen is the overproduction of ^{20}Ne relative to mixed ^{13}C, but the efficiency of recycling is slightly smaller since neutron capture reactions take place with ^{20}Ne as seeds.

Figure 2 summarizes the resultant abundances. The production of ^{16}O is only weakly dependent on the amount of mixed ^{13}C. On the other hand, the progress of neutron captures strongly depends on the latter, and for $^{13}C/^{12}C >\sim 0.01$, heavy s-process elements can be synthesized with including the heaviest elements of Pb and Bi even if stars are originally devoid of metals. Our model of $^{13}C/^{12}C = 0.02$ can reproduce, in good agreement, the observed abundances for light elements, such as Na, Mg, Al, and also for Sr from He 1327-2326 with including nitrogen produced by He-FDDM; our prediction on the Sr/Ba ratio is also consistent with the lower bound set by the observation, which increases with the temperature in the helium shell flashes. The abundances of HE 0107-5240 can also be explained with the dilution relative to ^{12}C, which increases via third dredge-up.

Nucleosynthesis by Type Ia Supernova for different Metallicity

Takuya Ohkubo*, Hideyuki Umeda*, Ken'ichi Nomoto* and Takashi Yoshida[†]

*Department of Astronomy, School of Science, The University of Tokyo, 7-3-1 Bunkyo-ku Hongo Tokyo Japan
[†]Astronomical Institute, Graduate School of Science, Tohoku University, Sendai 980-8578 Janan

Abstract. We calculate nucleosynthesis by type Ia supernova for various metallicity. We adopt two typical hydrodynamical models, carbon deflagration and delayed detonation. The two main points of this research are to see that (1)how the ejected mass of ^{56}Ni changes and (2)how abundance of each element (especially Fe-group elements) is influenced by varying metallicity. We find that (1)^{56}Ni mass changes about 15% in the range of $Z = 0.001 - 0.05$ and insufficient to explain all of the observed variety of SNe Ia peak luminosity, and (2)[Mn/Fe] and [Ni/Fe] show fairy dependence on metallicity (especially for delayed detonation model) while [Cr/Fe] or [α/Fe] do not.

Keywords: Nucleosynthesis

INTRODUCTION

Type Ia supernovae (SNe Ia) are observationally identified by the lack of H absorption in their spectra. Among several proposed physical models of SNe Ia, Chandrasekhar mass white dwarf model is accounted best based on the observational features such as photometric and spectroscopic features in early phases [1, 2] and is widely adopted for many applicational research.

Two major important meanings of researching SNe Ia are as follows. (1) They are used as the standard candles to measure distances for the cosmological research by using their luminosities (ejected ^{56}Ni mass — $M(^{56}$Ni) hereafter), because they have only small variations in their absolute magnitudes, shape of light curves, and spectra. (2) They are the major contributors of Fe-group elements such as Mn, Co, Fe, Ni in later phase of galactic chemical evolution.

For (1), there are some diversity in peak luminosities found from the observations of SNe Ia recently. For nearby SNe Ia, this is about 0.5 magnitude and $M(^{56}$Ni) is identified to be 0.4 – 0.7 M [3, 4, 5]. This means that SNe Ia are not completely standard candles, hence empirical laws such as distance indication need to be revised and we should investigate theoretically this diversity in detail the origin of this diversity.

For (2), the metallicity range in which SNe Ia begin to contribute to the galactic chemical evolution is a problem to be discussed. Generally, this is determined from the turning point of metallicity where the values [α/Fe] begin to decrease (in low metallicities [α/Fe] 0 4 for galactic bulge, disk stars, and dwarf spheroidal galactic stars). However, there are many abundance patterns which are not explained by the combination of conventional yields of SNe Ia and SNe II. While metallicity dependent SN II yields are proposed recently by [6], metallicity dependent SN Ia yields have not been researched systematically. The varieties of ejected Fe-group elements are expected as well as M(^{56}Ni), so we need to evaluate metallicity dependent SNe Ia yields.

METHOD

There are two major hydrodynamical models for explosion of WD; Carbon Deflagration model and Delayed Detonation model. The former is that the flame front propagates at a subsonic speed in the whole range [7] -W7. The latter one is that the deflagration wave is accelerated and changes to detonation [8, 9, 10, 11, 12] -CDD.

we calculate explosive nucleosynthsis of SNe Ia for different metallicities by using the two hydrodynamical models above. We take hydridynamical data (temperature and density history) by [7] for W7, and [12] for CDD. For initial chemical composition, we take the results of evolution by [12]. Although there are multi-dimensional hydrodynamical approach is proceeding [13, 14], we performed on sherical symmetry, because this research is focused on the nuclear feature by post process depending on metallicity.

RESULTS

What we find in this research are summarized as follows.

The ejected mass of ^{56}Ni, which determines the peak luminosity of a SN, varies about 16 % in the range of $Z = 0.001 - 0.05$ (Figure 1). This corresponds to the variation of peak luminosities $\triangle M \sim 0.2$. This is insufficient to explain all the diversity by observation, $\triangle M \sim 1.0$. In order to reproduce this diversity only by metallicity difference, it has to cover $Z = 0.001 - 0.2$, which corresponds to [Fe/H] $= -1.3 - 1.0$. This conclusion corresponds to the results of [15], who performed the similar calculation for W7. The difference of the absolute ^{56}Ni mass between theirs and ours is because we used updated electron capture rates by [16].

Among Fe-group elements, the yields of Ni and Mn strongly depend on metallicity (Figure 2). The extents of the dependences vary on different hydrodynamical models. Both [Ni/Fe] and [Mn/Fe] increase with metallicity through the difference of Y_e. The bulk of stable Ni is ^{58}Ni and only one stable Mn, ^{55}Mn, is originally synthesized as ^{55}Co. Since ^{58}Ni and ^{55}Co are neutron-rich, they are produced more in SNe Ia from smaller initial Y_e, which have metal-richer progenitors. Both [Ni/Fe] and [Mn/Fe] changes with metallicity for CDD models much more than for W7 model.

[Cr/Fe] is little affected by metallicity. Most of stable Cr is ^{52}Cr, which is originally produced as ^{52}Fe. ^{52}Fe is a α nucleus, so its abundance behaves similar to ^{56}Ni. The Cr yield strongly depends on hydrodynamical models. It is oversolar for CDD1, nearly solar value for CDD2, and undersolar for W7. Precise observations of [Cr/Fe] may indicate SN Ia explosion models appropriate for reproducing chemical evolution.

Cu and Zn are very underubandant even if metallicity changes. Therefore, we attribute these two elements to SNe II. Similarly, Na and Al are little produced.

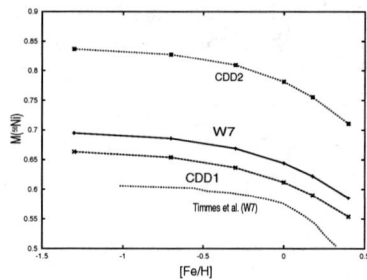

FIGURE 1. Ejected ^{56}Ni for each hydrodynamical model. The data by [15] for W7 are also plotted for comparoson

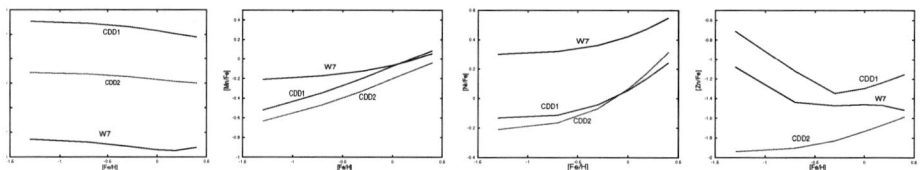

FIGURE 2. Mass fractions of each element (Cr and Mn, after decay) to ^{56}Fe normalized by solar values in logarithmic scale.

REFERENCES

1. P. Höflich, and A. Khokhlov, *ApJ* **457**, 500–528 (1996).
2. P. Nugent, E. Baron, D. Branch, A. Fisher, and P. H. Hauschildt, *ApJ* **485**, 812–819 (1997).
3. A. V. Filippenko, *ARA&A* **35**, 309 (1997).
4. P. A. Mazzali, E. Cappellaro, I. J. Danziger, M. Turrato, and S. Benetti, *ApJ* **499**, 49–52 (1998).
5. A. Saha, A. Sandage, G. A. Tamman, L. Labhardt, F. D. Macchetto, and N. Panagia, *ApJ* **522**, 802–838 (1999).
6. S. E. Woosley, and T. A. Weaver, *ApJ* **101**, 181–235 (1995).
7. K. Nomoto, F. K. Thielemann, and K. Yokoi, *ApJ* **286**, 644–658 (1984).
8. A. M. Khokhlov, *A&A* **245**, 114–128 (1991).
9. S. E. Woosley, and T. A. Weaver, *ApJ* **423**, 371–379 (1994).
10. K. Nomoto, et al. in *Thermonuclear Supernovae*, edited by P. Ruiz-Lapuente, R. Canal, and J. Isern, 1997, pp. 349.
11. K. Iwamoto, et al. *ApJ* **125**, 439–462 (1997).
12. H. Umeda, K. Nomoto, H. Yamaoka, and S. Wanajo, *ApJ* **513**, 861–868 (1999).
13. F. K. Röpke, M. Gieseler, M. Reinecke, C. Travaglio, and W. Hillebrandt, *A&A* **431**, 635–645 (2005).
14. V. N. Gamezo, A. M. Khokhlov, E. S. Oran, *ApJ* **623**, 337–346 (2005).
15. F. X. Timmes, E. F. Brown, and and J. W. Truran, *ApJ* **590**, L83–L86 (2003).
16. K. Langanke, and G. Martínez-Pinedo, *Nucl. Phys. A* **673**, 481–508 (2000).

SNe feedback and the formation of elliptical galaxies

Antonio Pipino* and Francesca Matteucci*

*Dipartimento di Astronomia, Università di Trieste

Abstract. The processes governing both the formation and evolution of elliptical galaxies are discussed by means of a new multi-zone photo-chemical evolution model for elliptical galaxies, taking into account detailed nucleosynthetic yields, feedback from supernovae, Pop III stars and an initial infall episode.

By comparing model predictions with observations, we derive a picture of galaxy formation in which the higher is the mass of the galaxy, the shorter are the infall and the star formation timescales. In particular, by means of our model, we are able to reproduce the overabundance of Mg relative to Fe, observed in the nuclei of bright ellipticals, and its increase with galactic mass.

This is a clear sign of an anti-hierarchical formation process. Therefore, in this scenario, the most massive objects are older than the less massive ones, in the sense that larger galaxies stop forming stars at earlier times.

Each galaxy is created outside-in, i.e. the outermost regions accrete gas, form stars and develop a galactic wind very quickly, compared to the central core in which the star formation can last up to ~ 1.3 Gyr. This finding will be discussed at the light of recent observations of the galaxy NGC 4697 which clearly show a strong radial gradient in the mean stellar $[<Mg/Fe>]$ ratio.

The role of galactic winds in the IGM/ICM enrichment will also be discussed.

Keywords: galaxies: ellipticals, lenticular and cD - galaxies: chemical evolution
PACS: 98.35.Ac, 98.35.Bd, 98.52.Eh, 98.56.Ew

INTRODUCTION

Any model of galaxy evolution presented so far had to overcome the strong challenge represented by the observational fact that elliptical galaxies show a remarkable uniformity in their photometric and chemical properties. Metallicity gradients are characteristic of the stellar populations inside elliptical galaxies. Evidences come from the increase of line-strength indices and the reddening of the colours towards the centre of the galaxies (for details and references see Pipino & Matteucci 2004, PM04). The study of such gradients provide insights into the mechanism of galaxy formation, particularly on the duration of the chemical enrichment process at each radius. Metallicity indices, in fact, contain information on the chemical composition and the age of the single stellar populations (SSP) inhabiting a given galactic zone. In particular, by comparing indices related mainly to Mg to others representative of the Fe abundance, it is possible to derive the [Mg/Fe] abundance ratio, which is a very strong constraint for the formation timescale of a galaxy. In fact, the common interpretation of the α-element (O, Mg, Ca, Si) overabundance relative to Fe, and its decrease with increasing metallicity in the solar neighbourhood is due to the different origin of these elements (time-delay model, Matteucci & Greggio, 1986), being the former promptly released by type II supernovae (SNII) and the latter mainly produced by type Ia supernovae (SNIa) on longer timescales. For a very short and intense star burst, the [α/Fe] ratios decrease at higher metallicity than in the solar vicinity. PM04 showed that a galaxy formation process in which the most massive objects form faster and more efficiently than the less massive ones can explain the photo-chemical properties of ellipticals, in particular the increase of [Mg/Fe] ratio in stars with galactic mass (see PM04). Moreover, from an extended analysis of metallicity and colour gradients, Pipino, Matteucci & Chiappini (2006, PMC) suggest that a single galaxy should form outside-in, namely the outermost regions form earlier and faster with respect to the central parts. A natural consequence of this model and of the time-delay between the production of Fe and that of Mg is that the mean [Mg/Fe] abundance ratio in the stars should increase with radius.

THE MODEL

The adopted chemical evolution model is based on that presented by PM04. In this particular case we consider our model galaxies as a multi-zone extending out to 10 effective radii, with instantaneous mixing of gas. Moreover we take explicitly into account a possible mass flow due to the galactic wind and a possible secondary episode of gas accretion in order to model late time gas accretion and/or interactions with the environment. The chemical code features a new self-consistent energy treatment which supersedes the previous one adopted by PM04 (see Pipino et al., 2005). Particular care is dedicated to a detailed calculations of Type Ia and II SN rates. The minimum SN efficiency required to develop a galactic wind is 10%.

RESULTS AND CONCLUSIONS

We find that SF and infall timescales decreasing with galactic mass are needed to explain the optical properties of elliptical galaxies (PM04). At the same time we reproduce the $L_X - L_B$ relation in the ISM of bright ellipticals (Pipino et al., 2005). Our best model satisfies the main constraint represented by the observed $[<\alpha/\text{Fe}>_V] - \sigma$ relation (see Fig. 1, left panel).

In Fig. 1 (right panel) we show the predicted radial trends of both the [< Mg/Fe >] (solid line) and [< Mg/Fe >$_V$] (dotted line) abundance ratios versus the *observed* one in NGC 4697. The latter is obtained by Mendez et al. (2005) by converting the line-strength indices into abundances. The agreement is remarkable, especially because we did not tune the input parameters (i.e. radius, mass) to exactly match NGC 4697. The observed increase of [Mg/Fe] with radius confirm PM04's model predictions, namely an outside-in formation process in which the central part of the galaxy form stars for a longer period compared to the most external regions. This can be explained in terms of galactic winds developing earlier where the local potential well is shallower.

By comparing the radial trend of [< Z/H >] with the *observed* one, we notice a discrepancy which is due to the fact that a CSP behaves in a different way with respect to a SSP. In particular the predicted gradient of [< Z/H >] is flatter than the observed one at large radii. Therefore, this should be taken into account when estimates for the metallicity of a galaxy are derived from the simple comparison between the observed

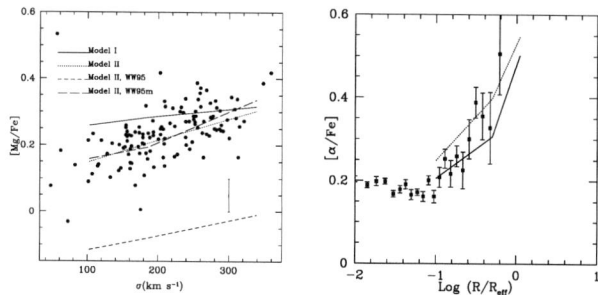

FIGURE 1. *Left:* [Mg/Fe] as a function of galactic velocity dispersion predicted by Model I (solid) and II (dotted) compared to the data from Thomas, Maraston, & Bender (2002). The typical error is shown in the bottom-right corner. For comparison we show the theoretical curves obtained with the same input parameters of Model II, but different yields (see text). WW95: yields by Woosley & Weaver (1995). WW95m: modified yields by Woosley & Weaver (1995), see PM04. *Right:* PM04's model IIb predictions for the mean mass-weighted [$<\alpha/\text{Fe}>$] (solid) and luminosity-weighted [$<\alpha/\text{Fe}>_V$] (dotted) abundance ratios in stars as a function of radius compared to the [α/Fe] derived for the galaxy NGC 4697 (Mendez et al. 2005, full squares).

line-strength index and the prediction for a SSP, a method currently adopted in the literature (see PMC).

The new energy formalism implemented in the chemical evolution code allows us to follow in a more detailed way the evolution of mass and energy flow into the ICM with respect to previous works. The predicted amount of Fe ejected by ellipticals into the ICM match the observations, and new data on the [α/Fe] ratios are in better agreement with our results (Pipino et al., 2005). Therefore, we confirm that SNe Ia are fundamental in providing energy and iron to the ICM.

ACKNOWLEDGMENTS

The work was supported by MIUR under COFIN03 prot. 2003028039. A.P. thanks the Organizers for having provided financial support for attending the conference.

REFERENCES

- Matteucci, F., & Greggio, L., 1986, A&A, 154, 279
- Mendez et al . 2005, ApJ, 627, 767
- Pipino, A., Matteucci, F., 2004, MNRAS, 347, 968 (PM04)
- Pipino, A., Matteucci, F., Chiappini, C., 2006, ApJ in press (astro-ph/0510556) (PMC)
- Pipino, A., Kawata, D., Gibson, B.K., Matteucci, F., 2005, A&A, 434, 553
- Thomas, D., Maraston, C., & Bender, R., 2002, Ap&SS, 281, 371
- Woosley, S.E., & Weaver, T.A., 1995, ApJS, 101, 181 (WW95)

Zinc Abundances in Metal-Poor Stars

Y.-j. Saito*, M. Takada-Hidai[†], Y. Takeda**, S. Honda** and M. Katsumata*

Department of Physics, Tokai University, Hiratsuka, Japan 25 9-1292
[†]*Liberal Arts Education Center, Tokai University, Hiratsuka, Japan 259-1292*
**National Astronomical Observatory, Mitaka, Tokyo 181-8588*

Abstract. We obtained high resolution (50000) spectra of 38 stars with $-3.0 <$ [Fe/H] ≤ 0 using HIDES at Okayama Astrophysical Observatory in order to clarify the behavior of zinc abundances and to obtain the clue to the origin of zinc. We estimated effective temperatures by the color indices based on the IRFM, and surface gravities by the basic relation based on luminosity and mass estimated from Hipparcos parallaxes and evolutionary tracks, on the HR diagram. Microturbulences and Fe abundances were determined from FeI and FeII lines, respectively. We measured the equivalent widths of ZnI lines at 4722.2, 4810.5 Å. We confirmed that [Zn/Fe] shows flat trend in the range of $-2.0 <$ [Fe/H] ≤ 0 with almost the solar value, and changes into an increasing trend at [Fe/H] $= -2.0$ with decreasing [Fe/H].

Keywords: Chemical evolution, Elemental abundance, zinc

INTRODUCTION

Zinc is one of very important elements because it is used as a tracer of metallicity in Damped Ly α systems. However the behavior of [Zn/Fe] to [Fe/H] is not clear, and the origin is not clearly understood either. A Si-burning is suggested to be the origin by Umeda & Nomoto (2002) and Cayrel et al. (2004), but the argument is also made about contribution of neutron capture process (especially s-process).

Behavior of zinc abundances ([Zn/Fe]) has attracted attention in these several years and various observational results have been reported, and we can divide their results roughly into two kinds of conclusions. One is flat trend with a solar value in $-2.5 <$ [Fe/H] < 0 (e.g., Sneden et al. 1991; Mishenina et al. 2002; Nissen et al. 2004). And another is increasing trend with decreasing [Fe/H] in the range of [Fe/H]< -2.0 (e.g., Blake et al. 2001; Johnson 2002; Cayrel et al. 2004).

We expect that [Zn/Fe] behavior changes from flat to an increase trend. However, the changing point ([Fe/H]) has never been reported. Therefore we investigate the behavior of zinc in the range of [Fe/H]$=-1.0$ to -1.5 to confirm the metallicity that [Zn/Fe] behavior changes to an increase trend.

OBSERVATIONS AND DATA REDUNCTION

High resolution spectra of 38 stars were obtained using the High Dispersion Echelle Spectrograph (HIDES) on the Okayama Astrophysical Observatory (OAO) 1.88-m telescope during October 2003 – April 2005. These spectra were obtained with a resolution of R=50000, to cover the wavelength range from 4650 Å to 5840 Å.

We executed data reductions by the standard procedure of echelle data using IRAF.

STELLAR PARAMETERS

The effective temperatures of the sample stars were estimated using the photometric data $(V - K)$ and [Fe/H], employing the empirical calibration obtained by Alonso et al.(1996a, 1999a), which is based on the infrared flux method (IRFM). According to Alonso et al.(1999a), the method of estimating the effective temperatures by using $(V - K)$ in some color indices has the lowest uncertainty, therefore, we estimated the T_{eff} by using the method.

We estimated the surface gravities following the standard procedures, based on data of T_{eff}, V magnitude, parallax, $E(B - V)$, bolometric magnitude, and the theoretical evolutionary track.

The microturbulent velocities, ξ, were estimated based on the condition that the Fe I abundances were constant even if equivalent widths changed.

ANALYSIS

We measured W_λ zinc lines at 4722.16Å, 4810.54Å. Both of the Zn I lines are very clear and unblended, and also strong enough to be measured in 38 all stars. All measurements of W_λ were carried out by Gaussian fitting, and the W_λ values were 12-120 Å(most were 40-60 Å). We measured the equivalent widths of Fe I and Fe II absorption lines between 4700 and 5700 Å to determine for the microturbulent velocities and the Fe abundances, respectively.

The analyses of Zn and Fe abundance were executed by using SPTOOL developed by Y. Takeda (2003, private communication). We adopted the gf-values and the lower excitation potentials (χ) by Biémont & Godefroid (1980) for zinc lines, and our adopting value are $\log gf = -0.390$, $\chi = 4.030$ eV for the 4722.16 Å line and $\log gf = -0.170$, $\chi = 4.080$ eV for the 4810.54 Å line. We selected the line list to determine the Fe I and Fe II abundances from Westin et al.(2000). The solar Zn value were adopted $\log \varepsilon$ (Zn)= 4.60 from Asplund et al.(2004)

RESULTS AND DISCUSSION

The [Zn/Fe] behavior of 38 our sample stars were shown in figure 1(left) as a function of [Fe/H]. As for our Zn abundances, we were able to confirm two different trends existed. One shows the trend that [Zn/Fe] behavior is flat trend with almost solar value in the range of $-2.0 <$ [Fe/H] ≤ 0. Other shows that [Zn/Fe] behavior is increasing trend with decreasing [Fe/H] in the range of [Fe/H] ≤ -2.0.

In the range of $-2.0 <$ [Fe/H] ≤ 0, [Zn/Fe] has large dispersion of approximately 0.4 dex. However the averaged value of [Zn/Fe] in this metallicity range is [Zn/Fe]=-0.02 ± 0.09, so that we should consider [Zn/Fe] in this metallicity range to be the solar value independent of iron abundances.

In the range of [Fe/H] ≤ -2.0, [Zn/Fe] values of all seven stars are more excessive than the solar value. As can be seen in figure 1(left), [Zn/Fe] behavior shows an increas-

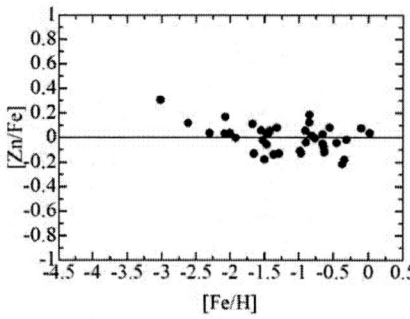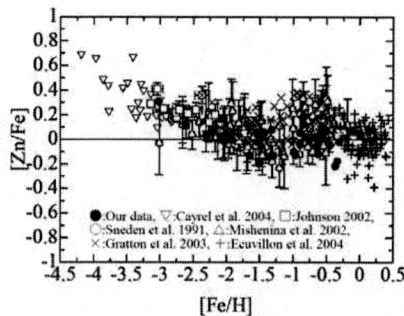

FIGURE 1. *Left*: Behavior of [Zn/Fe] in our 38 sample stars *Right*: Behaviors of [Zn/Fe] in our data and others including metal-rich stars.

ing trend with decreasing [Fe/H] in this metallicity range. In a word, it was confirmed that [Zn/Fe] became more excessive than the solar value as a boundary at [Fe/H]=-2.0 with decreasing [Fe/H], and keep increasing up to at least [Fe/H] ~ -3.0.

We showed the zinc abundance in metal-poor stars that had been examined before in figure 1(right) with our data. Our data and others were corresponding well for the abundance value and the behavior, so that we have understood an interpretation from our data can be trusted enough. In addition, the slope of our data in the range of [Fe/H] ≤ -2.0 is also corresponding to Cayrel et al.(2004) and Johonson(2002) well, therefore we were able to confirm [Zn/Fe] kept increasing up to [Fe/H] ~ -4.0 with a similar slope.

As a result of our investigation, the [Zn/Fe] has been understood it is the solar value in $-2.0 <$ [Fe/H] ≤ 0 and keeps doing enhancement in [Fe/H] ≤ -2.0. Such a large amount of zinc production can be explained by hypernovae suggested by Umeda & Nomoto (2005).

REFERENCES

(1) Alonso, A. et al. 1996a, A&A, 313, 873. **(2)** Alonso, A. et al. 1999a, A&AS, 140, 261. **(3)** Biémont, E., & Godefroid, M. 1980, A&A, 84, 361B. **(4)** Blake, L. A. J. et al. 2001, Nucl. Phys.A, 688, 502c. **(5)** Cayrel, R., et al. 2004, A&A, 416, 1117. **(6)** Chen, Y.Q. et al. 2002, A&A, 390, 225. **(7)** Girardi, L. et al. 2000, A&AS, 141, 371. **(8)** Johnson, J. A. 2002, ApJS, 139, 219. **(9)** Mishenina, T. V. et al. 2002, A&A, 396, 189. **(10)** Nissen, P.E. et al. 2004, A&A, 415, 993. **(11)** Schlegel, D. J. et al. 1998, ApJ, 500, 525. **(12)** Sneden, C. et al. 1991, A&A, 246, 354. **(13)** Takada-Hidai, M. et al. 2002, ApJ, 573, 614. **(14)** Takada-Hidai, M. et al. 2005, PASJ, 57, 347. **(15)** Takeda, Y. et al. 2002, PASJ, 54, 765. **(16)** Takeda, Y. et al. 2005, PASJ, 57, 751. **(17)** Umeda, H., & Nomoto, K. 2005, ApJ, 619, 427.

Astrophysics at RIA (ARIA) Working Group

Michael S. Smith[*], Hendrik Schatz[†], Frank X. Timmes[**], Michael Wiescher[‡] and Uwe Greife[§]

[*]Physics Division, Oak Ridge National Laboratory, Oak Ridge, TN, 37831-6354, USA
[†]National Superconducting Cyclotron Laboratory, Michigan State Univ., 1 Cyclotron Laboratory, East Lansing, MI, 48824-1321, USA
[**]Theoretical Division, MS B227, Los Alamos National Laboratory, Los Alamos, NM, 87545, USA
[‡]Dept. of Physics, Univ. Notre Dame, 225 Nieuwland Hall, Notre Dame, IN 46556-5670, USA
[§]Dept. of Physics, Colorado School of Mines, 1523 Illinois St., Golden, CO, 80401, USA

Abstract. The Astrophysics at RIA (ARIA) Working Group has been established to develop and promote the nuclear astrophysics research anticipated at the Rare Isotope Accelerator (RIA). RIA is a proposed next-generation nuclear science facility in the U.S. that will enable significant progress in studies of core collapse supernovae, thermonuclear supernovae, X-ray bursts, novae, and other astrophysical sites. Many of the topics addressed by the Working Group are relevant for the RIKEN RI Beam Factory, the planned GSI-Fair facility, and other advanced radioactive beam facilities.

Keywords: nuclear astrophysics, nucleosynthesis, radioactive beam, Rare Isotope Accelerator, RIA, nova, supernova, x-ray burst, neutron star
PACS: 26.30.+k, 26.50.+x, 26.60.+c, 95.30.-k, 97.10.Cv, 97.30.-b, 97.30.Qt, 97.60.Bw, 97.60.Jd, 97.80.Gm, 97.80.Jp, 98.38.Mz, 98.58.Mj

UNSTABLE NUCLEI IN ASTROPHYSICS

Unstable nuclei play an influential, and in some cases dominant, role in many exciting astrophysical phenomena [1]. From stellar explosions at the core of some massive stars - supernovae - to the surface of others - novae and X-ray bursts, from ultra-dense neutron stars to bloated red giant stars, and from the very early universe to the indicators of its current acceleration and eventual fate, information on unstable nuclei is needed to improve our understanding of the processes that shape our world. Some of the most exciting questions now being asked about the universe at its two extremes in length scales – the very large and the very small – are inextricably intertwined: What creates the elements from Iron to Uranium? Why do stars explode? What is the nature of neutron star matter?

These interdisciplinary questions have a broad appeal that captures the imagination of researchers and laymen alike. For example, the creation of the heavy elements was cited as one of the eleven most important questions for the next century by a National Academies study [2]. Recently, sophisticated and expensive platforms and missions such as the Hubble, Chandra, Spitzer, Sloan Digital Sky Survey, and others have provided incredibly detailed information on astrophysical phenomena over a wide range of wavelengths. Unfortunately, many theoretical models striving to explain these observations lack a firm empirical foundation because of the paucity of information on the properties of and reactions involving unstable nuclei.

In core collapse supernova, for example, it is important: to determine weak reactions on unstable nuclei near Fe [3]; to measure reactions that create and destroy long-lived radionuclides like ^{26}Al, ^{44}Ti, and ^{60}Fe that can help diagnose these explosions; and to measure the properties of and reactions on neutron-rich unstable nuclei that form heavy elements via the rapid neutron capture process [4]. In novae, it is important to measure capture reactions on proton-rich unstable nuclei to determine the energy generated and the nuclei synthesized in the outburst. This will help address qualitative issues in nova models such as the peak temperature, the heaviest nuclei synthesized, and the amount of material ejected [5]. For X-ray bursts, it is crucial to measure positron decay lifetimes and capture reaction rates on proton-rich unstable nuclei to make calculations of the X-ray output and subsequent neutron star evolution more realistic [6]. For thermonuclear supernovae, there is a need to determine electron capture rates on unstable nuclei to better understand explosion energetics and element synthesis [3], and the viability of these events as "standard candles" for cosmology research.

The Rare Isotope Accelerator (RIA) [7] is a proposed next-generation U.S. nuclear science facility that uniquely promises to provide access to the vast majority of nuclei in *all* astrophysical processes. RIA will usher in a new era where advances in observations, theory, *and nuclear science* synergistically improve our understanding of the cosmos. RIA will produce intense beams of unstable nuclei via three complementary techniques: fast beams via projectile fragmentation, reaccelerated beams by the Isotope Separator on Line (ISOL) technique, and a hybrid technique involving fragmentation, gas stopping, ionization, and reacceleration. There will be experimental halls devoted to Low Energy (less than 2 MeV/u), Medium Energy (less than 15 MeV/u), and Fast (greater than 15 MeV/u) beams. A schematic layout of the RIA facility is shown in Figure 1, while a possible arrangement of experimental equipment in the Low Energy Hall is shown in Figure 2. Astrophysics experiments will be pursued in all experimental halls.

ARIA WORKING GROUP

The Astrophysics at RIA (ARIA) Working Group was established to develop and promote the nuclear astrophysics research anticipated at the Rare Isotope Accelerator. Topics of discussion include the science motivation, types of experiments, facility issues, detection systems, the layout of experimental halls (e.g., Figure 2), as well as observational and theoretical needs for a successful program. Many of the issues are relevant for advanced radioactive beam facilities such as GSI-Fair and the RIKEN RI Beam Factory. We encourage membership of anyone interested in the scientific or technical aspects of the radioactive beam nuclear astrophysics program at RIA. For more information and to become a member, visit the working group website **ariaweb.org** [8].

ACKNOWLEDGMENTS

ORNL is managed by UT-Battelle, LLC, for the U.S. Department of Energy under contract DE-AC05-00OR22725. H. S. ackowledges support through NSF grants PHY 02-16783 and PHY 01-10253.

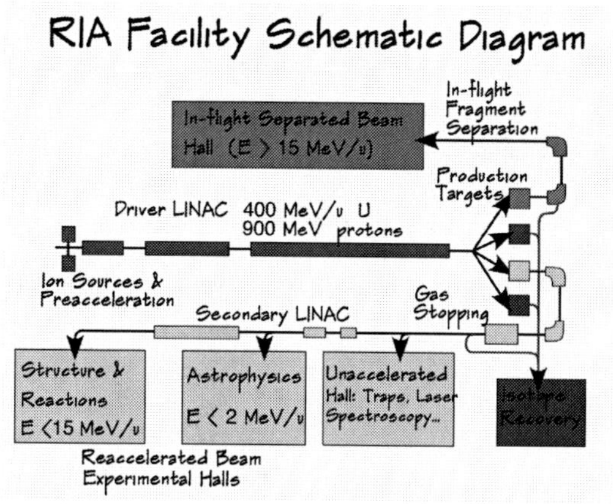

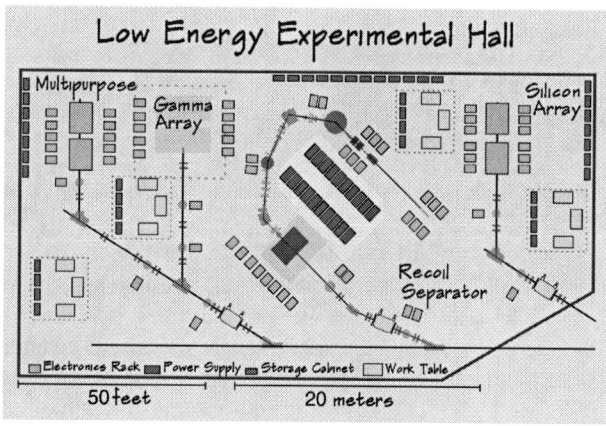

FIGURE 1. Schematic Diagram of the RIA Facility

FIGURE 2. Possible Low Energy Experimental Hall Layout at RIA

REFERENCES

1. M.S. Smith and K.E. Rehm, *Ann. Rev. Nucl. Part. Sci.*, **51**, 91–130 (2001).
2. M. Turner et al., "Connecting Quarks with the Cosmos: Eleven Science Questions for the New Century," *National Academies Press*, 2003.
3. K. Langanke, G. Martinez-Pinedo, *Rev. Mod. Phys.*, **75**, 819–862 (2003).
4. R. Surman, J. Engel, *Phys. Rev. C*, **64**, 035801 (2003).
5. J. Jose, in *Proc. Classical Nova Explosions, AIP Conf. Proc.* **637**, 104 – 113 (2002).
6. S.E. Woosley et al., *Astrophys. J. Suppl.*, **151**, 75 – 102 (2004).
7. http://www.orau.org/ria
8. http://ariaweb.org

Computational Infrastructure for Nuclear Astrophysics

Michael S. Smith*, Eric J. Lingerfelt*,†, Jason P. Scott*,†, Caroline D. Nesaraja*,†, W. Raphael Hix*, Kyungyuk Chae†,*, Hiroyuki Koura**, Richard A. Meyer‡, Daniel W. Bardayan*, Jeffery C. Blackmon* and Michael W. Guidry†,*

*Physics Division, Oak Ridge National Laboratory, Oak Ridge, TN, 37831-6354, USA
†Dept. Physics & Astronomy, Univ. of Tennessee, Knoxville, TN 37996-1200, USA
**Japan Atomic Energy Agency, Tokai, Naka-gun, Ibaraki 319-1195, JAPAN
‡RAME' Inc., Teaticket, MA, 02536, USA

Abstract. A Computational Infrastructure for Nuclear Astrophysics has been developed to streamline the inclusion of the latest nuclear physics data in astrophysics simulations. The infrastructure consists of a platform-independent suite of computer codes that is freely available online at **nucastrodata.org**. Features of, and future plans for, this software suite are given.

Keywords: nuclear astrophysics, thermonuclear reaction rates, nuclear data, visualization, abundances, nuclear masses, cross sections, s-factors, stellar explosions, simulations
PACS: 26.20.+f, 26.30.+k, 29.87.+g, 95.30.-k, 97.10.Cv, 97.30.Qt, 97.60.Bw, 97.80.Gm, 97.80.Jp

GENERAL DESCRIPTION

The **Computational Infrastructure for Nuclear Astrophysics** is a unique suite of computer codes, freely available online at **nucastrodata.org**, that is designed to greatly speed up the process of incorporating the latest nuclear physics results into astrophysical simulations and determining the astrophysical impact of these results. With a few mouse clicks, the suite enables creation, manipulation, and visualization of nuclear data sets as well as sample post-processing element synthesis calculations. The graphical interface enables users to easily employ all, or just a few, of the features, and save and export their results. The suite facilitates an online community where large datasets can quickly be shared with others, comments exchanged, and consensus on best results reached. Reaction rates are stored in the format of the most widely used rate collection [1], while the nucleosynthesis calculations utilize the reaction network code of Hix and Thielemann [2].

SOFTWARE FEATURES

The user-friendly interface enables users to:

- upload their cross section or s-factor data
- perform simple data evaluation tasks such as renormalizing or gain shifting

- extrapolate a cross section or s-factor with a linear function or with a theoretical cross section dataset
- compare theoretical nuclear mass models (their own or others) with other models or with experimental mass data (see Figure 1)
- examine separation energies of mass models and their relation to the r-process path
- create reaction rates from cross sections and s-factors
- calculate an inverse reaction rate by detailed balance
- scale up, down, or otherwise modify a reaction rate
- quickly find, plot, and compare all distinct rates for a given reaction
- generate reaction rates on a temperature grid and export
- read comments by experts on their latest reaction rates and post your own comments
- create, access, merge, and manage libraries of reaction rates
- replace old rates in a library with newer versions
- automatically generate a "recipe" of how a library was created for future reference
- share rates with others, or keep them private
- setup and run post-processing element synthesis calculations with new rate libraries
- compare simulations with different rate libraries
- vary input temperature / density histories and initial abundances in simulations
- plot simulation results – abundances and derivatives, reaction fluxes (Figure 2)
- visualize the simulation results with 2d animated nuclide charts (Figure 3)
- render movies and export for use in research and in presentations
- save your work in a dedicated disk space for future use
- share some or all of your work with colleagues around the world

FUTURE DEVELOPMENT

Features are continuously being added to this software suite, many in response to requests from users. In the future, there are plans to incorporate rate calculations from nuclear resonance parameters, theoretical cross section calculations, more simulation types, and more visualization routines.

ACKNOWLEDGMENTS

ORNL is managed by UT-Battelle, LLC, for the U.S. Department of Energy under contract DE-AC05-00OR22725.

REFERENCES

1. T. Rauscher, F.-K. Thielemann, *At. Data Nucl. Data Tables*, **75**, 1 – 351 (2000).
2. W.R. Hix, F.-K. Thielemann, *J. Comp. Appl. Math*, **109**, 321 (1999).

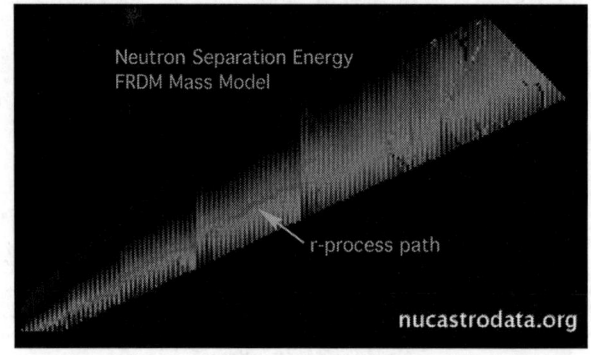

FIGURE 1. Neutron Separation Energy for the FRDM Mass Model

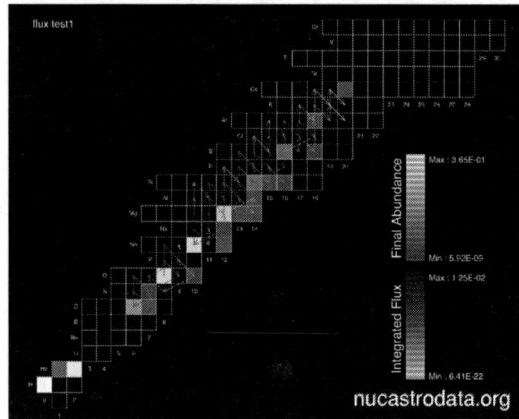

FIGURE 2. Integrated Reaction Flux and Final Abundances for a Nova Outburst

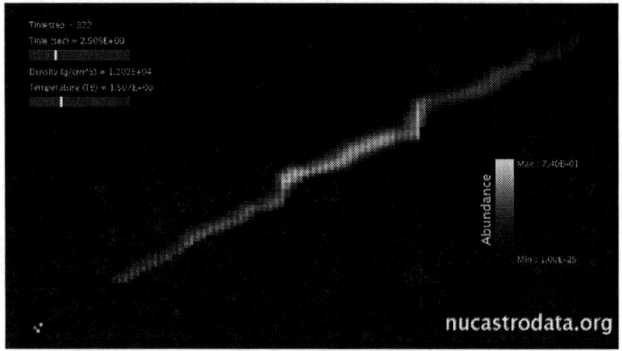

FIGURE 3. Abundances at one time-step in an r-process calculation

Fate of core-collapse supernovae: formation of neutron star and black hole

K. Sumiyoshi[*,†], H. Suzuki[**] and S. Yamada[‡]

[*]*Numazu College of Technology, Ooka 3600, Numazu, Shizuoka 410-8501, Japan*
[†]*National Astronomical Observatory of Japan, 2-21-1 Osawa, Mitaka, Tokyo 181-8588, Japan*
[**]*Faculty of Science and Technology, Tokyo University of Science, Yamazaki 2641, Noda, Chiba 278-8510, Japan*
[‡]*Science and Engineering & Advanced Research Institute for Science and Engineering, Waseda University, Okubo, 3-4-1, Shinjuku, Tokyo 169-8555, Japan*

Abstract. We study the fate of core-collapse supernovae from the massive stars by numerical simulations of neutrino-radiation hydrodynamics. We follow the long term evolution over 1 sec after the core bounce from the initial gravitational collapse to examine the explosion mechanism and to reveal the formation of neutron star and black hole. We explore the effects of equation of state (EOS) in these simulations by adopting the two sets of realistic EOS for supernovae. For the case of a $40M_\odot$ star, we find that the formation of black hole occurs in different timing depending on the softness of EOS and resulting properties of emergent neutrinos appear differently from ordinary supernova neutrinos.

Keywords: supernova, neutrino, black hole, neutron star, equation of state
PACS: 26.50.+x, 97.60.Bw, 97.60.Jd, 97.60.Lf

INTRODUCTION

Core-collapse supernovae originate from the gravitational collapse of massive stars having more than 10 solar masses ($M_\odot$). For the massive stars of $\sim$10–20$M_\odot$, supernova explosions occur due to the launch of shock wave by the bounce of core at high density, leaving a neutron star. Stars more massive than $\sim$20$M_\odot$ may have different fates since they usually have large iron cores. They are too massive to have the stellar explosion and the outcome will be the formation of black hole. The detection of neutrinos is clear and unique identification of such events and it is important to quantitatively predict the neutrino signals from the compact objects.

In order to to examine the explosion mechanism and to study the formation of neutron star and black hole, we perform numerical simulations of general relativistic neutrino-radiation hydrodynamics [1]. We follow the evolution from the beginning of gravitational core-collapse of $15M_\odot$ and $40M_\odot$ stars to explore the final outcome of massive stars. We evaluate the time profile of neutrino burst and the energy spectrum of neutrinos during the evolution. In order to determine the final fate, we perform the simulations for a long time scale ($\sim$1 s), which has not been explored before [1, 2]. For the case of $40M_\odot$ star, it is exciting to find when and how the black hole is formed from the massive star. In order to assess the influence of dense matter, we adopt two sets of EOS for the numerical simulations. We make comparisons of the evolutions of central core from the initial collapse up to the formation of compact object for the first time.

NUMERICAL SIMULATIONS

We perform numerical simulations by solving general relativistic neutrino-radiation hydrodynamics under the spherical symmetry [1]. We solve the Boltzmann equation for neutrinos together with the equations of hydrodynamics simultaneously. The initial models are taken from the iron core of the presupernova models by Woosley and Weaver [3]. Neutrino reaction rates are implemented as in the *standard* rates by Bruenn [4] together with the improved rates of pair processes and nucleon-nucleon bremsstrahlung.

We adopt two sets of equation of state (EOS) by Lattimer and Swesty (LS-EOS) [5] and by Shen et al. (SH-EOS) [6, 7]. The LS-EOS, which has been used as a *standard* in supernova simulations these years, is obtained by the extension of mass formula using the parameterized function of energy density. The SH-EOS, which is a new complete set of EOS, is constructed by the relativistic mean field (RMF) theory with a local density approximation. The RMF theory is based on the relativistic Brückner-Hartree-Fock theory [8] and is checked by the modern experimental data of neutron-rich unstable nuclei [9]. The characteristics of the relativistic SH-EOS as compared with the conventional LS-EOS, which is non-relativistic, appear mainly in the stiffness and the symmetry energy [10].

For the case of $15M_\odot$ star, we refer [1] to see the outcome of propagation of shock wave long after the bounce and the long term evolution of proto-neutron star. We have found that the influence of EOS appears clearly in late stage at ~ 1 sec after the bounce.

ν-SIGNAL FROM BLACK HOLE FORMATION

We report here the numerical results for the case of $40M_\odot$ star. After the core bounce, the shock wave is launched above 100 km and it immediately recedes down due to the dominance of accretion of material from the outer core of the massive star. The proto-neutron star is formed at center and its mass increases due to the accretion. The density and temperature of proto-neutron star increase due to the contraction according to the increasing mass. At 1.35 sec after the bounce, the proto-neutron star mass exceeds the maximum mass for hot and lepton-rich configurations determined by SH-EOS. The central part collapses dynamically and a black hole is formed.

The timing of the black hole formation is determined by the maximum mass for proto-neutron stars and depends substantially on EOS. In the case of LS-EOS, the black hole formation occurs at 0.57 sec, which is earlier than the case of SH-EOS, having the smaller maximum mass with LS-EOS. During the thermal evolution of proto-neutron star toward the black hole formation, neutrinos trapped in the central part gradually diffuse outward and are emitted as supernova neutrinos.

The average energies of neutrinos emitted from the central core in two simulations with LS-EOS and SH-EOS are shown as a function of time after the core bounce in Fig. 1. The rise of average energies comes from the temperature increase by the contraction of proto-neutron stars due to the increase of mass. The end points in the figure correspond to the births of black hole. It is interesting that the different timing of black hole formation for models with LS-EOS and SH-EOS is reflected in the different timing of the termination of neutrino emissions. The LS-EOS is softer than SH-EOS,

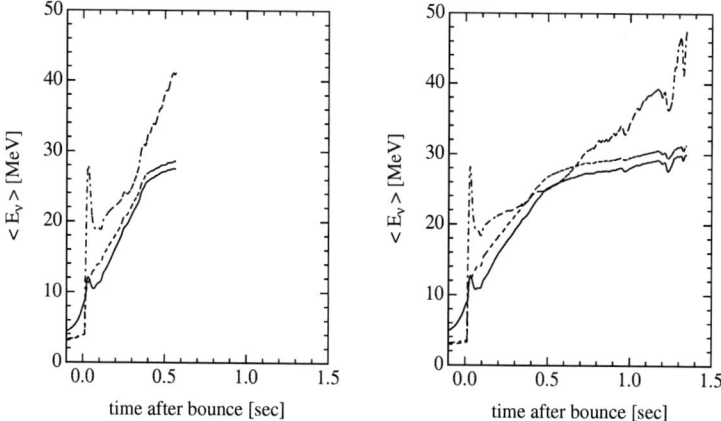

FIGURE 1. Average energies of ν_e (solid), $\bar{\nu}_e$ (dashed) and $\nu_{\mu/\tau}$ (dot-dashed) as a function of time after bounce in the cases of LS-EOS (left) and SH-EOS (right).

therefore, it leads to the smaller maximum mass, the earlier formation of black hole and the shorter neutrino burst. This difference in neutrino signals will provide us with the information on EOS. A terrestrial detection of neutrino bursts from the black hole formation in future will put a new constraint on the EOS at high density and temperature.

ACKNOWLEDGMENTS

The authors are grateful to H. Shen, K. Oyamatsu, A. Ohnishi, C. Ishizuka, S. Chiba and H. Toki for the collaborations on the supernova simulations and the table of equation of state. The numerical simulations were performed at NAO/ADAC (wks06a, wkn10b) and JAERI, and partially at RIKEN. This work is partially supported by the Grants-in-Aid for the Scientific Research (14039210, 14079202, 15540243, 15740160, 17540267) of the MOESSC of Japan and for the 21st century COE program "Holistic Research and Education Center for Physics of Self-organizing Systems".

REFERENCES

1. K. Sumiyoshi, S. Yamada, H. Suzuki, H. Shen, S. Chiba, and H. Toki, *Astrophys. J.* **629**, 922 (2005).
2. K. Sumiyoshi, S. Yamada, H. Suzuki, and S. Chiba, *Phys. Rev. Lett.* (2005), submitted.
3. S. E. Woosley, and T. Weaver, *Astrophys. J. Suppl.* **101**, 181 (1995).
4. S. W. Bruenn, *Astrophys. J. Suppl.* **62**, 331 (1986).
5. J. M. Lattimer, and F. D. Swesty, *Nucl. Phys.* **A535**, 331 (1991).
6. H. Shen, H. Toki, K. Oyamatsu, and K. Sumiyoshi, *Nucl. Phys.* **A637**, 435 (1998).
7. H. Shen, H. Toki, K. Oyamatsu, and K. Sumiyoshi, *Prog. Thoer. Phys.* **100**, 1013 (1998).
8. R. Brockmann, and R. Machleidt, *Phys. Rev.* **C42**, 1965 (1990).
9. Y. Sugahara, and H. Toki, *Nucl. Phys.* **A579**, 557 (1994).
10. K. Sumiyoshi, H. Suzuki, S. Yamada, and H. Toki, *Nucl. Phys.* **A730**, 227 (2004).

Magnetorotational Collapse of Very Massive Stars: Formation of Jets and Black Holes

Yudai Suwa*, Tomoya Takiwaki*, Kei Kotake[†] and Katsuhiko Sato*,**

Department of Physics, School of Science, the University of Tokyo, 7-3-1 Hongo, Bunkyo-ku,Tokyo 113-0033, Japan
[†]*Science & Engineering, Waseda University, 3-4-1 Okubo, Shinjyuku, Tokyo, 169-8555, Japan*
**Research Center for the Early Universe, School of Science, the University of Tokyo,7-3-1 Hongo, Bunkyo-ku, Tokyo 113-0033, Japan*

Abstract.
Population III stars are thought to be very massive stars. However, their properties are not clarified yet. In this research, we investigate the features of magnetorotational dynamics of such stars with computer simulations. We find jet-like explosion in a few models and reveal what type of initial model undergo explosion. In addition, we inquire features of newborn black holes as a remnant of core-collapse.

Keywords: population III, magnetohydrodynamics, jet, black hole

INTRODUCTION

Recently, great attention has been paid to the first stars, so-called Population III (hereafter Pop III) due to the discovery of hyper metal poor stars such as HE 0107-5240 and HE 1327-2326, which contain less than 1/100,000 of the iron observed in the Sun. These Pop III are predicted to have been predominantly very massive with $M \gtrsim 100 M_\odot$. In this research we study the evolution of such very massive stars using a two-dimensional magnetohydrodynamics code.

METHOD

All simulations were performed with a modified version of the explicit magnetohydrodynamic (MHD) ZEUS-2D code [1], which is an Eulerian code based on the finite-difference method and employs an artificial viscosity of von Neumann and Richtmyer to capture shocks. In so doing, the code utilizes the so-called constrained transport (CT) method, which ensures the divergence free($\nabla \cdot \vec{B} = 0$) of the numerically evolved magnetic fields at all times. Furthermore, the method of characteristics (MOC) is implemented to propagate accurately all modes of MHD waves. The self-gravity is managed by solving the Poisson equation with the incomplete Cholesky decomposition conjugate gradient (ICCG) method. Axial symmetry and reflection symmetry across the equatorial plane are assumed. Spherical coordinates (r, θ) are employed with logarithmic zoning in the radial direction and regular zoning in θ. One quadrant of the meridian section is covered with 300 $(r) \times$ 30 (θ) mesh points. We made several major changes to the base code to include microphysics. First we added an equation for electron fraction to treat

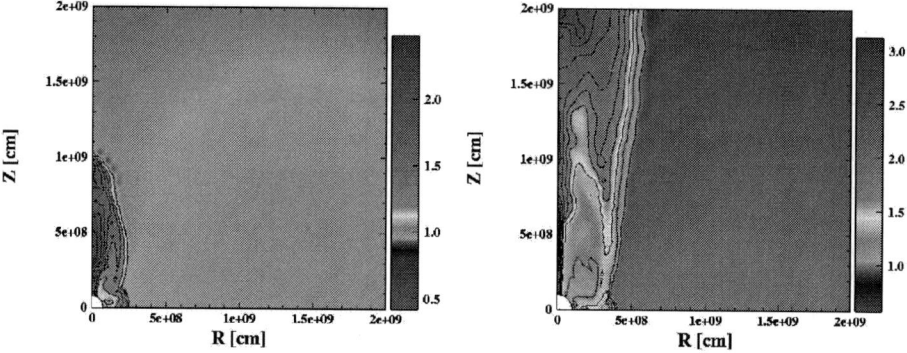

FIGURE 1. Time evolution of shock wave. They show the color coded contour plots of logarithm of entropy (k_B) per nucleon.

electron captures and neutrino transport by the so-called leakage scheme [2]. We extend the scheme to include all 6 species of neutrino ($\nu_e, \bar{\nu}_e, \nu_x$). ν_x means $\nu_\mu, \bar{\nu}_\mu, \nu_\tau$ and $\bar{\nu}_\tau$. Neutrino losses are included with thermal losses taken from [3]. The cooling rate, L_ν, is also estimated by the scheme. Second we have incorporated the tabulated equation of state (EOS) based on relativistic mean field theory instead of the ideal gas EOS assumed in the original code [4].

We start simulations with unstable $180 M_\odot$ He core of a $300 M_\odot$ star. We prepare polytropic star for density, whose polytrope index is $n = 3$, and calculate the energy on the assumption that core is isentropic and electron fraction $Ye = 0.5$ with Shen EOS. We assumed the core's entropy is $\sim 10 k_B$ per nucleon [5]. With these procedures, we calculate numerically the hydrostatic density and energy distribution of a $300 M_\odot$ star. From this star we produce a He core, $180 M_\odot$. In this paper we treat this core as the initial model. We assume in this study the differential rotation law. In addition, we assume that the initial magnetic field is nearly uniform (B_0 Gauss) in the core and dipole on the outside. We compute 9 models changing the total rotational energy and strength of magnetic field by varying the value of angular velocity and central magnetic fields.

NUMERICAL RESULTS

All models of very strong initial magnetic field ($B_0 = 10^{12}$G) form jets because the magnetic pressure exceeds the gas pressure by field wrapping and compression in the collapsing core. Such jets propagate to the outside of the core. Fig.1 shows the evolution of jet. The jet is obviously magneto-driven jet and leaves a very high entropy region behind the surface of shock. The more rapidly rotating models broaden the jet due to centrifugal force. Other models, which initially have weak magnetic fields, don't explode because the collapsed materials form a black hole before exploding.

We also investigate the initial black hole mass of each model. We determine black

TABLE 1. Initial Mass of Black Holes

B_0(Gauss) \ $T/|W|$(%)	1%	2%	4%
10^{10}G	70.4	87.3	106.6
10^{11}G	70.4	87.3	106.6
10^{12}G	57.9	75.8	96.6

hole formation with marginally stable orbit of Schwarzschild black hole. The initial mass of black holes for each model is summarized in Table 1. In this table, the mass of black holes are normalized in solar mass. Table 1 shows that the more rapid initial rotation, the bigger the black hole. As for the rapidly rotating star, the matter stop falling because of centrifugal force, thus more mass is necessary to form black hole. In addition to the rotation, the initial magnetic field affects the initial mass of black holes. The stronger magnetic field makes the initial black hole's mass smaller. This feature is due to the generation of the jet. When the jet arises, back-reaction overwhelm the matter behind the jet. Such a phenomenon quickens black hole formation time. Consequently, the initial black hole's mass gets smaller when the initial magnetic field is strong and jet-like explosions occur.

SUMMARY

We have performed time-dependent two-dimensional MHD simulations of the rotational core collapse of magnetized very massive stars. In this study, we systematically investigated how strong magnetic field and rapid rotation affect dynamics from the onset of core collapse to shock propagation in the core and black hole formation. We find that very massive stars can eject materials by the effects of rotation and magnetic field. The formation of black holes are also investigated. Both rotation and magnetic fields affect the initial mass of newborn black holes. Rapid rotation makes black holes big due to centrifugal forces and strong magnetic fields make black holes smaller due to back-reaction of jet.

REFERENCES

1. J. M. Stone, and M. L. Norman, *ApJS* **80**, 753–790 (1992).
2. R. I. Epstein, and C. J. Pethick, *ApJ* **243**, 1003–1012 (1981).
3. N. Itoh, T. Adachi, M. Nakagawa, Y. Kohyama, and H. Munakata, *ApJ* **339**, 354–364 (1989).
4. H. Shen, H. Toki, K. Oyamatsu, and K. Sumiyoshi, *Nucl. Phys.* **A637**, 435–450 (1998).
5. C. L. Fryer, S. E. Woosley, and A. Heger, *ApJ* **550**, 372–382 (2001).

Neutrino-Nucleus Reactions Induced by Supernova Neutrinos

Toshio Suzuki*, S. Chiba†, O. Iwamoto††, and T. Kajino¶

*Department of Physics, College of Humanities and Sciences, Nihon University, Sakurajosui 3-25-40, Setagaya-ku, Tokyo 156-8550, Japan
†Advanced Science Research Center, Japan Atomic Energy Agency, 2-4 Shirakata-shirane, Tokai, Naka-gun, Ibaraki 319-1195, Japan
††Nuclear Science and Engineering Directorate, Japan Atomic Energy Agency, 2-4 Shirakata-shirane, Tokai, Naka-gun, Ibaraki 319-1195, Japan
¶National Astronomical Observatory, Mitaka, Tokyo 181-8588, and Department of Astronomy, Graduate School of Science, University of Tokyo, Bunkyo-ku, Tokyo 113-003, Japan

Abstract. Neutrino-nucleus reactions induced by supernova neutrinos are investigated for ^{12}C and ^{4}He. Both of charge-exchange and neutral current processes are treated. Branching ratios to various decay channels are calculated by the Hauser-Feshbach theory taking into account the levels excited by ν's with a specific isospin and residual ones assuming isospin conservation.

Keywords: Neutrinos, Supernovae, Shell model, Nuclear reactions
PACS: 21.60.Cs; 25.30.-c; 25.30.Pt; 25.40.Kv; 26.50.+x; 24.60.Dr; 27.20.+n; 27.10.+h

INTRODUCTION

Recently, a modified shell model Hamiltonian for p-shell is obtained by properly taking into account the important aspects of the spin-isospin interaction [1], that is, the shell evolution as well as the change of magic numbers toward the drip-lines [2]. The modified Hamiltonian (SFO) can explain magnetic properties of the p-shell nuclei such as Gamow-Teller (GT) transitions better than conventional shell model Hamiltonians. In particular, agreements of calculated and observed magnetic moments are found to be systematically improved for the p-shell nuclei. These improvements can be traced to come from favorable behavior of monopole terms in the new Hamiltonan, especially those from the tensor part of the interaction [3].

Here, we study new ingredients of these developments on neutrino-nucleus reactions, which are dominantly induced by spin dependent excitations of nuclei. First, we show results of (v_e, e^-) reactions on ^{12}C for decay-at-rest (DAR) neutrinos. Calculated cross section for the exclusive GT transition to ^{12}N ($1^+_{g.s.}$) is found to be 9.96×10^{-42} cm^2 for the SFO Hamiltonian, which is enhanced compared with that of the Millener-Kurath (MK [4]) case; 8.48×10^{-42} cm^2. When g_A is quenched to $g_A^{eff} = 0.95$ g_A, that reproduces the observed B(GT) value, the value for the SFO is reduced to 9.06×10^{-42} cm^2. These values are consistent with the observed value; $8.9 \pm 1.2 \times 10^{-42}$

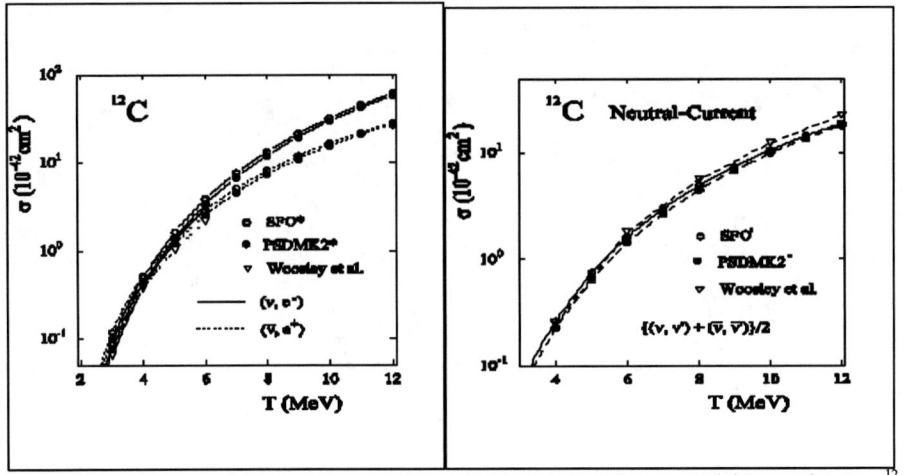

FIGURE 1. Cross sections for charge-exchange (left) and neutral-current (right) reactions on ^{12}C induced by supernova neutrinos.

cm^2 obtained by the liquid scintillator neutrino detector (LSND) experiment [5]. Calculated cross section to excited states in ^{12}N induced by spin-dipole as well as various multipole transitions is 8.35 (7.14) ×10^{-42} cm^2 for the SFO (MK) case, which needs to be reduced to 5.22 (4.87) ×10^{-42} cm^2 with the use of g_A^{eff} = 0.70 (0.75) g_A to explain the experimental value; 4.3 ± 1.0×10^{-42} cm^2 [5].

REACTIONS ON ^{12}C AND ^{4}HE

Charge-exchange and neutral-current reactions on ^{12}C for supernova neutrinos are obtained for the SFO and MK (PSDMK2 [6]) Hamiltonians. They are calculated for the ν spectra of the Fermi distribution with the temperature T = 2∼12 MeV by using

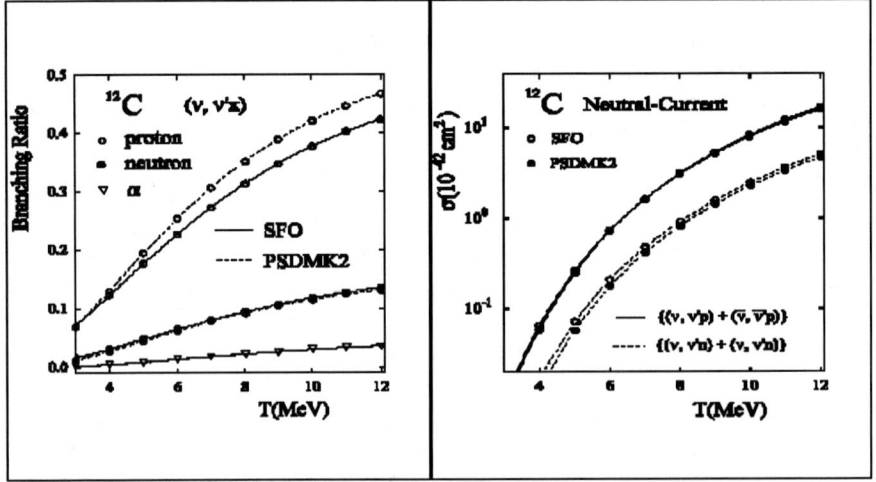

FIGURE 2. Calculated branching ratios for p, n, α and γ emission channels (left) and proton and neutron emission cross sections induced by neutral current for supernova neutrinos (right).

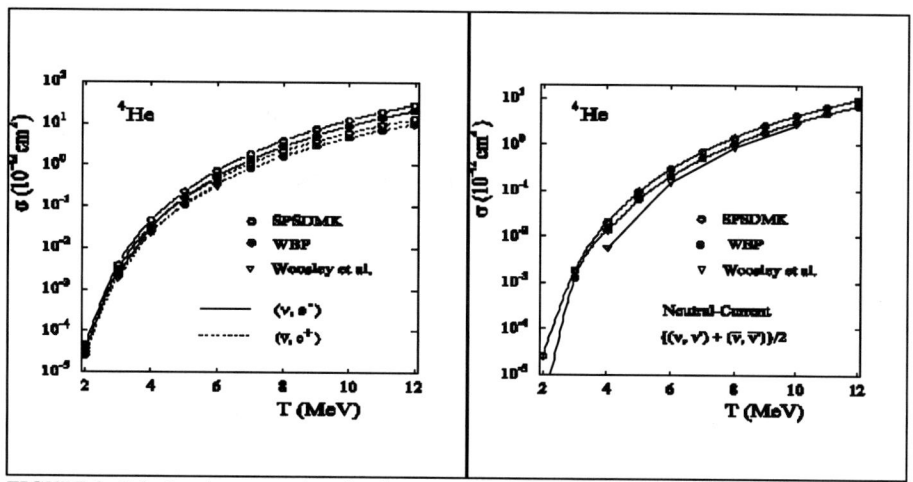

FIGURE 3. Calculated cross sections for charge-exchange (left) and neutral-current (right) reactions on ^{4}He induced by supernova neutrinos.

g_A^{eff} adopted above for the DAR case to explain the observed cross sections. The results are shown in Figure 1 showing some enhancement of the charge-exchange cross sections compared with the previous ones [7].

As for the neutral-current reactions, branching ratios to proton, neutron, α and γ emission chanels are calculated by the Hauser-Feshbach theory taking into account the levels excited by ν's and the residual ones assuming isospin conservation. Calculated branching ratios as well as the proton and neutron emission cross sections are shown in Fig. 2. The branching ratios for the proton emissions depend on the Hamiltonians, SFO or MK, while the neutron emission cross sections are found to be enhanced for the SFO case.

Calculated results for reactions on ^{4}He for supernova neutrinos obtained for the Warburton-Brown (WBP) [8] and MK (SPSDMK [6]) Hamiltonians with bare g_A are shown in Fig. 3. The reactions are induced dominantly by excitations of spin-dipole states. They are enhanced compared to the previous ones [7]. It woud be interesting to study possible implications for nuclear reactions and nucleosynthesis processes during the supernova explosions.

REFERENCES

1. T. Suzuki, R. Fujimoto and T. Otsuka, *Phys. Rev.* **C67**, 044302 (2003).
2. T. Otsuka, R. Fujimoto, Y. Utsuno, B. A. Brown, M. Honma and T. Mizusaki, *Phys Rev.Lett.* **87**, 082502 (2001).
3. T. Otsuka, T. Suzuki, R. Fujimoto, H. Grawe and Y. Akaishi, *Phys Rev.Lett.* **95**, 232502 (2005).
4. D. J. Millener and D. Kurath, *Nucl. Phys.* **A255**, 315 (1975).
5. LSND Collaboration, L. B. Auerbach et al., *Phys. Rev.* **C64**, 065501 (2001).
6. OXBASH, the Oxford, Buenos-Aires, Michigan State Shell Model Program, B. A. Brown, A. Etchegoyan and W. D. M. Rae, *MSU Cyclotron Laboratory Report*, No.524 (1986).
7. S. E. Woosley, D. H. Hartmann, R. D. Hoffman and W. C. Haxton, *Ap. J.* **356**, 272 (1990).
8. E. K. Warburton and B. A. Brown, *Phys. Rev.* **C46**, 923 (1992).

Chemical Evolution of Sulfur in the Metallicity Range of $-4 <$ [Fe/H]$< +0.5$

M. Takada-Hidai[*], M. Katsumata[†] and Y.-j. Saito[†]

[*]*Liberal Arts Education Center, Tokai University, Hiratsuka, Japan 259-1292*
[†]*Department of Physics, Tokai University, Hiratsuka, Japan 259-1292*

Abstract. We investigated a chemical evolution of sulfur in the Galaxy based on abundance results in the metallicity range of $-4 <$ [Fe/H]$< +0.5$. Trends and dispersions of [S/Fe] of 562 stars are discussed.

Keywords: Chemical evolution, Elemental abundance, Sulfur
PACS: 32.30.-r, 97.10.Cv, 97.10.Ex, 97.10.Tk

INTRODUCTION

Controversial results of a behavior of sulfure (S) have been obtained for recent years: one is an increasing trend with decreasing metallicity, the other a nearly flat trend in an [Fe/H]< -1 region. In a range of $-1 \leq$ [Fe/H]$< +0.5$, abundance analyses of S have been carried out to study whether a decreasing trend with increasing [Fe/H] from [Fe/H]$= -1$ continues still in the range of $0 \leq$ [Fe/H]$< +0.5$. Especially the S behavior in the range of $-0.5 <$ [Fe/H]$< +0.5$ has been examined based on the large sample of both stars with and without exoplanets in connection with finding a clue to a planet formation.

In this paper, we investigated the behavior of [S/Fe] in the range of $-4 <$ [Fe/H]$< +0.5$ to get new information of a chemical evolution of S in the Galaxy.

DATA AND ANALYSIS

The abundance data of 562 stars were collected from the literature listed in Table 1, which are mostly based on LTE analyses. In Figure 1, an overall trend of their [S/Fe] is shown in the range of $-4 <$ [Fe/H]$< +0.5$.

In Figure 2 (*Left*), we compared the behaviors of [S/Fe] between 34 giants and 57 dwarfs in the range of $-4 <$ [Fe/H]< -1. The sample of 50 dwarfs analyzed by Caffau et al. (2005) was also plotted for reference. The least-squares fits are given for giants and dwarfs, respectively. In Figure 2 (*Right*), the behaviors of [S/Fe] of dwarfs are shown, and the least-squares fits are given for stars with and without exoplanets of the samples of THKS(2005) and Ecuvillon et al. (2004), respectively.

TABLE 1. Literature for S abundances. A total number of stars collected is 562, allowing duplication. RMT stands for numbers of the Revised Multiplet Table.

References [Abbreviation]	RMT	S I lines (Å)	Star
Takada-Hidai & Sargent (2005) [TH-S(2005)]	1, 6	9212-37, 8694	16
Takada-Hidai et al. (2005b) [THKS(2005)]	6, 8, 10	8694, 6748-57, 6046-52	149
Takada-Hidai et al. (2005a) [TH et al(2005)]	1, 6	9212-37, 8694	21
Caffau et al. (2005)	1, 6, 8	9212-37, 8694, 6748-57	50
Ecuvillon et al. (2004)	8	6473-57	143
Nissen et al. (2004)	1, 6	9212-37, 8694	34
Ryde & Lambert (2004)	1	9212-37	10
Takada-Hidai et al. (2002) [TH et al(2002)]	6	8694	26
Chen et al. (2002)	6, 8, 10	8694, 6757, 6046-52	26
Sadakane et al. (2002)	8	6743-57	11
Takeda et al. (2001)	6, 10	8694, 6052	18
Israelian & Rebolo (2001)	6	8694	6
Santos et al. (2000)	10	6046-52	8
François (1987, 1988) [François-Dwarf]	6	8694	24
Clegg et al. (1981) [Clegg et al-Dwarf]	6	8694	20

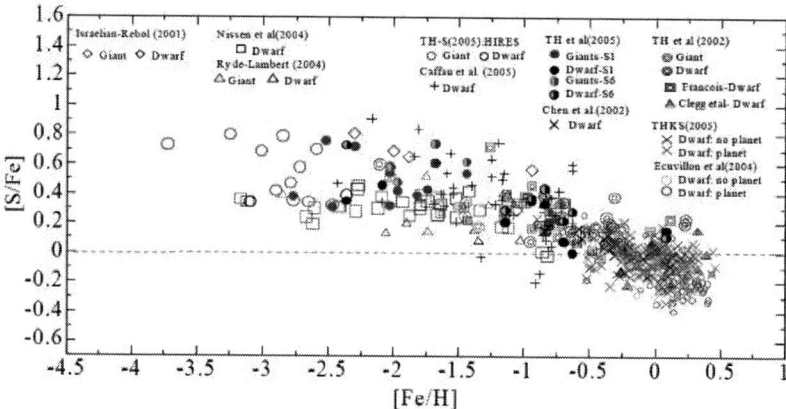

FIGURE 1. Overall behavior of [S/Fe] vs [Fe/H] in the range of −4 <[Fe/H]< +0.5.

RESULTS AND DISCUSSION

As seen from Fig.1, the overall behavior of [S/Fe] seems to be consistent with those of other α-elements, while a large dispersion of ~ 0.6 dex is demonstrated in the range of [Fe/H]$<\sim -1$, which is caused mainly by the results of giants.

The left panel of Fig.2 illustrates that [S/Fe] of giants seems to show a moderately increasing trend (slope $= -0.15$) with decreasing [Fe/H], as well as a large dispersion. On the contrary, dwarfs shows a nearly flat trend with a smaller dispersion. The flat trend may be explained by chemical evolution models based on usual supernovae, while the increasing trend may require hypernovae nucleosynthesis. The right panel of Fig.2

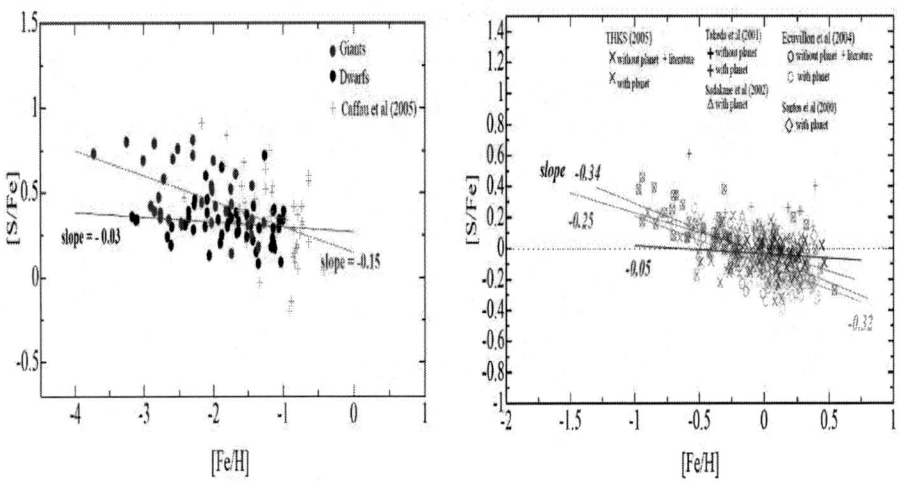

FIGURE 2. *Left*: Behaviors of [S/Fe] in giants and dwarfs with [Fe/H]< −1, and Caffau et al.'s dwarfs sample. *Right*: Behavior of [S/Fe] in dwarfs with −1 <[Fe/H]< +0.5.

demonstrates that there is a smaller dispersion of < 0.5 dex in −1 <[Fe/H]< 0, while a larger one of ∼ 0.6 dex in 0 ≤[Fe/H]< 0.5. The decreasing trend with increasing [Fe/H] of stars *without* planets inferred from THKS(2005) seems to be consistent with that from Ecuvillon et al. (2004), but there is a certain difference of trends between THKS(2005) (nearly flat) and them (decreasing with a slope of -0.32) for stars *with* planets. This issue should be examined in further studies.

REFERENCES

Caffau, E. et al. 2005, A&A, 441, 533
Chen, Y.-Q. et al. 2002, A&A, 390, 281
Clegg, R.E.S. et al. 1981, ApJ, 250, 262
Ecuvillon, A. et al. 2004, A&A, 426, 619
François, P. 1987, A&A, 176, 294; 1988, A&A, 195, 226
Israelian, G., & Rebolo, R. 2001, ApJ, 557, L43
Nissen, P.E. et al. 2004, A&A, 415, 993
Ryde, N., & Lambert, D.L. A&A, 415, 559
Sadakane, K. et al. 2002, PASJ, 54, 911
Santos, N.C. et al. 2000, A&A, 363, 228
Takada-Hidai, M. et al. 2002, ApJ, 573, 614
Takada-Hidai, M. & Sargent, W.L.W. 2005, IAU Symp. 228, 277
Takada-Hidai, M. et al. 2005a; 2005b, in preparation
Takeda, Y. et al. 2001, PASJ, 53, 1211.

Aspherical Ejecta of Type Ia Supernovae Inferred From High Velocity Features

M. Tanaka[*], P. A. Mazzali[†,**], K. Maeda[‡] and K. Nomoto[*]

[*]*Department of Astronomy, Graduate School of Science, University of Tokyo, Hongo 7-3-1, Bunkyo-ku, Tokyo 113-0003, Japan; mtanaka@astron.s.u-tokyo.ac.jp*
[†]*Max-Planck-Institute für Astrophysik, Karl-Schwarzschild-Str., 1, D-85741 Garching bei München, Germany*
[**]*Osservatorio Astronomico di Trieste, Via Tiepolo 11, I-34131 Trieste, Italy*
[‡]*Department of Earth Science and Astronomy, Graduate School of Arts and Science, University of Tokyo, Meguro-ku, Tokyo 153-8902, Japan*

Abstract. Spectral synthesis in 3-dimensional space for the early phase spectra of Type Ia supernovae is presented. In particular, the high velocity absorption features that ubiquitously exist at the earliest epochs ($\sim$ 10 days before maximum light) are investigated. The increasing number of early spectra available allows statistical study on the geometry of the ejecta. The observed diversity in the strength of the high velocity features (HVFs) can be reproduced through a "covering factor", which represents the fraction of the projected photosphere that is concealed by high velocity material. Various geometrical models involving high velocity material with a clumpy structure or a thick torus can naturally account for the observed statistics of HVFs. Models with 1 or 2 blobs, as well as a thin torus or disk-like enhancement seem to be unlikely as a standard situation.

Keywords: supernovae: general — radiative transfer — line: profiles
PACS: 97.60.Bw, 96.25.Tg

INTRODUCTION

Type Ia supernovae (SNe Ia) have been used as a distance indicator in cosmology after the relation between their maximum luminosity and the shape of the light curve (LC) was found. The origin of this empirical relation is, however, not fully understood mainly because of uncertainties regarding the properties of the explosion mechanism.

Detailed observations of SNe Ia indicate that SNe Ia with similar LCs may have different absorption line velocities in the early phase spectra [1]. In particular, the high velocity features (HVFs) in the Ca II IR triplet have been the subject of interest. Here, HVFs are defined as the absorption features that have a higher velocity than the photospheric velocity. Mazzali et al. [2] showed they are commonly seen at the earliest epochs ($\sim$ 10 days before maximum light) and there is a diversity in the strength and the velocity range of HVFs. Simultaneously, observations of spectropolarimetry [3] and multi-dimensional numerical simulations [4] have suggested that the explosion is aspherical. Although the origin of HVFs is still under debate, understanding HVFs can cast light on the explosion mechanism.

We present here synthetic spectra computed in 3-dimensional (3D) space with various geometries and show how HVFs are affected by different geometrical configurations and line-of-sight effects. Although all previous studies in multi-dimensional space performed modeling for each SN [5], no systematic study has been made. A comparison

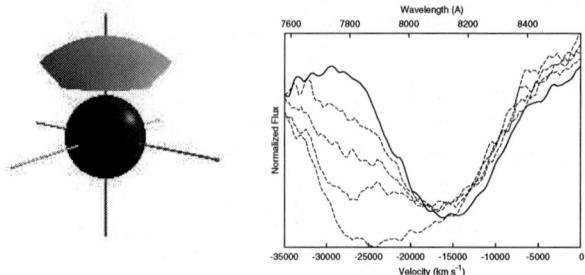

FIGURE 1. The geometry of the model with one blob (left) and synthetic spectra (right). The dashed lines in the right panel show the synthetic spectra with various line-of-sights, which correspond to the viewing angles of $0°, 21°, 33°$ and $42°$, respectively, going from deep to shallow absorption. The angle is measured from z-axis. The solid line shows the profile computed with the original W7 model.

with the statistical properties of HVFs enables us to constrain the geometry of the ejecta and possibly the nature of the explosion.

METHOD AND MODELS

We have developed a 3-dimensional Monte Carlo radiative transfer code based on the 1D Monte Carlo code [6]. The input parameters are the luminosity $L(\theta, \phi)$, the photospheric velocity $v_{ph}(\theta, \phi)$ and the epoch since the explosion t. With these parameters, the temperature in the ejecta, excitation and ionization of each element are computed in all zones. Since the path of each packet is affected by the density and temperature structure, the emergent spectrum depends on the orientation.

Our models are based on the spherical deflagration model W7 [7]. We introduced additional material in the outer layer of the ejecta to produce HVFs. We are not concerned here with the origin of this material, which may come from fluctuations of the explosion or from interaction with CSM or an accretion disk. The velocity range and the degree of the density enhancement are determined by fitting one of the strongest CaII HVF, which was observed in SN 2002dj, with the 1D code. We mapped this density enhancement into 3D space conserving the range and the degree of the enhancement. We tested various morphologies that may be realistic, including one or two large blobs, a small number of discrete blobs, a crowded clumpy structure, and tori of the various opening angles.

RESULTS AND DISCUSSION

Figure 1 shows the geometry of a model with a large blob (left) and the synthetic line profile (right). In the left panel, the sphere and the detached region show the photosphere and the density enhancement, respectively. The high velocity absorption in the synthetic spectra becomes deepest when seen on the z-axis, which is defined as the direction of the blob. If we move our line-of-sight to the equatorial plane, the high velocity absorption becomes weaker and it disappears when the line-of-sight reaches the edge of the blob.

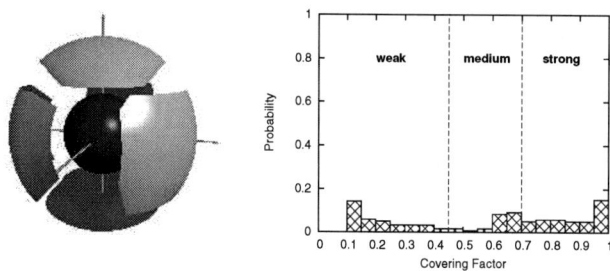

FIGURE 2. The geometry of the model with 5 large blobs (left) and the distribution of the covering factor (right).

We find the strength of the HVFs is determined by the fraction of the photosphere that is concealed by the dense blob, which we define as the "covering factor (f)".

If such a blob exists, the observed variation of the strength of the HVFs can be reproduced by changing lines-of-sight. We roughly classify the HVFs by their strength into three groups, strong HVFs ($f \gtrsim 0.7$), medium HVFs ($f \sim 0.45 - 0.7$) and weak HVFs ($f \lesssim 0.45$). However, we have to consider the statistical properties of the observed HVFs. Recent observations suggest that a considerable fraction of the earliest spectra have high velocity absorption [2].

We investigate several geometrical models and the expected statistical properties of the HVFs by means of the distribution of covering factors. As a result, we find that the model with relatively large blobs (Fig. 2) or a thick torus whose thickness is comparable to the size of the photosphere can reproduce the observed statistics. Models with 1 or 2 blobs, as well as a thin torus seem to be unlikely as a standard situation because of a large fraction of "weak" HVFs. Furthermore, since the model with many small blobs cannot produce the diversity, it is also unlikely as a standard model.

While a torus model sounds appealing, the main argument against it is the absence of hydrogen lines and the variation of the HVF velocity. The blob models may result from the explosion if the mushroom structure of the deflagration is not completely washed away in a delayed detonation [8].

A statistical study presented in this paper can be useful to distinguish the various 3D hydrodynamical models as well as the analysis of the spectropolarimetry. More early observations can constrain the structure more accurately.

REFERENCES

1. Benetti, S., et al. 2005, *ApJ*, **623**, 1011–1016
2. Mazzali, P.A., et al. 2005, *ApJ Letters*, **623**, 37–40
3. Wang, L., et al. 2003, *ApJ*, **591**, 1110–1128
4. Röpke, F., K. & Hillebrandt, W. 2005, *A&A*, **431**, 635–645
5. Kasen, D., et al. 2003, *ApJ*, **593**, 788–808
6. Mazzali, P.A. & Lucy, L.B. 1993, *A&A*, **279**, 447–456
7. Nomoto, K., Thielemann, F.-K. & Yokoi, K. 1984, *ApJ*, **286**, 644–658
8. Gamezo, V.N., Khokhlov, A. M. & Oran, E. S. 2005, *ApJ*, **623**, 337–346

Population III Core-Collapse Supernova Yields and Extremely Metal-Poor Star Abundance Pattern

N. Tominaga*, H. Umeda* and K. Nomoto*,†

*Department of Astronomy, School of Science, University of Tokyo, Bunkyo-ku, Tokyo 113-0033, Japan
†Research Center for the Early Universe, School of Science, University of Tokyo, Bunkyo-ku, Tokyo 113-0033, Japan

Abstract. Recently, very accurate observations of abundances of extremely metal-poor (EMP) stars have been performed with VLT and SUBARU. They determined the abundance patterns of many EMP stars with $-4.2 \lesssim $ [Fe/H] $\lesssim -2$. In the result, it is confirmed with small dispersions that the abundance ratios of these stars are obtained as a function of the metallicity. For example, the EMP stars with smaller [Fe/H] have larger [Zn,Co/Fe] and smaller [Cr/Fe]. Therefore Cayrel et al. (2004) and François et al. (2004) suggested homogeneous mixing in the Galaxy even below [Fe/H] ~ -3, although this conflicts with the Galactic chemical evolution models. We calculate explosions and nucleosynthesis of population III supernovae (SNe) with mixing-fallback model (Umeda & Nomoto 2003; 2005; Tominaga et al. 2006) and show that the trends of the abundance ratios with the small dispersions can be understood by the variations of explosion energies and progenitors' masses in inhomogeneous mixing.

Keywords: Galaxy: halo — nuclear reactions, nucleosynthesis, abundances — stars: abundances — stars: Population III — supernovae: general
PACS: 26.20.+f, 26.30.+k, 26.50.+x, 97.10.Cv, 97.10.Tk, 97.20.Tr, 97.20.Wt, 97.60.Bw, 98.80.Ft

INTRODUCTION

In the early universe, the enrichment by a single SN can dominate the preexisting metal contents and the abundance pattern of the enriched gas may reflect nucleosynthesis in the SN (e.g., [8, 9, 10]). The second stars formed in the enriched gas should reflect nucleosynthesis in a single SN and can constrain the yield of the SN. Among the second generation stars, low mass ($\sim 1 M_\odot$) stars have long life-times and might be observed as low [Fe/H] ($\lesssim -3$) stars, called extremely metal-poor (EMP) stars.

A recent observation ($-4 \lesssim $ [Fe/H] $\lesssim -2$: [1]) provided trends with small dispersions, especially in [Cr/Fe], different from [2] ($-3 \lesssim $ [Fe/H] $\lesssim -2$). Much flatter trends of [(Mg, Mn)/Fe] than the previous studies are also shown in [1], while the other trends are almost similar to those found earlier. Cayrel et al. (2004) and François et al. (2004) suggested that the elements have been already mixed homogeneously even below [Fe/H] $\lesssim -3$ and that the trends can be reproduced by the difference of the life time of progenitors with different masses.

However, this suggestion conflicts with the Galactic chemical evolution models that suggest inhomogeneous mixing in such early phases [4]. Additionally, r- and p-process nuclei observed in EMP stars show large scatter that can not be reproduced under the

TABLE 1. The explosion models.

$M_{\rm MS}/M_\odot$	13	15	18	20 (A)	25 (A)	25 (A)
$E_{\rm K}/10^{51}$ergs	1	1	1	10	5	10
$M(^{56}{\rm Ni})/M_\odot$	0.07	0.07	0.07	0.08	0.11	0.10
[Fe/H]	−2.35	−2.35	−2.35	−3.28	−2.85	−3.20
$M_{\rm MS}/M_\odot$	30 (A)	30 (B)	40 (A)	40 (B)	50 (A)	50 (B)
$E_{\rm K}/10^{51}$ergs	20	20	30	30	40	40
$M(^{56}{\rm Ni})/M_\odot$	0.16	0.05	0.28	0.12	0.37	0.26
[Fe/H]	−3.30	−3.85	−3.23	−3.60	−3.23	−3.39

assumption of homogeneous mixing, except for finding the other sites of r- and p-process nuclei instead of SN explosions.

MODELS

We construct core-collapse SNe models listed in Table 1 for the Pop III $13-50\,M_\odot$ stars. For normal SN models ($M_{\rm MS} = 13, 15, 18 M_\odot$), the mass cuts, the boundary between the central remnant and the ejecta, are determined to yield $M(^{56}{\rm Ni}) = 0.07 M_\odot$, while for hypernova (HN) models ($M_{\rm MS} = 20-50 M_\odot$), the mixing-fallback model is applied and the parameters are determined so that [O/Fe] = 0.5 (A) or [Mg/Fe] = 0.2 (B).

Addtionally, in order to obtain better agreements with the observations, we consider two additional effects. One is a neutron excess. Recent studies (e.g., [11, 12]) have suggested that Y_e may be significantly varied by the neutrino process during explosion. The region, where the neutrino absorption and the Y_e variation occurs, is Rayleigh-Taylor unstable, thus having a large uncertainty in Y_e. Another is an expansion before the main explosion. We reduced the density of the pre-supernova progenitor artificially without changing the total mass as [6]. We assumed that such a low density would be realized if the explosion is induced by multiple jets including the relatively weak jets prior to the main strong SNe jets. The weak jets expand the interior of the progenitor before the SN explosions due to the main jets (as described in the appendix in [6]).

TRENDS WITH METALLICITY

The comparisons with the observed abundance patterns of individual stars are shown in [13]. In this paper, we focus on the trends of [X/Fe] vs. [Fe/H]. According to the SN-induced star formation model ([14, 9, 10]; $[{\rm Fe/H}] \simeq \log_{10}(M({\rm Fe})/E_{51}) - C$, where C is a constant value, and take as $C = 1.2$ in this paper), normal SN and HN models have larger [Fe/H] ~ -2.4 and smaller [Fe/H] ~ -3.5, respectively (Table 1).

Figures 1 show the comparisons between the observed abundance ratios [X/Fe] against [Fe/H] and yields of individual SN models in Table 1 and the IMF integrated yield over all SNe and HNe models. [Fe/H] of the IMF integrated abundance ratios are

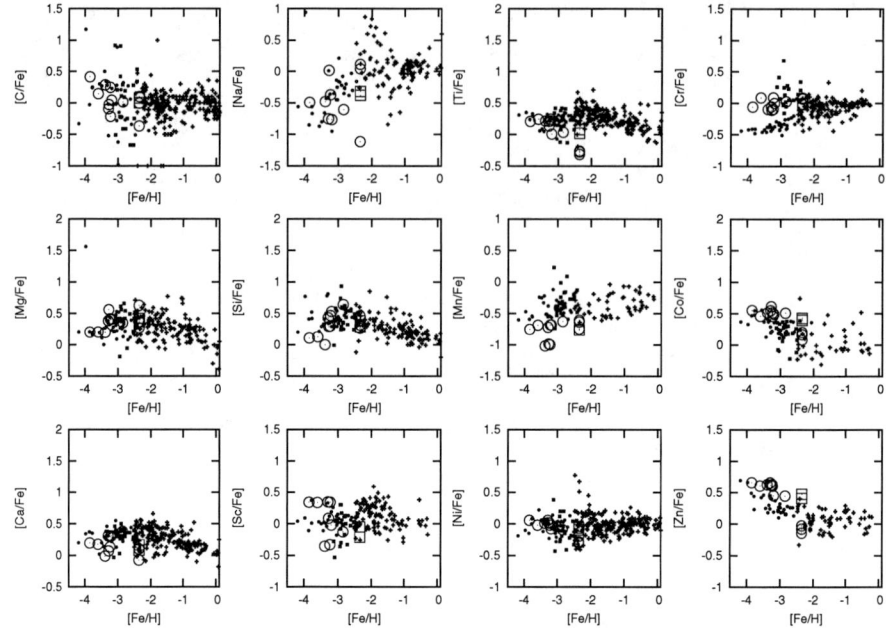

FIGURE 1. The comparison between the [X/Fe] trends of observed stars (the previous studies; *cross*, [1]; *filled circle*, [2]; *filled square*) and those of individual stars models (*open circle*) and IMF integration (*open square*). See [7] for detail and the references of the previous studies are therein.

assumed to be same as normal SNe models ([Fe/H] ~ -2.4). The yields show good agreements with the abundance patterns and trends of recent observations [1, 2].

We note that the observed abundance of most elements are roughly constant for $-2.5 \lesssim$ [Fe/H] $\lesssim -1$. This can be interpreted as that the SN ejecta had been mixed homogeneously at $-2.5 \lesssim$ [Fe/H] and consistent with chemical evolution model in [4].

REFERENCES

1. R. Cayrel, et al., *Astron. Astrophys.*, **416**, 1117–1138 (2004).
2. S. Honda, et al., *Astrophys. J.*, **607**, 474–498 (2004).
3. P. François, et al., *Astron. Astrophys.*, **421**, 613–621 (2004).
4. J. Tumlinson, *Astrophys. J.*, **641**, 1–20 (2006).
5. H. Umeda, and K. Nomoto, *Nature*, **422**, 871–873 (2003).
6. H. Umeda, and K. Nomoto, *Astrophys. J.*, **619**, 427–445 (2005).
7. N. Tominaga, H. Umeda, & K. Nomoto, in preparation (2006).
8. J. Audouze, and J. Silk, *Astrophys. J.*, **451**, L49–L52 (1995).
9. S. G. Ryan, J. E. Norris, and T. C. Beers, *Astrophys. J.*, **471**, 254–278 (1996).
10. T. Shigeyama, and T. Tsujimoto, *Astrophys. J.*, **507**, L135–L139 (1998).
11. H.-Th. Janka, R. Buras, & M. Rampp, *Nucl. Phys. A*, **718**, 269–276 (2003).
12. M. Liebendörfer, et al., *Nucl. Phys. A*, **719**, C144–C152 (2003).
13. H. Umeda, N. Iwamoto, N. Tominaga, K. Nomoto, and K. Maeda 2006, this volume.
14. D. F. Cioffi, C. F. McKee, and E. Bertschinger, *Astrophys. J.*, **334**, 252–265 (1988).

Electric Field Strength Of Coherent Radio Emission In Rock Salt Concerning Ultra High-Energy Neutrino Detection

Y.Watanabe, M.Chiba, O.Yasuda, Y.Shibasaki, T.Kamijo,
Y.Chikashige[†], T.Kon[†], Y.Shimizu[†], A.Amano[†], Y.Takeoka[†],
S.Ninomiya[†], S.Mori[†]

Tokyo Metropolitan University, 1-1 Minami-Osawa Hachioji-shi, Tokyo, Japan
Faculty of Science and Technology, Seikei University,
3-3-1 Kichijyoji Kitamachi Musasino-shi, Tokyo, 180-8633, Japan [†]

Abstract. Detection possibility of ultra high-energy (UHE) neutrino (E $>10^{15}$ eV) in natural huge rock salt formation has been studied. Collision between the UHE neutrino and the rock salt produces electromagnetic (EM) shower. Charge difference (excess electrons) between electrons and positrons in EM shower radiates radio wave coherently (Askar'yan effect). Angular distribution and frequency spectrum of electric field strength of radio wave radiated from 3-dimensional EM shower in rock salt are presented.

Keywords: Neutrino detectors; Cosmic ray detectors; Askar'yan effect; Rock salt
PACS: 95.55.Vj

INTRODUCTION

Ultra high-energy (UHE) neutrinos (E$>10^{15}$ eV) can travel without energy loss over astronomical distance. UHE neutrinos give us information about early stage of the universe. Salt Neutrino Detector (SND) aims to detect UHE neutrinos, especially produced by decay of charged pions originated from interactions between UHE cosmic rays and the cosmic microwave background (Greisen, Zatsepin, Kuz'min) [1].

The UHE neutrino can be detected using a natural huge rock salt. Interaction between an UHE neutrino and rock salt produces a hadron shower in which many π^0s are included. The π^0s decay into 2 γ s. Thus, a gigantic electromagnetic (EM) shower is generated. Charge difference (excess electrons) between electrons and positrons in an EM shower radiates Cherenkov radio wave coherently (Askar'yan effect).

F.Halzen, E.Zas, and T.Stanev calculated the electric field strength of radio wave radiated from high-energy EM shower in ice [3]. We calculated electric field strength in rock salt. For the calculation of electric field strength, EM shower in rock salt was simulated by using a program of Geant4 included LPM effect [4]. This paper presents angular distribution against EM shower axis and frequency spectrum of the electric field strength of radio wave radiated from an EM shower with energy 10^{12}eV-10^{15}eV.

CALCULATION OF ELECTRIC FIELD STRENGTH

For calculation of electric field strength (Eq. (5)), we derived electric field strength with frequency spectrum using the Lienard-Wiechert in ref. [3]. Electric field strength is shown Eq. (1), which is called "acceleration field" in ref. [5].

$$\mathbf{E}(R,t) = \frac{e}{c}\left[\frac{\mathbf{n}\times\{(\mathbf{n}-\boldsymbol{\beta})\times\dot{\boldsymbol{\beta}}\}}{(1-n\boldsymbol{\beta}\cdot\mathbf{n})R}\right]_{ret} \quad (1)$$

Where e is elementary electric charge, n is refractive index, c is velocity of light, R(t') is the distance from a charged particle to the observation point, and **n** is a unit vector for the observation direction from the charged particle in EM shower (Figure.1). The velocity and acceleration of a charged particle are $\boldsymbol{\beta}$ and $\dot{\boldsymbol{\beta}}$, respectively. The square brackets with subscript "ret" mean that the quantity in the brackets is evaluated at the retarded time, t=t'+R(t')/c. In order to calculate electric field strength of frequency spectrum, we used Fourier transformation in Eq. (2). When R is large, we can use *Fraunhoffer approximation* in Eq. (3).

$$E_\omega = \frac{1}{\sqrt{2\pi}}\int_{-\infty}^{+\infty} E(t)e^{i\omega t}dt \quad (2)$$

$$R(t') = R_0 - \mathbf{n}\cdot\mathbf{r}(t') \quad (3)$$

Where ω is angular frequency, R_0(=150 m) is distance from the origin of EM shower to the observation point, **r**(t') is position vector of the charged particle from the point O. Substitution of Eq. (1) and (3) in Eq. (2) and integration leads to Eq. (4).

$$\mathbf{E}(R,\omega) = \frac{ie\omega}{cR\sqrt{2\pi}}\exp\left[i\frac{\omega R_0}{c}\right]\int_{-\infty}^{\infty} e^{i\omega\left(t'-\frac{\mathbf{n}\cdot\mathbf{r}(t')}{c}\right)}\mathbf{n}\times(\mathbf{n}\times\boldsymbol{\beta})dt' \quad (4)$$

We assumed $\boldsymbol{\beta}$ is constant for flight duration ($\delta t = t_2 - t_1$) of a charged particle. By using $\mathbf{r}(t') = \mathbf{r}_1 + c\boldsymbol{\beta}(t-t_1)$, we could integrate Eq.(4). Where t_1, t_2, r_1 and r_2 are the generated point (t_1, r_1) and disappearing point (t_2, r_2).

$$RE(R,\omega) = \frac{en^2}{\sqrt{2\pi}c^2}\mathbf{v}_\perp\exp\left[i\frac{nR\omega}{c}\right]\exp\left[i\omega\left(t_1-\frac{\mathbf{r}_1\cdot\mathbf{n}}{c}\right)\right]\left(\frac{\exp[i\omega(1-n\boldsymbol{\beta}\cdot\mathbf{n})\delta t]-1}{1-n\boldsymbol{\beta}\cdot\mathbf{n}}\right) \quad (5)$$

Where $\mathbf{v}_\perp$ is the velocity of perpendicular direction toward observation point as $-\mathbf{n}\times(\mathbf{n}\times\mathbf{v})$. By summing electric field strength (Eq. (5)) of all charged particle, we calculated the coherent electric field strength:

$$RE(R,\omega) = \sum_{i=1}^{Total\_charged\_particle}(RE)_i \quad (6)$$

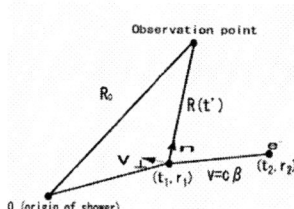

FIGURE 1. The geometry of electric field strength from a charged particle.

Angular distribution (left) and frequency spectrum (right) is shown Fig 2. Angular distribution shows interference pattern around Cherenkov angle. The electric field strength around Cherenkov angle (=65°) is higher than that of black body radiation above 10^{13}eV. The electric field strength has a maximum at about 4 GHz, above which it decreases gradually. Because destructive interference effect becomes stronger above 4 GHz.

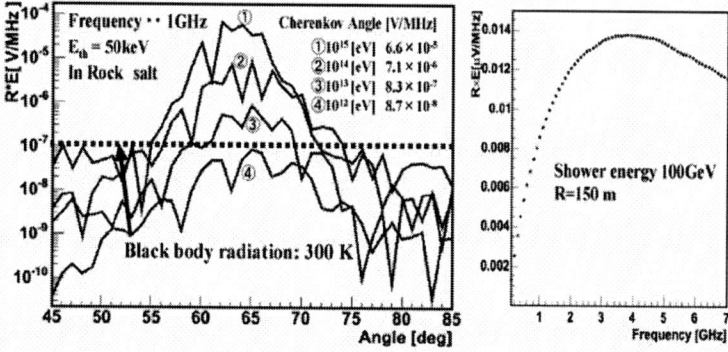

FIGURE 2. These figure show electric field strength of angular distribution (left) and frequency spectrum at Cherenkov angle (right). Dot line is 300K Black body radiation with band width 100 MHz.

In conclusion, it is possible to detect radio wave above 10^{13} eV in case of no radio wave absorption in rock salt. In order to make plan of SND, We should calculate electric field strength, including attenuation length of radio wave measured in refs [6,7,8,9] and realistic antenna array.

REFERENCES

1. K. Greisen, Phys. Rev. Letters **16**, 748 (1966).
2. G. A. Askar'yan, Soviet Physics JETP **14** 441 (1952).
3. E.Zas, F.Halzen, T.Stanev, Phys. Rev. D **45**,362(1992)
4. S.Agostinelli *et al*, Nuclear Instruments and Methods in Physics Research, NIM **A 506**(2003), 250-303
5. J.D. Jacson, Classical Electrodynamics (Wiley, NewYork, 1975)
6. M.Chiba, T.Kamijo, M.Kawaki, H.Athar, M.Inuzuka, M.Ikeda, O.Yasuda, Proc. 1st Int. Workshop for Radio detection if High Energy Particles [RADHEP-2000], UCLA, AIP Conf. Proc. **579**, p.204 (2000)
7. T.Kamijo, M.Chiba, Memoirs of Faculty of Tech., Tokyo Metropolitan University, No.**51 2001**, 139(2002)
8. M.Chiba etal., Proc .of the First NCTS Workshop Astroparticle Physics, Taiwan, World Scientific Publishing Co. Ltd. P.99 (2002)
9. T.Kamijo, M.Chiba, in Proc. Of SPIE **4858** Particle Astroparticle Physics Instrumentation, edited by Peter W.Gorhm, (SPIE, Bellingham, WA) p.151 (2003)

The extraction of fractions of the resonant component from analyzing powers in ^{6}Li(d, α)^{4}He and ^{6}Li(d, p$_0$)^{7}Li reactions at very low incident energies

M. Yamaguchi*, Y. Tagishi*, Y. Aoki[†], T. Iizuka[†], T. Nagatomo[†], T. Shinba[†], N. Yoshimaru[†], Y. Yamato[†], T. Katabuchi** and M. Tanifuji[‡]

*The Institute of Physical and Chemical Research (RIKEN), Wako, Saitama, Japan
[†]Institute of Physics and Tandem Accelerator Center, University of Tsukuba, Ibaraki, Japan
**Gunma University Graduate School of Medicine, Maebashi, Gunma, Japan
[‡]Department of Physics, Hosei University, Tokyo 102-8160, Japan

Keywords: Nuclear reaction, Polarization
PACS: 24.10.-i

Recently, several groups deduced the astrophysical $S(E)$ factor for ^{6}Li+d reactions at zero energy using cross section data for astrophysical interests [1, 2]. Generally astrophysical $S(E)$ factor at zero energy was extrapolated from data which is not affected by electron screening. The extrapolated $S(E)$ factor depends on the reaction model. The model dependence especially appears when the cross section has two components, resonant component and nonresonant one. In this work, we extracted the fraction of the resonant component in the ^{6}Li(d,α)^{4}He and ^{6}Li(d,p)^{7}Li reactions using polarization observables.

We measured the vector and tensor analyzing powers of the ^{6}Li(d,p$_0$) and and ^{6}Li(d, α) reactions with polarized deuteron beams at the incident energy of 90 keV. The experiment was performed using a Lamb shift-type polarization ion source [3] at the University of Tsukuba Tandem Accelerator Center (UTTAC). Figure 1 shows the layout in the scattering chamber. The target was a layer of lithium carbonate, Li$_2$CO$_3$, having a thickness of about 10 μg/cm^2 on an aluminum backing. Polarized deuteron beam was introduced to the lithium carbonate layer and was intercepted by the aluminum backing. A slit with a diameter of 5 mm was placed at a distance of 150 mm upstream from the target. We placed twelve Si-SSD's around the target at every 15° scattering angles from 0° to 165° to detect the emitted proton and α-particles.

Figures 2 and 3 show the measured analyzing powers for (d, p$_0$) and (d, α) reactions. The predictions by IMA (Invariant Amplitude Method) [4] are also shown. The predictions well reproduce the measured analyzing powers. From these data, we confirmed the spin and parity of the sub threshold resonance to be 2^+ and obtained the fraction of the resonant component for the (d, p$_0$) reaction as 0.90 ± 0.05 which is consistent with the earlier value about 0.85 extracted from the cross section data [5]. As the fraction of the resonant component for the (d, α) reaction, we obtained 0.998 ± 0.003, which is much larger than the earlier value about 0.5 extracted from the cross section data [2].

We also attempted to reproduce the analyzing-power data by the DWBA calculation. However, at present, we did not obtain successful results. This implies also negligible or very small contribution from nonresonant processes which are not accounted in the DWBA calculation.

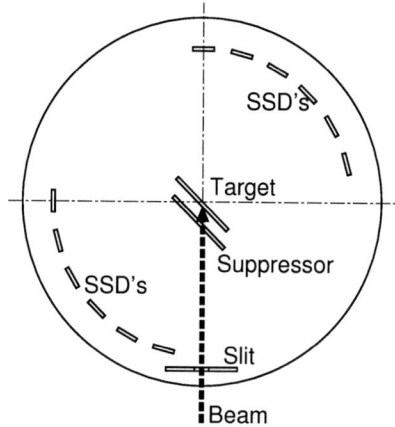

FIGURE 1. Layout in the scattering chamber.

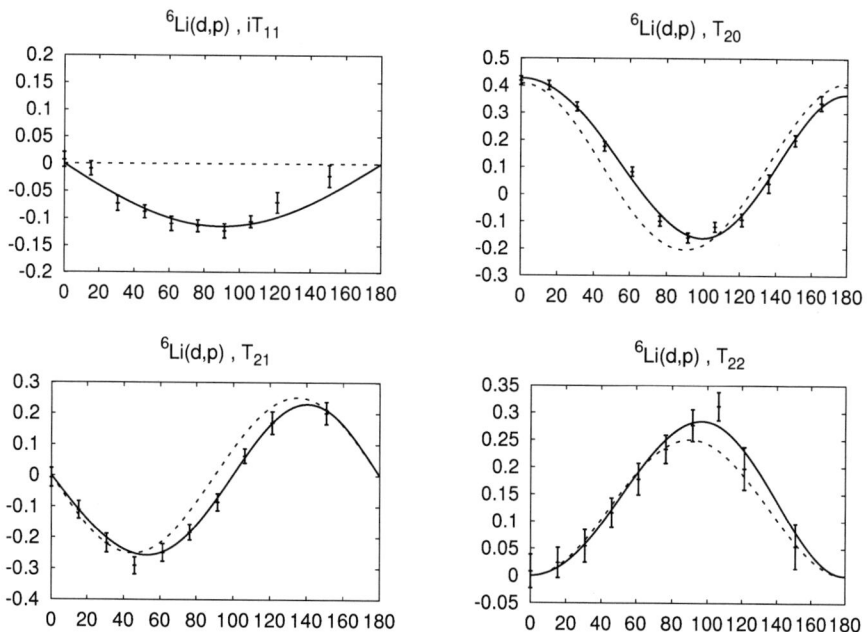

FIGURE 2. Experimental result for ^{6}Li$(d,p)^7$Li. The dashed and solid lines represent the prediction of IMA. The dashed lines are the results in which the angular momentum of the incident channel is restricted to s-wave. The abscissas represent the scattering angle in the C.M. system.

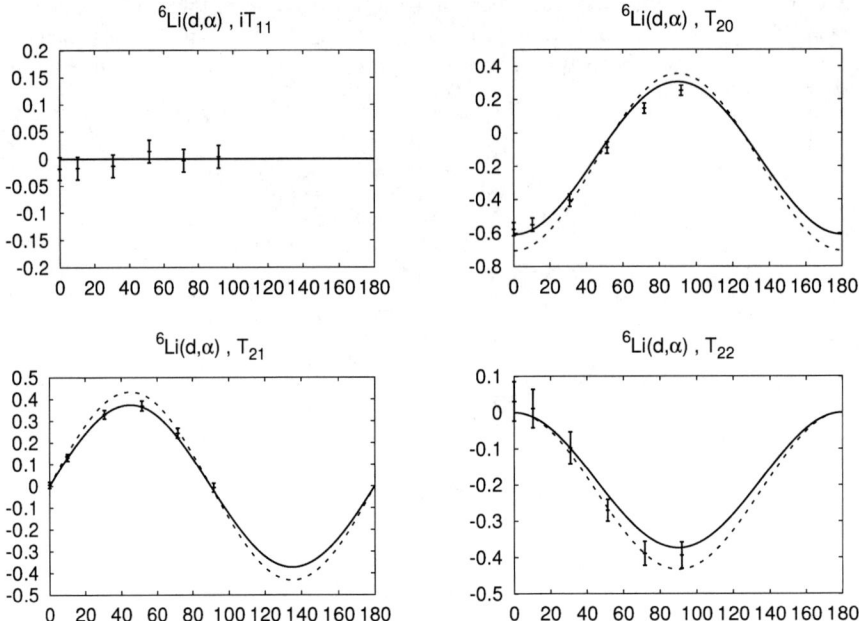

FIGURE 3. Experimental result for ^{6}Li$(d,\alpha)^4$He. The dashed and solid lines represent the prediction of IMA. The dashed lines are the results in which total angular momentum are restricted to 2^+. The abscissas represent the scattering angle in the C.M. system.

REFERENCES

1. J. E. Monahan, A. J. Elwyn, and F. J. D. Serduke, *Nucl. Phys. A* **269**, 61 (1976).
2. K. Czerski, A. Huke, H. Bucka, P. Heide, G. Ruprecht, and B. Unrau, *Phys. Rev. C* **55**, 1517 (1997).
3. Y.Tagishi, and J.Sanada, *Nucl. Instr. and Meth.* **164**, 411 (1979).
4. M. Tanifuji, and H. Kameyama, *Phys. Rev. C* **60**, 034607 (1999).
5. K. Czerski, H. Bucka, P. Heide, and T. Makubire, *Phys. Lett. B* **307**, 20 (1993).

Study on the dominant reaction path in nucleosynthesis during stellar evolution by means of the Monte Carlo method

K. Yamamoto, K. Hashizume, T. Wada, M. Ohta, T. Nishimura[1][†], M. Y. Fujimoto[†], K. Katō[†], T. Suda and M. Aikawa[2][‡]

Department of Physics, Konan University, 8-9-1 Okamoto, Kobe 653-8501, Japan
[†]*Division of Science, Hokkaido University, Sapporo 060-0810, Japan*
Meme Media Laboratory, Hokkaido University, Sapporo 060-0813, Japan
[‡]*Institut d'Astronomie et d'Astrophysique, C.P.226, Université Libre de Bruxelles, B-1050 Brussels, Belgium*

Abstract. We propose a Monte Carlo method to study the reaction paths in nucleosynthesis during stellar evolution. Determination of reaction paths is important to obtain the physical picture of stellar evolution. The combination of network calculation and our method gives us a better understanding of physical picture. We apply our method to the case of the helium shell flash model in the extremely metal poor star.

Keywords: Reaction path, Nucleosynthesis
PACS: 26.50.+f;26.30.+k

INTRODUCTION

The abundance variation of nuclear elements in stellar evolution is generally obatined by solving a set of differential equations for nuclear reaction network. When we assume an initial abundance and a scenario of stellar evolution, the reaction network calculation gives the abundance of an arbitrary element at arbitrary time. However, it is difficult to extract the information on the reaction paths by which a specific element is produced from the reaction network calculation, in particular, when various reaction paths compete in the late evolutionary stages.

Detailed knowledge of the reaction paths to a specific element is important to investigate the astrophysical model for stellar evolution and to improve the evaluation of nuclear reaction database in the astrophysical energy region.

In order to investigate quantitatively the contribution of various reaction paths to a specific element, we propose the Monte Carlo method for treating the set of diffetential equations for nuclear reaction network. In combining with the usual reaction network calculation, this method enables us to elucidate the transmutation paths of a certain element under the stellar evolution. We can investigate also the variation of the dominant reaction path under a modification in the reaction rate of a specific nuclear reaction.

---

[1] JSPS Reserch Fellow.
[2] FNRS Research Fellow.

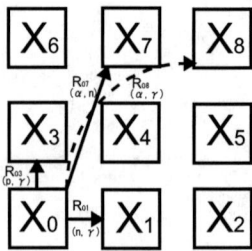

FIGURE 1. Schematic diagram of the Monte Carlo method. The transration from element X_0 to X_i is determinated stochastically according to the reaction rate R_j attached to the arrows.

In this study, we apply our Monte Carlo method to the cases of the nucleosynthesis in the convective zone; the case of the helium shell flashes in the extremely metal poor stars experiencing ^{13}C mixing[1]. In particular, we study the competition of the reaction paths in the helium flash convective zone shown in Fig:3 which is explained later.

CALCULATION METHOD AND APPLICATION

For a reaction i(j,l)k, the reaction rate R_j is denoted as

$$R_j(t) = Y_j(t)\rho(t)N_A \langle\sigma v\rangle_{ik} \quad (1)$$

where Y_j[mol/g], ρ[g/cm^3] and $N_A \langle\sigma v\rangle_{ik}$ [cm^3/sec/mol] mean the number abundance of projectile j, average density and average reaction rate, respectively[4, 5, 6]. We then calculate the residual probability of nucleus i, P_i, after the time interval $t - t_0$ as

$$P_i(t) = \exp\left[-\int_{t_0}^{t}\sum_j R_j(\tau)d\tau\right] \quad (2)$$

where t_0 is the initial time. The schematic diagram of our Monte Carlo method is illustrated in Fig:1. Element X_i decays with the probability $1 - P_i$ according to Eq. (2).

We apply our Monte Carlo method to the case of the heilum shell flashes in the extremely metal poor stars experiencing ^{13}C mixing[1]. Result of this model with network calculation is illustrated in Fig:2.

RESULT AND SUMMARY

Result of our Monte Carlo method is illustrated in Fig: 3. In the left pannel, the paths from ^{12}C to ^{20}Ne are shown. Suppose we have 100 ^{12}C, it is found that 2.03 of them capture neutron and only 0.37 of them capture α-particle. The dominant path is the one by the reaction (n γ). The importance of the neutron production by ^{13}C(α n)^{16}O and ^{17}O(α n)^{20}Ne reactions is confirmed.

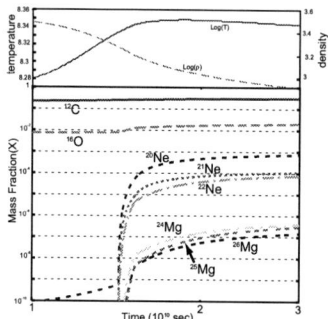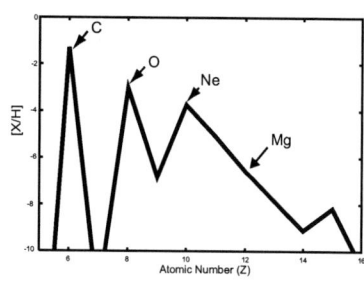

FIGURE 2. Result of network calculation. The temperature and the density case shown in the left pannel. The light element abundance is plotted in the right. There are produced during one cycle of convection [1, 2, 3].

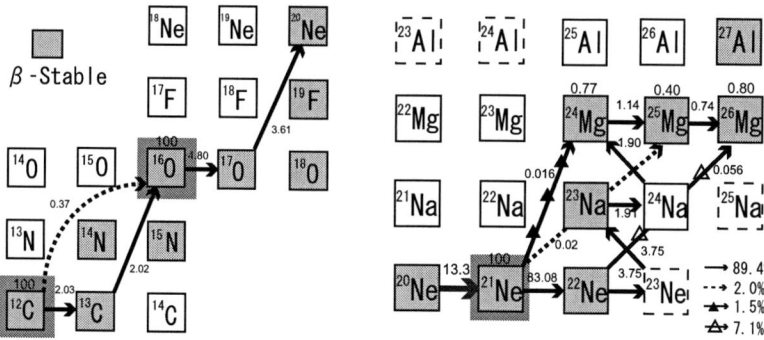

FIGURE 3. Two expamples of the reaction paths with Monte Carlo method. The left pannel shows the paths from ^{12}C to ^{20}Ne, and the right pannel shows those from ^{21}Ne to Mg isotopes.

In the right pannel, the paths from ^{21}Ne to Mg isotopes are shown. The paths look more complcated. Suppose we have 100 ^{21}Ne, we can see that 83 of them capture neutron and only about 0.01 of them capture α-particle. The dominant path to magnesium isotopes is the series of neutron capture reaction and β-decay.

With our Monte Carlo method, we can extract various informations from nucleosynthesis models for the astrophysicis.

REFERENCES

1. T. Suda et al., *Astrophys. J*, **611**, 476–493 (2004).
2. M. Aikawa et al., *Astrophys. J*, **608**, 983–988 (2004).
3. M. Aikawa et al., *Astrophys. J*, **560**, 937–956 (2001).
4. Bao, Z. Y. et al., *At. Data Nucl. Data Tables*, **76**, 70–154 (2000).
5. Y. Toyoshima, Master's Thesis, Dept. of phys, Konan Univ., **No.234**, (2005)
6. C. Angulo. et al., *Nucl Phys*, **A656**, 3–183, (1999)

Seven-Layer Supernova Mixtures Reproducing Isotopic Ratios of Presolar Grains

Takashi Yoshida

Astronomical Institute, Graduate School of Science, Tohoku University, Miyagi 980-8578, Japan

Abstract. We investigate supernova mixtures reproducing isotopic ratios of individual presolar grains from supernovae. We divide the supernova ejecta of 3.3, 4, 6, and 8 $M_\odot$ He star models into seven layers, i.e., the Ni, Si/S, O/Ne, C/O (3.3 and 4 $M_\odot$ models) or O/C (6 and 8 $M_\odot$ models), He/C, and He/N layers. Then, we seek the mixing ratios of the mixtures reproducing the isotopic ratios of ^{12}C/^{13}C, ^{14}N/^{15}N, ^{16}O/^{17}O, ^{16}O/^{18}O, ^{26}Al/^{27}Al, ^{29}Si/^{28}Si, ^{30}Si/^{28}Si, and ^{44}Ti/^{48}Ti as many as possible measured in a low density graphite grain KE3a-322. Six isotopic ratios of the grain are reproduced by two mixtures of 4 $M_\odot$ model and a mixture of 6 $M_\odot$ model. The main component of the mixtures is the He/N layer to reproduce ^{12}C/^{13}C ratio and to be the C/O ratio of the mixtures slightly larger than unity. However, ^{14}N/^{15}N is not reproduced by the mixtures. Isotopic ratios of ^{16}O/^{17}O and ^{16}O/^{18}O are reproduced by containing the He/C- and He/N-layer components in the mixtures. The Si/S layer should be contained to reproduce the excess of ^{28}Si and the mixing ratio is smaller than 1×10^{-4} for this grain. The Ni layer can be contained to reproduce ^{26}Al/^{27}Al. However, too much contribution of the Ni layer raises ^{44}Ti/^{48}Ti in mixtures.

Keywords: supernovae, nucleosynthesis, presolar grains, isotopic ratios
PACS: 26.30.+k, 96.50.Mt, 97.60.Bw

INTRODUCTION

Silicon carbide type X and low density graphite are considered to be supernova origin (e.g., [1]). Most of these grains have excesses of ^{28}Si [2] and some of them have evidence for the original presence of ^{44}Ti [3]. Low density graphite indicates the excesses of ^{18}O produced in the He layer in massive stars [4]. In order to reproduce isotopic and compositional signatures of the grains, large-scale inhomogeneous mixing in supernova ejecta has been required because the composition of the whole or a part of supernova ejecta does not reproduce both of the signatures [2]. Recently, quantitative comparison of isotopic ratios of presolar grains from supernovae with those of the mixtures evaluated using supernova nucleosynthesis models has been carried out. One procedure is evaluating the ranges of isotopic ratios of mixtures with various mixing ratios [5, 6, 7]. However, it is not determined with this procedure that how many isotopic ratios of individual single grains are reproduced by mixtures. The other procedure is to seek the mixtures reproducing isotopic ratios of individual grains [8] but only four-layer mixing in seven-divided layers has been taken into account. In this proceeding, we seek the mixtures considering seven-layer mixing of supernova ejecta reproducing isotopic ratios of individual presolar grains from supernovae. We show the isotopic ratios and the mixing ratios of the mixtures reproducing isotopic ratios of a low density graphite KE3a-322.

SUPERNOVA EJECTA AND MIXING MODELS

We adopted the abundance distributions of supernova ejecta of 3.3, 4, 6, and 8 $M_\odot$ He star models in [7] to reproduce isotopic ratios of KE3a-322. The 3.3, 4, 6, and 8 $M_\odot$ He star models correspond to 13, 15, 20, and 25 $M_\odot$ star at the zero-age main sequence [9]. We divided the supernova ejecta into seven layers; the Ni, Si/S, O/Si, O/Ne, C/O (3.3 and 4 $M_\odot$ models) or O/C (6 and 8 $M_\odot$ models), He/C, and He/N layers.

We evaluated the isotopic ratio of species i and species j of a mixture as a function of the mixing ratio x as follows: $r_{\text{iso,mix}}(x) = Y_i/Y_j = (\sum_a Y_{ia}x_a)/(\sum_a Y_{ja}x_a)$, where Y_{ia} is the abundance of species i averaged in layer a, x_a is the mixing ratio of layer a, Y_i is the abundance of species i of the mixture. Then, we determined the mixing ratios x of the mixtures reproducing isotopic ratios of individual presolar grains from supernovae using χ^2-value evaluation;

$$\chi^2 = \sum_{\text{iso}} \frac{\{r_{\text{iso,grain}} - r_{\text{iso,mix}}(x)\}^2}{\sigma^2_{\text{iso,grain}}}, \tag{1}$$

where $r_{\text{iso,grain}}$ is isotopic ratio "iso" of a grain, $\sigma_{\text{iso,grain}}$ is its 1σ value. We sought the minimum value of χ^2 and the corresponding mixing ratios of the mixtures. We considered that N_{isotopes} isotopes of the grain are reproduced by the mixture with the obtained mixing ratios when $\chi^2 \leq 4N_{\text{isotopes}}$.

RESULTS AND DISCUSSION

We investigated the mixtures reproducing eight isotopic ratios, i.e., $^{12}C/^{13}C$, $^{14}N/^{15}N$, $^{16}O/^{17}O$, $^{16}O/^{18}O$, $^{29}Si/^{28}Si$, $^{30}Si/^{28}Si$, and $^{44}Ti/^{48}Ti$ of KE3a-322. We found two mixtures of the 4 $M_\odot$ model and one mixture of the 6 $M_\odot$ model reproducing six isotopic ratios of KE3a-322. Figure 1 shows the mixing ratios, the isotopic ratios, χ^2-values, and the C/O ratios $r(C/O)$ of the mixtures. The mixture of 4 $M_\odot$ (1) model reproduces the isotopic ratios except $^{14}N/^{15}N$ and $^{26}Al/^{27}Al$. The mixture of 4 $M_\odot$ (2) model reproduces except $^{14}N/^{15}N$ and $^{44}Ti/^{48}Ti$. The mixture of 6 $M_\odot$ model reproduces the isotopic ratios except $^{12}C/^{13}C$ and $^{14}N/^{15}N$. We show main characteristics of the mixtures reproducing the isotopic ratios of KE3a-322.

The main composition of the three mixtures is the He/N layer; the mixing ratio is more than 90%. The He/N layer indicates small $^{12}C/^{13}C$ and $^{16}O/^{17}O$, large $^{26}Al/^{27}Al$, and $r(C/O)$ slightly larger than unity. The He/N layer is important to reproduce $^{12}C/^{13}C$ and to be $r(C/O) \geq 1$ in the mixtures. However, $^{14}N/^{15}N$ is very large, so that $^{14}N/^{15}N$ of the grain is not reproduced by the mixtures.

The mixing ratio of the He/C layer is $3 \sim 4 \times 10^{-2}$. The He/C layer indicates large $^{12}C/^{13}C$ and $r(C/O)$. Thus, too large mixing ratio of the He/C layer increases $^{12}C/^{13}C$ and $r(C/O)$ of mixtures. The He/C layer is also important to reproduce $^{16}O/^{18}O$ because only the He/C layer is rich in ^{18}O in supernova ejecta.

The mixing ratio of the Si/S layer is $5 \sim 8 \times 10^{-5}$. The Si/S layer is rich in ^{28}Si so that the layer is necessary to reproduce the excess of ^{28}Si of the grain. However, too much

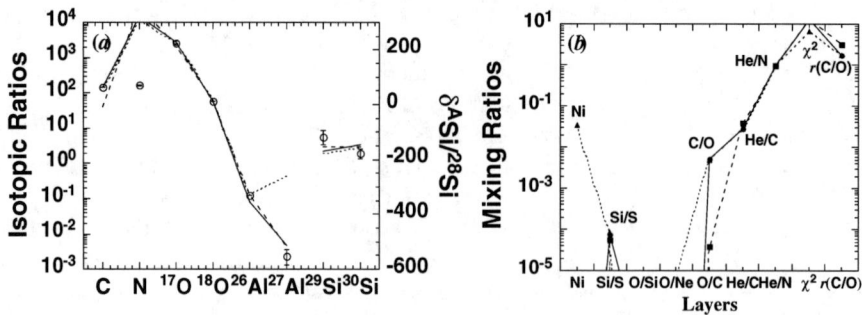

FIGURE 1. (*a*)The isotopic ratios of the mixtures reproducing the isotopic ratios of KE3a-322. (*b*)The mixing ratios, χ^2, and the C/O ratios of the mixtures. Open circles with error bars indicate measured isotopic ratios of KE3a-322. Solid, dotted, and dashed lines correspond to the isotopic ratios, mixing ratios, χ^2, and the C/O ratios of the mixtures of 4 $M_\odot$ (1), 4 $M_\odot$ (2), and 6 $M_\odot$ He star models.

Si/S layer becomes $\delta^{29}\text{Si}/^{28}\text{Si}$ and $\delta^{30}\text{Si}/^{28}\text{Si}$ close to -1000. The C/O ratio of this layer is much smaller than unity. So, the Si/S layer reduces the C/O ratio of the mixture.

About 3% of the Ni layer is included in the mixture of 4 $M_\odot$ (2) model which reproduces the isotopic ratios except $^{14}\text{N}/^{15}\text{N}$ and $^{44}\text{Ti}/^{48}\text{Ti}$. The Ni layer indicates small $^{14}\text{N}/^{15}\text{N}$ and $r(\text{C/O})$ and large $^{26}\text{Al}/^{27}\text{Al}$ and $^{44}\text{Ti}/^{48}\text{Ti}$. In this case, the Ni layer contributes to reproduce $^{26}\text{Al}/^{27}\text{Al}$ but brings about too large $^{44}\text{Ti}/^{48}\text{Ti}$. The Ni layer is important to reproduce $^{44}\text{Ti}/^{48}\text{Ti} \sim 0.4$ shown in some presolar grains from supernovae [6, 8]. The other two mixtures do not contain the Ni layer component. They reproduce $^{44}\text{Ti}/^{48}\text{Ti}$ of 0.0024 in KE3a-322.

The O/Si layer should not be contained in the mixtures to reproduce $\delta^{29}\text{Si}/^{28}\text{Si} > \delta^{30}\text{Si}/^{28}\text{Si}$. The O/Si layer indicates an excess of ^{30}Si. For the grains indicating ^{30}Si excesses rather than ^{29}Si such as KE3a-141, the O/Si layer can be contained in mixtures.

This work has been supported in part by the Ministry of Education, Culture, Sports, Science and Technology, Grants-in-Aid for Young Scientist (B) (17740130).

REFERENCES

1. K. Lodders, and S. Amari, *Chemie der Erde Geochemistry* **65**, 93–166 (2005).
2. S. Amari, P. Hoppe, E. Zinner, and R. S. Lewis, *ApJ* **394**, L43–L46 (1992).
3. L. R. Nittler, S. Amari, E. Zinner, S. E. Woosley, and R. S. Lewis, *ApJ* **462**, L31–L34 (1996).
4. S. Amari, E. Zinner, and R. S. Lewis, *ApJ* **447**, L147–L150 (1995).
5. C. Travaglio, R. Gallino, S. Amari, E. Zinner, S. E. Woosley, and R. S. Lewis, *ApJ* **510**, 325–354 (1999).
6. P. Hoppe, R. Strebel, P. Eberhardt, S. Amari, and R. S. Lewis, *Meteoritics Planet. Sci.* **35**, 1157–1176 (2000).
7. T. Yoshida, H. Umeda, and K. Nomoto, *ApJ* **631**, 1039–1050 (2005).
8. T. Yoshida, and M. Hashimoto, *ApJ* **606**, 592–604 (2004).
9. K. Nomoto, and M. Hashimoto, *Phys. Rep.* **163**, 13–36 (1988).

Accretion in Sirius binary system

Alexander Yushchenko[*,†] and Vera Gopka[†]

[*]*Astrophysical Research center for the Structure and Evolution of the Cosmos, Sejong University, Seoul, Korea*
[†]*Astronomical observatory, Odessa National University, Odessa, Ukraine*

Abstract. The chemical composition of Sirius A was calculated using Copernicus ultraviolet spectrum. Our results and earliar investigations of this star permit to find the abundances of 50 chemical elements in the atmosphere of Sirius A. We founf that the abundance pattern of Sirius A is strongly contaminated by s-process enriched matter. It can be explained by mass transfer from Sirius B at previous stage of its evolution. Three different estimates of the age of Sirius B from 10^3 to 10^8 years are possible.

Keywords: nucleosynthesis, stars: abundances, stars: chemically peculiar, stars: individual (Sirius)
PACS: 26.20+f 95.75.Fg 97.10.Cv 97.10.Tk 97.30.Fi 97.80-d

INTRODUCTION

A significant number of upper main sequence star show spectral peculiarities, which reflect abundance. It is difficult to find the star without anomalies in this part of HR diagram. As an example Yushchenko et al. [10] found that the majority of binary systems with A4-F1 components which considered to be normal ten years before, appeared to be chemically peculiar stars after detailed examination of abundance patterns. The main theory which can explain these anomalies is diffusion theory proposed by Michaud in 1970. Recent review of this theory was made by Michaud [2]. However, the simplest diffusion model lead to anomalies which are different from the observations. That is why additional mechanism is necessary to fit the observed abundance patterns.

Accretion phenomena was first discussed as an only reason, which affect the abundances. In 1989 the interplay between diffusion and accretion was proposed by Proffitt & Michaud [4]. They compared barium stars (binary systems: red giant + white dwarf) and main sequence binaries (main sequence + white dwarf) and concluded that mass transfer from an AGB star (now white dwarf) can have occurred in, at most, five percents of Am stars. Atomic diffusion, however, might still be able to modify the surface abundances. They showed that time scale for such diffusion to occur might be only a few millions years if the envelope is sufficiently stable. But during this few millions years we should be able to observe s-process enriched matter at the surfaces of main sequence stars. We tried to find these type anomalies in three binary systems. Two of these systems consist of main sequence star and a white dwarf. The brightness of the systems permit to have an excellent spectra at 2-meter class telescopes. These are the systems of Sirius and Procyon. The third system is a prototype of mild barium stars ζ Cyg (HD202109). ζ Cyg B is also a white dwarf. The orbital periods of above mentioned three binary systems are 50, 40, and 18 years respectively. Spectral classes of main components are A1V, F5IV-V, and G8III.

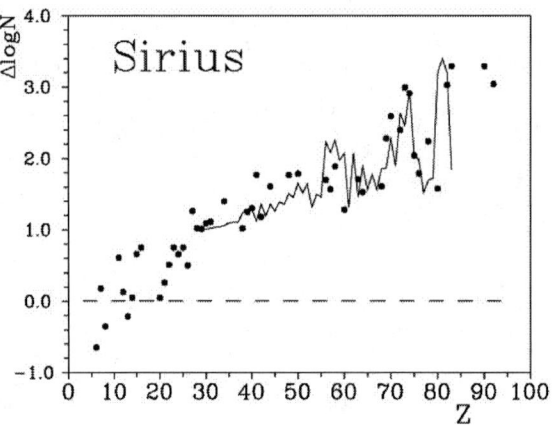

FIGURE 1. Abundances of chemical elements in the photosphere of Sirius A. The axes are atomic numbers and logarithmic abundances with respect to solar or Solar system values. Observed values are marked by points. Line - calculated abundances, scaled wind accretion model for ζ Cyg A [9]. This is s-process enriched abundance pattern.

ABUNDANCE PATTERN OF SIRIUS A

In this poster we present the results obtained using Copernicus ultraviolet spectral atlas of Sirius A published by Rogerson [4]. The wavelength region is from 1,649 to 3,170 Å, resolution 0.1 Å, signal to noise ratio S/N>100. The atlas is in excellent agreement with with recent HST observations of Sirius. The rotational profile of Sirius is wider than the instrumental profile of Copernicus and HST spectrographs, and the brightness of the star permit to have high quality spectrum using both instruments. Here after we will discuss the abundances, obtained from Copernicus spectrum and previously published results of different authors. We used URAN software [8] to fit the observed spectrum by calculated one in semiautomatic mode. Line lists were taken mainly from Kurucz CD-roms. Newer determinations of oscillator strengths for heavy elements were used also. More detailed description of used method can be found in our last papers [9-11]. In several words it is the identification of spectral lines on the base of comparison of synthetic and observed spectra and calculations of abundances of all elements using the spectrum synthesis method.

As a result of comparison of ultraviolet spectrum of Sirius A with synthetic spectrum we identified lines of 18 chemical elements with atomic numbers Z>29. These are Cu, Ga, Y, Zr, Mo, Cd, Sn, Hf, Ta, W, Re, Os, Hg, Pb, Th, U. Abundances of other elements were taken from previous investigations of Sirius. The total abundance pattern of Sirius A consists of 50 elements.

Figure 1 shows the abundance pattern of Sirius A with respect to the solar photosphere abundances. For several elements, with unknown abundances in the solar photosphere, meteoritic abundances were used.

DISCUSSION

As it was shown by Yushchenko et al. [9], barium stars with long orbital period (P>1600 days) can be formed through wind accretion. Yushchenko & Gopka [7] showed, that all abundances of r-, s-process elements in the spectrum of Procyon A are near solar values. The only exception is tellurium (Z=52). It means that the contamination of photosphere of Procyon A by s-process enriched matter can not be detected at present time. The convective photosphere of F5 type star can not conserve the overabundances of s-process enriched matter. These overabundances should be destroyed in short time. Fig. 1 is an illustration of the fact that the abundance pattern of Sirius A can be explained as a result of diffusion and wind accretion. The scaled abundance pattern of mild barium star ζ Cygni A can fit the maximums and minimums in the abundance pattern of Sirius A. The stable atmosphere of A-type star conserve the overabundances of s-process elements for a longer time, than in the case of Procyon. As it was pointed in the introduction diffusion should make these overabundances undetectable few millions years after the contamination event. We can point three possible estimates of the time, when the atmosphere of Sirius A was contaminated by s-process enriched matter. Sirius B became a white dwarf:

- near $160 \cdot 10^6$ years ago - Holberg et al. [1] (standard theory of white dwarf).
- less than $1-5 \cdot 10^6$ years ago - our abundances and diffusion theory.
- $1-2 \cdot 10^3$ years ago - from historical researches, see [5, 6] and references therein.

Standard white dwarf theory, diffusion theory and the ancient observations of Ptolemy two thousands years ago are in conflict. We can not resolve the problem at this moment. Now we are preparing new homogeneous abundance pattern of Sirius A using high quality spectrum in visual and near infrared wavelength region and Copernicus observations of Sirius A shortward of atlas [4].

REFERENCES

1. J. B. Holberg, M. A. Barstow, F. C. Bruhweiler, A. M. Cruise, and A. J. Penny, *Astrophys. J.*, **497**, 935–942 (1998).
2. G. Michaud, "Atomic diffusion in stellar surfaces and interiors" in *The A-Star Puzzle*, Edited by J. Zverko, J. Ziznovsky, S. J. Adelman, and W. W. Weiss IAU Symp. 210, Cambridge University Press, Cambridge, UK, 2004, pp. 173–183.
3. C. R. Proffitt, and G. Michaud, *Astrophys. J.*, **345**, 998–1007 (1989).
4. J. B. Rogerson, *Astrophys. J. Suppl.*, 163, 369–486 (1987).
5. T. J. See, *Astr. Nachr.*, **229**, 245–272 (1927).
6. D. C. Whittet, *Mon. Not. Royal Astr. Soc.*, **310**, 355–359 (1999).
7. A. V. Yushchenko, V. F. Gopka, *Astronomy Letterrs*, **22**, 412–421 (1996).
8. A. V. Yushchenko, "URAN: a Software System for the Analysis of Stellar Spectra", in *Proceedings of the 20th Stellar Conference of the Czech and Slovak Astronomical Institutes*, Edited by J. Dusek, Brno, Czech Republic, 1998, pp. 201–203.
9. A. V. Yushchenko, V. F. Gopka, C. Kim, Y. C. Liang, F. A. Musaev, G. A. Galazutdinov, *Astron. Astrophys.*, **413**, 1105–1114 (2004).
10. A. V. Yushchenko, V. F. Gopka, V. L. Khokhlova, D. L. Lambert, C. Kim, and Y.W. Kang, *Astron. Astrophys.*, **425**, 171–177 (2004).
11. A. Yushchenko, V. Gopka, Chulhee Kim, F. Musaev, Y.W. Kang, V. Kovtyukh, and C. Soubiran, *Mon. Not. Royal Astr. Soc.*, **369**, 865–873 (2005).

The Exponential S-Factor in the PP Chain

Norhasliza Yusof, Ithnin Abdul Jalil and Hasan Abu Kassim

Department of Physics, University of Malaya, 50603 Kuala Lumpur, Malaysia

Abstract. The astrophysical S-factor widely used in the literature is a linear or a second order polynomial fit in terms of energy. In this work, we assume an exponential form of the S-factor. The S-factor can then be written in the form $\ell n S_0 + f(E)$ where $S_0 = 5.523$ MeV and $f(E) = ax + bx^2 + cx^3$. By fitting the experimental cross section for $He^3 - He^3$, we obtain $a = -0.88, b = 0.59$ and $c = -0.11$. We evolve a solar model and compare with those of the standard cross section compilation. Most of the physical and chemical profiles differ at most by a few percent but the He^3 abundance is higher by 20 %. This might affect the Be^7 and B^8 neutrino productions.

Keywords: Nuclear astrophysics, solar interior, solar evolution
PACS: 26.20+f; 96.60.Jw; 97.10 Cv

DETERMINATION OF NEW S-FACTOR

The astrophysical reaction cross section as a function of energy E is defined as

$$\sigma(E) = \frac{S(E)}{E} exp(-b/E^{\frac{1}{2}}). \tag{1}$$

This equation indicates that the astrophysical S-factor $S(E)$ contains all the nuclear effects while the exponential term is the tunneling probability between the reacting particles, known as Gamow factor. Due to large Coulomb barrier, $S(E)$ is approximated as a smoothly varying function of energy either in linear or second order polynomial fits. These type of functions can be easily implemented and extrapolated to low energies when running stellar evolution codes, since the kinetic energies of particles are small in stellar interior.

Nuclei must be close to each other in order for the nuclear wavefunctions to overlap. Penetration of the Coulomb barrier permits the interaction between nuclei to occur in stellar interior at low energies. The exchange of mesons among nucleons might happen during the interaction between the nucleons in nuclei. We investigate this idea by using the Yukawa-type potential. The simplest form of the Yukawa potential can be represented in form $\frac{1}{R}exp-(\frac{r}{R})$. This leads us to deduce that the appropriate form for the S-factor should be an exponential function of energy $S(E) \propto e^{f(E)}$ where

$$f(E) = \ell n \sigma + b/E^{\frac{1}{2}} + \ell n E - \ell n S_0. \tag{2}$$

A complete S-factor might include the Coulomb penetration probability.

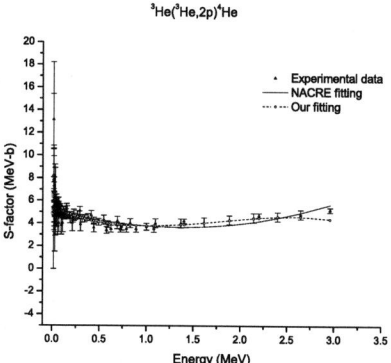

FIGURE 1. S-factor for the Experimental data, NACRE and Our Fitting

$HE^3(HE^3, 2P)HE^4$ REACTION

The cross section or the S-factor of $He^3(He^3, 2p)He^4$ reaction has special interest in determining solar neutrino flux through the production. The experimental values of the S-factor for the $He^3(He^3, 2p)He^4$ reaction vary very rapidly with energy. The value of S-factor is a constant at energy above 2.969 MeV (Figure 1). By fitting the available experimental data for the reaction, the complete S-factor is

$$\ell n S_0 = 1.7092 - 0.86E + 0.59E^2 - 0.11E^3. \tag{3}$$

From this equation, we calculated the S-factor and compare it with NACRE compilation as shown in the Figure 2.

SOLAR MODEL

We evolve a solar model using Equation (3) and NACRE compilation. From Figure 2, the abundance is higher by 20% at 0.3 solar radius when compared to NACRE.

ACKNOWLEDGMENTS

This work is funded by the Intensification of Research in Priority Areas (IRPA) grant 09-02-03-0149 from the Malaysian Government and a short term research grant F0204/2004D from University of Malaya. The authors would like to thank Raphael Hirsch for useful discussions on the solar model.

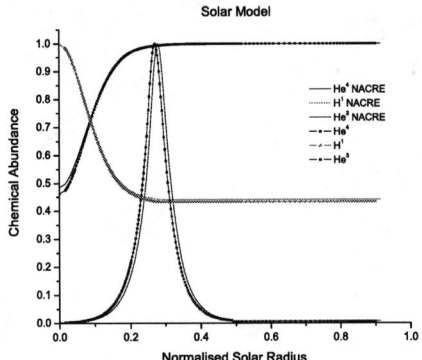

FIGURE 2. Chemical Abundances Profile from Our Solar Model.

REFERENCES

1. D.Clayton, *Pinciples of Stellar Evolution and Nucleosynthesis*, The University of Chicago Press, Chicago, 1983, pp. 296-309.
2. C.E. Rolfs and W.S. Rodney, *Cauldrons in the Cosmos : Nuclear Astrophysics*, The University of Chicago Press, Chicago, 1988, pp. 296-309.
3. C. Angulo et al, *Nucl. Phys. A* **656**, 3 (1999).

List of Participants

Aikawa, Masayuki
Institut d'Astronomie et d'Astrophysique
Universite Libre de Bruxelles,
Campus de la Plaine,
B-1050 Brusseles
Belgium
aikawa@astro.ulb.ac.be

Amadio Guilherme
CNS, Wako Branch at RIKEN
University of Tokyo
2-1 Hirosawa, Wako
Saitama 351-0198
Japan
amadio@cns.s.u-tokyo.ac.jp

Amanik, Philip
Department of Physics
University of California San Diego
Mail Code 0319, 9500 Gilman Drive,
La Jolla, CA 92093
USA
pamanik@physics.ucsd.edu

Amari, Sachiko
Box 1105, Physics Department
Washington University
One Brookings Drive
St. Louis, MO 63130
USA
sa@wuphys.wustl.edu

Aoki, Wako
National Astronomical Observatory
2-21-1 Osawa, Mitaka,
Tokyo 181-8588
Japan
aoki.wako@nao.ac.jp

Balantekin, Baha
Physics Department
University of Wisconsin
Madison, WI 53706
USA
baha@physics.wisc.edu

Bishop, Shawn
Heavy Ion Physics Laboratory
RIKEN
2-1 Hirosawa, Wako
Saitama 351-0198
Japan
bishop@rarfaxp.riken.jp

Bisterzo, Sara
Dipartimento di Fisica Generale
University di Torino
Via Pietro Giuria 1
10125 Torino
Italy
bisterzo@to.infn.it

Calura, Francesco
Dipartimento di Astronomia
Universita di Trieste
Via G.B. Tiepolo 11
34131 Trieste
Italy
fcalura@ts.astro.it

Chen, Alan
Department of Physics and Astronomy
McMaster University
1280 Main St. West
Hamilton ON, L8S4M1
Canada
Chenal@macmaster.ca

Cherubini, Silvio
INFN
Laboratori Nozionali del Sud
Via Santa Sofia 62
I-95123, Catania
Italy
cherubini@lns.infn.it

Chiba Masami
Tokyo Metropolitan University
1-1 Minamiosawa, Hachioji
Tokyo 192-0397
Japan

Coc, Alain
CSNSM
Bat. 104
Orsay Campus, 91405
France
coc@csnsm.in2p3.fr

Das, Kumar
Osaka Electro-Communication University
18-8 Hatsucho, Nayagawa,
Osaka 572-6830
Japan
suranjan@isc.osakac.ac.jp

Depagne, Eric
European Southern Observatory
Alonso de Cordova 3107
Santiago
Chili
edepagne@eso.org

Diehl, Roland
Max Plank Institute fuer Extraterrestrische Physik
D-85748 Garching
Germany
rod@mpe.mpg.de

Ejiri, Hiro
RCNP
Osaka University
Mihongaoka, Ibaraki
Osaka 567-0047
Japan
ejiri@rcnp.osaka-u.ac.jp

Franchini, Mariagrazia
INAF
Osservatorio Astronomico di Trieste
Via G.B. Tiepololl
I-34131 Trieste,
Italy
franchini@ts.astro.it

Francois, Patrick
Paris Observatory
61 Avenue de l' Observatoire
F-750114 Paris
France
patrick.francois@obspm.fr

Fujimoto, Shin' ichiro
Kumamoto National College of Technology
2659-2 Suya, Gousi,
Kumamoto 861-1102
Japan
fujimoto@ec.knct.ac.jp

Fukuda, Tomokazu
Osaka Electro-Communication University
18-8 Hatsu-cho, Neyagawa
Osaka 572-8530
Japan

Fuller, George
University of California San Diego
Mail Code 0319, 9500 Gilman Drive,
La Jolla, CA 92093
USA
gfuller@ucsd.edu

Fulop, Zsolt
ATOMKI
POB 51
H-4001 Debrecen
Hungary
fulop@atomki.hu

Gallino, Roberto
General Physics
University of Torino
I -10125 Torino
Italy
gallino@ph.unito.i

Gopka, Cera
Astronomical Observatory
Odessa National University
Debrovolskogo 98/34,
Odessa 65069,
Ukraine
yua@tm.odessa.ua

Guidry, Mike
Department of Physics and Astronomy
University of Tennessee
Knoxvill
TN 37996-1200
USA
Guidry@utk.edu

Gupta, S.N.P
Bhilai Steel Plant
1B /Street-57 /Sector-8,
Bhilai,C. G. 490008
India
snp.gupta@gmail.com

Hahn, Kevin
Ewha Womans University
11-1 Daehyun-dong, Seodaemun-gu
Seoul 120-750
Korea
ishahn@ewha.ac.kr

Hayakawa, Takehito
Japan Atomic Energy Agency
8-1 Kunimidai, Kizu
Kyoto 619-0215
Japan
hayakawa.takehito@jaea.go.j

He, Jianjun
CNS, Wako Branch at RIKEN
University of Tokyo
2-1 Hirosawa, Wako
Saitama 351-0198
Japan
he@cns.s.u-tokyo.ac.jp

Hirschi, Raphael
Department of Physics and Astronomy
Universitaet Basel
Klingelbergstrasse 82,
4056 Basel
Switzerland
RAPHAEL.HIRSCHI@UNIBAS.CH

Honda, Satoshi
National Astronomical Observatory
2-21-1 Osawa, Mitaka,
Tokyo 181-8588
Japan
honda@optik.mtk.nao.ac.jp

Ichikawa, Kazuhide
Institute for Cosmic Ray Research
University of Tokyo
5-1-5 Kashiwa-no-Ha, Kashiwa,
Chiba 277-8582
Japan
kazuhide@icrr.u-tokyo.ac.jp

Ichiki, Kiyotomo
National Astronomical Observatory
2-21-1 Osawa, Mitaka,
Tokyo 181-8588
Japan
ichiki@th.nao.ac.jp

Inoue, Susumu
National Astronomical Observatory
2-21-1 Osawa, Mitaka
Tokyo 181-8588
Japan
inoue@mpi-hd.mpg.de

Inoue, Kunio
Department of physics
Tohoku University
6-3 Aramaki, Aoba, Sendai
Miyagi 980-8578
Japan
inoue@awa.tohoku.ac.jp

Ishimaru, Yuhri
Academic Support Center
Kogakuin University
2665-1 Nakanomachi, Hachioji
Tokyo 192-0015
Japan
kt13121@ns.kogakuin.ac.jp

Ishiyama, Hironobu
Instiute of Particle and Nuclear Studies
High Energy Accelerator Research Organization
1-1 Oho, Tsukuba
Ibaraki 305-0801
Japan
hironobu.ishiyama@kek.jp

Iwamoto, Koichi
College of Science and Technology
Nihon University
1-8-14 Surugadai, Chiyoda
Tokyo 101-8308
Japan
wamoto@phys.cst.nihon-u.ac.jp

Iwamoto, Nobuyuki
Nuclear Data Center
Japan Atomic Energy Research Institute
2-4 Shirakata-shirane, Tokai
Ibaraki 319-1195
Japan
Niwamoto@ndc.tokai.jaeri.go.jp

Kaeppeler, Franz
Forschungszentrum Karlsruhe
Institut fuer Kernphysik
Hermann-von-Helmholtz-Platz 1
D-76344 Eggenstein-Leopoldshafen
Germany
franz.kaeppeler@ik.fzk.de

Kajino, Taka
National Astronomical Observatory
2-21-1 Osawa, Mitaka
Tokyo 181-8588
Japan
kajino@nao.ac.jp

Kassim, Hasan
Department of Physics
University of Malaya
50603 Kuala Lumpur
Malaysia
hasanak@um.edu.my

Kawagoe, Shiou
The Graduate University for Advanced Studies (SOKENDAI)
2-21-1 Osawa, Mitaka
Tokyo 181-8588
Japan
shiou@th.nao.ac.jp

Kishimoto, Chad
Department of Physics
University of California San Diego
9500 Gilman Dr. MC 0350, La Jolla
CA 92093
USA
ckishimo@physics.ucsd.edu

Kobayashi, Chiaki
National Astronomical Observatory
2-21-1 Osawa, Mitaka
Tokyo 181-8588
chiaki@MPA-Garching.MPG.DE

Kodaira, Keiichi
The Graduate University for Advanced Studies
Shonan Village, Hayama
Kanagawa 240-0193
Japan

Kotake, Kei
Science & Engineering
Waseda University
3-4-1 Okubo, Shinjuku
Tokyo 169-8555
Japan
kkotake@heap.phys.waseda.ac.jp

Koura, Hiroyuki
Advanced Science Research Center
Japan Atomic Energy Agency
Tokai, Naka-gun
Ibaraki 319-1195
Japan

Kubono, Shigeru
Center for Nuclear Study (CNS)
University of Tokyo, Wako Branch at RIKEN,
2-1 Hirosawa, Wako
Saitama 351-0198
Japan
kubono@cns.s.u-tokyo.ac.jp

Kusakabe, Motohiko
National Astronomical Observatory
2-21-1 Osawa, Mitaka
Tokyo 181-8588
Japan
kusakabe@th.nao.ac.jp

Kwon, Young Kwan
Department of Physics
Chung-Ang University
Huksuk-Dong, Dongjak-Gu,
Seoul 156-756
Korea.
Ykkwon11@hanmail.net

Lattimer, James
Department of Physics and Astronomy
Stony Brook University
Stony Brook
NY 11794-3900
USA
lattimer@astro.sunysb.edu

Le, Hong Khiem
Center for Nuclear Study
The University of Tokyo, Wako branch at
Riken campus
2-1 Hirosawa, Wako
Saitama 351-0198
Japan
lhkhiem@iop.vast.ac.vn

Li, Zhihong
Department of Nuclear Physics
China Institute of Atomic Energy
P.O. Box 275(46)
Beijing 102413, P.R. China
zhli@iris.ciae.ac.cn

Liebendoerfer, Matthias
Departement fuer Physik und Astronomie
Universitaet Basel
Klingelbergstrasse 82,
CH-4056 BASEL
SWITZERLAND
Matthias.Liebendoerfer@unibas.ch

Limongi, Marco
Osservatorio Astronomico di Roma
National Institute for Astrophysics
Via Frascati, 33
Monteporzio Carone (Rome), I-00040
Italy
marco@mporzio.astro.it

Liu, WeiPing
China Institute of Atomic Energy
P.O. BOX 275(46),
Beijing 102413
China
wpliu@iris.ciae.ac.cn

Madokoro, Hideki
Cyclotron Center
RIKEN
2-1 Hirosawa, Wako
Saitama 351-0198
Japan
madokoro@postman.riken.jp

Maeda, Keiichi
Department of Earth Science and
Astronomy
University of Tokyo
3-8-1 Komaba, Meguro
Tokyo 153-8902
Japan
maeda@esa.c.u-tokyo.ac.jp

Matteucci, Francesca
Astronomy Department
University of Trieste
Via G. B. Tiepolo 11
34124 Trieste
Italy
matteucci@ts.astro.it

Mezzacappa, Anthony
Physics Division
Oak Ridge National Laboratory
Bldg. 6025, PO Box 2008,
Oak Ridge, TN 37831-6354
USA
mezzacappaa@ornl.gov

Minato, Futoshi
Department of Physics
Nuclear Theory Group
Tohoku University
6-3 Aoba, Aramakiji, Aoba-ku, Sendai
Miyagi 980-8578
Japan

Misawa, Toru
Department of Astronomy and Astrophysics
Pennsylvania State University
525 Davey Lab
University Park, PA 16802
USA
misawa@astro.psu.edu

Mizoi, Yutaka
Osaka Electro-Communication University
18-8 Hatsu-cho, Neyagawa
Osaka 572-8530
Japan

Mizutani, Kohei
Department of Physics
Saitama University
255 Shimo-Okubo
Saitama 338-8570
Japan
mizutani@cr.phy.saitama-u.ac.jp

Morossi, Carlo
INAF
Osservatorio Astronomico di Trieste
Via G. B. Tiepolo 11
I-34131 Trieste
Italy
morossi@ts.astro.it

Motizuki, Yuko
Nishina Center
RIKEN
2-1 Hirosawa, Wako
Saitama 351-0198
Japan
motizuki@rarf.riken.jp

Motobayashi, Tohru
Nishina Center
RIKEN
2-1 Hirosawa, Wako
Saitama 351-0198
Japan

Mukhamedzanov, Akram
Cyclotron Institute
Texas A&M University
Collage Station, TX 77843
USA
akram@comp.tamu.edu

Muto, Takumi
Department of Physics
Chiba Institute of Technology
2-1-1 Shibazono, Narashino
Chiba 275-0023
Japan
takumi.muto@it-chiba.ac.jp

Nagashima, Masahiro
Department of Physics
Kyoto University,
Oiwakechou, Kitashirakwa, Sakyo
Kyoto 606-8502
Japan
masa@scphys.kyoto-u.ac.jp

Nagataki, Shigehiro
Department of Physics
Kyoto University,
Oiwakechou, Kitashirakawa, Sakyo
Kyoto 606-8502
Japan
nagataki@yukawa.kyoto-u.ac.jp

Nakamura, Riou
Department of Physics
Kyushu University
4-2-1 Ropponmatsu, Chuo
Fukuoka 810-8560
Japan
riou@gemini.rc.kyushu-u.ac.jp

Nakamura, Ko
Department of Physics
University of Tokyo
7-3-1 Hongo, Bunkyo
Tokyo 113-0033
Japan
nakamura@resceu.s.u-tokyo.ac.jp

Nishikawa, Yoshihisa
Department of Physics
Konan University
8-9-1 Okamoto, Higashi-nada, Kobe
Hyogo 658-8501
Japan

Nishimura, Nobuya
Department of Physics
Kyushu University
4-2-1 Ropponmatsu, Chuo
Fukuoka 810-8560
Japna
nobuya@gemini.rc.kyushu-u.ac.jp

Nishimura, Takanori
Department of Physics
Hokkaido University
Kita 10 Nishi 8, Kita
Sapporo 030-0810
Japan
nishimura@astro1.sci.hokudai.ac.jp

Nomoto, Ken'ichi
Department of Astronomy
The University of Tokyo
7-3-1 Hongo, Bunkyo
Tokyo 113-0033
Japan
nomoto@astron.s.u-tokyo.ac.jp

Nordstrom, Birgitta
Niels Bohr Institute
Rockefeller Building, Juliane Maries
Vej 30
DK 2100 Copenhagen
Denmark
birgitta@astro.ku.dk

Norris, John
Research School of Astronomy &
Astrophysics
ANU Mount Stromlo Observatory
Cotter Road, Weston Creek Act 2611
Australia
jen@mso.anu.edu.au

Nugent, Peter
Lawrence Berkeley National Laboratory
1 Cyclotron Rd, MS50F-1650, LBNL,
Berkeley, CA 94720
USA
penugent@LBL.gov

Ogata, Kazuyuki
Department of Physics
Kyushu University
6-10-1 Hakozaki, Higashi, Fukuoka
Fukuoka 812-8581
Japan
kazu2scp@mbox.nc.kyushu-u.ac.jp

Ohkubo, Takuya
Department of Astronomy
University of Tokyo
7-3-1 Hongo, Bunkyo
Tokyo 113-0033
Japan
ohkubo@astron.s.u-tokyo.ac.jp

Ohno, Shin-ichi
Theoretical Radiation Research Laboratory
12-5 Shiratori-dai, Aoba-ku, Yokohama
Kanagawa 227-0054
Japan

Ohta, Masahisa
Department of Physics
Konan University
8-9-1 Okamoto Higashinada Kobe
Hyogo 658-8501
Japan

Ot s uka, takaharu
Department of Physics
University of Tokyo
7-3-1 Hongo, Bunkyo
Tokyo 113-0033
Japan
otsuka@phys.s.u-tokyo.ac.jp

Okamura, Sadanori
Department of Astronomy
University of Tokyo,
7-3-1 Hongo, Bunkyo
Tokyo 113-0033
Japan

Omiya, Masashi
Department of Physics
Tokai University
1117 Kitakaname, Hiratuka
Kanagawa 259-1293
Japan

Ono, Hiroyuki
Faculty of Science and Technology
Tokyo University of Science,
2641 Yamazaki, Noda
Chiba 278-8510
Japan

Otsuki, Kaori
Department of Astronomy and
Astrophysics
University of Chicago
5640 South Ellis Ave, LASR103
Chicago, IL60637
USA
otsuki@uchicago.edu

Oura, Yasuji
Faculty of Science, Physics
Tokyo Metropolitan University
1-1 Minamiosawa, Hachioji
Tokyo 192-0397
Japan

Paul, Biswajit
TATA
Institute of Fundamental Research
Homi Bhabha Road, Colaba
Mumbai 400005
India
bpaul@tifr.res.in

Pipino, Antonio
Dipartimento di Astronomia
Universita' di Trieste
Via G.B. Tiepolo 11
34100 Trieste
Italy
antonio@ts.astro.it

Ropke, Friedrich
Max-Planck-Institut fuer Astrophysik,
Karl-Schwarzschild-Street 1
D-85741 Garching
Germany
fritz@mpa-garching.mpg.de

Rolfs, Claus
Department of Physics
Ruhr-Universitaet Bochum
Universitaetstrasse
44780 Bochum
Germany
rolfs@ep3.ruhr-uni-bochum.de

Sagara, Kenshi
Department of Physics
Kyushu University
6-10-1 Hakozaki, Higashi, Fukuoka
Fukuoka 812-8581
Japan

Saito, Akito
Center for Nuclear Study (CNS)
University of Tokyo, RIKEN Campus
2-1 Hirosawa, Wako
Saitama 351-0198
Japan
akito@cns.s.u-tokyo.ac.jp

Saito, Yu-ji
Department of Physics
Tokai University
1117 Kitakaname, Hiratuka
Kanagawa 259-1293
Japan
saitoyj@lynx.rh.u-tokai.ac.jp

Shigeyama, Toshikazu
Research Center for the Early Universe
University of Tokyo
7-3-1 Hongo, Bunkyo
Tokyo 113-0033
Japan
shigeyama@resceu.s.u-tokyo.ac.jp

Shima, Tatsushi
Research Center for Nuclear Physics
Osaka University
10-1 Mihogaoka, Ibaraki
Osaka 567-0047
Japan
shima@rcnp.osaka-u.ac.jp

Smith, Michael
Physics Division
Oak Ridge National Laboratory
MS-6354, Bldg. 6025, PO Box 2008
Oak Ridge, TN 37831-6354
USA
msmith@mail.phy.ornl.gov

Spitaleri, Claudio
INFN-Laboratori Nazionali del Sud
Via S. Sofia 62
I-95123 Catania
Italy

Su, Jun
China Institute of Atomic Energy
P.O.BOX 275(46)
Beijing 102413
China

Suda, Takuma
Department of Physics
Hokkaido University
Kita 10, Nishi 8, Kita
Sapporo 030-0810
Japan
suda@astro1.sci.hokudai.ac.jp

Suda, Akihiro
Department of Physics
Tohoku University
6-3 Aoba, Aramaki, Aoba-ku, Sendai
Miyagi 980-8578
Japan

Sumiyoshi, Kohsuke
Numazu College of Technology
3600 Ooka, Numazu
Shizuoka 410-8501
Japan
sumi@numazu-ct.ac.jp

Suwa, Yudai
Department of Physics
University of Tokyo
7-3-1 Hongo, Bunkyo
Tokyo 113-0033
Japan
suwa@utap.phys.s.u-tokyo.ac.jp

Suzuki, Hideyuki
Department of Physics
Faculty of Science and Technology
Tokyo University of Science
2641 Yamazaki, Noda
Chiba 278-8510
Japan
suzuki@chs.nihon-u.ac.jp

Suzuki, Toshio
Department of Physics
Nihon University
2641 Yamazaki, Noda
Chiba 278-8510
Japan
suzuki@phys.chs.nihon-u.ac.jp

Tachibana, Shogo
Department of Physics
University of Tokyo
7-3-1 Hongo, Bunkyo
Tokyo, 113-0033
Japan
tachi@eps.s.u-tokyo.ac.jp

Takada-Hdiai, Masahide
Liberal Arts Education Center
Tokai University
1117 Kitakaname, Hiratsuka
Kanagawa 259-1292
Japan
hidai@apus.rh.u-tokai.ac.jp

Takeda, Yoichi
National Astronomical Observatory
2-21-1 Osawa, Mitaka
Tokyo 181-8588
Japan

Tanaka, Masaomi
Department of physics
University of Tokyo
7-3-1 Hongo, Bunkyo
Tokyo 113-0033
Japan
mtanaka@astron.s.u-tokyo.ac.jp

Tatsumi, Toshitaka
Department of Physics
Kyoto University
Oiwakechyou, Kitashirakawa, Sakyou
Kyoto 606-8502
Japan
tatsumi@ruby.scphys.kyoto-u.ac.jp

Terada, Kentaro
Department of earth and Planetary
Systems Science
Hiroshima University
1-3-1 Kagami-yama, Higashi-Hiroshima
Hiroshima 739-8526
JAPAN
terada@sci.hiroshima-u.ac.jp

Togano, Yasuhiro
Department of Physics
Rikkyo University
3-34-1 Nishi-Ikebukuro, Toshima
Tokyo 171-8501
Japan
toga@ne.rikkyo.ac.jp

Tominaga, Nozomu
Department of Astronomy
University of Tokyo
7-3-1 Hongo, Bunkyo
Tokyo 113-0033
Japan
tominaga@astron.s.u-tokyo.ac.jp

Turuta, Sachiko
Department of Physics
Montana State University
Bozeman
Montana 59717
USA
uphst@gemini.msu.montana.edu

Tytler, David
University of California San Diego
9500 Gilman Drive, MC 0350
La Jolla, CA 92093-0424
USA
tytler@ucsd.edu

Umeda, Hideyuki
Department of Astronomy
Univerity of Tokyo
7-3-1 Hongo, Bunkyo
Tokyo 113-0033
Japan

Vergados, Ioannis J.D.
Department of Physics
University of Ioannina
Ioannina, GR 451 10
Greece
vergados@cc.uoi.gr

Wada, Takehiro
Department of Physics
Konan University
8-9-1 Okamoto, Kobe
Hyogo 658-8501
Japan

Wanajo, Shinya
Research Center for the Early Universe
University of Tokyo
7-3-1 Hongo, Bunkyo
Tokyo 113-0033
Japan
wanajo@resceu.s.u-tokyo.ac.jp

Watanabe, Yusuke
Department of Physics
Tokyo Metropolitan University
1-1 Minamiohawa, Hachioji
Tokyo 192-0397
Japan
watanabe@hakone.phys.metro-u.ac.jp

Woods, Philip
School of Physics
University of Edinburgh
Edinburgh, EH9 JJZ
UK

Yamada, Shoichi
Departement of Physics, School of Science
and Enginnering
Waseda University
3-4-1 Okubo, Shinjuku
Tokyo 169-8555
Japan
shoichi@waseda.jp

Yamaguchi, Hidetoshi
CNS, Wako Branch at RIKEN
University of Tokyo
2-1 Hirosawa, Wako
Saitama 351-0198
Japan
yamag@cns.s.u-tokyo.ac.jp

Yamaguchi, Mitsutaka
Heavy Ion Nuclear Physics Laboratory
RIKEN
2-1 Hirosawa, Wako
Saitama 351-0106
Japan
myamagu@rarfaxp.riken.jp

Yamamoto, Kazuyuki
Department of Physics
Konan University
8-9-1 Okamoto, Kobe
Hyogo 653-8501
Japan
dn621003@center.konan-u.ac.jp

Yoneda, Ken-ichiro
Nishina Center
RIKEN
2-1 Hirosawa, Wako
Saitama 351-0198
Japan
kyoneda@riken.jp

Yong, David
Department of Physics and Astronomy
University of North Carolina
Phillips Hall, CB #3255,
Chapel Hill, NC 27599-3255
USA
yong@physics.unc.edu

Yoshida, Takashi
Astronomical Institute
Tohoku University
Aramaki, Aoba, Sendai
Miyagi 980-8578
Japan
tyoshida@astr.tohoku.ac.jp

Yoshihara, Kazuhisa
Department of Physics
Faculty of Science and Technology
Tokyo University of Science
2641 Yamazaki, Noda
Chiba 278-8510
Japan

Yurimoto, Hisayoshi
Department of Natural History of Science
Hokkaido University
Kita 10 Nishi 8, Kita, Sapporo
Hokkaido 060-0810
Japan
yuri@geo.titech.ac.jp

Yushcenko, Alexander
Astrophysical Research Center for the
Structure and Evolution of the Cosmos
Sejong University
Kunja-Dong 98, Kwangjin-gu
Seoul, 143-747
Republic of Korea
yua@sejong.ac.kr

Yusof, Norhasliza
Department of Physics
University of Malaya
50603 Kuala Lumpur
Malaysia
norhasliza@perdana.um.edu.my

SCIENTIFIC PROGRAM

* Invited speaker

November 8 (Tuesday)

Opening Session

Chair: Nomoto, K. (Tokyo)

(15m) Kodaira, K. (President of GUAS) Opening address
(5m) Kajino, T. (Chair person) Opening address

Big Bang Cosmology and Particle Astrophysics

Chair: Nomoto, K. (Tokyo)

(30m) Nugent, P.* (LBL) Supernovae and Dark Energy
(20m) Ichiki, K. (Chicago) On the origin of dark energy in brane world cosmology
(20m) Yong, D. (North Carolina) Li abundances in halo subgiants

Big Bang Cosmology and Particle Astrophysics

Chair: Mukhamedzanov, A. (Texas A&M)

(30m) Tytler, D.* (UC San Diego) The Tension in Standard Big Bang Nucleosynthesis: the baryon density, D, He and Li
(20m) Coc, A. (CNRS Paris) Recent results in Big-Bang Nucleosynthesis
(20m) Shima, T. (Osaka RCNP) Experimental study of photonuclear reactions relevant to nuclear astrophysics
(20m) Li, Z. H. (CIAE) Determination of the astrophysical reaction rate for $^8Li(n,\gamma)^9Li$ reaction from the measurement of $^2H(^8Li,^9Li)^1H$ reaction

Topical Session: Most Metal-Deficient Stars: Observation and Theory

Chair: Francois, P. (Paris)

(30m) Norris, J.* (ANU) Metal abundances from first generation stars to the solar
(20m) Aoki, W. (NAO/GUAS) Abundance study of the most iron-poor star HE1327-2326 with Subaru/HDS
(20m) Suda, T. (Hokkaido) Nucleosynthetic Signatures of Pop.III Survivors and the Origin of HE0107-5240 and HE1327-2326
(20m) Umeda, H. (Tokyo) The first chemical enrichment in the universe and the formation of hyper metal-poor stars
(20m) Hirschi, R. (Basel) Models of rotating stars at very low Z: High carbon and nitrogen production

Poster 3min-Presentation (PP1)

Chair: Yoneda, K. (RIKEN)

Poster Session (PS1)

November 9 (Wednesday)

Cosmic and Galactic Chemical Evolution and Structure Formation
Chair: Nugent, P. (LBL)
- (30m) Matteucci, F.* (Trieste)   The chemical evolution of the Milky Way: From light to very heavy elements
- (20m) Nordstrom, B. (Niels Bohr)   Chemical evolution in the Milky Way Disk
- (20m) Ishimaru, Y. (Kogakuin)   Galactic chemical evolution with heavy metals produced by the first generation stars
- (20m) Kobayashi, C. (MPA/NAO) Galactic and Cosmic Chemical Evolution with Hypernovae

Explosive Nucleosynthesis in Supernovae
Chair: Fuller, G. (UCSD)
- (20m) Limongi, M. (Rome)   Core Collapse Supernova Nucleosyntyesis
- (20m) Shigeyama T. (Tokyo)   Light element production in Type Ic SNe
- (20m) Maeda, K. (Tokyo)   Optical and high energy emission from multi-dimensional supernovae as a tracer of explosive nucleosynthesis

Weak Interaction and Neutrino Physics (I)
Chair: Tytler, D. (UCSD)
- (30m) Fuller, G.* (UCSD)   Neutrino flavor transformation and neutrino-nucleus   interactions in stellar collapse and cosmology
- (30m) Inoue, K.* (Tohoku)   Results from KAMLAND
- (20m) Balantekin, A. B. (Wisconsin)   Does neutrino magnetic moment play a role in astrophysics?
- (20m) Yoshida, T. (Tohoku) Neutrino oscillation effect on supernova light element synthesis

Weak Interaction and Neutrino Physics (II)
Chair: Mezzacappa, A. (Oak Ridge)
- (20m) Vergados, J. (Ioannina)   Detecting earth and sky neutrinos by measuring electron and coherent nuclear recoils with large spherical TPC's
- (20m) Ejiri, H. (ICU)   Two novel approaches for direct dark matter searches: Detection of ionization electrons and hard X-rays

Supernovae, Neutron Stars and High Density Matter
Chair: Mezzacappa, A. (Oak Ridge)
- (30m) Lattimer, J.* (Stony Brook)   Constraints on the dense matter equation of state from observations
- (20m) Tsuruta, S.* (Montana)   Recent developments in neutron star thermal evolution theories and observation
- (20m) Tatsumi, T. (Kyoto)   Ferromagnetism in quark matter and origin of magnetic field in compact stars

Poster 3min-Presentation (PP2)
Chair: Shigeyama, T. (Tokyo)

November 10 (Thursday)

Supernova Explosion Mechanism
Chair: Lattimer, J. (Stony Brook)

- (30m) Mezzacappa, A.* (Oak Ridge) Ascertaining the core collapse supernova mechanism: The road ahead
- (20m) Roepke, F.* (MPA) Multi-dimensional models for Type Ia supernova explosions
- (20m) Yamada, S. (Waseda) Gravitational collapse of massive stars
- (20m) Sagara, K. (Kyushu) Experiment of ^{4}He(^{12}C,^{16}O)gamma reaction at Kyushu University

Neutron-Capture and r-Process Nucleosynthesis
Chair: Norris, J. (ANU)

- (30m) Francois, P.* (Paris) Abundance of heavy elements in extremely metal-poor stars
- (30m) Smith, M.* (Oak Ridge) Radioactive beams and exploding stars at ORNL
- (20m) Honda, S. (NAO) Subaru/HDS studies of r-process elements in metal-poor stars from near UV spectra
- (20m) Otsuki, K. (Chicago) Origin of main r-process elements

Neutron-Capture and s-Process Nucleosynthesis
Chair: Fulop, Z. (ATOMKI)

- (30m) Kaeppeler, F.* (Karlsruhe) Neutron capture in massive stars? the challenge of the weak s-process
- (20m) Gallino, R. (Torino) Carbon-rich stars with and without s-process enhancements at very low metallicity: Lead stars and the problem of Sr, Y, and Zr

Nuclear Astrophysics and Nucleosynthesis (I)
Chair: Kubono, S. (CNS, Tokyo)

- (40m) Rolfs, C.* (Bochum) Frontiers of experimental nuclear astrophysics
- (20m) Ishiyama, H. (KEK) Study of astrophysical (α, n) reactions on light neutron-rich nuclei using low-energy radioactive nuclear beams

Nuclear Astrophysics and Nucleosynthesis (II)
Chair: Smith, M. (Oak Ridge)

- (30m) Mukhamedzanov, A.* (Texas A&M) Indirect methods in nuclear astrophysics
- (20m) Cherubini, S. (Catania) Trojan horse method: Experimental results
- (20m) Ogata, K. (Kyushu) Determination of S_{17} from ^{8}B breakup by means of the method of continuum-discretized coupled-channels
- (20m) Yamaguchi, H. (CNS, Tokyo) Proton resonance scattering of ^{7}Be
- (20m) Togano, Y. (Rikkyo) Coulomb dissociation of ^{27}P for study of ^{26}Si(p,γ)^{27}P reaction

Poster Session (PS2)

November (11 Friday)

X-Ray, Gamma Ray, and Cosmic Rays
                                                Chair: Kaeppeler, F. (Karlsruhe)
- (30m)   Diehl, R.* (MPA) Studies of isotopic abundances through gamma-ray lines
- (20m)   Chen, A. (McMaster) Probing galactic ^{26}Al with radioactive ion beams
- (20m)   Tsunemi, H. (Osaka)   X-rays from SNR
- (20m)   Tachibana, S. (Tokyo) The abundance of live ^{60}Fe in the early solar system

Meteoritic Abundances
                                              Chair: Matteucci, F. (Trieste)
- (30m)   Amari, S.* (Washington) Presolar graphite from the Murchison meteorite: Imprint of nucleosynthesis and grain formation
- (30m)   Yurimoto, H.* (Hokkaido)   Meteoritic abundances
- (20m)   Terada, K. (Hiroshima)   Eu isotope analyses of SiC Grains from the Murchison Meteorite

Nuclear Astrophysics and Nucleosynthesis (III)
                                        Chair: Balantekin, A. B. (Wisconsin)
- (30m)   Liebendoerfer, M.* (Basel)   Composition of the innermost core collapse supernova ejecta and the vp-process
- (20m)   Hayakawa, T. (JAERI) Universality of the p-process nucleosynthesis in supernova explosions and scaling laws for p- and s-process nuclei in the solar system abundances
- (20m)   Wanajo, S. (Tokyo) The r-process and p-process in core collapse supernovae
- (20m)   Fulop, Z. (ATOMKI)   Radiative capture reactions and the astrophysical p-process

Closing

- (10m)   Nomoto, K./Kubono, S.   Closing Remark

AUTHOR INDEX

A

Aikawa, M., 59, 359, 455, 497
Amadio, G., 275, 362
Amanik, P. S., 365
Amano, A., 491
Amari, S., 311
Anderson, J., 205
Ando, H., 53
Ando, Y., 281
Aoi, N., 281
Aoki, W., 21, 53, 92, 221, 324, 368
Aoki, Y., 494
Arai, K., 359, 449
Arnould, M., 359
Asplund, M., 53

B

Baba, H., 281, 395
Bai, X., 37, 427
Balantekin, A. B., 128
Barbuy, B., 205
Bardayan, D. W., 470
Barklem, P. S., 53
Bartlett, A., 227
Baugh, C. M., 442
Beers, T. C., 53, 205, 368
Bisterzo, S., 368
Blackmon, J. C., 470
Blokhintsev, L. D., 255
Blondin, J. M., 179
Bonifacio, P., 205
Bravo, E., 333
Brown, S., 255
Bruenn, S. W., 179
Burjan, V., 255

C

Calura, F., 371
Carney, B. W., 21
Cayrel, R., 205
Chae, K., 470
Charlton, J. C., 433

Chavez, M., 383, 436
Chen, A. A., 298
Chen, Y., 37, 427
Cherubini, S., 255, 263
Chiba, M., 491
Chiba, S., 380, 479
Chieffi, A., 99, 311
Chikashige, Y., 491
Christlieb, N., 53
Coc, A., 25
Cole, S., 442
Crucilla, V., 263

D

Das, S. K., 249, 374
Deliyannis, C. P., 53
Demichi, K., 281
Depagne, É., 205, 377
Diehl, R., 289
Di Marcantonio, P., 436

E

Ejiri, H., 147
Elekes, Z., 281
Endo, T., 380
Eracleous, M., 433
Eriksson, K., 53

F

Franchini, C., M., 383
Franchini, M., 436
François, P., 205
Frebel, A., 53
Frekers, D., 227
Frenk, C. S., 442
Fröhlich, C., 333
Fuchi, Y., 249, 374
Fujikawa, H., 275, 362
Fujimoto, M. Y., 53, 59, 455, 497
Fujimoto, S., 386, 452
Fukuda, N., 281

Fukuda, T., 249, 374
Fuller, G. M., 365, 418
Fülöp, Z., 281, 351
Furukawa, T., 374
Futakami, U., 281

G

Galaviz, D., 351
Gallino, R., 311, 368
Giomataris, Y., 140
Gol'dberg, V. Z., 255
Gomi, T., 281
Gono, Y., 395
Gopka, V., 389, 503
Goriely, S., 389
Gorres, J., 227
Greife, U., 467
Guidry, M. W., 470
Gulino, M., 263
Guo, B., 37, 427
Gupta, S. N. P., 392
Gyürky, G., 351

H

Hanayama, H., 403
Hartmann, D. H., 134
Hasegawa, H., 281
Hashimoto, M., 386, 449, 452
Hashimoto, S., 269
Hashimoto, T., 249, 374
Hashizume, K., 497
Hayakawa, T., 339
He, J. J., 275, 362, 395
Heil, M., 235
Higurashi, Y., 281
Hill, V., 205
Hillebrandt, W., 190
Hirayama, Y., 249, 374
Hirschi, R., 71
Hix, W. R., 333, 409, 470
Hokoiwa, N., 395
Honda, S., 53, 221, 464
Huss, G. R., 304

I

Iben, Jr., I., 59, 455
Ichikawa, K., 400
Ichikawa, S., 249, 374
Ichiki, K., 15, 403
Ieki, K., 281
Iizuka, T., 494
Ikezoe, H., 249, 374
Imai, N., 249, 281
Inafuku, K., 275
Inafuku, S., 362
Inoue, K., 119
Inoue, S., 105, 446
Irgaziev, B. F., 255
Ishihara, M., 281
Ishikawa, K., 281
Ishikawa, T., 249, 374
Ishimaru, Y., 92, 221
Ishiyama, H., 249, 374
Iwamoto, K., 406
Iwamoto, N., 59, 65, 324, 339, 409, 424, 455
Iwamoto, O., 479
Iwasa, N., 275, 281, 362
Iwasaki, H., 281, 395
Izumi, H., 374

J

Jalil, I. A., 412, 506
Jeong, S. C., 249, 374
Johnson, E., 255

K

Kajino, T., 5, 15, 37, 53, 134, 339, 415, 479
Kamijo, T., 491
Kamimura, M., 269
Kang, Y. W., 389
Kanno, S., 281
Käppeler, F., 235
Kassim, H. A., 412, 506
Katabuchi, T., 494
Katayama, I., 249, 374
Kato, K., 497
Kato, S., 395

Katsuma, M., 359
Katsumata, M., 464, 482
Kawagoe, S., 415
Kemper, K., 255
Khiem, L. H., 275, 362
Kibe, M., 395
Kimura, K., 134
Kishimoto, C. T., 418
Kiss, G. G., 351
Kita, N. T., 304
Kodaira, K., 3
Kon, T., 491
Kondo, Y., 281
Kotake, K., 386, 421, 452, 476
Koura, H., 470
Kroha, V., 255
Kubo, T., 281
Kubono, S., 275, 281, 362, 395
Kunibu, M., 281
Kunugise, T., 406
Kurita, K., 281
Kusakabe, M., 424
Kwon, Y. K., 275, 362

L

Lacey, C. G., 442
La Cognata, M., 263
Lamia, L., 263
Langanke, K., 333
Lattimer, J. M., 155
Lee, C. S., 395
Lee, J. H., 395
Li, Z., 37, 427
Lian, G., 37, 427
Liebendörfer, M., 333
Limongi, M., 99, 311
Lingerfelt, E. J., 470
Liu, W., 37, 427
Lu, Y., 37

M

Madokoro, H., 430
Maeda, K., 65, 111, 485
Malagnini, M. L., 383, 436
Marcantonio, P. D., 383
Martínez-Pinedo, G., 333

Maruyama, T., 380
Máté, Z., 351
Mathews, G. J., 15, 227
Matsuda, M., 249, 374
Matsuyama, Y. U., 281
Matteucci, F., 79, 371, 461
Mazzali, P. A., 485
McWilliam, A., 21
Mengoni, A., 227
Messer, O. E. B., 179
Meyer, R. A., 470
Mezzacappa, A., 179, 333
Michimasa, S., 281, 395
Minemura, T., 281
Minezaki, T., 53
Misawa, T., 433
Mitsuoka, S., 249, 374
Miura, M., 281
Miyatake, H., 249, 374
Mizoi, Y., 249, 374
Mizuta, A., 445
Mohr, P., 351
Molaro, P., 205
Momotyuk, A., 255
Moon, J. Y., 395
Mori, S., 491
Morishita, Y., 304
Morossi, M., 383
Morrosi, C., 436
Motizuki, Y., 430
Motobayashi, T., 281
Moustakidis, C. C., 147
Mukhamedzhanov, A. M., 255, 263
Murakami, H., 281
Muto, T., 439

N

Nagashima, M., 442
Nagataki, S., 445
Nagatomo, T., 494
Nakai, K., 374
Nakamura, K., 105, 446
Nakamura, R., 15, 449
Nakmura, T., 281
Nesaraja, C. D., 470
Niikura, M., 275, 362
Ninomiya, S., 491
Nishimura, M., 395

Nishimura, N., 452
Nishimura, S., 275, 362, 395
Nishimura, T., 59, 455, 497
Nishio, K., 249, 374
Nomoto, K., 53, 65, 333, 339, 409, 424, 458, 485, 488
Nomura, T., 249, 374
Nordström, B., 87, 205
Norris, J. E., 45, 53
Notani, M., 281, 395
Nugent, P., 9

O

Ogata, K., 269
Ohkubo, T., 458
Ohnishi, N., 421
Ohno, H., 403
Ohta, M., 497
Okamoto, T., 442
Ota, S., 281
Otsuki, K., 227

P

Pipino, A., 461
Pizzone, R. G., 255, 263
Plez, B., 205
Prantzos, N., 92
Primas, F., 205, 377

R

Rauscher, T., 351
Roeder, B., 255
Rogachev, G., 255
Rolfs, C., 245, 263
Romano, S., 255, 263
Röpke, F. K., 190
Ryan, S. G., 53, 92, 221, 368

S

Saha, P. K., 374
Saito, A., 275, 281, 362
Saito, Y., 464, 482

Sasaqui, T., 374
Sato, K., 421, 445, 476
Schatz, H., 467
Schuster, W. J., 21
Scott, J. P., 470
Serata, M., 281
Shavrina, A., 389
Shibasaki, Y., 491
Shigeyama, T., 105, 446
Shima, T., 31
Shimizu, T., 430
Shimizu, Y., 491
Shimoda, G., 304
Shimoda, T., 249, 374
Shimoura, S., 281
Shinba, T., 494
Shizuma, T., 339
Shu, N., 37, 427
Smith, M. S., 213, 467, 470
Somorjai, E., 351
Spitaleri, C., 255, 263
Spite, F., 205
Spite, M., 205
Steinhauer, A., 53
Straniero, O., 368
Su, J., 37, 427
Suda, T., 59, 455, 497
Sugimoto, T., 281
Sugiyama, N., 403
Sumiyoshi, K., 415, 473
Suwa, Y., 476
Suzuki, H., 415, 473
Suzuki, T., 446, 479

T

Tachibana, S., 304
Tagishi, Y., 494
Tajitsu, A., 433
Takada-Hidai, M., 53, 464, 482
Takahashi, K., 403
Takamura, A., 134
Takeda, Y., 467
Takeoka, Y., 491
Takeshita, E., 281
Takeuchi, S., 281
Takiwaki, T., 476
Tanaka, M., 249, 485
Tanaka, M. H., 374

Tanifuji, M., 494
Tatsumi, T., 171, 380
Terada, K., 324
Teranishi, T., 275, 362, 395
Thielemann, F.-K., 333, 409
Timmes, F. X., 467
Togano, Y., 275, 281, 362
Tominaga, N., 65, 409, 488
Tostevin, J., 227
Trache, L., 263
Tribble, R. E., 255, 263
Truran, J., 227
Tsangarides, S., 53
Tsuruta, S., 163
Tudisco, S., 263
Tumino, A., 255, 263
Typel, S., 263

U

Ue, K., 281
Umeda, H., 65, 339, 409, 458, 488
Umezu, K., 15
Utsunomoiya, H., 359

V

Vergados, J. D., 140, 147

W

Wada, T., 497
Wakabayashi, Y., 275, 362
Wanajo, S., 92, 105, 221, 345, 446
Wang, B., 37, 427
Wang, Y., 427

Watanabe, Y., 249, 491
Watanabe, Y. X., 374
Wiescher, M., 227, 467
Williams, I. S., 324
Wu, K., 37

Y

Yahiro, M., 15, 269
Yamada, K., 281
Yamada, S., 196, 386, 415, 421, 452, 473
Yamaguchi, H., 275, 362
Yamaguchi, M., 494
Yamamoto, K., 497
Yamato, Y., 494
Yan, S., 37, 427
Yanagisawa, Y., 281, 395
Yasuda, O., 491
Yokomakura, H., 134
Yoneda, K., 281
Yong, D., 21
Yoshida, A., 281
Yoshida, T., 134, 324, 458, 500
Yoshihara, K., 415
Yoshii, Y., 53
Yoshikawa, N., 249, 374
Yoshimaru, N., 494
Yurimoto, H., 319
Yushchenko, A., 389, 503
Yusof, N., 412, 506

Z

Zeng, S., 37, 427
Zilges, A., 351
Zinner, N. T., 333